Unifying Themes
in Complex Systems

VOLUME I

NEW ENGLAND COMPLEX SYSTEMS INSTITUTE SERIES ON COMPLEXITY

YANEER BAR-YAM, EDITOR-IN-CHIEF

Unifying Themes in Complex Systems, Volume I
Proceedings of the First International Conference on Complex Systems
Edited by Yaneer Bar-Yam

Unifying Themes in Complex Systems, Volume II
Proceedings of the Second International Conference on Complex Systems
Edited by Yaneer Bar-Yam and Ali Minai

Virtual Worlds: Synthetic Universes, Digital Life, and Complexity
Edited by Jean-Claude Heudin

Unifying Themes in Complex Systems

VOLUME I

Proceedings of the First International Conference on Complex Systems

Edited by Yaneer Bar-Yam

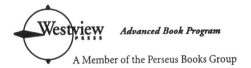

Westview PRESS *Advanced Book Program*

A Member of the Perseus Books Group

Copyright © 2000 by the New England Complex Systems Institute

Published in the United States of America by Westview Press, A Member of the Perseus Books Group, 5500 Central Avenue, Boulder, Colorado 80301–2877, and in the United Kingdom by Westview Press, 12 Hid's Copse Road, Cumnor Hill, Oxford OX2 9JJ.

Find us on the world wide web at www.westviewpress.com

Westview Press books are available at special discounts for bulk purchases in the United States by corporations, institutions, and other organizations. For more information, please contact the Special Markets Department at the Perseus Books Group, 11 Cambridge Center, Cambridge, MA 02142, or call (617) 252–5298, (800) 255–1514 or email j.mccrary@perseusbooks.com.

A Cataloging-in-Publication data record for this book is available from the Library of Congress.
ISBN 0–8133–4122-1

The paper used in this publication meets the requirements of the American National Standard for Permanence of Paper for Printed Library Materials Z39.48–1984.

Contents

INTRODUCTION

The integration of ideas and methods from many disciplines into the study of complex systems has, in recent years, generated excitement within the scientific community as well as in the general public. The conceptual shift from reductionist strategies of inquiry to integrative cross disciplinary and unified descriptions has had profound impacts on our understanding of environmental, social, economic, medical, biological, and physical aspects of our world. The real world teaches us that events and actions have multiple causes and consequences. Scientists are developing new concepts and methods that can be used to address questions about the world which cannot be answered by focusing on individual forces or parts, thus building our understanding of the functional and coincidental relationships in and between systems. These concepts provide a new unity and beauty, simplifying the otherwise bewildering properties of diverse complex systems into a common framework. They allow an individual to develop an appreciation for the essential questions that must be asked and the value of and relationships between different perspectives.

The first International Conference on Complex Systems, hosted by the New England Complex Systems Institute, took place Sept. 26-30, 1997 in Nashua, NH with over 250 participants and to general acclaim. It had two major aims: first, to investigate those properties or characteristics that appear to be common to the very different complex systems now under study; and second, to encourage cross fertilization among the many disciplines involved. Highlights of the conference (program attached) included a well attended pedagogical session, extended plenary sessions, breakout sessions and a poster session which ran throughout the conference. Due to the wide ranging interdisciplinary nature of the conference, speakers devoted substantial time to introducing essential concepts. The conference dealt with issues of significant public interest, including: individual psychology, dynamics of social and economic change, the human genome, and ecology. The conference success led to a second conference Oct. 25-30, 1998 and to plans for future conferences. Further information about upcoming conferences can be found on the World Wide Web at http://necsi.org/

A central contribution of the ICCS conferences is building an international community that is devoted to the unified study of complex systems. This is a new discipline and creating a common foundation, knowledge base, language and context for the discipline is an essential part of its growth. At this critical stage in the development of the field, the conference served a central role in advancing these goals. The choice of an isolated location promoted active discussions.

A listserv discussion group originating from participants in ICCS, continues to grow in number of participants and in activity.

A conference devoted to complex systems provides unique opportunities for transdiciplinary communication and collaboration. The study of complex systems in a unified framework is the ultimate of interdisciplinary fields. It draws inspiration from the diverse approaches and strategies that have been developed in many disciplines, ranging from physics to anthropology, and in return is relevant to advancing disciplinary objectives. Feedback from participants at the conference pointed out that many were exposed to a wealth of new ideas relevant to their own disciplines and interests. Among the many collaborations that arose from the conference was a collaboration that involved researchers in meteorology, molecular biology, genomics, and computer science. This collaboration is building a unified representational framework to be applied to each of the disciplines.

The field of complex systems has attracted not only the attention of scientists and engineers but also corporate managers and management researchers, doctors and medical researchers, policy makers and political scientists. ICCS has provided an opportunity for them to develop an understanding of both the science and application of complex systems concepts. Both professionals and researchers also have an opportunity to contribute their real world experience to the study of complex systems. The application of complex systems to management has been discussed at ICCS as well as in a large number of recently published books and topical conferences. Specific applications range from corporate planning, to human resource management, to design of scheduling and control systems. Applications of complex systems to medicine have been discussed including topics ranging from patient monitoring and diagnosis to the immune system and cancer.

The development of such basic new concepts carries with it an educational priority. Since the fundamental concepts of complex systems unify and integrate across the disciplines, they should be introduced to students early. Counter to the ongoing specialization of professions and splintering of disciplines, complex systems provides an opportunity for developing coherent educational experiences with diverse problems within a common framework, helping students build the conceptual tools to understand not only their field of specialization but more generally their ever more complex and rapidly changing world.

This suggests that building educational programs in complex systems should be given a high priority. The ICCS was itself an important educational event. Remarkably, the pedagogical sessions were attended by over half of the conference participants. Since most plenary lectures were designed to be comprehensible to nonspecialists, the conference served as a lecture and discussion equivalent of well over a full semester courses for attending graduate students. Approximately 60 graduate students and a few undergraduate students attended with reduced registration, and additional financial aid was given using NSF support. The impact of the conference in encouraging students to seek further education in the field of complex systems has also become apparent.

Moreover, presentations at the conference addressed the problem of communicating complex systems concepts using diverse media (new text, graphics, interactive simulations). Other presentations described the use of complex systems concepts to build educational programs using the insights of complex systems to achieve better educational outcomes. Collaborations arising from the conference include several that are devoted to education of complex systems concepts and the use of complex systems concepts in education and education reform. These collaborations address education at all levels of schooling, ranging from elementary school to postgraduate and continuing education. A national initiative to bring complex systems concepts into K-16 education is being actively planned.

The International Conference on Complex Systems was the launching event of the New England Complex Systems Institute (NECSI), a new and growing academic educational and research institution dedicated to advancing the study of complex systems. NECSI was established as a joint effort of faculty of New England academic institutions for the advancement of communication and collaboration outside of institutional and departmental boundaries. The academic and educational objectives of this institute are commensurate with the objectives of the conference.

In this collection, I have tried to capture some of the spirit of the conference using a few transcripts of the talks. Technical difficulties with the transcription process eventually precluded including many of the talks, nevertheless, the sprit and outstanding content can be seen from the talks that are included. The transcripts are contained in Part I of this book. Two of the plenary talks whose transcription were not possible are represented by formal articles (Goldstein and Fell). Part II contains the formal peer-reviewed papers of the conference proceedings which are also published on-line in InterJournal (http://interjournal.org/).

I would like to thank the people and institutions who contributed time, effort and financial support to this conference. Temple Smith was originally responsible for suggesting the importance of holding the conference and his on-going advice was essential in making it a success. Sean Pidgeon, then at Oxford University Press, had the vision to recognize the importance of this event and provide partial financial and administrative support. A key part of his support was in the form of his remarkable assistant Merilee Johnson who managed the conference communications of thousands of e-mail messages. Sean's involvement is to be credited to Irv Epstein who provided the referral. Michael Lissack joined the organizational effort just before the conference, but had an immediate and large impact. Tom Toffoli and Hiroki Sayama are to be thanked for their assistance in formatting these proceedings. Jeff Robbins of Perseus Press is the editor responsible for publisher support for this book series. Partial financial support was also provided by the National Science Foundation, Coopers and Lybrand Consulting, McKinsey and Company, Addison Wesley Longman Advanced Book Division (now part of Perseus Press) and John Benjamin Publishing. Academic institutions co-sponsoring the ICCS conference are the Santa Fe Institute, the Society for Chaos Theory in Psychology & Life Sciences, and the College of Engineering of Boston University.

The organizing committee and session chairs (listed on the following pages) played an essential role in making this conference represent a wide range of views and in attracting outstanding speakers. Given the many and diverse roots of the study of complex systems, no one individual can know the people and subjects that should be represented in such a gathering. The collective behavior—the conference—emerged from the efforts of the outstanding people involved and they are to be credited with the outcome.

Yaneer Bar-Yam
Newton, Massachusetts

"Significant Points" in the Study of Complex Systems

Yaneer Bar-Yam

In order to help establish a backdrop for the conference, I compiled (with much feedback and contributions from others) a list of "significant points" in the study of complex systems. These are supposed to represent key conceptual insights coupled with mathematical tools for the analysis and discussion of complex systems in general. Feedback and additions are welcome. The points are provided only in brief. In general some familiarity is assumed. Some controversial points are included (what field has no controversy?).

(1) Multi-scale descriptions are needed to understand complex systems. Relevant mathematical tools are scaling laws, fractals & trees, renormalization, multigrid. These specific methods are not exclusive of the more general issue of relating finer scale descriptions to larger scale descriptions (e.g. which fine scale parameters are relevant on larger scales, etc.). Examples: weather—patterns on all scales (cyclones, tornadoes, dust devils); proteins—secondary, tertiary, quaternary structure; physiology—molecules, cells, tissues, systems; brain—hemispheres, lobes, functional regions, etc.; economy/society—similar.

(2) Fine scales influence large scale behavior. Relevant mathematical tools are nonlinear feedback iterative maps, mathematics of deterministic chaos, amplification & dissipation. The specific methods of deterministic chaos are not exclusive of more general issue of fine scale effects on large scale behavior. Examples: weather—"butterfly effect", proteins—enzymatic activity is amplification, physiology—neuromuscular control (a nerve cell action triggering a muscle), economy/society—the relevance of individuals to larger scale behaviors (how many people watch Michael Jordan).

(3) Pattern formation: Prominent among simple mathematical models that capture pattern formation are local activation/long range inhibition models. e.g. Turing patterns, and the work of Prigogine. Examples: weather—cells of airflow, protein—alpha and beta structure, physiology—processes of pattern formation in development, brain/mind—various patterns of interconnection and pattern recognition mechanisms (on-center off-surround), magnetic bubble memories, patterns of species in phenome or genome space, economy/society—patterns of industrial/residential/commercial areas.

(4) Multiple (meta) stable states. Small displacements (perturbations) lead to recovery, larger ones can lead to radical changes of properties. Dynamics on such a landscape do not average simply. Mathematical models are generally based upon local frustration e.g.. spin glasses, random Boolean nets. Attractor networks use local minima as memories. Examples: weather—persistent structures, proteins—results of displacements in sequence or physical space, physiology—the effect of shocks, dynamics of e.g. the heart, brain/mind—memory, recovery from damage, economy/society—e.g. suggested by dynamics of market responses.

(5) Complexity—answer to question "How complex is it?" There is much discussion of this question. A general answer: The amount of information necessary to describe the system. There are various important issues that require clarification. One of these relates to the use of inference to obtain the description from a seemingly smaller amount of information. This leads the concept of algorithmic complexity. Another relevant point: The apparent complexity depends on the scale at which the system is described, however, once a particular scale is chosen the complexity should be well defined and bounded (at a particular instant) by the information necessary to describe the microstate of the system (the entropy). Also note that complexity on a large scale requires correlations on a small scale, which reduces the smaller scale complexity. Example: random motion (high small scale complexity) averages out on a larger scale.

(6) Behavior (response) complexity. To describe the behavior (actions) of a system acting in response to its environment, where the complexity of the environmental variables are $C(e)$ and of the action is $C(a)$, we often try to describe the response function f, where $a = f(e)$. However, unless simplifying assumptions are made, specifying the response to each environment requires an amount of information that grows exponentially with the complexity of the environment (a response must be specified for each possible environment). Specifically $C(f) = C(a) \times 2^{C(e)}$. This is impossible for all but simple environments (e.g. less than a few tens of bits). This means that behaviorism in psychology, or strict phenomenology in any field, or testing the effects of multiple drugs, or testing computer chips with many input bits, is fundamentally impossible.

(7) Emergence. Related to the dependence of the whole on parts, the interdependence of parts, and specialization of parts. This is directly relevant to questions about how we study systems both theoretically and experimentally. Parts must be studied "in vivo". For example—"If you remove a vacuum tube from a radio and the radio squeals do not conclude that the purpose of the tube is to suppress squeals." While studying the parts in isolation does not work, the nature of complex systems can be probed by investigating how changes in one part affect the others, and the behavior of the whole.

(8) Evolution. Related to the dynamics of change of the collection of complex systems that are present as part of a larger system, for example the earth. Traditional views emphasize reproduction (trait heredity) with variation, selection with competition. Newer perspectives (e.g. Kauffman) emphasize dynamic landscapes and co-evolution in ecosystems. Multiscale perspectives emphasize

the relationship between cooperation and competition at different scales. Among the candidates for evolutionary dynamics are biological organisms, immune system maturation, artificial evolution of drugs, genetic algorithms/evolutionary computation, linguistic and cultural evolution.

(9) The relationship of descriptions and systems. This is relevant to our understanding of theory and simulations, the recognition of systems in their models, encoding and decoding (compression), and the subject of algorithmic complexity. Specific applications are apparent in biological development (genome vs. physiology), engineering design, and memory vs. experience.

(10) Selection is information (a la Shannon theory). The amount of information necessary to specify a system is obtained by enumerating the possible states and comparing them with the possible states of the description e.g. a bit string, or e.g. English language (at about 1 bit/character). This enables the systems to be enumerated and one of them specified. Selection as information is relevant to the issue of multiple selection: replication (reproduction) with variation, and comparative selection (competition) as a mechanism for POSSIBLE increase in complexity. Consistent with modern biological views of evolution it is essential to emphasize that selection does not have to increase complexity.

(11) Composites. To form a new complex system take parts (aspects) of other complex systems and recombine them. For this to work parts must be partially independent. Examples—sexual reproduction, creativity (e.g. seeing a person walking and a bird flying and imagining a person flying by combining information of shape and motion represented in different parts of the brain), and modular construction (building blocks) in artificial systems. The purpose of composites is to allow rapid evolution.

(12) Control hierarchy. When (if) a single component controls the collective behavior (not the individual behaviors of all the components) of a system, then the collective behavior cannot be more complex than the individual behavior. i.e. there is no emergent complexity. Examples: muscle (since the muscle is controlled by a single neuron, its collective behavior is no more complex than the neuron behavior), society/economy: corporate hierarchies/dictatorships/etc. (to the extent that central control is exercised complexity of collective behavior is bounded by the complexity of the controlling individual).

(13) Modeling and simulation: There are a number of simulation methodologies that have arisen as having general application in the study of complex systems. These include: Monte Carlo, simulated annealing, cellular automata. Other methods have been mentioned above.

International Conference on Complex Systems: Organization and Program

Organization:

Hosts:

New England Complex Systems Institute

Oxford University Press

Partial financial support:

National Science Foundation

Coopers and Lybrand Consulting

McKinsey and Company

Addison Wesley Longman Publisher

John Benjamin Publisher

Cosponsors:

Santa Fe Institute

Society for Chaos Theory in Psychology & Life Sciences

College of Engineering, Boston University

Conference Chair:

Yaneer Bar-Yam - NECSI

Executive Committee:

Temple Smith - Boston University

Sean Pidgeon - Oxford University Press

Organizing Committee:

Philip W. Anderson - Princeton University

Kenneth J. Arrow - Stanford University

Michel Baranger - MIT

Per Bak - Niels Bohr Institute

Charles H. Bennett - IBM

William A. Brock - University of Wisconsin

Charles R. Cantor - Boston University

Noam A. Chomsky - MIT

Leon Cooper - Brown University

Daniel Dennett - Tufts University

Irving Epstein - Brandeis University

Michael S. Gazzaniga - Dartmouth College

William Gelbart - Harvard University

Murray Gell-Mann - CalTech/Santa Fe Institute

Pierre-Gilles de Gennes - ESPCI

Stephen Grossberg - Boston University

Michael Hammer - Hammer & Co

John Holland - University of Michigan

John Hopfield - Princeton University

Jerome Kagan - Harvard University

Stuart A. Kauffman - Santa Fe Institute

Chris Langton - Santa Fe Institute

Roger Lewin - Harvard University

Richard C. Lewontin - Harvard University

Albert J. Libchaber - Rockefeller University

Seth Lloyd - MIT

Andrew W. Lo - MIT

Daniel W. McShea - Duke University

Marvin Minsky - MIT

Harold J. Morowitz - George Mason University

Alan Perelson - Los Alamos National Lab

Claudio Rebbi - Boston Unversity

Herbert A. Simon - Carnegie-Mellon University

Temple F. Smith - Boston University

H. Eugene Stanley - Boston University

John Sterman - MIT

James H. Stock - Harvard University

Gerald J. Sussman - MIT

Edward O. Wilson - Harvard University

Shuguang Zhang - MIT

Session Chairs:

Michael Lissack - NECSI

Sean Pidgeon - Oxford University Press

Charles Cantor - Boston University

Seth Lloyd - MIT

Michel Baranger - MIT

Melanie Mitchell - Santa Fe Institute

Stuart Kauffman - Santa Fe Institute, BIOS

Mehran Kardar - MIT

Jim Crutchfield - Santa Fe Institute

Robert Axtell - Brookings Institute

Kristian Lindgren - Institute of Physical Resource Theory

Shuguang Zhang - MIT

Günter Wagner - Yale University

Piet Hut - Institute for Advanced Study

Ken Showalter - West Virginia University

Marshall VanAlstyne - MIT

Richard Strohman - UC Berkeley

Jim Collins - Boston University

Bruce Boghosian - Boston University

Harold Morowitz - George Washington University

Jim Stock - Harvard University

Lynn Stein - MIT

Larry Rudolph - MIT

Bruce Boghosian - Boston University

Gerald J. Sussman - MIT

Subject areas: Unifying themes in complex systems

The themes are:

EMERGENCE, STRUCTURE AND FUNCTION: substructure, the relationship of component to collective behavior, the relationship of internal structure to external influence.

INFORMATICS: structuring, storing, accessing, and distributing information describing complex systems.

COMPLEXITY: characterizing the amount of information necessary to describe complex systems, and the dynamics of this information.

DYNAMICS: time series analysis and prediction, chaos, temporal correlations, the time scale of dynamic processes.

SELF-ORGANIZATION: pattern formation, evolution, development and adaptation.

The system categories are:

FUNDAMENTALS, PHYSICAL & CHEMICAL SYSTEMS: spatio-temporal patterns and chaos, fractals, dynamic scaling, non-equilibrium processes, hydrodynamics, glasses, non-linear chemical dynamics, complex fluids, molecular self- organization, information and computation in physical systems.

BIO-MOLECULAR & CELLULAR SYSTEMS: protein and DNA folding, bio-molecular informatics, membranes, cellular response and communication, genetic regulation, gene- cytoplasm interactions, development, cellular differentiation, primitive multicellular organisms, the immune system.

CELLULAR SYSTEMS: cellular response and communication, genetic regulation, gene-cytoplasm interactions, development, cellular differentiation, primitive multicellular organisms, the immune system.

PHYSIOLOGICAL SYSTEMS: nervous system, neuro-muscular control, neural network models of brain, cognition, psychofunction, pattern recognition, man-machine interactions.

HUMAN SOCIAL AND ECONOMIC SYSTEMS: corporate and social structures, markets, the global economy, the Internet.

Program:

Sunday Sept. 21

PEDAGOGICAL SESSION - **Michael Lissack** session chair

Atlee Jackson - nonlinear dynamics

Richard J. Gaylord - cellular automata with mathematica

Norm Margolus - cellular automata machines

Richard Lewontin - evolution

David Fogel - evolutionary computation

Mark Tramo - neurobiology

Jeffrey Goldstein - psychology and corporations

Seth Lloyd - complexity

RECEPTION SESSION - **Sean Pidgeon** - session chair

Matthew Golombeck - Science Results from the Mars Pathfinder Mission

Monday Sept. 22

Yaneer Bar-Yam - welcome

EMERGENCE - **Charles Cantor** - session chair

Daniel Keyser - emergent properties and behavior of the atmosphere

David Fell - systems properties of metabolic networks

Steve Lansing - social emergence in anthropology

Hayward Alker - political emergents at the world level

COMPLEXITY - **Seth Lloyd** - session chair

Charles Bennett - information and complexity

Greg Chaitin - the limits of mathematics

Murray Gell-Mann - patterns and complexity

Dan McShea - a hypothesis about hierarchies

SCALING - Michel Baranger - session chair

> H. Eugene Stanley - scaling in medicine and biology
> Per Bak - self-organization and scaling

Tuesday September 23

SELF-ORGANIZATION - Melanie Mitchell - session chair

> Ken Showalter - chemical waves and patterns
> Albert Goldbeter - modeling biological patterns
> Günter Wagner - evolution
> Robert Berwick - evolution of language
> Tom Ray - evolution of artificial systems
> Margaret Geller - astrophysical patterns

AFTERNOON BREAKOUT SESSIONS

> Stuart Kauffman - evolution and webs in economics: session chair
>
> > Mark Newman, Per Bak, Stuart Pimm
>
> Mehran Kardar - scaling: session chair
>
> > Geoffrey West, Daniel H. Rothman, Didier Sornette, Vincent A. Billock/J. A. Scott Kelso/Gonzalo C. de Guzman
>
> Jim Crutchfield - measures of structural complexity: session chair
>
> > Dave Feldman, William Macready/David Wolpert, Ing Ren Tsang/Ing Jyh Tsang, Gad Yagil, Martin Zwick
>
> Robert Axtell/Kristian Lindgren - modeling social systems: session chair
>
> > Jeff Schank, James Uber/Ali Minai

EVENING BREAKOUT SESSIONS

> Shuguang Zhang - biomolecular & cellular systems: session chair
>
> > Don Ingber, Hyman Hartman, Christian Forst, Catherine Willet
>
> Günter Wagner - evolution: session chair

Elena Budrene, David Gutnick/Eshel Ben-Jacob, Kei Tokita, Lee Altenberg, Andreas Wagner, Raffaele Calabretta

Piet Hut - modeling modeling: session chair

Evelyn Fox Keller, Javier Dolado, Stuart Kauffman, Christian Reidys

Ken Showalter - controlling chaotic systems: session chair

F. Tito Arecchi, Mingzhou Ding, Jim Collins, Atlee Jackson

Marshall VanAlstyne - organizational structure and behavior: session chair

Jim Hines, Michael Lissack, Steve Maguire

Wednesday September 24

INFORMATICS - **Richard Strohman** - session chair

David Kingsbury - information management

William Gelbart - Drosophila

Temple Smith - whole genome bioinformatics

ENGINEERING COMPLEXITY - **Seth Lloyd** - session chair

Alex d'Arbeloff - Engineering management at Teradyne

Frans Kaashoek - complexity and software design

Thursday September 25

TIME SERIES - **Jim Collins** - session chair

Jim Stock - economic time series

Blake LeBaron - markets

Richard Cohen - chaotic hearts

Ary Goldberger - fractal dynamics of the heartbeat in health and disease

COMPUTATIONAL METHODS - **Bruce Boghosian** - session chair

Claudio Rebbi - computer simulations

Melanie Mitchell - genetic algorithms, cellular automata and emergent computation

AFTERNOON BREAKOUT SESSIONS

Harold Morowitz - origin of life: session chair

Roland Somogyi, Shuguang Zhang, Peter Wills, Stuart Kauffman

Jim Stock/Jim Collins - time series: session chair

Tim Sauer, Frank Moss, Lars Hansen, Frank Diebold

Lynn Stein/Larry Rudolph - computational systems: session chair

Mitchel Resnick, Kazuhiro Saitou, Yasuo Kuniyoshi

Bruce Boghosian - simulation and modeling: session chair

Frank Alexander, David Meyer

BANQUET SESSION - Jerome Kagan - session chair

Herbert Simon - can there be a theory of complexity?

Friday September 26

EMERGENCE IN MIND AND BRAIN - Gerald J. Sussman - session chair

Michael Gazzaniga - evidence from split brains

Marvin Minsky - specialization

Stephen Grossberg - how does the brain self-organize attentive recognition codes

Steve Pinker - language acquisition / how the mind works

EXTENDED DISCUSSION: MIND AND BRAIN - Daniel Dennett, William Sulis - chairs

Poster session

Petra Ahrweiler Modeling a theory agency as a toolkit for Science Studies SiSiFOS: Simulating studies on the internal formation and the organization of science

Philippe Binder Droplet breakdown in cellular automata

Bart de Boeck From inductive inference to the fundamental equation of measurement

Stefan Boettcher Aging in a model of self-organized criticality

Valery K. Bykovsky Man-made structures encoded with molecular information (computer-engineered physical reality: A synthesized complexity)

Tuan Cao-Huu High performance computing and the developments of positron emission tomography at the Massachusetts General Hospital

Michael Chechelnitsky Error estimation in ocean circulation models

Chang H. Choi The empirical relationship of self-organizing system to organizational effectiveness

Marshall Clemens Toward a model of the complex systems domain

David R. Collins Stochastic bifurcations in biological coordination

T. Gregory Dewey Algorithmic complexity and thermodynamics of molecular evolution

Michael Doebeli Controlling spatial chaos in metapopulation models with long-range interactions

Keith E. Donnelly Hazards, self-organization, and risk compensation: A view of life at the edge

Udo Erdmann Structure formation by Active Brownian particles with nonlinear friction

David Fell Systems properties of metabolic networks

Andrew M. Fraser Chaos and detection

Stefanie Fuhrman Genetic network modeling and inference

Mario Giampietro Biophysical analyses of socioeconomic processes in relation to the sustainability issue

Paul Halpern Emergent behavior in structurally dynamic cellular automata

Jeff Hausdorff Fractal dynamics of human walking rhythm

Collin Hill Transition to chaos and models of genetic networks

Vasant Honavar Kolmogorov complexity, simple distributions, and inductive learning

Jason G. Mezey and **Günter Wagner** Canalization by directional selection through genetic piracy

Kenneth J. Moriarty Parallel computational complexity in statistical physics

Jeff Morrow Estimating product line architecture complexity: Application to 'sunset' technology gracefully

Danko Nikolic A dual processing theory of brain and mind: Where is the limited processing capacity coming from?

Partha Niyogi and **Robert C. Berwick** Language change and dynamical systems

Mark Poolman and **David Fell** – Emergent behaviour in a computer simulation of photosynthesis

Michael A. Riley Chaotic dynamics of rhythmic movements

Andrew J. Rixon Pattern generation in simulated biological systems

Luis Mateus Rocha Matter and symbols: Requirements for evolving complexity

James N. Rose Robust non-fractal complexity

Helge Rose Evolutionary strategies of optimization and the complexity of fitness landscapes

Matthias Ruth Dynamic modeling of economy-environment interactions: The case of metals and fuels in the US

Kazuhiro Saitou Conformational switching as assembly instructions in self-assembling mechanical systems

Marie-Vee Santana Small scale oscillatory dynamics of the haptic perceptual system

Bruce K. Sawhill Statistical mechanics of boolean nets

Jeff Schank Self-organized huddles of rat pups modeled by simple rules of individual behavior

Bernard Sendhoff, Clemens Potter and **Werner von Seelen** The role of information in evolving biological systems

Lawrence Kai Shih Simulating the effects of knowledge policies

Jacqueline Signorini Folded and immersed cellular automata

Robin Snyder Cluster size in the contact process: A step toward understanding predator-prey individual-based models

Didier Sornette Emergence in earthquakes: From atomic scale transformations to mechanical rupture

W. Sulis and **I. N. Trofimova** TIGoRS as an associative memory in complex systems

Yasuhiro Suzuki and **Hiroshi Tanaka** Symbolic chemical system based on abstract rewriting system and its behavioral pattern

I. N. Trofimova Functional differentiation in developmental systems

Stephen C. Upton Warfare as a complex adaptive system

Burton Voorhees Reasoning about complex systems: Towards an epistemology of complexity

Lipo Wang Ordering chaos in a neural network using linear feedback

Usami Yoshiyuki Regeneration of extinct animals in computer

NOTE: Uri **Wilensky** and **Mitchel Resnick** provided an interactive demonstration during the conference.

Publications:

Proceedings:

Conference proceedings (this volume)
Video proceedings are available to be ordered through the New England Complex Systems Institute.

Journal articles:

Individual conference articles were published through the refereed on-line journal InterJournal and are available on-line (http://interjournal.org/) as manuscripts numbered 94-150.

Other products:

An active listserv discussion group has resulted from the conference. Access and archives are available through links from http://necsi.org/.

Web pages:

http://necsi.org/
Home page of the New England Complex Systems Institute with links to the conference pages.
http://necsi.org/html/iccs.html
International Conference on Complex Systems.
http://necsi.org/html/iccs_program.html
Conference program.
http://necsi.org/html/iccs2.html
Second International Conference.

http://interjournal.org/
Interjournal: refereed papers from the conference are published here.

Part I

Transcripts

Chapter 1

Can there be a science of complex systems?

Herbert A. Simon
Carnegie Mellon University

At the outset of my remarks, I must apologize for appearing before you in this insubstantial form. We should expect that one important chapter in a theory of complexity, perhaps more than one chapter, will be devoted to describing the ways in which we may avoid complexity. Now the chief complexity in travel lies in the journey to and from the airport—threading your way through streets, highways, and airport ticket counters and passageways populated by other human beings and their vehicles, each intent on his or her own mission; meanwhile managing to keep track of your baggage and other possessions. That complexity is best avoided by not traveling.

Many years ago, when trains were the chief vehicle of inter-city travel, I had a fantasy of living in Cleveland, with an apartment in the city's main railroad terminal (which was, in fact, a combined terminal-hotel—perhaps still is). I would catch my train to New York, disembark in Penn Station, take a subway to my hotel or office destination, transact my business and return home: all without ever emerging into the open air of either city. Trips to Boston and many other cities could be managed in the same way. A truly ant-like, but quite simple, existence. Today I can realize my dream without moving to Cleveland, but with the help of PicTel, and the non-existent, hence simple, ether that transmits electromagnetic waves. True, I did have to walk to the Carnegie Mellon Campus, but that is only a half mile from my apartment, a pleasant leisurely trip, without baggage except for my lecture notes.

But let me get on with the task. If my topic—the possibility of a science of complexity—has caused you any anxiety, let me relieve you of it immediately. I am going to argue that there can be such a science—that the beginnings of one already exists—and I am going to try to sketch out some of its content, present and prospective. As you will see, the theory of complex systems is much concerned with the avoidance of complexity; or, to put the matter more precisely, the theory is concerned with how systems can be designed, by us or by nature, to be as simple as possible in structure and process for a given complexity of function. The task is to defeat complexity by removing it from mechanism.

1. General systems theory?

But before I turn to complex systems, let me say a few words about systems in general. The idea that there are systems in the world, and that somehow, they permit and even require special attention, is an old idea. In rather recent times, since the First World War, the idea has undergone three transformations. First, there was the holism of South Africa's President Smuts, with its insistence that new properties emerge with the growth of systems and that systems are not reducible to their parts.

One kind of emergence is hard to deny. If one doesn't have a system of at least two atoms, one can't have molecular forces. Emergence should have no terror for us, for we can recognize emergent properties as no more than new theoretical terms. Thus, if we are concerned with the temperature equilibria of mixtures of liquids, we simply average the temperatures of the component liquids multiplied by their masses. But if we now introduce two different liquids, generalizing our law of temperature equilibrium calls for a new property: specific heat. If we now include the specific heat of each kind of liquid as one of its properties, then there is nothing irreducible about the system. That property simply becomes an uninteresting parameter if we limit ourselves to a single kind of liquid.

My impression is that almost everyone today accepts reducibility in principle. Reducibility in practice is another matter; often we are quite unable to perform the computations that would represent a system in terms of its components, or to understand the reduced system without the aid of simplification by aggregation at the higher level. Hence, I think most biologists are quite unconcerned (and rightly so) at the possibility that they will be put out of work by particle physicists, or even by organic chemists. Later on, I will have more to say about "simplification by aggregation." By the time of the Second World War, the study of complex systems was taking on the form and name of general systems theory, which was built around such ideas as homeostasis, and around Wiener's cybernetic notions of information and control. After a promising start, general systems theory also began to die on the vine, through lack of nourishment from tangible scientific results. Homeostasis was largely absorbed by control theory. Information theory and control theory remain healthy enterprises, but they do not seem to have produced a general theory of systems. Instead, they became something much more specific, the former concerned with channel capacities of communications systems and the latter mainly with the design of feedback mechanisms to maintain the stability of dynamic systems.

My own diagnosis of the demise of general systems theory is that it attempted to be just that: a theory that described all systems. But as there is very little that is true for all known systems, or even all known large systems, the theory found little content. It became little more than an injunction that in designing large systems, one should not try to design the individual parts without full consideration of their contexts. That is good advice, but hardly earns the honorific title of "Theory." Today, interest in complex systems begins with some new ideas that can be added to those of general systems theory. In recent years, we have learned a great deal about a large and important class of systems, called chaotic. Then we have had the birth of genetic algorithms for the investigation of the origins of complexity, and the game of life for the study of self-replication. To these, I would add my own favorite candidate: nearly completely decomposable systems, that bear a kinship with the renormalizations of particle physics.

What is new in the situation is that we are no longer talking about systems in general, but about particular classes of systems that have specific properties (e.g., chaos or near-decomposability), or specific mechanisms for their generation (e.g., genetic algorithms or rules for a game of life). An interesting theory can arise out of these special cases. Or, if we don't like special cases, we can put the matter as follows:

What properties of a system are conducive to an ability to perform complex functions, or to rapid development toward such an ability? The former might be termed the "efficiency question," the latter the "attainability question." To survive in difficult environments, systems must be capable of performing a wide range of adaptive functions, hence of economizing on the energy they devote to each. To acquire this capability in competition with other metamorphosing systems, they must be able to increase their efficiency, to evolve, relatively rapidly.

Putting matters still more simply, we are not really interested in large systems so much as in the mechanisms that allow them to manage multifunctionality efficiently and to increase their capacity to do so. When I express confidence about the prospects of a theory of complex systems, it is this kind of a system and the mechanisms that support it that I have in mind.

2. Some principles of complex system design

What, then, are the principles that must inform the design of a system of the sort I have just described? I will organize this part of my discussion under four headings: homeostasis, membranes, specialization, and temporal specialization. There are undoubtedly others, but I should like to begin by examining these four.

2.1. Homeostasis

Homeostasis is a venerable term in discussions of complex systems. By means of feedback mechanisms or by other methods (e.g., by screening inputs for homogeneity), a system may be able to hold the values of some of its important properties within narrow limits, and thereby greatly simplify internal processes that are sensitive to these properties. However complex the external environment, the internal environment becomes much simpler. A familiar example of homeostasis is temperature control. As the rates of various chemical reactions are differentially sensitive to temperature, maintenance of a system becomes very much simpler if the internal temperature is constant.

Maintenance of a stable internal temperature no doubt greatly facilitated the adaptation of birds and mammals to life outside the relatively uniform environment of a large body of water. (I'm afraid that I'll have to turn to others to explain how reptiles manage on land without that stability.) As our knowledge of genomes advances, it will be interesting to learn whether there is a connection between the very large amount of DNA possessed by most amphibians and their need to adapt to a wide range of habitats that have different and changing temperatures.

Homeostasis is facilitated if the system possesses some kind of skin that greatly attenuates the transmission of environmental changes into the interior of the system, and the same principle can be applied to insulate the various subsystems of a system from each other. I will have more to say about membranes in a moment.

A third mechanism of homeostasis is maintaining inventories of the inputs to internal processes. Inventories decrease the need for precision in timing inputs that are obtained from the external environment, or obtained from intermediate internal processes. Inventories are important to the degree that there is uncertainty about the time required to obtain required substances from the environment, and to the degree that the substances can be stored efficiently. Thus, biological organisms typically carry sizable inventories of food (for example, in the form of fat), but not of oxygen, which is used at too rapid a rate to be readily stored.

We see that both variability of the environment and its unpredictability impose a need for homeostasis, and we might conjecture, therefore, that chaotic environments would call for especially powerful homeostatic mechanisms. Here again, feedback comes to the rescue, for appropriately contrived feedback devices can sometimes maintain a system within a specific small subregion of the strange attractor of a chaotic system; whereas without feedback, the system would be doomed to range over the whole area of the strange attractor, with consequent wide variability of its environment.

In general, we can think of homeostasis as a method of reducing system complexity at the cost of some new complexities in the form of the homeostatic mechanisms themselves. As in virtually all questions of design, there are tradeoffs here that put bounds on how much net reduction in complexity can be provided by homeostasis.

2.2. Membranes

I have already referred to the insulation of a system from its environment: by means of skins in organisms, or their counterparts, like walls in houses, or the ozone layer of our Earth. But there is much more to membranes than insulation. In particular, membranes may contain specialized transport mechanisms that move particular substances or information from the external to the internal environment, or vice versa. Sensory organs can be regarded as one form of transfer agent, motor organs as another, and specialized systems for transport of substances between cells and between organelles within cells, a third.

All of these transfer agents represent specializations (and consequent complications) in the boundary surfaces. Instead of adjusting uniform surfaces to handle the complex mix of substances that have to enter and leave the system, particular regions of the surfaces are specialized for the transport of specific substances. Thus each of the sense organs has very specific sensitivities to particular kinds of stimuli. Each of the excretory surfaces (urinary, rectal, pores) are specialized to the removal of particular classes of substances. Perhaps the motor system appears less specialized than the other transfer mechanism—its task is to exert physical force on the environment—but we need only recall such specializations as the opposed thumb or the use of the tongue in speech to recognize that it is very specialized indeed.

The British biochemist, Peter Mitchell, and others following him have demonstrated a most remarkable variety of very specific membrane transport systems between and within cells. The mind boggles at what a homogeneous membrane would be like that could effect all of these kinds of transfer at any point in its surface without destroying the homeostasis of the interior. The design problem is simplified to the point of realizability by requiring each transport mechanism to transfer only one or a small number of substances, at the expense of restricting specific kinds of transfer to particular membrane locations. Again we see a tradeoff between a simplification, on the one hand, in boundary structures and complexity, on the other hand, in the additional mechanisms that are required to perform specific functions.

2.3. Specialization

We have already seen several examples of specializations that at the same time contribute to simplification of a system, viewed as a whole, but add new complexities in the form of the specialized mechanisms themselves. Can anything of a general kind be said about the tradeoff?

We can think of specialization within a system as an application of the principle of divide-and-conquer. If the task is to design a system that must perform many functions in order to operate successfully in its environment, then design a set of subsystems, each capable of performing one (or a few) of these functions, and connect them appropriately so that they can cooperate with each other. Clearly, the ease with which this strategy can be executed depends on the number and nature of the connections among the functions. If there are few connections, it should be easy both to design the components and to connect them. As the numbers of connections increases, the task becomes more difficult.

A first requirement for any successful system is that it be dynamically stable. R. L. May has shown that, if we represent the presence or absence of connections between pairs of elements in a linear system by a matrix of 1's and 0's and choose these elements at random, the system will very probably be stable if the percentage of 1's is sufficiently small, and highly probably unstable if the percentage of 1's is sufficiently large. Moreover, the shift from the stable to the unstable condition occurs quite suddenly as the percentage of 1's increases. This simple result gives us an important cue to the feasibility of specialization: specialization should be carried out in such a way as to keep the interactions between the specialized components at as low a level as possible.

Effective specialization is a central topic in the literature on human organizations. In fact, in the so-called "classical" literature of the 1930's (and in some of books and papers published even now), the advice is given to specialize: by function, by process, by area, and by clientele (and perhaps along other dimensions as well). As a checklist of possible bases of specialization, the advice is useful, but only as a first step, for it is internally inconsistent. If we put all of the marketing activities of a company in one department, and all of the financial activities in another (specialization by function), then we will not put all of the stenographers (specialization by process) in the same department. Hence choices must be made, at each level of the organization, as to which dimensions of specialization are the more important for the sets of heterogeneous tasks that are grouped together at that level.

2.4. Near-decomposability

Determining the degree of interconnection among various subsets of the elements of a system is not a trivial task. We must consider not only the number of different kinds of interactions, but also with the complexity of each, as well as its frequency and duration. One principle that has emerged from observation of the kinds of complex systems that actually occur in the world is that most such systems have a hierarchical structure. By hierarchy I do not mean a structure of power, although that may be present, but a boxes-within-boxes arrangement of subsystems and sub-subsystems This kind of structure is as visible in physical and biological systems as it is in human organizations. A much higher frequency and intensity of interaction takes place between components belonging to a single sub-system than between components belonging to different sub-systems; and this principle holds for all levels of the hierarchy.

The property of near-decomposability has important consequences for the behavior of a system that possesses it. Suppose that a system has a number of layers of subsystems. Because of the hierarchical arrangement of interactions, if the system is disturbed, subsystems at the lowest level will come to their internal steady states before the systems of which they are components at the next level above. Because their subsystems return rapidly to a steady state, the system above can be described in terms of the average behavior of the subsystems—specifically their principal eigenvalues. Broadly speaking, the principal eigenvalues at the various levels of the hierarchy will represent the dynamic behavior of the system in different temporal ranges: the eigenvalues at the lowest level, determine the very short-term behavior; at the successive levels above, the eigenvalues determine the dynamics over longer time intervals; and the principal eigenvalue on the highest level determines the system's long-term dynamic behavior.

This mathematical structure of a nearly decomposable system allows us to fix our attention on particular system levels if we are interested in dynamics within a specified frequency range. Even more important, it allows us to factor the system, so that we do not have to deal with all of its complexity at once. Having determined the behavior of subunits at one level, we can replace the details of these subunits by a small number of aggregate parameters, and use these to represent the system at the next level above. Or, looking from the top down, we can say that the behavior of the units at any given level does not depend on the detail of structure at the next level below, but only upon the steady state behavior, in which the detail can be replaced by a few aggregated parameters.

It is also easy to show that systems composed of stable subsystems can be composed into larger systems orders of magnitude more rapidly than can systems lacking this underlying structure. We can use this fact to explain, for example, the evolution of the hierarchy that leads from quarks to elementary particles, to atoms to molecules, and further up the biological hierarchy at least to unicellular organisms. There is no obvious way, however, in which the argument explains how multi-celled organisms achieved similar hierarchical structure. They are not formed by composition of simpler organisms, but through specialization of cells during maturation. In trying to understand the development of multi-celled organisms as specialized hierarchical structures we will need to invoke new principles that are not yet understood.

One direction the exploration can take is to ask whether hierarchy allows the various components of the system to evolve relatively independently of each other, obtaining feedback, through natural selection of the entire organism, of the particular contribution to increased fitness that is provided by each component. Near-independence of the several component functions of the system should greatly simplify its fitness landscape, so that, at least in the small, the optimum (or good) values of parameters of one subsystem would be rather insensitive to the values for the other subsystems This insensitivity, in turn, should accelerate improvement of fitness through natural selection.

Perhaps these remarks are sufficient to persuade you that near-decomposability is a property that supports complexity of function, and that nearly-decomposable systems are important objects for study in a theory of complexity.

3. Organizations and markets

Finally, I would like to discuss two kinds of systems that play a dominant role in the complex human systems we call economies: organizations (chiefly business and governmental) and markets. Between them, organizations and markets handle substantially all of the economic production and exchange that takes place, as well as the decision making that controls it. Because their components are human beings, these institutions have peculiarities, particularly in the motivations of their participants. Motivation is not usually a topic we think about in connection with the interactions of parts of organisms or the behavior of automated mechanical or electronic systems. Hence some issues arise here that are not prominent in discussions of other kinds of complexity.

3.1. The market bias of contemporary economic theory

So-called neoclassical theory devotes most of its attention to markets and puts forth several arguments as to why they generally perform their functions better than organizations do. The first argument for markets is that they are thrifty in their use of information. Each actor (consumer or business firm) can make its decisions with a minimum of information about the other actors—essentially, it needs only a knowledge of the market prices of its inputs and outputs. This is Adam Smith's "Invisible Hand." The second argument is that they make effective use of the most reliable of human motivations (some would say, "the only motivation"): self-interest.

Modern economics usually gets along with a minimalist theory of organizations, or if it enlarges this theory (as does the so-called "new institutional economics"), it does so by postulating a large array of contracts to characterize the relations between the owners of organizations and their employees—that is, it redescribes intraorganizational relations as contractual market relations., and tries to compare organizations with markets largely in terms of the relative costs of transactions in the two systems under various circumstances.

In the light of this subordination of organizations to markets in much economic theorizing, it might surprise an observer of a modern economy to note that most of the activities of members of organizations, perhaps eighty per cent as a rough guess, are not market transactions at all, but decision-making and communication activities that involve extensive problem solving, using data that go far beyond market prices, and often also incorporating authority relations among the participants. It might surprise the same observer to note that many of the interactions between different firms involve the exchange of far more information than prices, and carrying out transactions quite different from contracting for purchase or sale. In fact, these interactions often resemble the within-organization interactions to an extent that makes the boundary between the organizations fuzzy. It is a small step from a franchised retailer to a branch of a large retail chain. The interactions in real markets are something quite distinct from arms-length market transactions.

Given the important role of markets and organizations in economies, a high priority needs to be given, in research on complex systems, to deepening our understanding of the real nature of these two kinds of structure. Many topics will appear on this agenda. We will have to learn why the exchanges in markets frequently call for information about many things besides prices. We will have to take account of the motives besides self-interest, especially organizational identification and loyalty, that play a central role in the decisions of members of organizations. We will have to reassess the circumstances under which markets exhibit greater effectiveness than organizations, and the circumstances under which they are less effective. In particular, we will have to understand how bounded rationality—limits on knowledge and computation (whether by humans or computers)—affects these relative advantages and disadvantages. In sum, the theory of markets and organizations, and of their mutual relations, deserves a high place on the agenda of the study of complexity.

3.2. Motivations in organizations

I will not try to survey this whole domain, but will focus on the motivational issues; for without a correct, empirically grounded picture of the goals and values that direct human choices in organizational settings, it is hard to understand why so much of our economic activity should take place inside the skins of organizations. Notice that the advantages of specialization do not necessarily give organizations an advantage over markets.

Adam Smith was quite clear on this matter, for although he was a staunch advocate of specialization (witness his celebrated example of pin making), he conceived it as being accomplished through something like the putting-out system, where the farmer grows the flax, which he sells to the spinner, who sells the yarn to the weaver, who sells the cloth to the mercer. In fact, Adam Smith was wholly skeptical that corporations could be relied upon to represent their owners' interests, and confident that they would be operated (inefficiently) as to line the pockets of their managers and employees. He took Oxford and Cambridge Universities as prime examples of the gross inefficiencies that inevitably accompanied corporate organization.

What Adam Smith did not take into account (partly for lack of experience in his era with large organizational forms other than armies), was that human selfishness is far from synonymous with the desire to maximize personal wealth. On the contrary, there is voluminous evidence that humans, in any situation, tend to identify with a "we," and to make their decisions in terms of the consequences (including economic consequences) for the "we," and not just the "I." In some circumstances, the unit of identification is the family, in other circumstances, the nation or an ethnic group. And in many circumstances in modern society, the unit is the company or government agency that employs the individual. The decisions, and the behaviors they lead to, that we may expect from someone who is employed by and identified with an organization are often totally different from the decisions and behaviors the same person exhibits in roles that evoke a different "we"—the family, say.

It is not hard to build the case that loyalty to groups, even to the point of many sacrifices of the "I" to the "we," are exactly what we would expect as a result of the workings of evolution. I made that case, in one form, in a paper that appeared in Science in December 1990, which employs a standard neo-Darwinian argument that does not challenge the doctrine of the "selfish gene."

The introduction of group loyalties instantly changes all discussion of incentives in organization. Direct economic incentives may continue to play an important role (and they undoubtedly do in fact); but the organization where loyalties are present can be far less preoccupied with the problems of measuring the marginal contribution of employees or guarding against their opportunism that it would have to be if personal economic motives were the only bond of the employee to the organization. Personal motives play a much larger role in the decisions to join or leave organizations than they do in behavior while employed.

But the issue goes far beyond motivation. There is not merely organizational loyalty, but also organizational identification, a more inclusive mechanism. Working in an organization exposes the employee to a daily environment of beliefs and information that is vastly different from what would be encountered in other environments, including other organizations and other positions in the same organization. As creatures of bounded rationality, the decisions we reach are strongly influenced by the social environment of information that surrounds us. Most of the time, in our daily jobs, the tasks we are doing are tasks that take their meaning directly from organizational objectives (objectives of the whole organization or of the particular units in which we work), and only very indirectly (in terms of the wages and other perquisites of the employment relation) from our personal objectives.

Of course we should not be so naive as to believe that self interest does not influence the way in which employees handle organizational matters, but, in the context of a stable employment relation, one can surely make better predictions of the daily behaviors of a manager or employee from knowing what his or her job is than from knowing his or her private needs or wants. We make such predictions all the time: the bus driver will collect our fares and drive the bus along the prescribed route, and which driver is on the bus on a particular day makes only a marginal difference.

The human propensity to acquire organizational loyalties and identifications, which change both motivation and cognitive outlook, is a powerful force toward enabling organizations to accomplish certain kinds of tasks that markets perform badly or not at all. In building our theories of that complex system called the economy, we will need to incorporate identification as an important component in the explanation of organizational performance, thereby changing substantially our view of the relative roles of markets and organizations.

In introducing this exceedingly complicated topic, my aim is to warn against excessive generality in our theories of complexity. The complexity of biological systems is not going to resemble, in all aspects, the complexity of geophysical structures and processes; and both of these are going to differ in many fundamental ways from the complexity of human institutions. The theory of complex systems is perhaps going to look more like biology, with its myriad of species and of proteins, than physics, with its overreaching generalizations.

3.3. Adaptive production systems

The most common way, today, in computer science for representing processes is in the form of production systems: systems of if-then, or condition-action, relations. Whenever the conditions of a production are satisfied, its action is taken.

(There must also be rules of precedence to choose between several productions whose conditions are satisfied simultaneously.) It has been shown that production systems are as general as Turing Machines. It would be a good exercise for a biologist to show how the Krebs cycle could be represented as a production system ("If such and such proteins, etcetera, are present, then synthesize the following protein: —".) In fact, the MECHEM program of Raul Valdes-Perez represents chemical reaction paths in precisely this way, by sets of productions.

Production systems may be adaptive, that is, they may have the capability of constructing new productions (new condition-action pairs) and adding them to the system—to themselves. One way this learning can be accomplished is by presenting to the production system an example of the steps in a process and allowing it to determine what actions were taken at each step, and what conditions were present that triggered the actions. The system then forms the conditions into the left-hand side of a new production, and the actions into the right-hand (action) side. The system must have available a set of detectable features that can be encoded as conditions, and a set of executable actions that can be encoded as actions. The complete conditions and actions for any production can be assembled, in tinker-toy fashion, from members of the sets of primitive features and actions.

Learning in this manner, from worked-out examples, is widely used by human learners, and often even incorporated explicitly in such instructional procedures as homework or classroom exercises. It is an extremely powerful and very general learning technique, and probably at the core of most human learning that is more complex than the acquisition of simple reflexes (and perhaps even of that). Adoptive production systems could provide a promising model of emerging biological systems, with the introduction of mutation and crossover. I am not aware that any models of this kind have been constructed.

4. Conclusion

Perhaps I have surveyed enough specific topics that are highly relevant to the behavior of complex systems to show that the theory of complex systems, if pursued correctly, is unlikely to suffer the fate of general systems theory. Complex systems are not just any old systems. They possess a number of characteristic properties, because without these properties they would be unlikely to come into existence, or to survive even if they were born. We would expect them to be homeostatic; to have membranes separating them from their environments, and internal membranes between their parts; to specialize, so that complex functions tend to be performed in few locations; and generally, we would expect them to be nearly decomposable. When they exist in chaotic environments, we would expect them to possess special mechanisms for dealing with that chaos. We would expect to identify a number of generic forms of complexity, of which markets and organizations are examples, and systems that learn (e.g., adaptive production systems) as another example.

At the present state of our knowledge, all of these expectations are possibilities, whose correctness, and whose causal and supporting mechanisms if they are correct, remain to be verified. And of course this is a very partial menu, for I should not like to claim that I have identified, much less described, all of the facets of complexity. For those of us who are engaged in research on complexity, life ahead looks very exciting indeed.

Chapter 2

Evolution

Richard Lewontin
Museum of Comparative Zoology Laboratories
Harvard University

I took seriously the charge this is a pedagogical session, and so what I have done is to suppress my tendency to be assertive and argumentative on substantive issues, and instead try to be pedagogical. That is to say try to outline for you those elements of evolutionary processes that I think are essential to understand if one is going to engage in research on complexity in evolution, on evolution, or to use, for example, genetic algorithms and other so-called evolutionary processes to solve other problems. What is there that is particular to evolutionary processes, what's really going on? I think I should begin by saying that my prejudice is that biology is like the law; that is to say everything depends on the jurisdiction. There are no absolute general principles that are very interesting, and one really needs to know the specifics of biological organisms or the nexus of a very large number of weakly determining forces. And that's why biology is interesting: Precisely because, except when one is sick, and indeed the definition of illness is that an organism is dominated by one force, rather than being the nexus of a large number of weakly determined forces, and that's why complexity is relevant. So I've got to make a lot of distinctions.

The first thing I want to do is distinguish for our purposes between two kinds of evolutionary processes, because the word "evolution" is used rather loosely by people in general. One is what I'll call a transformational evolutionary process. The evolution of the stars is a classic example of a transformational evolutionary process. That is to say that the ensemble of objects changes as an ensemble because each object in the ensemble is going through a transformational process, and all of the objects are going through a similar generalized transformational process. The people in this room as an ensemble over the years will become grayer and their centers of gravity will drop and so on, and the ensemble changes.

The description of the ensemble in state space will change because each object is undergoing the same developmental process. And cosmic evolution, I mean the theory of cosmic evolution is a classic example of the transformational theory of evolution.

There is another form of evolution, and that's the one we regard as applying to biological organisms, and that's variational evolution. Variational evolution has the mechanism that the ensemble changes its position, the cloud of points changes position in the state space not because each object is undergoing a motion in the state space, but because the ensemble itself varies in its content to be formed of different individuals with different properties, and the ensemble moves in the state space because there is a change in the relative proportions of the different objects. That's the theory of organic evolution, that you have a population of organisms that differ one from another, there is some process which differentially enriches the ensemble with some kinds as opposed to other kinds, and therefore the ensemble itself changes in the space. The individual objects are not themselves undergoing a transformational process. Well, they do because we all develop from the egg to death, but that transformational process is superimposed—is a much shorter time scale. That's not the evolutionary process. And it's very important to distinguish between these two because often people say they would like to study the evolution of language, the evolution of culture, the evolution of consciousness, the evolution of mathematics, the evolution of automobiles. They all are thinking about this process because organic evolution is so trendy. And much of the story of putting evolutionary scenarios on things like language involves trying to map and make an isomorphism between the ensembles and the objects in organic evolution as an ensemble of objects in linguistic evolution. So, for example, you have phonemes and allophones which are supposed to be isomorphic to genes and allelic forms of genes and so on. I don't know whether linguistic evolution is transformational or variational or some complicated mixture of both at the same time constants. I don't know. But it's extremely important if you're talking about studying the evolution of some process—which is not well known to be transformational, like the stars or well known to be variational, like organic evolution—that you're not making an assumption about the underlying form of that evolution. I think a lot of trouble has arisen by the attempt to homologize the evolution of other things with the variation mystique. Now given that, and this is not the place to go through the history of this, except to say that briefly, that before Darwin people thought that organic evolution was also transformational and Darwin's contribution was to envision it as a variational process, i.e., as a process of changing the relative proportions of already existent types within the ensemble.

1. Selection and production

Now the schema, then, which we have to confront is the following one. We have some ensemble which consists of a mixture of different types: genotypes, phenotypes. By the way, almost everything I'm going to say has already been said by the first speaker. I'm in a sense focusing on a number of different topics; he covered everything. I really resent that. We have this ensemble of objects. For example, triangles and circles and plusses and minuses and so on. And then there's a process of selection, a process of differential survivorship of the different kinds of objects such that the ensemble then becomes, a new ensemble in which there are a lot more triangles and fewer zeroes and one plus and one minus. So the proportions change in the ensemble. And that's roughly speaking what's called selection, or there are other reasons why they could change, and I want to deal with that in a moment, but it is a selection process. That's the essence of a variational process: some culling process. However, organic evolution has a second feature which is central to our understanding of evolutionary process, and that is that the objects themselves are mortal. And because they are mortal, this culling which has changed the ensemble of distribution is in danger of being completely nullified in the production of another ensemble of also mortal objects. That is to say these mortal objects must in turn produce, I will not say reproduce because that's a glaring illusion. There is no reproduction of organisms. Nobody in this room is a reproduction of his or her parents. That's absurd. There is a production of organisms. And that production of organisms then produces a new ensemble, and that new ensemble could be anything. It could, for example, look just like the original one. And if it looks just like the original one, then the selection process would have been wasted, so to speak, or you'd be at some kind of an equilibrium, which we have, and indeed there are equilibria in evolutionary biology in which the production process reproduces the ensemble that had been altered by the selection process. And it's here that genetics is relevant. This is all the rest of biology. This is genetics. So that now genetics produces a new ensemble which is not, and the issue is, is it a copy of the original? A perfect copy? Or is it a perfect copy of the one after selection? Or is it something in between? Or what other possibilities are there? To put it another way, the essence of an evolutionary process is you have a distribution of frequencies over genotype. The selection process carries you to a different distribution. And then the productive process carries you to a new distribution which could have different properties. One is that it could, alas, look like the original distribution, in which case you have completely destroyed the process of selection. The other possibility is it could perfectly copy the selected distribution, in which case the productive process has cost nothing. And the third possibility, the one that usually happens, is that the distribution produced in the production process has moved somewhat but not as much as the selection process would have moved it. And the ratio of the change after production and the change after selection in known in population genetics and evolution genetics as the heritability of the traits being looked at. Heritability in this sense is defined as that proportion of the selection difference that you've produced which is itself maintained after the production process. That you would call the reproduction process.

There is another possibility, alas, which is that the ensemble produced by the production process could be even farther than the selection process carried you. Let's say there's something about the production process which is pushing you in the same direction as the selection process, in which case the evolution could occur without selection at all, but more slowly; or, worse, from some standpoint, it could move you back in the opposite direction.

Fundamental to the evolutionary process as we understand it in biology is that heredity itself is unbiased. That is to say that the mechanism of transmission is such that by itself it does not cause a movement. We ultimately think of hereditary as unbiased because we have Mendel's laws, and this is why I say it depends on the jurisdiction. Mendel's laws say that if I have a heterozygote, the sperm or the eggs produced are 50/50 big A and little a. This means that a deterministic system, forgetting for the moment about a stochastic system, if there's no selection, you'll have a perfect reproduction of the ensemble. But the fact of the matter is, that Mendel's laws are a mechanism, a particular machine which doesn't always operate. There are systems and circumstances under which the production process itself is biased and drives an evolutionary process in the absence of selection or against selection. This is known in biology as mitotic drive, or other phenomena, in which in the segregation process of heterozygotes, far more of one is produced than the other.

This brings me to the point that I really want to emphasize. That is that the theory of evolution by natural selection is structurally unstable to the assumption of symmetrical heredity. And that asymmetrical heredity, when you write out the equations, turns out to be extremely powerful and to dominate any selection process. The result is if it were common that heredity were asymmetrical; i.e., that the particular DNA sequence that makes big A, for reasons only of its sequence and having nothing to do with the way it influences and is translated into organismic properties, if that sequence, for some strange reason, went into 90sperm, then in fact the population would go very rapidly to 100big A, even if that were bad news for the organism. Indeed we have many such cases known in nature: famous cases of alleles in wild populations of house mice in which a lethal gene is maintained in very high frequency in the population, precisely because the selection process which is getting rid of the gene is reversed by the asymmetrical process of production, and you get in fact an equilibrium.

Q: So in which of those two processes do you place errors?
A: You mean stochastic?
Q: I mean like mutations.
A: I'll get to mutations.
Q: OK.
A: I'm trying to make a particular point. A characteristic of the organic evolutionary process is that by and large-heredity is symmetrical and therefore any directional movement is a consequence of the sieving process and not a consequence of the productive process. But the productive process can be asymmetrical, and in that case it dominates the selection process and can even nullify the selection process because the actual dynamics are such that it's like an infection, an infectious gene that sweeps through even though it's bad for the organism.

So if you're going to think about analogs of the evolutionary process—and your objects are mortal then you have to be aware of the fact that the productive process is important. For example, cultural evolution has been modeled not using genes as the causes of heritability, but the transmission of ideas from generation to generation, from parents to offspring, and so on. If the transmissional process which is modeled is not exactly symmetrical, then the transmission process itself drives the direction. You have to be very careful of this in your model. That's the main point I wanted to make.

2. Variation

Now let's talk about the variations. The second feature of such a selection process is that the selection process itself negates the process. That is to say if you're going to have a directional selection process which changes the ensemble frequencies in a given direction, then pretty soon the system goes completely through one type and it comes to a grinding halt. So we have the question of the generation of the variation. Why is the population varied? And that means that we have to consider the production of the variations.

By the way, just a quick example of what happens when you don't have mortality. The best example of a natural selectional process is panning for gold. If you pan for gold, the water carries away the lighter particles, leaving behind the heavy ones. Pretty soon you have nothing but heavy particles. Well, not nothing, but you have great enrichment. However, gold is immortal, there is no production process that has to be considered. There are no genetics. You again have to consider whether the system that you want to analogize to a biological evolution process has the property that you have to consider the production process or whether the objects are immortal.

Now I want to come to the question of mutation. First, let me emphasize that all that is required for heritability is that the bias of the distribution introduced by the selection process be partially or completely reproduced. Only the statistical bias. All that's required is statistical heritability; i.e., that the distribution is biased. It does not require that any individual object in the ensemble have any similarity to its parent object. Any system of reproduction which will perpetuate some fraction of the distributional bias will do. This is quite important. You don't have to have genes, you don't have to have DNA reproduced, you don't have to have anything so long as you have some mechanism of the reproduction of part of the motion of the ensemble. Now in organic evolution we have genes. Of course we also have other passages of similarity from parent to offspring, including where you are born, which is passed from parent to offspring for most organisms who actually move a finite distance, and also various kinds of nurture and also cellular components, like mitochondria. I mean there's an awful lot of physical heredity which is not genes. So, by the way, cloned people would not truly be clones because they wouldn't have the right mitochondria. So not very many genes are in the mitochondria, but they're very important genes.

Now this brings us to the question of what the basic properties of the mutational process are when we're talking about organic evolution. Mutational process have three properties in organisms. The first is that the probability of a change of any one nucleotide is extremely low: about 10^{-8} per cell division per nucleotide. It's a very weak process. That turns out to be important because a lot has been said about the essential feature of genetics, namely there's no directed mutation; i.e., if an organism is living in a hot place, there is no greater tendency for mutations to occur that make it possible to survive in the hot place. It turns out the thing is not really unstable for that assumption because the frequency of the mutation is so low anyway that mutation as a process cannot drive evolution. Even if there were selective mutations, they're still too rare to drive the evolutionary process. So that's an irrelevancy. So it's very low frequency, about 10^{-8} per nucleotide.

Second, there are a hell of a lot of nucleotide. And that means that any individual may be mutant in an immense variety of ways. That is to say there's a very high dimensionality and you may take steps almost any direction. Therefore two individuals are almost never new variants of the same kind. This has very important consequences for a stochastic model because in a stochastic model we have finite populations, and not only that, but in the segregation process there is a randomness in the segregation of genes. A new mutation is occurring in essentially in any population of any size, of any reasonable size once and once only in one individual, heterozygous individual in the entire ensemble. The probability that that new variant will be lost to the population by a random step is very high. In fact, it's N^{-1} over N, where N is the total number of genes in the population, twice the population size. So almost inevitably every new mutation that actually occurs is lost, even when that mutation is favored. We have a random walk on the line, and the absorbing barrier is right next door. Remember that. Even if you take one step to the right away, the next time you are likely to take two steps. And that's very important. New variation of a particular kind is very rare. Therefore the effective rate of appearance of any particular change per unit time, the effective rate, by which I mean getting to high enough frequency so that it matters, is extremely slow. You have to remember that.

The third thing is that, despite the fact that there are a very large number of nucleotides and therefore the directions in which mutations can occur are extremely large, the measure of the sphere of accessible states from where you are in the total space is extremely small; i.e., the problem of the states of accessibility of that mutation process, even when you discount the fact that most mutations are lost, is that you can't get there from here in one step or even two steps or a few steps mutationally if you just choose some arbitrary place where you want to get. The space of accessibility has a very small measure.

Fourth, many of the mutated steps that can occur fall in the same functionally equivalent set. But this has a very odd consequence, which is important to recognize. I'm really overlapping on the selection part, but I need to make the point now: Here's where you are, there's some sphere of accessibility around you by single steps, by mutational steps. Some subset of those will be functionally equivalent from the standpoint of natural selection. Selection won't care whether it's this one, this one, or this one. All will do, so to speak. So all of those are OK. But what you have is then, drawing this, we move to there or there or there, so there are new spheres around each of these of accessibility by mutation. But although these three are functionally equivalent, it does not follow that the next step is functionally equivalent. As the sphere expands, the mutations which can occur from here now are not necessarily functionally equivalent to the ones that can occur from here. So although these fall in the same functional set, these mutational ones do not. The consequence of that is that the actual evolutionary process under the influence of selection and mutation is a process of going through a maze in which at one step three different walks in the maze may be OK, but most of them are dead ends because what you've got to do is to follow either one or a small set of paths, a very small measure through the space, and get to there, and even though selection has said, "Yes, that's OK," and may even delude you into going a step further, you will find that you're headed in the wrong direction. And this is not just theoretical. We have experiments in which the attempt to change the function of an enzyme in three steps has shown this very nicely. In experiments in which you make mutations, you select for the first step, you get five different mutations, all of which do the trick, you then try to go to the second step. Four of the five are unable to mutate to the next step, even though they were functionally equivalent after the first step. You get some mutations from here and you get several from here, but again, only one of those turns out to lead onto the next functionally equivalent state. So it's a feature of the topology of the accessibility of mutation that the selection process over multiple mutations is threading your way through a space of very high dimensionality along a pathway of very small measure. And that has to be understood if you're going to analogize or if you're interested in this process. This means that most things that could happen and would be a good thing to happen don't happen because first of all mutations are effectively very rare, and secondly the organism will make the mistake of accepting, let me put it that way, or choosing or favoring mutations that look good now but they are on the primrose path.

Q: Could I just ask about new variations being difficult to get off the ground? Is that the situation that there has to be a small population to get off the ground?
A: No, you don't have to have a small population. The small population of course has the property that a new mutation is now at high frequency because it is one out of a small number, but could I get on to that at another point? I mean it would involve me going on.
Q: I was just wondering how new variations get off the ground at all.
A: Originally new variations get off the ground by a purely stochastic process. That is to say that it is almost never the case in real organisms that the bias in favor of reproduction of a new variant is sufficiently high that it can overcome the random walk element. It's a random walk of a drunkard who is not exactly on a flat surface, but the surface is so slightly tilted. In fact, it's the signal to noise ratio that matters. The distribution can actually be shown to be a signal to noise distribution and therefore it's a product of the selection intensity times

population size that matters, and most selection biases are very small, so it's really a stochastic matter to get high enough in frequency. The actual life history of a new mutation: most new mutations appear and disappear immediately. Another one may actually go up a little, come down, then go out. Another one goes up and comes down. And then one by sheer chance reaches high enough frequency to protect it from the immediate loss, and now it may in fact eventually go to fixation. It's a typical random walk process with a very slight drift from a physicist's standpoint, a very slight potential.

Q: If population has room to grow exponentially temporarily, then if the mutations are going to be a couple of generations, you'll have a huge number of individuals who have the mutation, which doesn't seem quite consistent.

A: Well, there are two points to make. What determines the probability of fixation in the random walk is the initial frequency, not the initial absolute number. But second, speaking as a biologist, the famous populations that grow exponentially are a curiosity of the bacteriological laboratory. Or, well, there are sometimes extreme reductions in population size, then a chance that for a few generations they grow larger, but what matters is the relative frequency. The drift term depends on, if you're one out of 1000 or ten out of 10,000, the fixation property is the same. It's proportional. If you're one out of 10,000, of course, that's a different matter. And it might easily turn out that in the population expansion, expansion does not include that one individual. We have to see the first few steps. So this is the point about mutations. I'd like to go on if I could. Yes.

Q: Could that be viewed as some sort of a threshold for the selection process?

A: Well, there's no appearance of any threshold, there really isn't. It's just that it's a random walk with a very slight potential function in it, and most of the loss occurs here, and very little of the loss occurs once you get here. I think that's the best way to describe it. There's no threshold. The next thing I want to say is that mutation's not the only source of variation. The second source of variation, genetic variation, is the recombination of mutations of alleles, the mixing in of new combinations. Now if there were no interactions between individual genes, that wouldn't be particularly interesting, but still be of some interest. It's the recombinational process which is interesting. That is to say it takes an ensemble which has a certain distribution of types in it and produces from it by a purely recombination process new types. And the rate of recombination between adjacent nucleotide pairs is exactly the same, essentially the same, as mutational rate, about 10^{-8} per adjacent nucleotide pair, but recombinations over longer distances are much higher in frequency and recombination of the total process is fairly common.

Now it's extremely important to understand that when we come to the selection process we cannot consider the selection process in the absence of this recombinational process because recombinational process is part of the reason why the difference created by the selection is not reproduced after the production process. Because recombination—sexual recombination—is orthogonal to the selection process, and therefore enters as a noise term. That is to say you put some information in and then you destroy a lot of that information accumulated by taking random steps because recombination process simply is an orthogonal selection. A proper understanding of the way in which the population moves through the state space requires that you simultaneously consider the selection process and the recombinational process, and when you write out the equations for that—you make any kind of formal model—you find that the recombinational process and the selection process can be and often are of the same order of magnitude with the same time constant, and therefore they have to be treated as coupled. However, sometimes the recombination process is much larger, faster, than the selection process, in which case you can decouple it. You can consider the population is on a slow-moving equilibrium manifold with respect to the combination, or the combination has occurred and it's on equilibrium, and then the selection process moves that equilibrium fully through space, always staying at equilibrium for the combinational process. The issue, as I say, depends on the jurisdiction—for any particular system that you want to model, are recombination and selection on the same time scale or on different time scales? It's exactly like the molecular problem: you rapidly reach equilibrium through molecular action, and then move slowly along that equilibrium, or both are occurring simultaneously . For many years it was assumed by evolutionary geneticists that they were on different time scales, that recombination was so large that the population simply went to a recombinational equilibrium and then selection moved slowly along that equilibrium. But it turns out that's not true. For a lot of recombination, it's a complex process which both have to be considered simultaneously. It turns out that the very simple equations for this process, even in the case of two genes, are not soluble generally and have to be treated by numerical methods, and I want to say a quick word about bifurcations here.

Lest you think that bifurcations are a phenomenon of complex relations, I would like to point out to you that at one time it was thought that the structure of equilibria under selection and recombination had built-in bifurcations—with many different state of equilibria possible simultaneously for one of the parameters. It turns out that that was a complete artifact of symmetries put into the selection parameters, and the moment you break the symmetries, all of those bifurcations disappear and the trajectory through the space of the equilibria is a single, continuous string which can be represented on an infinite torus. Bifurcations don't really exist. Moreover it turns out that all the selection models can be transformed by continuous deformation, one into the other, and the outcomes are continuously transformed. It just turns out to be the nature of the equation. There's nothing universal about that. That's the way it is.

And this brings me to another pedagogical point about work. This is a space of gene frequencies, X between 0 and 1, no alleles have a frequency less than 0 or greater than 1. And there is some recombination parameter which can be as small as 0 and biologically not much larger than .5. It can be .51 or something like that. And when you look at the result of some selection model, you will get some picture of the frequency of a type against the recombination. And you say,

"Ah. Below a certain selection intensity there are no solutions at all." But that's because as biologists you forget that what you have are a set of equations, not a set of organisms, that in fact these equations behave that way in a biologically unrealistic frame. And the correct way to do the modeling is not to consider the biological constraints on the parameters and the variables to begin with. Once you've got a set of equations they behave in a certain way. Examine their topology and then put on them the frame of biological reality. And then you'll discover that all kinds of situations which appear to be discontinuous, situations in which equilibria disappear or appear suddenly out of nowhere are in fact nothing but the result of continuous deformation such that the solutions appear in the biological frame of reality. And that's a way of doing your business which is tremendously revealing, and which it makes it possible to unify what appear to be completely disparate results from different models of selection, and that's something that people have realized only recently.

I have no time left at all. And so I want to say one more thing and stop. And that is that the standard model of selection is a model in which there is an environment in which the organisms are being selected, and therefore the organisms can be regarded as moving on some kind of surface, so-called adaptive surface, an objective function, and the rules of genetic evolution are such that you can only do local hill climbing. There's nothing in genetics which enables a population to see the global maximum, where it is. You can only see the local maximum. And usually there are many local maxima of more or less equivalent height, and it depends on where you start, where you wind up. The problem of using genetic algorithms to solve engineering problems suffers from the same problem that biology suffers from: namely that we don't have a mechanism for solving problems at a distance. In fact, the problems don't exist for the organism if they're at a distance. An organism does not know that it would be a good thing to fly if it doesn't have any wings. And one of the chief problems in evolutionary reconstruction is how we select for novelties. The most famous case is wings on insects. Very small paddle-like wings don't give you any lift, and you can prove that yourself with a couple of ping pong paddles and see what it does for you. It's because of the nature of the physics of aerodynamics. Only when they get big enough do you get the lift. And this brings us to the possibility that selection is increasing the size of the wing until it's large enough, so now flying becomes a potential problem, namely it happens to incidentally give you lift. This is the notion of pre-adaptation, the idea that you get unexpected consequences.

Finally, and I wish I had another hour, the whole model is wrong. Because organisms do not get selected in an environment. That is to say the equations for change are not the dO/dt, where O is organism, is some function of organism in environment. That's the classic generalized picture of evolution, of evolutionary genetics. You have an organism, it's variable, you have an environment, it may be fluctuating or it may be constant, it doesn't matter, and you can write the equation, the dO/dt as a function of O and E. The fact of the matter is that organisms are themselves adaptive machines or functioning machines who are changing their environment in which they live in every instant, which are selecting the environment, which they are in fact, just as there is no organism without an environment, there is no environment without an organism. There's a physical world, but there's no environment without something to be environed. You have to be around something. And the thing determines what's around it. There's a sphere created. So the correct set of equations is the dO/dt is a function–F–of O and E, and dE/dt is a function–G–of O and E and we have almost no information about G. F is given to us by Mendel's rules, by physiology and so on. But G requires an understanding of the natural history of organisms and the way in which they remake their local world, which we don't have at all. And unfortunately, any conclusion you come to about generalities in the evolutionary process which do not take account of the fact that this is happening, namely that they're not moving on a fixed surface, a fixed objective function, they're moving on a trampoline in which every step that the ensemble takes deforms the surface, and without that knowledge, I don't know what to say.

Chapter 3

Psychology and corporations: A complex systems perspective

Jeffrey Goldstein
Adelphi University

(See Chapter 24.)

Chapter 4

Genome complexity (Session introduction: Emergence)

Charles Cantor
Boston University

I don't work in the field of complex systems analysis per se, although I work on a complex system. So this gives me the opportunity to relate a story. The man who was my advisor as a freshman in college, a Professor named Jack Miller, a well known nuclear chemist, told me two things which were very influential in my life. The first, which was very reassuring, is that every professor only gives one lecture; they change it to conform to the circumstances, but they really only know one lecture, so I don't have to feel guilty telling you about just what I know about. The second thing he did was taught me was not to become a biologist, which wasn't permanent, but it was helpful at the time. So I work on the human genome, which I guess is a good example of a complex system, and I'll relate two anecdotes, to try to give you some focus for this meeting. We have 100,000 genes. Each one is a set of instructions to make a protein. And we today essentially know at least a snippet of the information, about almost every one of them. This has gone much more rapidly than people thought when the genome project was started because the technology got better. Within a couple of years we'll probably know all of those genes in their entirety. So we have a complete parts list almost to the human, and we do have a complete parts list for about 10 to 15 simple organisms. However, we don't know what most of those parts do. So it's a very embarrassing situation to be in, although we'll find out eventually. But we do know that extremely subtle differences in those parts have dramatic consequences, and that the spectrum of these differences is much more elaborate and complicated than anybody thought in 1985 when all this started. For instance, there's a gene called p53 (it's a terrible name) that is altered in many cancers. We know something about what this gene does—but the thing that's frightening about it is there are already 6,000 different point changes that have been detected in that gene in different people. And those changes lead to different cancer presentation. It's not clear which changes are cause and which are effect, so one has to be careful. But just in this one case alone, sorting out the mess is going to be remarkably time-consuming.

The other example I want to give you is a very simple one. We know in some cases diseases are simpler; we know the exact molecular changes that lead to certain kinds of disease. We can absolutely mimic those changes perfectly in experimental animals. We can make the precise change in their precise gene, and they don't get the same symptoms. Now that's very discouraging for people who want to build animal models, if you think of what happens in humans there are things we call a certain disease, but the symptoms varies enormously from person to person. The reason for that is that we, unlike laboratory animals, are very outbred and diverse. A number which I can tell you is that any two in the room differ from each other at the level of DNA sequence quite substantially. Each of you have 6 billion base pairs of DNA. You all know by looking around you that everybody is different. Current estimates are that at the DNA level there are 6×10^6 differences between any two people. The population of the earth is around 5×10^9. So the total DNA sequence diversity on the planet is something like 3×10^{16}. That's a pretty big number. In contemplating how we're going to untangle this diversity, the thing that's reassuring is that biological function, what do organisms do, is remarkably similar, whether it's simple bacterium or a person they breathe oxygen, make carbon dioxide, have a very similar small molecules that should carry out most functions. So there's an enormous commonality. There are even people who are argue, as David Botstein does, for example, that we're going to be able to work out biology of most human diseases in simple organisms like yeast but I disagree with this notion. I don't think he could find a manic depressive disease, even find a cardiovascular model in yeast. So that while certain kinds of problems will be very easy to solve in simple organisms, most problems are going to require at the level of mammals if not humans. I've tried to give you just a little bit of a flavor for where I'm coming from, and I'm sure that occasional biologists throughout this meeting will probably touch on other aspects of the field which we call genomics today, which is the study of whole genomes. Suffice it to say that to make a working model, a systems model of even the simplest genomes with a few thousand genes, a few thousand is completely beyond anything that's possible today, but hopefully it will be possible in the future. Let me add one cautionary note to all the speakers that follow. I've never in my life have seen such a diverse program and such a diverse audience, and so I tried in my comments to avoid almost any jargon, and I ask each of the speakers that follow in this morning's session, which is going to go from molecules to the planet, try to forget those specialized words that are used in discussions with colleagues, and use plain English insofar as it is possible so that the audience will be able to understand your problem. There are four speakers this morning. Each of them I hope will talk briefly enough so that we have a little bit of time for questions. That's what makes this fun. And the general topic is emergence, which Yaneer tried to define it for you his way, but I didn't quite get it. So I guess I would define it as emergence means that the whole is more than the sum of the parts. That's what he was trying to get at. The first speaker, is Daniel Kaiser, and he's going to talk on emergent properties and behavior of the atmosphere.

Chapter 5

Emergent properties and behavior of the atmosphere

Daniel Keyser
State University of New York at Albany

The title of this talk is "Emergent Properties and Behavior of the Atmosphere." By way of introduction, I am an atmospheric scientist or meteorologist. I originally intended to become a weather forecaster, but I changed course in graduate school to become the academic equivalent of what I originally intended—a professor who performs weather-related research. Thus, I am visiting as an emissary or a representative of the weather-research community.

In his invitation, Professor Bar-Yam encouraged me to relate my knowledge of the weather to complex systems in general and to the concept of emergence in particular. By necessity, I will talk about my perceptions of complexity and emergence from the perspective of an outsider to the complex systems community. The talk will begin with a brief overview of the atmospheric system and its governing equations, which will be followed by a classification of atmospheric phenomena and processes according to scales of atmospheric motion and so-called "scale interactions." The talk proceeds to address the topics of phenomenological studies and atmospheric predictability, and concludes with several research issues related to these particular topics that are receiving the attention of the weather-research community at this time.

I start with several perceptions of complex systems from my meteorological vantage point. A quote from Simon [8, p.86] stating that complex systems are composed of "a large number of parts that interact in a nonsimple way," and that in complex systems "the whole is more than the sum of the parts," suggests to me that in complex systems we do not have pure superposition of phenomena and processes. These ideas are familiar in that complex systems lie somewhere between the extremes of strict linearity and full nonlinearity, that is in between circumstances where interactions are straightforward to analyze and understand, but aren't necessarily very interesting, and full nonlinearity, where interactions are too complicated to keep track of and comprehend from a deterministic point of view. This sense of intermediacy will come up when I discuss the concept of scale interactions. Another familiar theme is related to the question of what happens to the properties of a system if a component part is removed: this question comes up in the context of atmospheric predictability.

Next, some discussion of the atmospheric system and its governing equations. Paraphrasing Atkinson [1, pp.1-2], the connection between observed weather and climate and the concepts of fluid dynamics may be found in the motion of the air. The general goal is to understand the dynamical behavior of an atmosphere that comprises a continuous, stratified, turbulent fluid, of order 100 km deep, confined to a rotating spheroid possessing a highly irregular lower boundary (containing continents and oceans) and subject to differential heating and gravity. Atmospheric motions are described by a coupled nonlinear system comprising five partial differential equations and one algebraic equation: the vector equation for the three components of three-dimensional motion; the thermodynamic equation for temperature; the mass conservation equation for density; and the equation of state relating density and temperature to pressure. In actual weather prediction, where consideration of the hydrologic cycle is crucial, it is necessary to consider an additional mass conservation equation (actually a set of mass conservation equations) governing the evolution of water substance (more precisely its constituent phases). This system of equations is solvable, in principle, for future atmospheric states given suitable specifications of the initial state; boundary conditions; and physical processes, such as radiation, energy exchanges involving phase changes of water substance, and frictional dissipation.

Realization of the goal of understanding and predicting atmospheric motion requires us to confront a number of complications. The first of these is that the governing equations, as formulated, apply to abstract entities referred to as "air parcels," rather than to fixed points in space. Air parcels are sufficiently large for statistical properties of molecular motion such as pressure and density to be well-defined, but are sufficiently small for these properties to be homogeneous throughout the air parcel [3, p.26]. Use of these equations for weather prediction requires that they be transformed from parcels to fixed points and that they be averaged to apply to grid volumes of manageable size; i.e., the atmosphere needs to be discretized in some form. The transformation from parcels to fixed points and the subsequent discretization process exposes the inherent nonlinearity of the governing equations, and the averaging procedure connected with the discretization procedure introduces the notion of subgrid-scale processes, which must be related to the resolvable-scale variables through a procedure referred to as "parameterization." Limitations in the predictive skill of the governing equations arise from imperfections in the specification of the initial state, which never can be known exactly, even for the resolvable-scale variables applicable to finite-size grid volumes; from uncertainties in the representation of physical processes; and from uncertainties in the parameterization of subgrid-scale processes.

And, finally, the nonlinearity of the governing equations ensures that, given sufficient time, small-scale errors in the initial state will infect and contaminate all other scales of a forecast, however accurate the model formulation. This consequence of nonlinearity—the growth, propagation, and dispersion of errors in atmospheric prediction models—is a central theme of atmospheric predictability.

Next, some perspectives on the concepts of scales and scale interactions. Atmospheric phenomena and processes occur over a vast range of scales, from approximately 40,000 km, the circumference of the Earth (planetary-scale waves), to centimeters (small-scale turbulent eddies). Despite the well-known property of atmospheric motions to vary on all scales ranging from the largest planetary-scale waves to the smallest turbulent eddies, it is conventional practice to subdivide the atmosphere into discrete size ranges or scales of motion. The question arises as to why particular phenomena and processes occur in the first place, and, given that they occur, explaining their characteristic scales and lifetimes. The answer to this question and the accompanying explanations may be sought in the characteristic geometry of the atmosphere (i.e., spherical shell of finite depth); in the characteristic spatial and temporal scales of instabilities of various atmospheric flows; and in the fundamental frequencies of oscillation of the atmosphere, involving stratification (i.e., the buoyancy frequency), rotation (i.e., the inertial frequency), and the latitudinal gradient of rotation (i.e., stationary planetary Rossby waves). Although it is conceptually convenient to categorize phenomena and processes according to scale, all scales eventually will influence all other scales because the governing equations are nonlinear [3, p.13].

The interdependence of atmospheric motions of all scales is referred to as "scale interaction." To quote Kerry Emanuel [4, p.13], "Scale interaction' is at once one of the most popular and one of the most ambiguous phrases in all the meteorological literature." Please allow to me to continue by quoting directly from Emanuel [4, p.14], since I believe that you will find complex systems residing within his hierarchy of progressively general scale interactions:

"In a very simple system consisting of a slowly varying mean flow with a very weak disturbance upon it, the main scale interaction can be considered to be the influence of the mean flow upon the disturbance. As the perturbation becomes more substantial it exerts an increasing influence on the mean flow, and other scales of motion develop because of secondary instabilities. The scale interactions become more and more numerous, and the general degree of disorder becomes greater. In the most highly nonlinear systems such as fully turbulent flow the scales interacting in a complex way are extremely numerous, and an explicit description of their interaction becomes problematic; one then resorts to stochastic treatments of the interactions. Perhaps the most difficult flows to deal with are ones for which the degree of disorder is too great to allow explicit treatment of the dynamics but too small to allow a simple stochastic treatment."

I now move on to the topic of phenomenological studies. Here one turns to the discipline of synoptic meteorology, which may be defined as the use of observations from a variety of sources to describe the structure and dynamics of atmospheric phenomena through their life cycles and to identify their interrelationships with other phenomena, especially those of different scales [7, p.307]. This definition emphasizes understanding atmospheric phenomena in terms of their interrelationships with each other rather than in isolation, which points to one of the defining characteristics of emergent behavior. As an example of how phenomenological studies are conducted, I will spend a few minutes recalling the so-called "Superstorm" of 12-14 March 1993, which influenced a large portion of the eastern United States and which should be very familiar to many members of the audience. A key to understanding this historic meteorological event is its multiscale nature manifested in the multitude of weather phenomena that occurred during the course of its lifetime. This event did not simply feature a record-breaking snowstorm in terms of amounts and areal coverage of snowfall, but also contained severe thunderstorms and a tornado outbreak over Florida and a storm surge affecting the northwestern coast of Florida.

Stepping back and examining the planetary-scale flow evolution preceding the Superstorm indicates major changes leading up to its development: as you examine the midtropospheric flow for the Northern Hemisphere on a daily basis between 8 March and 13 March 1993, you can see how the flow becomes increasingly well-organized as the hemispheric wave pattern amplifies and a deep trough forms in the eastern United States. You also can confirm the extent to which the gradients of geopotential height increase in midlatitudes over the five-day period shown here (note how the contours delineating the wave pattern come closer together over time); this increase is manifested as an increase in kinetic energy. At the same time that the kinetic energy is increasing, a different type of energy, the so-called "available potential energy," a diagnostic indicative of the amount of potential energy available to be converted into atmospheric motion, is decreasing dramatically. A time series of available potential energy for the Northern Hemisphere, calculated by Bosart et al. [2], shows the evolution of this diagnostic from mid-February through mid-March 1993. One sees the gradual buildup of available potential energy, reaching a maximum about 9-10 March, and then the sudden, dramatic decrease over the next five days as the Superstorm forms. Another fascinating perspective of the antecedent flow evolution in the month preceding the Superstorm is found in the trajectories of several upper-tropospheric coherent structures that contributed to the development of the storm. One of these trajectories may be traced back to 24 February 1993 to a point of apparent origin over southwestern Asia. Another feature originated near Seattle, WA, on 19 February 1993, traveled all the way around the Earth, nearly intersecting its point of origin, and then continued to merge with the aforementioned feature over the eastern United States on 14 March 1993, when the Superstorm was at or near peak intensity. This evolution holds potentially important ramifications for atmospheric predictability, because it suggests that some forecasts of cyclogenesis may be limited by the accuracy with which antecedent features, such as the two coherent structures considered here, are observed and predicted.

Atmospheric predictability addresses, among other topics, the perennial question of why weather forecasters cannot achieve the same accuracy as astronomers in their predictions of the motion of celestial bodies, such as comets, planets, and stars. To respond to this question, I refer to the chapter entitled, "The Butterfly Effect," in the popular book, "Chaos" [6, pp.9-32], which draws upon the pioneering work on predictability by Edward Lorenz. Astronomers do not achieve perfection, but they receive a lot more credit from the general public than atmospheric scientists because they are dealing with a much more periodic system than the atmosphere. Weather forecasters assume that the periodic part (the annual and diurnal cycles) are well known and not especially interesting. We know that there are well-defined seasons in New England and that fall has just begun (there was a touch of frost on the cars and grass this morning). But weather forecasters grapple with the perturbations upon the periodic part of the flow—the so-called nonperiodic part. Gleick [6, p.8] proceeds to introduce one of the most important concepts in atmospheric predictability—small differences between given atmospheric states may amplify dramatically in very short periods of time—referred to as "sensitive dependence on initial conditions" and the "Butterfly Effect." As noted by Dutton [3, p.14], this behavior should be distinguished from traditional linear thinking, where small causes produce small effects such that if flows start from slightly different initial conditions, they should remain relatively similar over long periods of time. When considering atmospheric predictability, one not only is concerned with the sensitivity of flows to differences in initial conditions, but also to boundary conditions, and to the approximations used in simplifying the governing equations.

Another perspective on atmospheric predictability is that some flows and their constituent phenomena are more predictable than others, which is sometimes referred to as the "regime dependence" of predictability. Given the importance of and intrinsic scientific interest in identifying which flow regimes are more predictable than others justifies continuing research on the structure and behavior of weather phenomena, especially those that are hazardous to society. Such phenomena include extratropical cyclones (the same type of system that organized the record-breaking snowfall in the Superstorm of 1993) and tropical cyclones (referred to as hurricanes when occurring over the North Atlantic and eastern North Pacific Oceans). It has become increasingly appreciated that extratropical cyclogenesis often results from the interaction between wavelike and vortical features found on the tropopause (the boundary separating the troposphere and stratosphere) and comparable structures based at or near the Earth's surface. These vertical (and also lateral) interactions can explain the existence and behavior of the anticyclones and cyclones, as well as the frontal systems, routinely displayed on surface weather charts. (In regard to tropopause-based vortical features, recall the example of the history of the predecessor disturbances that contributed to the development of the Superstorm of 1993.) Incidentally, the tropopause plays a fundamental role in the exchange of dynamical and chemical properties between the stratosphere and troposphere, so that its behavior is of great interest from an environmental, as well as predictive, point of view. With respect to tropical cyclones, despite recent improvements in the predictive skill of the motion and tracks of these systems, there presently is little skill in forecasting changes in their intensity, which often may fluctuate rapidly. As in the case of extratropical cyclones, the intensity of tropical cyclones may be strongly affected by interactions with their environment, including structures and processes involving the tropopause—evidence, yet again, that atmospheric systems and flows cannot be understood and predicted in isolation, but depend critically on interactions with other systems and the environment.

I conclude with some remarks on present research investigations in atmospheric predictability being conducted by the weather research community, drawn from the paper by Emanuel et al. [5]. One issue is how the predictability of certain phenomena and flow regimes may be optimized for a given amount of cost and effort in observing the initial conditions of the atmosphere in areas deemed critical to the success of a given forecast. A prime example of where lack of observations in data-sparse regions is believed to make a significant difference in forecast quality is tropical-cyclone track and intensity, referred to above. Furthermore, an overall lack of data over the North Pacific Ocean and northern Canada is believed to compromise forecast quality over Alaska and the West Coast at all forecast ranges, and to affect the remainder of the United States over the short and medium range (2-7 days). Two technological innovations are especially promising in filling these so-called data voids in a cost-effective manner: The first is the advent of programmable observing platforms, which offer the potential of monitoring the atmosphere "adaptively," when and where observations are required to make a difference in forecast accuracy. The second is the operational production of ensembles of numerical forecasts in real time based on slightly differing initial conditions, which provide the means to determine quantitatively those regions of the atmosphere that are susceptible to rapid error growth and thus need to be observed selectively with greater accuracy than other regions. (The preceding statement presumes that the distribution of analysis uncertainty is known, since an adaptive observing strategy needs to consider not only regions where errors in initial conditions will amplify rapidly, but also those regions where errors in the initial conditions are large.) These ongoing developments show how principles of atmospheric predictability, an understanding of the structure and dynamics of atmospheric phenomena, and the numerical weather prediction enterprise presently are being combined to refine and extend the skill of weather forecasts toward their limits of deterministic predictability.

I close here. Thank you very much.

Q: It's hard to understand exactly what you mean by sensitivity to initial conditions in a system that's always there.
A: Well, I am referring not to the atmosphere itself, but to numerical simulations of it where we don't have exact observations and we never know exactly the state of that system—which you say is always there. In other words, the notion of sensitivity to uncertainties or errors in the specification of initial conditions arises when we attempt to simulate the atmosphere with approximate models and imperfect data. Hope this helps.
Q: It helps a little bit. The notion that sometimes the sensitivity to initial conditions is there and then sometimes it's not, what determines that?

A: I don't believe that a general answer yet exists to this question. Because I am not a specialist in any sense of the word in atmospheric predictability I cannot say a lot more, but my impression is that while infrastructure is being developed to diagnose regional or local sensitivity, a dynamical or conceptual framework for understanding the results of such diagnoses remains for future research.

Q: OK. In connection with that, is the lack of predictability really due to your inability to get enough data, or is it really something wrong with the model itself, the basic paradigm you're using?

A: I believe that part of the answer lies in the nature of the equations of motion and their practical application, and part with the inability to observe the Earth—atmosphere system perfectly. From what I understand—and others could answer you with greater authority—the equations are fundamentally approximate for all the reasons I mentioned earlier. We have to go from parcels to points, we have to go to finite volumes, we have to average—that is, we have to go through a whole set of steps to obtain a predictive system that is tractable. And. finally, we don't have perfect observations down to the molecular scale. In regard to this last point, my recollection is that even if the predictive equations were a perfect model of atmospheric behavior and if these equations did not have to be approximated in any way to enable solving for future atmospheric states, the absence of an exact initial specification would remain as an insurmountable barrier to the realization of perfect deterministic weather forecasts.

Q: It looks like the picture that you've been painting is writing a set of equations and finite-differencing them and getting a larger and larger computer to solve them. That's kind of working your way from the bottom up, and I was wondering if people had worked out any other method that would be useful, instead of brute forcing.

A: I can't rule out alternative approaches to the numerical weather prediction paradigm, although I am not aware of any. Nevertheless, it is standard practice to use statistically based methodologies to refine and extend the utility of numerical weather forecasts.

Q: As to the previous question about sensitive dependence on initial conditions, it's actually a feature of physics having to do with our interpretation of it, but you start off with two systems which are slightly different and you ask how do they diverge in time, and the difference between the two states may sometimes decrease or sometimes grow exponentially, and because the exponential gets large after several e-folding times, there will be some physical conditions under which the smallest error will amplify to a very large error after a certain length of time.

Q: On this same question of sensitivity, I would guess from the descriptions you've been giving that differences in initial conditions do not amplify uniformly, and so what you would typically see if you started two systems with slightly different initial conditions is that the difference might grow only slowly for a while and then after a rather unpredictable length of time there would be a period of exponential divergence.

A: That's right. The error growth is nonuniform in space and time, and right now a fundamental research problem is to try to understand just those points you're making, as well as how to apply them operationally.

Q: When you talk about scale, are you talking mostly about spatial scale?

A: And time.

Q: Could you talk about the interaction between time scale and spatial scale, and give an example?

A: A hypothetical example of a scale interaction might involve the interplay between a hurricane and its environment over the tropical North Atlantic Ocean, resulting in a change in the strength of the hurricane over a short period of time. Over a longer period, the hurricane may travel into midlatitudes, become extratropical, and modify the planetary-scale wave pattern by triggering the formation of a blocked flow pattern over Europe, which, in turn, may persist for a month. This example considers interactions going both ways—not only from large to small scales in space and time, but upscale as well.

Q: I'm from the West Coast and we always worry about the jet stream. Now the point is the jet stream keeps going constantly, it just dips every once in a while. Can you essentially say when you will ever be able to predict the jet stream?

A: Well, we are predicting the location and the evolution of the jet stream on an operational basis. Recall the diagram that I showed of the postage-stamp-sized snapshots of the Northern Hemispheric circulation looking down on the North Pole. In these diagrams, the jet stream corresponds to localized regions of large gradients in geopotential height, which are predicted as substructure in the planetary-scale wave pattern.

Q: Is there any work being done on the impact of human systems on the regional level? I know we talked a lot about global warming and other environmental concerns, but what about the impact population centers may have on atmospheric conditions?

A: An example is large cities, which can be the source of thermal and pollution plumes. These respective plumes ultimately could influence the weather and climate in downwind locations.

Q: Could you say a little bit more about the measurement of the quantity referred to as available potential energy?

A: Available potential energy consists of a global integral based on the variance of the temperature field and on its vertical stratification. Also, available potential energy is calculated or diagnosed from temperature data rather than being measured directly.

Q: Could available potential energy actually be defined on a local scale?

A: It doesn't really make sense to define available potential energy in a pointwise manner because of its global property. To reiterate, this diagnostic applies to the behavior of a system rather than to a specific point.

[Applause]

Bibliography

[1] Atkinson, B. W., 1984: The meso-scale atmosphere. Inaugural lecture. Department of Geography and Earth Science, Queen Mary College, University of London, 30 pp.

[2] Bosart, L. F., G. J. Hakim, K. R. Tyle, M. A. Bedrick, W. E. Bracken, M. J. Dickinson, and D. M. Schultz, 1996: Large-scale antecedent conditions associated with the 12-14 March 1993 cyclone ("Superstorm '93") over eastern North America. Mon. Wea. Rev., 124, 1865-1891.

[3] Dutton, J. A., 1986: The Ceaseless Wind: An Introduction to the Theory of Atmospheric Motion. Dover, 617 pp.

[4] Emanuel, K. A., 1986: Overview and definition of mesoscale meteorology. Mesoscale Meteorology and Forecasting, P. S. Ray, Ed., Amer. Meteor. Soc., 1-17.

[5] Emanuel, K., E. Kalnay, C. Bishop, R. Elsberry, R. Gelaro, D. Keyser, S. Lord, D. Rogers, M. Shapiro, C. Snyder, and C. Velden, 1997: Observations in aid of weather prediction for North America: Report of Prospectus Development Team Seven. Bull. Amer. Meteor. Soc., 78, 2859-2868.

[6] Gleick, J., 1987: Chaos: Making a New Science. Penguin Books, 352 pp.

[7] Keyser, D., and L. W. Uccellini, 1987: Regional models: Emerging research tools for synoptic meteorologists. Bull. Amer. Meteor. Soc., 68, 306-320.

[8] Simon, H. A., 1969: The Sciences of the Artificial. M.I.T. Press, 123 pp.

Chapter 6

Systems properties of metabolic networks

David A. Fell
School of Biological & Molecular Sciences
Oxford Brookes University

(See Chapter 19.)

Chapter 7

A hypothesis about hierarchies

Dan McShea

Duke University

Arguably, the emergence of higher levels of organization is a rare event in the history of life. Two billion years ago some prokaryotic cells joined to form the first eukaryotic cell, and then around seven hundred million years ago, give or take, some clones of eukaryotic cells joined to form the first multicellular organisms. It is not clear that very much has happened—hierarchically speaking—since then. We have had a few cases of multicellulars getting together to form (near) superorganisms: some insect colonies, some colonial, marine invertebrates, and perhaps some highly social vertebrate groups. But these are not as integrated at the superorganism level as the multicellulars are at theirs, and anyway a few bee hives, the Portuguese Man of War, and the United Nations is not much to show for seven hundred million years. So, apparently, this emergence of higher levels doesn't happen very easily. And maybe that's what makes it so interesting when it does happen.

There are many questions that could be addressed in this area. We would like to know under what conditions levels arise or fail to arise? In what groups of organisms did they arise and why those groups? Are levels ever interpolated? Do they dissipate as well as arise? Are they ever lost entirely? I am going to be addressing a somewhat different issue: specifically, what's the relationship between the emergence of higher levels and complexity, with complexity understood simply as number of different parts at a given level. I am going to propose a hypothesis about that relationship in a moment, but first let me give you an outline.

First, I am going to define some terms; organism, hierarchy, and bit of jargon I am creating for myself, the word "thing." Then I will explain the hypothesis and the predictions that it makes and show how I'll be testing it. I'll give you some early data, and finally discuss some possible additional tests that I am considering trying out.

The first term is "thing," and let me build up to what I mean by it here. Imagine a collection of parts, an assortment of molecules, if you like. Now consider an interactive collection of parts, maybe the gas molecules in the balloon. Now, a cohesive interactive collection of parts, say the molecules in a rock. A rock is a thing. A solar system is a thing. This talk, including all of you in the audience, is a thing.

Notice in these examples that there is an implicit further requirement, namely, a lack of cohesiveness and interactiveness at the edges or boundaries of a thing—think of it as the flip side of cohesiveness and interactiveness. A thing need not have a sharp boundary or a hard boundary, but it does have to be somewhat isolated. You can think of a thing as just a region of interactiveness or connectedness surrounded by a region of lesser interactiveness and connectednesss. These qualities come in degrees, what might be called degrees of "thingness." Some related terms: you can speak of the process of "thingification." (A terrible word—I'll try not to speak of it.) There are parts of things or "subthings." There is the unification of things to produce "superthings." In other contexts, what I am calling "things" have also been called entities or individuals, but the overlap in meaning is imperfect.

The second term is organism. I don't mean anything very precise yet by that term. I just mean a thing with functionality added. And, by functionality, I just mean the usual sort of self-serving activities that organisms engage in, like self-production, self-reproduction, internal division of labor, homeostasis, and so on. So, any internally interactive and externally isolated set of objects that can also self-produce, self-reproduce, self-preserve, and so on, is what I mean by an organism. By the way, it's not completely obvious that in order to be an organism, you also have to be a thing, though I will give you reasons later on to think that might be true.

The third term is hierarchy. By hierarchy I just mean physical nestedness of parts within wholes. Importantly, I am only talking about object hierarchies here, like the rooms in the building, or individuals in a society, not process hierarchies like an army chain of command. There are two kinds of object hierarchies: hierarchies of things and hierarchies of organisms, plus combinations of these. A given system could be things at one level, and organisms at the next level, and things at the next level, in any combination that you can think of. In all of these, there's a simple part/whole relationship, and you will hear me using the words part and whole often instead of thing and organism. I will do this when it doesn't matter whether the system is a thing or organism at that level.

Now the hypothesis: I call it the "hierarchy hypothesis," because the phrase is alliterative, and I couldn't think of a better one. The hypothesis is this: as functionality arises at higher levels—for example when a group of metazoans get together to form a society or a superorganism, or some free-living cells get together to form an multicellular organism—the number of different sub-part types is expected to decrease. Suppose we're talking about organisms getting together to form a superorganism. The logic is simply this. As the superorganism emerges, it takes over functions previously performed by the component organisms, and their parts—i.e., the subparts, become unnecessary. Then, natural selection is expected to eliminate those sub-parts in the interest of economy. An example: compare the life of the human skin cell with that of a free-living single cell organism, like an Amoeba. A skin cell has a pretty easy life. Food is delivered to its doorstep. There is no need for defense, no need to move. It gets central heating and central air. And because there are fewer functional requirements, the skin cell is expected to require fewer parts; the Amoeba, on the other hand, has to do it all for itself.

Notice that the hypothesis need not be true. There are a number of reasons it could turn out to be false. One of them is that it's possible that parts get locked in by developmental constraints, that selection is unable to eliminate parts because they are too integrated into the developmental program of the newly formed superorganism. Another reason the hypothesis could be wrong: it could also be that, in order to be part of a larger cooperative activity like a superorganism, you actually need more functionality. In other words, in order for a component organism to play its role in the superorganism's physiology and development and so on, it actually has to perform more functions. By the way, I have been speaking casually here in talking about loss of parts. I should always say number of different types of parts. What matters is number of types, not total numbers of parts.

This logic is related to, actually a generalization of, the standard explanation in evolutionary biology for the simplification of parasites. The host takes over many functions, so the parasite is expected to lose parts, the argument goes. Of course, conventionally, we don't think of the parasite-host relationship as an organism-superorganism relationship, but from the point of view of the parasite, it really is, especially when the relationship is an obligate one.

So, to summarize, "organismification" at an upper level results in a loss of parts two levels down. And a corollary would be that "thingification" at an upper level results in no loss of parts, because there's no functional take-over when the higher-level entity is just a thing and not an organism.

Some may have noticed an assumption, a kind of auxiliary hypothesis, here, namely that number of parts is correlated with number of functions. It would be nice if we didn't have to make that assumption. In other words, in order to test the hierarchy hypothesis, it would be nice if we could count functions directly. But we can't do that. We can only count physical parts. So what we need instead is a reason to think that parts are going to be correlated to functions, and I am going to digress here to make that argument.

The argument actually has three parts. First, I need to show why you would expect parts and functions to be related to each other at all. Second, assuming that you are convinced that they should be related to each other, I want to show that it is not a simple one-to-one relationship; and, third, I want to show why the correlation is expected to be good anyway, even though it is not one-to-one, given certain assumptions.

First, why are parts and functions expected to be correlated? The answer is a modification of an argument by Gunther Wagner and Lee Altenberg. First, for a system to perform a function, its internal activities must be coordinated, and coordination in turn requires internal connectedness. Second, performing a function requires minimal interference from other functions. In other words, it requires external isolation. So, for example, if you are a bird, and your functions are eating and flying, it had better not be the case that your wings convulse every time you open and close your beak. Functions need to be isolated from each other within the organism. This combination of internal connectedness, or coordination, and external isolation, is what makes a functional unit a thing, and the argument is that selection favors the isolation of functions in things, in parts, in order for them to work properly.

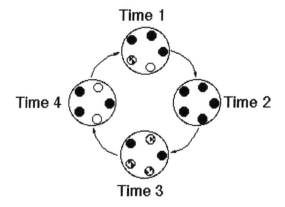

Figure 0.1:

Now, based on what I have just said, you might think that the expectation is that there should be a one-to-one correspondence between things and functions, that each part should have one function. But this is not so, and here's why: consider a function as a state cycle. The diagram (Fig. 0.1) shows a state cycle for some function, that is, for one part with a single function. The part has five sub-parts, the little circles drawn within it. And in the course of performing its function, the part goes through a series of four steps; time one, time two, time three, time four; and each of those sub-parts changes. For example, if you look in the upper right, there is a little circle there that's hatched at time one. It stays hatched in time two. It's shaded at time three, shaded still at time four, then returns to its original condition. All of the sub-parts go through state cycles like this, and in doing so perform some function—whatever it is: reproduction, homeostasis, etc.

So what happens when we try to merge functions together into the same part?

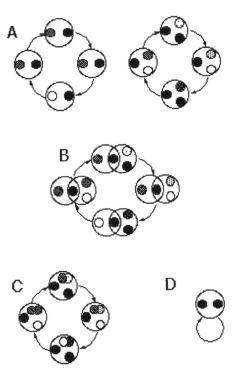

Figure 0.2:

Look at the next diagram (Fig. 0.2), and for the moment just at A. We are going to try this merging step wise. In A, we have got now two parts, each with its own separate function, and they are completely isolated from each other. Now look down at B. It's the same two parts, the same two functions, but here they are partly overlapping, that is, they share one sub-part. How is this possible? It's possible because the state cycle of the sub-part that they are sharing, the dark, black circle, is short and simple. It stays black throughout its state cycle. What that means is that both functions make the same, simple demand on that sub-part, namely that it always stay in the same state. In biological terms, the heart and brain share a part, the circulatory system, and both make the very simple demand on it, namely that it stay oxygenated all the time; that's why they can share it.

Now look at C. These two functions are attempting total overlap within the same part, and it's not going to work. It's not going to work, because we have got a sub-part here that's trying to do two incompatible state cycles at the same time. Now, this is not to say that you can't have two functions overlapping completely, sharing all of their sub-parts. You can, provided those sub-part state cycles are short and simple, as in D. A biological example would be a clam's shell that is able to perform the function of defense and burrowing at the same time. How is the shell able to do this? Well, both functions make the same demand on the shell, be rigid all the time. That's all. In other words, both functions demand a one-step state cycle, so it's easy for those functions to overlap entirely within the same part.

Here's the point: you can get partial overlap. You can even get complete overlap, but it becomes difficult or unlikely as state cycles become long and complex. Another way to put it is this: assuming that organismic functions have some distribution of state cycle lengths and complexities, and assuming that organisms have got to find ways to have these functions performed effectively by some collection of parts: in very few of those parts will it be possible for many functions to overlap significantly.

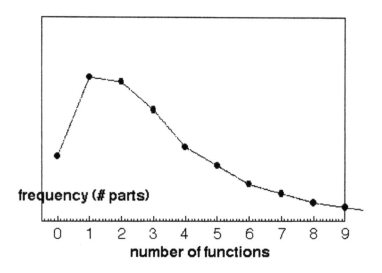

Figure 0.3:

So what can we say about the relationship between parts and functions? Consider this hypothetical curve (Fig. 0.3) showing the distribution of parts and functions in organisms. The X axis is number of functions. The Y axis is number of parts, and the curve shows number of parts having, or participating in, X number of functions in organisms generally. So the height of the curve here indicates how many parts tend to have one function, how many parts have two functions, how many have three, and so on?

I think we can claim with some confidence that the curve must be bounded at zero. We certainly can say it has a positive mean. Of course, the shape of this curve is uncertain; we don't even know that it's unimodal. But assuming that there are no strange interactive effects, it is clear that—and here's the whole point of the digression—the variance ought to be low. That is, as we move out to the right, we are going to see a quick drop-off in the number of functions that we can squeeze into a single part. In other words, the number of parts having or participating in three, four, five, six functions, and higher, is going to be very low. It follows that the variance of this curve, whatever its shape, is going to low, which means that most of the time when we sample from it, most of our samples are going to be near the mean, whatever the mean is. It follows that in sampling from this, the correlation between parts and functions is going to be relatively good. That is, when organisms sample from this curve, in structuring themselves in evolution, so to speak, the correlation between parts and functions is going to be relative good. So, again summarizing the digression, the relationship between parts and functions is not one-to-one. But number of parts is expected to be well correlated with the number of functions, and therefore we can use number of parts as a proxy for number of functions in comparative studies.

Now the test of the hierarchy hypothesis. I am going to test at a small scale, at the cellular level, comparing number of types of parts in two extreme cases: in free-living eukaryotic cells which are not part of any larger organism, and therefore are expected to have more parts, and in the cells of animals, metazoans, which are part of a larger organism, and therefore are expected to have fewer parts. I will be looking at modern organisms only, but there is no assumption of ancestry here, no assumption that the free-living cells, like Amoeba, are ancestral to the metazoan cells.

So what are the parts of a cell? Consider a generalized insect cell. It has a flagellum, it has endoplasmic reticulum, it has a Golgi apparatus, some mitochondria, and so on. Each of these is a type of part. I have to do better than giving you examples, so let me attempt an operational definition. What I mean by a part of a cell is the most inclusive sub-cellular unit that is still a "thing." Why "most inclusive sub-cellular?" That requirement is designed to eliminate all of the sub-sub-parts, and sub-sub-sub-parts. For example, consider the internal membranes of mitochondria, the cristae. Are those going to count . as parts? No, because they are more directly parts of the mitochondria, not of the cell. The functions they serve are most directly those of the mitochondria, and those of the cell are served only indirectly.

Still there are lots of problems. I'll quickly run through three of them. Parts occur at all scales, from the size of the cell nucleus down to molecule-sized parts. It would be impossible to count them all. So, I am going to limit my count of parts to those that are smaller than the cell, but still electron-microscope visible. Thus, the molecule-sized parts are going to be missed. I think this is fine provided that no bias is thereby introduced. I can think of one reason it would introduce a bias, however. It could be that cells in metazoans are communication specialists, as you might expect from a cell that is integrated into a larger whole, and communication might tend to require smaller, molecule-sized parts. I don't know that this is so. It could be that an Amoeba, in communicating with its environment, has the same number of communication requirements, and the same demand for molecule-sized parts. But it is something to worry about.

A second problem: cells pass through developmental stages, and their parts and functions often change radically in the course of that development. I am going to ignore these changes, instead counting parts in cell at a single moment in time, taking a single snapshot of the cell, so to speak.

The third problem is the most troublesome. I am not going to be using a microscope to look at these cells and to count the parts in them. I am going to be basing these counts on journal articles. And the problem is that the descriptive literature tends to focus on specializations; so, for example, a description of a retinal cell, is going to focus on the photon-receiving apparatus and may not mention the mitochondria in the cell. To solve this problem, I devised a list of what I call the "standard set" of cell parts, and I am going to assume every cell has this entire list (unless the literature actually specifies that some part is missing). Thus, all of the cell part counts that I'll show you are really part counts in addition to the standard set.

Now the results. I have eleven phyla, or eleven species representing eleven different phyla of free-living eukaryotic cells, plus ten metazoan cell types, all randomly chosen. In other words, they were not chosen based on their degree of "partedness." I picked them only because I found good cell-level descriptions. The free-living cells seemed to have more parts, an average of 5.5 compared to 3.0 for the metazoan cells. The sample size is small but the difference in means is highly significant. Actually, on account of phylogenetic relatedness among these organisms, the comparison has to be done somewhat differently than I've shown here, but doing it differently doesn't change the basic result.

Here's what I plan to do next. First, I need to increase the sample size. Then I want to test the hypothesis at the next level up. The prediction at the next level up would be that the zooids, castes, and polyps that are the parts in integrated functional colonies should have fewer internal parts than those in colonies that are less integrated and functional.

Well, let me close with some speculation. The supposed increase in complexity and evolution that Murray Gell-mann was referring to is usually understood to occur in two senses. One is the accumulation of parts at the single scale within organisms and the other is the accumulation or the addition of higher levels integration. Well, first suppose the second increase actually occurs, that higher and higher levels are added in evolution, albeit only rarely. Further, suppose that suppose the hierarchy hypothesis is true, that parts at certain lower levels are lost as higher levels emerge. If so, then as we add levels at a higher and higher scale, we must be constantly eliminating parts at a smaller and smaller scale. In other words, the only thing that is happening in evolution that complexity, is moving up in scale, leaving a trail of simplified levels below it.

Chapter 8

Session introduction: Informatics

Richard C. Strohman
Emeritus Professor
Department of Molecular and Cell Biology
University of California at Berkeley

The Human Genome Project (HGP) may be seen in terms of five distinct but overlapping phases of it's evolution. These phases evolve through technological development within HGP and from discovery from independent researchers who have been following a more complex vision of life for many years. These phases are: (I) Monogenic and (II) Polygenic determinism, (III) Protein phenotype as opposed to genotype analysis (Proteome), (IV) Model organism analysis (Transgenics) and (V) Complex adaptive systems analysis. One may detect in this series a strong movement away from strict genetic determinism where informational complexity in whole genomes is proving to be unmanageable. The transitions from stage I-V are driven by mis-assumptions and their resulting experimental anomalies discovered within each stage. As each stage develops it confronts increasing complexity and the need for some kind of complex systems analysis and non-linear theory capable of structuring the growing data bases and poviding for testable hypothesis of the system's behavior. However, for a variety of reasons, ranging from scientific narrowness within the community of experimental biologists, to the socioeconomic needs of an ever increasing dominance of technology over fundamental science, non-linear systems approaches capable of recognizing simple rules for adaptive behavior have not been embraced until stage V. Instead, modem biology has repeatedly relapsed into what has been called "blind reductionism" and the search for specific genes or gene programs that are thought to "cause" or otherwise control complex organismal behavior. However, with stage V we see the reemergence of interest in systems approachs which complement the traditional and essential process of a reductionism whose proper role is the description of agents and their dynamics within levels of organization each level with its own rules and laws of behavior. This is contrasted with a vulgar reductionism which insists on reducing complex behaiors at higher leals exclusixely to the laws governing the lower levels; for our purposes in this paper, to the laws and rules governing genes and molecular agents. it is this "levels" problem that generates the confusion now characterizing so much of what is seen as a failure in biological reductionism. Stage V recognizes the illegitimate nature of extending linear genetic causality, gene programs, etc. to the levels of organization and behavior above the genome and locates "the program" in environmentally open networks of agents, including genes (as circuits) , whose total behavior is not reducible to the agents themselves. Postgenomics will mostly be concerned with the rules gowming non-linear biological open systems: complex adaptive systems in which genes play a necessary but insufficient role in determining phenotype.

Chapter 9

Whole genome bioinformatics

Temple Smith
Boston University

For those of you who don't know me I will try to be out of character and be controversial. [Laughter] That was a joke for those who think they know me. I am going to try to do two things today. One, I want to make some attempt to relay the ideas from complexity to the examples in biology that are being talked about in as quantitative a way as possible on a couple of simple examples.

The other thing I want to do is take the extreme position that I am going to start out being as reductionist as possible and, in front of you, try to undergo a metamorphosis here. [Laughter] If I begin to sprout fly eyes, there is something to worry about.

Well, biology is over. [Laughter] We have the complete genetic sequence of a number of organisms. And the organism is completely described by the streams of these four letters, okay? At least at some level that dogma is a very useful dogma for raising money at this moment inside of the United States. [Laughter]

And if you had invested in at least three companies I can think of who are doing nothing but massive sequencing, your stock would have gone from 12 to 73 yesterday morning. So anybody who does not feel that this is the end-all, you are not involved in stock.

The genomes that are now completely and accurately sequenced are listed here, these are the ones in the public domain (slide–TIGR list). And the thing that is probably of most interest to biologists who are not complete reductionists is the fact that this information immediately gives us information about the evolutionary history. In fact, when we have yeast and M. Jannaschii, which is an archaeabacteria (the third kingdom of life that has only been recognized by the average biologist in the last probably 10 years, by Carl Wose), we now have a view of the oldest aspects of life back when there were no fossils, no classic information, only that based on this sequencing.

However, the position taken by most people is, let's start from the DNA sequence and derive everything. Let's suppose I give you the entire human chromosome—one, just one. What can you tell me about it? The reason I want to look at this problem is that it brings up one of the measures that would involve complexity that everybody sort of agrees on, but is not in general very well quantified.

We can ask a series of questions and it is almost clear to everybody, I think, that if that is all I give you and you are from my home planet of Alpha Centuri, you are not going to get very far.

Suppose I give you all of physical chemistry. You still won't get anywhere. If I actually gave you the three-dimensional structure of DNA, you are going to immediately get something. Because now you can combine the chemistry with this string of symbols because the symbol now is written in the language of chemistry. If I gave you the terrestrial environment and a temperature range, you still probably can't get very far. If I give you an entire bacterial genome, you know these two things are related. And they have different levels of complexity by any Shannon type measure, any type measure—they would have similarities and so on.

If I now tell you everything there is to know about E. coli—about where it lives, how it swims, and so on—there is a major breakthrough. You will now understand that this sequence I gave you probably codes for a living thing, and doors will begin to open.

If I gave you all of Drosophila, that is, 20 years from now I gave you everything about Drosophila, you would probably understand that this is a multicellular organism, that it underwent complex development—you'd know a lot of things about it. We can continue this until I give you the complete history and description of everything that happened on earth up to 25 million years ago. If I give you that complete context, plus your one thing. Well, you would know a lot of things. You would know what sex it was; you would probably be able to make inferences about its social behavior. But, language? I don't know. You might discover that it looks like it codes for expanded areas in the brain that are related to some of the other creatures that you know everything about—including the genome—that have to do with signaling. But does that tell you language? I doubt it. Would it tell you that these creatures are highly philosophical and religious, self-aware—oh boy, there is a controversial one, right? Consciousness? [Laughter]

Let's abandon this and go back to real science. This is the central dogma of modern biology and bioinformatics. We have the sequence that codes for structure and control of how that structure—where and when it is expressed. The structure in some ultimate sense encodes for at least biochemical function, generally on the surface of proteins, the complex surface controls which proteins bind with which proteins, which catalytic functions are carried out, and a combination of where and when. And these large complexes of many things come together so that the biological role is a part of the feedback of the environment. This part obviously is extremely complex.

All right, let's start. In the simplest case, there is a sequence written in an alphabet of the 20 letters of protein, and a really good 21st century biochemist would say "Oh yeah, it folds up like that." Protein structure prediction is one of the grand challenges listed by the NSF, and I will come back to that. Is that a complex problem or not, and what are the measures for it? But at the moment, the way we determine these structures is we go out and measure them. And I think that Kingsbury underestimated the rate at which genes are appearing, because they are now appearing—you can count all the micro-organisms—at 71 a day. That's how many genes are being sequenced. The estimate for the year 2005, when the human genome is to be completed and some of the new techniques of screening populations are in place, I have heard estimates as high as 10,000 genes.

And so the data alone is complex. But this coding from here to here—this is what we get, directly or indirectly, as the sequence information. This is the structure. But even given the structure at the moment, nobody is going to, by de novo techniques, predict the structure of this protein [see a beta propeller].

Even in very simple cases where we know a lot and the structures are not very complicated, and you can predict the structure reasonably well if I give you a hint. I give you this microcrystal, which is sort of unusual in chemistry and biology, as a sulfur iron cluster. Apparently when Nature discovered it [see slide], it bifurcated a number of times to use it in a slightly different context. Even in this case, when I tell you the microcrystals which is a very small protein, we can just barely solve it. The protein folding problem is considered one of the grand challenges by the NSF and the computer science world, and the altar upon which a large number of graduate students have been sacrificed. [Laughter]

That is not an exaggeration. That is, we know how proteins fold; we are all willing to bet almost everything. My first born particularly. [Laughter]

Not my second born, my first born. Given the theory of physical chemistry in an aqueous environment, we know how to do it. The computational power to carry out that algorithm—we have no hardware or software currently capable of doing it. But to describe how to do it, whatever your favorite measure of descriptive complexity of an algorithm, you do it for physical chemistry—you've got the protein fold. But Nature has lots of proteins for which that's probably not true for two reasons—two other measurements of complexity.

One of those essentially has to do with the Chomsky hierarchy, and though David Searl has credited me for doing this, it was he that noticed it on a slide of mine. I had never noticed it. He deserves the credit. That is, proteins are assembled as modules in nature. And in fact, evolution does its best to not re-invent the wheel, it takes whatever it has got on hand and sort of modifies and duplicates and shuffles the pieces around. How many people in here notice when you pass a black Volvo wagon older than four years that the back doors don't look right? Look closely. The top is painted black so it looks like the door is square but in fact the door is curved. They saved a lot of money; they took the doors off the coupe and put them on the station wagon, and they did that for six years. If you haven't noticed that, you will now. Once it is pointed out to you as it was to me, every Volvo you see that is older than about three years "That is really weird, the door looks like a coupe door." They just re-used the door. Nature does it all the time.

So in the single domain protein we apparently know how to deal with. In these kinds of proteins where you duplicate them and glue them together, like this one—I'll show you a number of beautiful examples like that—Nature does this all the time. Sometimes for a very simple, but elegant reason. She does this a lot too where she will actually take the sequence of one and drop another one in there. In the higher organisms this is very common because of the way the information is coded in the DNA with introns and exons, which is in a sense not terribly relevant. For interlacing, which would represent the top of the Chomsky language hierarchy, Nature doesn't do it very often, maybe she cannot. In fact, I only know of one case where people believe this is a verified sequence.

Now this hierarchy greatly complicates trying to pull proteins by your bio-chemistry algorithm just because the order in which they fold and how they fold is apparently so fundamental. In fact, the other measure of complexity is the measure that is always related to all of these entropy kinds of measures, which is the content. I hate when people use the entropy idea instead of missing information, because I'm a physicist and entropy is very well-defined, we have the zeroth law. We have the state from which to subtract the zeroth law 'P log P'. Without that, you don't have the measure. All of these measures are relative and it turns out that to be able to refold many proteins, you have to know the context of all of the machinery in the cell, all the heat shock proteins, all the foldases, and so on.

In fact, the proteins that are simple in this sense, not the high IGG immuno-hemoglobins of the immune system, which are effectively the same structure repeated four times or five times? Four times. If you try to fold those in a test tube, they get all tangled up and sink to the bottom. Because those four units sort of don't know that the two pieces of this unit go here, this one and this one. They get all confused. So in fact their information descriptive complexity is lower because it just repeats the same unit. But their context complexity is much higher because you've got to have all of this other information. In fact, they do not fold by themselves.

So there is in the protein folding area sort of three classic type measures of complexity with which we are well familiar. All of the algorithms that we know about in terms of what is called the inverse folding approach of protein structure and the class are all NP-complete. My colleague, Rick Lathrop, from MIT, who is now at the University of California at Irvine, did one of the many proofs of this. And that is a binary division that is hard and not a continuous measure of complexity.

There is a language hierarchy, which there is obviously more than the simple Chomsky hierarchy, and that clearly applies to the way we look at protein structure. You need to know where you are in the hierarchy to figure out what you are doing. And then there are all sorts of information measures that require that you have some relative measure, and obviously this has to be subtracted from this. I don't know what the measure is—they all tend to have this context—and that is effectively a continuous degree of complexity.

Here is a paper recently published in Science. I wish I had published this paper. Here is a set of genes which do a very, very wide variety of things: from controlling replication of chromosome ends to running the AIDS virus to get it into DNA and copy it back into the genomes, and you're stuck with it. A whole variety of very different functions that all have one thing in common— they manipulate the nucleic acids and are built out of a set—here's a dark blue module, a light blue. These modules have all been assembled with other pieces and in slightly different words to carry out an incredible variety of functions that were not originally recognized to be built out of these same Lego pieces.

In fact, map the Lego pieces onto the three-dimensional structure and you can see exactly how this was done. You can see how the things that fit into the DNA were assembled. And then the modifications were done so you can look at it. Now it's obvious that that line is RNA, that line is DNA.

To some extent this is where biology will be around 2005. And that is, we will have a database, not of sequences, not of organisms, but of the module tinker toys. We will teach biochemistry at this level. We will teach people "Here are the toys". Now let's see how Nature assembled them into very different, complex, integrated systems.

Now I am supposed to talk about genomes. Well there is an example of a collection of genomes I know about. [laughter]

1799—someplace in Egypt [see Rosetta Stone slide]. But this is how we know most of what we know, or what we think we know. We can't read this, by George, no more than you could read the human genome sequence I gave you without the context. You couldn't read this. And yet you knew a lot; you knew about the pyramids, you knew who the people were. You had Roman text about what they knew and who they slept with and when. You knew all of that information. Couldn't read it. Couldn't read it until you discovered that in sort of two forms of Greek, you had the same text. This is what the genome projects are about.

All of the genome projects are about sequencing pairs. Now the reason this works is the same reason the Rosetta Stone works. All of the organisms on earth, as far as we can tell, are related, and nature is conservative, in the sense that you heard the talk yesterday morning about how we can divide the complexity, if you like, of evolution into sort of these three divisions in adaptation and so on. Evolution is not going to waste the time to reinvent something totally new if it is already there.

Now I'm not a Darwinian, I am a Dobzhanskian. And Dobzhansky made, I think, one of the critical points—you can't select anything you don't already have. Which is interesting regarding the last point from yesterday's talk about innovation. It means you've got to have a light sensitive cell before you start selecting for eyes. That's all right. You are already tranducing light into ATP for energy—you are already partway there. So everything we know is by comparing sequences.

Now let me just first hit the databases hard. I am one of the founders of GenBank, so look on my back and see all of the self-flagellation I go through with apologizing. GenBank does not exist as an archival database. The archives are wonderful, particularly in basements covered with dust. They are not a living thing. As was pointed out by both speakers earlier, the annotations are dynamic, they change what we learn, and they literally capture the integrated complexities of biological systems.

But they're just simple things that are disasters. There is a wonderful gene in yeast, when the yeast sequence came out, labeled a topoisomerase. Topos are one of my favorite enzymes, they changed the topology of things. And this is labeled a topoisomerase. And yet when I looked at it, it's not a topo. It's not a topo by any stretch of the imagination. [laughter]

What had happened was this transitivity of the information in the database was a topo and this domain over here encoded that function. But it was linked to other things by local similarity. And another sequence matched it, but this region was labeled, so it was labeled a topo. So this gene matched that in a different region, so it inherited. This didn't have one amino acid in common with the topoisomerase. The database runs about, according to different people's estimates, 5% such errors. Actually, you can count all other kinds of errors in the database that would make as high as a 30% error rate. [laughter]

What does this mean as we go forward and sequence blindly a few hundred million nucleotides and all our analysis is done by the Rosetta Stone? Potential disaster. To the rescue? My graduate students! [laughter]

Lots of people are trying to figure out how can I recognize those module patterns that I showed you in the topoisomerase. That is, so I can actually say what it is about the structure or the function that Nature has conserved in evolution. 'Conserved' is the wrong word; 're-used' is the right word. It is not saving it, it is re-using it. These are not conservative modules, they are re-used and they are alive, they are not left over from the archives. And it turns out that when you fold up a protein in three-dimensional space, the things that carry out the key function are not all in one place in the linear sequence, but spread out. So these regions vary in all kinds of random ways and in ways that have nothing to do with the function you are trying to identify. They have what I love to say is that 'the devil is in the details'.

Often one change makes all the difference. One of the things that is so interesting is that exact same set of genes that control and lay down the fruit fly's wonderful compound eye, particularly in terms of a wave that passes through the cell, just like the chemical waves that we heard about the other day. A wave that passes through. Behind it you get a differentiation that divides both ways and you get the segmentation of the eye.

Those same genes and part of the same wave structure is involved in your eye. In fact, looking at two of those genes in one case in the functional region, there is only a four-amino acid difference. So if you see people whose eyes look a little bit segmented, don't be surprised. [laughter]

Oh, we are supposed to be talking about complexity.

Well, here are the algorithms that are necessary in outline to identify the little dispersed clumps of re-used information. And each one of these boxes is probably a few thousand lines of code. What they allow you to do though is to look at problems like this—this is a well-studied problem, it was originally studied by Doolittle by hand. Here is a set of proteins that have modules in them, but no single module in common. So if you search the databases by the simple-minded tech search, they often get all pulled out together, but they have different functions and, if you like, measures of complexity on graphs. There is a whole graph theory approach to dissecting complex multidomain proteins.

However, you do have to worry about the Chomsky hierarchy, because Nature does do this lots of times. She drops one functional phrase inside another. Wouldn't English be wonderful it did that? If you opened up a word like 'complexity' and drop it in 'bioinformatics'.

In any case, this Chomsky hierarchy problem actually gets very complicated because I can drop another domain into that domain. There are cases like that. Let me quickly show you some of these and relate them to the database problems. Here is a protein in yeast that was originally labeled by this function only, it was not recognized it had this function.

Here is an E. coli, the gene is a single gene, and this one is a single gene. And in M. Jannaschii, the archaebacteria that is the same, and in another bacteria is the same. So in this higher organism, it has taken genes that are separate genes and glued them together. That is an interesting form of regulation because you guarantee that whenever you make this, you make that. So it solves part of the complex regulation problem in bacteria that has to do with operons. This is a theme throughout yeast.

In addition, if you look at what the reactions carry out, you find that the two domains carry out reactions that are chemically linked from here to here, and from there to there. So this has a very simple, cute trick.

It gets more interesting. Here again is a yeast gene and an archaebacteria gene. Here is another yeast gene. Here is an E. coli case where the E. coli has the multidomains and the yeast does not. Here it is, but those aren't even warm yet. There we go—let's let them get really happy here. Here is something that was brought up that also complicates looking at the genome, let alone looking at sequences. Yeast in this case not only has all of these functions concatenated together, if you like, the engine, two transmissions, the rear wheels, and the brakes. Yeast also has each one of these separately coded for someplace else. In this case, we strictly understand why, and they are coded for separately. That is because they are transported into the mitochondria, which is one of the small internal symbiotic machines inside yeast and all higher cells. So to import them in there, you can't import this complex. Why? Unfold it to get it in there and you refold it on the other side, and you cannot refold this without all of the complex refolding context machinery in yeast.

Well, I will give you one more just to show—here is sort of an extreme case of a set of genes that sequentially catalyze and are related to a very complex phenomenon that goes on in bacteria where in yeast they are strung together. But not in the order that they are in the bacteria; they are labeled sort of in bacterial order; partly in the order that they were understood. And in this case the yeast does not have single duplicate copies.

Let me show you one last example of one of the things that we learned directly related to the genome. People are actually convinced that there are no operons in higher organisms—that is, a set of regulatory commands—and then a whole bunch of genes that are controlled by the single set which the bacteria use a lot. Wrong. These have operons—they are a very interesting kind; all the ones we know about are pairs. Two yeast genes, one running this way on the DNA, one going on the opposite strand this way. We looked at all of the genes that were less than 1,000 apart in that sense, and said just for fun, "Who are they?" Well, about half of them are unknown. So we said, "What about all the ones where somebody thinks they know what the two are?" And if you look at them, you find that in every case they are stoicheometrically related. And that is, it requires one copy of this and one copy of that for the complexes. So it has solved a trivial problem—don't make three of this and one of this—ever—but always make equal amounts of each. So it has stuck the two genes together and said the machinery to control them is common and they overlap.

Now we cannot read the machinery in this region. We can't read it. But we know what it says—sort of an interesting case. I'm going to try to end if I can with sort of a fascinating example of one of my favorite proteins. There is a whole family of proteins called the beta propellers. And if you actually schematically look at them, they look just like a prop on a boat. They are composed of a sequence that folds up like this, a single sequence that folds up like this. And it is made out of repetitive structural units that do not come from the same sequence. The structures are identical, and I mean identical. You can superimpose the beta carbon to a half an Angstrom RMS. A structural biologist will say, "That can't be, give me the data." Two minutes later he would say "Mm, interesting." [laughter]

They are not the same sequence, they are structurally the same.

So nature has re-used this module over and over. However, this module has no common function. It is part of signal transduction. It is part of RNA splicing. It is involved in regulation. There is a hint now that it is key in the mechanics you heard about last night, stacking together to absolutely force the cytoskeleton to be in the right form.

So here is a module that Nature has used over and over. If you look at this from a structural level, it has low information or Kumnogoroff kind of complexity that just repeats. Do it 8 times, or 4 times or 15. Proteins made out of 15 of these. There is one frog, or toad, that has 9 of these repeats where the sequence is identical in every repeat. Isn't that very interesting, in this boring sequence? However, do a knock out of it and the embryo doesn't develop, yet it has no known function.

The sequences are, at one level, everything you need—provided you have the entire context: ecology, the history of life on earth, and so on. So to some extent I think Wally Gilbert was correct. That is, the sequences are the raw material for the biology of the 21st century. And when you look at sequences, particularly in terms of structure or function, you can talk about measures of complexity which are required to understand and work with the problem. It seems to me in most cases people say "I could give you some measure of complexity, but will that do you any good? Is that going to change the way you approach the problem?"

Perhaps in this case it does. Sam Karlin, the wild mathematician from Stanford, has looked at sequences for a long time with various kinds of information measures, most of them sort of entropy or Shannon-type measures and found lots of interesting things that other people didn't notice, that some of it can be correlated to the real world, some of it may be totally incidental to the history of evolution.

There are cases in biological systems, as far as I can tell, where measures, binary measures, incomplete measures, discrete measures, various language measures; and there are various graph theory measures, and almost certainly continuous measures of complexity related to context or your reference data whenever you are doing a classic entropy or information measure.

And we do use them. We use them all the time to decide, how do I attack this problem? How do I dissect the problem with the pieces I can attack?

Now the one thing that I sort of haven't done anything about because I just had one slide, is the biggest challenge facing sequence analysis. And this relates to what I think underlines the idea that maybe you have to do the reductionism to get the pieces of the puzzle, but you can't put the puzzle back together until you get the whole measure. And the reason here is a trivial example of a minor gene and its regulatory signals in front of it. This gene can be turned on and off at probably 200 or 300 different levels in probably a dozen different tissues, and it probably behaves differently in hundreds of different organisms in taxonomic lines of descent by manipulating in trivial ways whether this protein has one phosphate on it or two; by manipulating whether this is here because then this won't be there; but if this is there then this will be there. These pieces which in a cell are just floating around. Let me end by giving you a physicist's view of complexity, at least when I was a graduate student.

I worked on the 3-body problem and most of my fellow people thought I had walked off the deep end, although we actually had people who worked up to the 5-body problem. [laughter]

And then we had people who worked on the infinite-body problem. Not everything between 5 and infinity was complex, okay? From 5 down it was simple and from infinity up, it's trivial. [laughter]

I mean, the uranium nucleus that was worked on, for instance, was trivial. What one has to understand is that, in the cell, the level of these networks and feedbacks of the regulation is in this range between 5 and a couple of hundred. And the numbers of each of these proteins is in that same range. You can't do kinetics. Somebody said this last night—you cannot do kinetics.

I worked on the lac operon in E. coli, the simplest case. We got under most circumstances five regulatory molecules. Can you do kinetics on 5 or use differential equations ?. We do not have mathematical tools to deal with this level. This is a graph theoretical circuit problem, we need the electrical engineers to work on algorithms to figure out how to put those layers on top of one another on a little chip. And they actually have interesting measures of complexity that tell them whether or not they're going to be able to build this chip in five layers or 30. 30 of course are too many.

Some of the complexity measures are probably applicable here. But I don't think, unless I am wrong, that there are even any databases that have used them. So we need lots of input; there is real data, real problems, small numbers, many interactions linked to an outside world that is difficult to describe. Thank you.

[Applause]

Q: ??.

A: Let me give you an example that many of young biologists today are familiar with—David Botstein's stuff using one of these new chip technologies which allows them to look at a few thousand genes at a time genes at a time being expressed. And there is dogma in the yeast community, a simple organism in what is called the Gal operon system, the Gal sugar system. And people would do mutations where they would effect one of these genes or two of these genes and see what happens. And there must be a few hundred papers published on when you change the sugar level or you change what is going on in this system, these are the only 6 or 7 genes that are effected.

So they did the experiment; they actually did this and looked at 3,000 genes simultaneously as they did these experiments. A lot of people had only looked at 3 or 4 and there were about 70 genes being turned on and regulated at the same rate as the key one that everybody else was looking at.

Q: Is this in bacteria?

A: No, this is in yeast. But in bacteria, according to Fred Blattner, we now see the same thing even in the lac system. Even when you turn on lactose, depending on what sugar you are going from, it now looks like the expression changes over about 100 genes. So the feedback networks are extremely complicated even in bacteria.

I do agree with you that the Rosetta Stone approach will help with much of this by sort of working up and seeing how evolution duplicated and patched things together. I am a Dobzhanskian in that sense.

Q: ??

A: Yes, I will repeat the question. If you take a gene out of humans, and change it slightly and move it into Drosophila, will it matter? In some cases, not at all. That is the fact that there are 300 amino acids different in this protein. And I cut it out and splice it in, and I replace the gene in Drosophila or in a couple of cases yeast, it makes not an iota of difference. So the details don't seem to matter.

On the other hand, sometimes a single amino acid makes all of the difference in the world. Remember, this is an evolutionary process. So that we both have lots of differences that are close to irrelevant, neutral in some sense; and you have differences that were selected for that difference and that difference alone. And both processes are all mixed together when you look at the sequence. The sequence alone can't do that.

And one of the things people want to do is they want to take out genes one at a time, two at a time, three at a time, and right there, you understand, experimentally, the complexity just goes to pot real quick. And so in a sense we need new ways of looking at this stuff. Another question? Yes.

Q: Is there anything known about this molecular aspect if you have a membrane adjacent which changes...

A: It is in some cases. To do the experiments, what we call the micro experiments in vivo is extremely hard. The one case where this has been done a lot is in nerves where we can actually take pipettes small enough to look at one channel at a time. We can ask what the ionic concentration is, what the PH is locally. And what you find is of course how this molecular complex behaves locally depends on, and what its structure is, depends on the local environment.

Obviously most of what we look at most of the structures we know and the ones that are more robust. If you're going to crystallize it, it has got to crystallize, so it is very hard. But today we have tools in NMR that are allowing us to look at some of these structures closer to in vivo. And one of the interesting things about beta propellers is they do not undergo allosteric change. As far as we can tell they are the most rigid protein structures known; very different Ph, very different ionic strength and bound and unbound to their complexes, there is no hint that the basic propeller structure changes a half an Angstrom Which again fundamentally violates everything we know about enzymes ...

Q: There is a question as to how to move to understand function...

A: Well, I think that is in fact what people are trying to do, and we had lots of hints about this and we talked about that. That is, this is an old thing out of Scientific American. You have a set of genes which are all related, they use the same module repeated over and over. They are involved in controlling various forms of differentiations in cell type determination. And the thing that was sort of interesting when they were first looked at was we found sort of this spatial sequence in mouse and fruit flies what is roughly the same, and they are sequenced, the way they are laid out the chromosome is the same. And that didn't seem like an accident although it may be an accident.

And what people tried to really do to attach to these genes then was these higher level things; their function and in what context is this module found? Because the function you want is not the biochemical function that binds DNA, you want this higher level function that hopefully you can go up another notch and we can do a real bottom-up, given the context. Not that you can start with the chromosome and get the organism, but I can start with the chromosome and work my way up as I look at the other data; the ecological data, the environmental data and so on, look at the context.

And we are doing this except that this data is not available, except in pieces, it is scattered. In fact, one of the most depressing things is nobody is sequencing one of the spiral cleavage organisms. These are organisms that undergo very different embryonics than the fruit fly or you and I. And in fact, they actually differentiate in a spiral rather than this thing fold very differently. And they were mapped in the early 1800's so that people know every single cell in there, by name in the order in which they differentiated and how they ended up. And there is a wealth of data. No molecular information on probably the richest developmental stuff. However you are not going to get funding for that because humans don't spiral cleavage from Alpha Centuri?? and therefore they can't get funding.

The other thing is the complexity of this problem is partly political.

Q: There are two problems there. There is the money matter and there is the scientific matter...

A: To pull the young people out of the other disciplines of graph theory and the mathematical areas that we need, and to integrate ecologists and people into modern biology first of all is going to take time, it is a cultural and social problem. And right now the funding is being driven by the fact that in the short term, if I want to solve malaria, I don't care about the holistic approach. I want to solve malaria, you give me that sequence, you give me a protein in there, if I can get a chemical that will bind to it and that won't bind anything in my body and kill that little guy dead as a hammer. And by George, that is perfectly logical, and science goes up and down, and I think that you just have to face it. The economy is going to drive the reductionist view for another ten years. That's not bad. We will have a lot of data to work with when we re-start. So what? [Laughter]

Q: Yes, but that is the short range view, right?

A: Yes, it is.

Q: So you don't need to understand the function of the organism in order to discover a tool to fix it. But that is perhaps why it takes so much money and so much time to bring a product onto the market because you don't really understand it. And what you've got is some sort of a search mechanism that is totally random that finds you ultimately some tool to use. If you have to rely on that process to get to the market you're right; we're in deep trouble. Now—

A: I don't think we are in trouble, I just think we have to wait, we have to be patient.

Q: Well of course we have to be patient.

A: I don't think there is any reason why the scientific community has to be able to understand what most of us already own. Kingsbury went to get an airplane, I have got a grant due in two days ... [Laughter]

And in a sense you might as well accept that it's an American disease, maybe a world-wide disease—we all have to be doing 25 things at one time and give up sleep. And I think one of the things that it has forced me to think about is, I'm not going to worry about it. Biology will recover in 25 or 30 years to be a great science and I am not going to worry about it. You can worry about it, you're retired. [Laughter]

Can I just put in a few subversive words? I may have a point that I can make. I just thought I would mention words like holistic medicine, homeopathy and so-on which have always looked at the holistic level and may have some interesting things to say. Now think we probably have a long way to go before the connection will be made. But it is really just a thought that occurred to me that there have been some rather unscientific ways of looking at the whole body which have come up with ideas which we are perhaps working towards from the other end now.

In fact I went and looked at Aristotle's description of plants. And he had, as far as I can tell, particularly if I can give him a little leeway, a very useful description of plants. They took one of the elements—water, they took another one of the elements, air. And he didn't quite understand light but he knew if you covered them up they died. So I will give him the benefit of the doubt that if I talked to him for ten minutes he would have understood light.

He had a model of a uniform thing, the leaf, which took water and air, mixed them together somehow in the presence of light, and created more leaf. And it's crap. And it is a good description and it sure tells a farmer that you have to water and that you can't grow things in Northern Michigan in the Upper Peninsula because it's dark most of the year.

So in fact what I would call empirical holistic descriptions at the top level are often very useful, they give you information. They generally make it difficult for intervention or modification. But again, I'm not worried about it. These things will eventually integrate. Why can't they? Why do they have to be integrated next week? I mean, you don't want to die, but you are. [Laughter]

I am immortal. [Laughter]

Q: But I mean, you gave us the prediction that in the year 2005 or 2010 that this genome information project is going to evolve to the point where we've got all the toys. So what are these toys? These toys are simply larger aggregates of the genetic elements you are talking about, right? And then what is going to happen after that, you said, is that we are going to come to some understanding about how the cell puts the toys together to get to the function. Now that is 2005, that is not too long to wait, I agree with you.

But it seems to me that that is not much of a progress because what we need to know now is not about the contingency element having to do with how all of these things may or may not be put together, but the robust, the generic, robust properties that do put them together as a matter of fact right now.

A: I don't think that is going to happen, I don't think we have the data to do it.

No. But we know something about the fact that these dynamical prophecies that are responsible for integrating function. And in some way, I have forgotten which complexity book I read this in, but I think he was going for that perhaps you didn't really need to know in great detail what the elements were if you had no information about the generic organizing processes, you wouldn't have to in fact completely go into this reductionistic scheme for your understanding. You could get it out of generic process. Now that is a possibility, but for some reason we can't free ourselves up to the point of taking it seriously, because we don't have the data but that's is the Catch-22.

Q: It is not clear to me the posture that you are taking. The whole business of identifying blocks and modularity is in fact a very big step away from the early reductionistic approach...

A: You can define 'reductionism' in lots of ways. I am interested at this particular moment in time with the particular graduate students who happen to be in the lab—we are very interested in developing tools that would allow us to come up with the 1000 building blocks that basically build all of life on earth. I would like to have that list. That list, without a doubt we will have on or before 2005. If I get all the funding I have requested, we can have it in 3 years... [Laughter]

The reason you don't ?? me is because I'm not taking a stance at that level. I mean, I agree with all these positions when I say that each of us has a role to play right now and the tools that I have that I can use, I would love to be the Renaissance man. I try to read Aristotle and Plato, I try to read modern medicine, I even read my wife's journal on holistic medicine. But let's face it. The data is ?? and if I try to encode the data on just yeast, that is, if I look at all the environments of how yeast behaves—that integrating system. I try to keep track of all of the ways I can trigger the mating response in yeast. And you may not think this is very complicated, but I am writing a wonderful story about the humans moving those genes into us. So that when you go into the bar as a 19-year old and you look at who is sitting on the stools you say "Well tonight, should I move my cassette over to be female, or should I be male? ??. You can't go to Switzerland, but this you can do in a few minutes. [Laughter]

This kind of understanding—even in yeast, we have no way—and I have talked to the yeast geneticists—they don't even know how to record the holistic data that's available on yeast. They don't even know how to collect it—group it together. So I think that in a sense it is a futile effort at this point in time to say that I can do this, that I could take all this data. If you talk to the experts, and they say "I don't know how to take ??? data", which would be interesting to me.

Q: What about exeriments that record activity patterns. Once you get these activity patterns, at least then you don't have the rules yet by which these were generated, but at least you know what the output of the system is, and then maybe we can bring that in agreement will knowledge like your modules and so forth and begin to understand how to engineer this thing.

A: I think that is correct. But that is ten years away before it's is done. I am working with David Botstein and we are a long ways away. And then unless you know something that nobody else in the field knows, you don't know how to represent that data. When I went to look up the sexual cassette information on yeast, there are 1,100 papers—I'm going to read 1,100 papers? [laughter]

[Applause]

Chapter 10

Session introduction: Computational methods

Bruce M. Boghosian
Center for Computational Science
Boston University

Good afternoon and welcome to the session on Computation and Complexity. My name is Bruce Boghosian and I'm from the Center for Computational Science at Boston University. I wanted to say a few words about the relationship between computation and complex systems research at the outset of the session in the way of introducing the two speakers. There's been a close co-dependence between computation and complex systems research. They've existed symbiotically for many decades now. There are some interesting reasons for that, and I wanted to bring up four of them in preparation for the session.

With regard to the first of these reasons, there's been much discussion here about the definition of a complex system, I'd like to borrow from Richard Dawkin's description in his book "The Blind Watchmaker' when he describes a complex system as something that is heterogeneous first of all, and something that we try to understand in layers. This description is sometimes called "hierarchical reductionism" where one tries to understand the whole by trying to understand a number of its parts. And then you look at those parts and break them down. And you try to go down several levels—at least you go down until you don't feel a need to go down further any more. So if you're a biologist you might look at an organism and you might break that down to cells. And then you might go dawn to the molecular level. If you're a physicist you might begin at the atomic level and go down to the level of an atomic nucleus and work your way down to quarks and strings, But the point is that there are several levels—multiple scales of length and time going on in all these situations. And so you wind up with a situation in which the problems that you encounter in doing this may be somewhat similar or at least reminiscent of each other from one field of study to another.

So the first and most obvious connection with computers is the fact that computer programs—or at least good computer programs—are written in precisely this way. The most fashionable style of programming that is done these days is what is called object-oriented programming. This is characterized by a number of different features, one of which is that the program should be constructed of sub-programs and those in turn should be constructed of sub-programs. And each of these subprograms should be what's called "encapsulated." They should be "black boxes" in and of themselves. And the communication between one level and the next should be very controlled. Computer scientists talk disparagingly about what they call "spaghetti" computer code which is laced with "GO TO" statements and other things that break down that kind of hierarchical structure. Another feature of object-oriented programming is called "inheritance." This means that you can have computer subroutines and data structures that inherit properties from one level of the hierarchy to the next. So, to push this analogy probably further than it should be pushed, the problems faced by a software engineer who's trying to maintain a computer program of half a million to a million lines of code are analogous to those of a biologist, in the sense that they're both dealing with a system that has many, many interacting pieces on many, many different levels.

The second feature that I wanted to mention in the way of the connection between computation and complex systems is that the discrete nature of computation on the computer has provided new physical models that have allowed us to more easily glimpse generic features of complex behavior. One example of that is the class of dynamical systems called cellular automata that the second speaker in this section will be telling us more about. They are a set of dynamical systems that were introduced by John Van Neumann who was one of the pioneers of the theory of computation earlier in this century. They've been used in many different applications throughout the natural sciences, in physics, biology, and other areas. They tend to capture the universal properties of many different dynamical systems in a way that's particularly well suited for simulation on the computer—and so certain subtle features of which dynamical systems are capable, such as the ability to compute and the ability to self-replicate, that would be very, very difficult to see in the context of a real dynamical system built out of physical parts become very easy to construct in the context of a cellular automaton.

The third area in which I think computers have had a tremendous impact in the field of complex systems is In the area of computer visualization. And my comments in this area will be made from my perspective as a physicist on this subject. The past thirty years have seen a reintroduction of geometry— and concomitantly of pictures—into the study of physics. This is something that's not to be taken for granted. If you open a physics textbook today you will see plenty of pictures in it. This was not always the case. There was a time, hundreds of years ago, when geometry was considered an indispensable part of the education of any natural philosopher, But after the analyticization of geometry by Descartes there began a period of decline over centuries. So that by the end of the 19th century the goal was to make things as analytic as possible and to dispense with the pictures. You can pick up a copy of Whittaker's 1904 Treatise on the Analytical Dynamics of Particles and Rigid Bodies, which educated an entire generation of physicists in classical mechanics, and you will find five hundred pages of very excellent text on classical dynamics and classical mechanics—and almost no illustrations in the entire five hundred pages! That would be unthinkable in any classical mechanics text today. So the situation has changed, especially from the 1960's to the present day.

Several things have brought about this reversal, and I would like to claim that the computer has been a principal one of those. The first pictures that we saw, maybe thirty years ago. of the phase space dynamics of chaotic Hamiltonian systems made us realize that such systems are generic in physics. The very simple examples like the harmonic oscillator and the pendulum, and so forth, that introductory courses tend to emphasize, turned out to be the exception rather than the rule. Systems as simple as a driven pendulum or a charged particle moving in the field of an electromagnetic wave turned out to be capable of very complex behavior, The phase-space structure is sometimes called a homoclinic tangle. And the only way really to appreciate what it looks like is with computer-generated pictures.

Fractals, which have also been mentioned numerous times throughout the talks here, are another thing that we really began to appreciate well. I think they have been appreciated in many different sciences for over a decade now. But it was really the work of Mandelbrot in the 1960's and 70's that clarified their role and ubiquity in science, and provided a mathematical foundation unifying them across several different fields of study. And perhaps it's not surprising, given Mandelbrot's association with the IBM Watson Research Center, that these ideas involved an interplay between computers and physics.

The final area that I'd like to mention before I introduce the first speaker is that of multi-scale simulation which is kind of a buzz-word in the computational sciences today. Ironically in that same book that I mentioned a moment a ago, "The Blind Watchmaker' by Dawkins, he refers to physics as the study of simple systems, as being in contrast with biology. From a biologist's perspective I suppose physics is the study of simple systems; although if you're a physicist banging your head away at these problems they don't seem so simple. Partisanship aside, I disagree with this statement for the following reason: Physics does indeed run into problems of multiple scales in length and time. One classic problem that physicists have looked at for decades, without success to this day, is that of fluid turbulence. The equations for viscous fluid flow have been known for over a century now. But the solutions when that flow is at all irregular or the fluid is moving fast are maddeningly difficult to characterize; even the gross, statistical properties about those solutions are very, very difficult to ascertain. There's a hierarchy of eddies in turbulent flow—from the smallest scales, heavily damped by viscosity, all the way to the way of to the size of the system. There are power laws governing the size of the eddies that are present in the system. But no reliable method of taking advantage of that scaling and of truncating that spectrum of eddies has never been achieved. And this remains an active area of research to this day.

Another area where physicists come against multiple scales of length is in what is sometimes called "soft" condensed-matter physics, This is the physics of soft materials like emulsions and colloids, polymers, liquid crystals, foams, and so forth, In most of these cases no hydrodynamic description is known. Here you're at even more of a disadvantage than you were for viscous fluids; you can't even write down the equation in many of these cases—they are often unknown. So what people will oftentimes do is go down to the molecular level. And this is something that has been made possible by computers to simulate the motion of the molecules in these kinds of materials, and then try to bootstrap your way up the length scales. And that's easier said than done because the molecular scales are on the order of Angstroms, and the structures that you want to be able to simulate in these things—if it's an emulsion for example—you might be looking at emulsion droplets that are fifty or a hundred nanometers across. That's very much larger than the size of a molecule and so if you have to build that up from its component molecular motions it's runs into a very intractable computational problem very quickly, So there's been a great deal of work over the past few years to find physically motivated computer algorithms that allow you to talk about both kinds of molecular dynamics. For example, people investigate molecular dynamics where you lump particles into bigger macro-particles that move about. And you try to understand how you can measure some of those dynamics and then work your way up the length scales. And you will hear more about that and about applications to physics in the first talk as well in the break-out session this afternoon.

Part II

Papers

Theories in (inter)action: A complex dynamic system for theory evaluation in Science Studies

Petra Ahrweiler
University of Bielefeld, Germany
petra.ahrweiler.@post.uni-bielefeld.de
(Collaborators: Rolf Wolkenhauer, Stefan Woermann)

Science Studies is an interdisciplinary enterprise of philosophy, sociology and history of science which tries to investigate the genesis and the development of scientific knowledge both with respect to "internal" epistemic factors and to "external", mainly social and historical factors. Due to the fact that the participating disciplines follow heterogenous concepts, the various theories of Science Studies are not only supplementary but partly competing. What is the respective explanation power of any participating theories? Where are possible links, opportunities for cooperation and integration between them? These questions should be answered in order to formulate concrete interdisciplinary projects.

SiSiFOS (abbreviation of "simulating studies on the internal formation and the organization of science") is a multiagent system for testing, evaluating, combining, and integrating theories from the field of Science Studies. Every theory can be characterized by a notion system which in some way refers to an empirical system only accessible through this description. In SiSiFOS, book versions of theories are converted into computational theories using common notion systems as an interface. The computational theories act as autonomous "agents" in an environment which consists of other agents and data that is to be explained: all agents are engaged in contributing to solve a global problem, namely to explain as many units of data as possible. SiSiFOS' theory-agents are "reactive" agents trying to apply their respective patterns to the "environment".

The output of SiSiFOS gives information not only about the explanations and the competing chains of explanations that have been found by the community of agents — it also gives information about: which agents have suggested equivalent explanations for certain aspects, which agents offer alternative explanations, which agents contribute to which aspects of the explanations, and which agents have cooperated. The participating theories are evaluated through the procedures of the system instead of finding an optimal solution for a defined problem (the normal application of multiagent systems).

The special task of SiSiFOS is mirrored in the characterization of its agents and its interaction mechanism. SiSiFOS is a production system which creates an artificial explanation network of interacting (cooperating or competing) theories from Science Studies.

To constitute SiSiFOS' theory agency, theoretical works of important authors in the field of Science Studies are selected and read, such as Popper's "The Logic of Scientific Discovery" ([7]) or Kuhn's "The Structure of Scientific Revolutions" ([5]). For every chosen theory, two kinds of clauses are extracted from the original sources: concepts and theses. The first ones help to illustrate the specific theoretical terms and to avoid misunderstandings due to their common use in colloquial language (e.g.: What is "normal" science? — cf. [5]). The latter ones representing the core of the translation process constitute the "rule-set", i.e. the qualitative relations that are put forward by the theory.

There are two problems that arise when a theory is translated into a formal language. On the one hand, it has to be taken into account that the necessity to fix the procedures of a program without indetermination does not have the effect that the program formulates concepts and theses of the theory in a more definite way than the theory itself and so gets to statements the theory itself would not state. But one also has to avoid, on the other hand, that the program formulates the concepts and theses fuzzier or more restrained than the theory, and therefore is not able to deduce certain statements about the given data which are well statements of the theory.

For SiSiFOS' theory-agents there is one global task: to explain as many aspects of the given data as possible. The agents work on the global problem in a sequence of structurally equivalent steps. In these steps the agents do not work on agent-specific partial problems; in every step each agent instead works on the entire problem modified by its actual specification. So all agents work on the same problem, but they use different notions and hypotheses for finding explanations and therefore they possibly get to different results.

Agents may cooperate or compete: Every agent takes note of the results the other agents have found and takes them into consideration in the following steps, in so far as they differ from its own results but do not contradict them. The contradictions between the agents are not eliminated: The respective solutions that have been found by the agents remain, even if they contradict each other partly or completely.

In SiSiFOS, data is available for all agents in the same way. The "shared memory" of the agents, where they can find the actual sets of data is called "blackboard" in the following passages. Using the blackboard methodology, messages are not addressed to other agents but only deposited — speaking within the blackboard metaphor, "written on the blackboard". Every agent may read every message being on the blackboard; this reading happens every now and then following a certain timing.

When data concerning the development of a certain academic discipline in a certain period (with respect to "internal" and "external" factors) has been collected, it is up to the SiSiFOS-user to divide the whole of the data in single sets; for these sets an order is fixed. Then the data structured this way is treated by the theory-agents of SiSiFOS as is shown in the algorithm (table 0.1), aiming to explain as many aspects of the data as possible.

SiSiFOS is no problem solving system, although each implemented theory is a problem solving entity for itself. The purpose of SiSiFOS as a whole is not to search for the optimal theory (the best problem solution, the ultimate explanation etc.) concerning a given problem. It is a theory agency, where theories are allowed to form competing "chains of explanations" and to persist in their explanation alternatives without being forced to decide between them. SiSiFOS is a "production system" performing an interaction process between problem solving theories, which create a virtual explanation network. The essential point for the SiSiFOS user is, to analyze and to evaluate the remarks, which have been stored on the back of the blackboard and which document the explanatory process of the agents.

At this time, SiSiFOS is still not fully implemented. One of the two basic features of the interaction mechanism is completely programmed, namely the process where theory agents try to explain data. In a first attempt, we tested this mechanism with very simple theory agents each able to "explain" the shape of different geometrical figures. Then, we applied the working interaction mechanism to our domain of scientific theories while still using simple fictitious theory agents acting on a small artificial data base which we created for this purpose. The first theory agent, Kuhn's "Structure of Scientific Revolutions", is now fully implemented and is able to perform its task as a single-agent system. The second theory agent, Philip Kitcher's "The Advancement of Science", is partly implemented. Others already exist as a notion system but wait for implementation.

Enabling cooperation between the agents is the second important feature of the interaction mechanism. We have already implemented a mechanism for identifying "identity" and "contradiction" which are the two easiest concepts supporting or preventing cooperation. At this time, we look for concepts of "similarity", "neighborhood", or "compatibility": their implementation may not only include a mapping process within a given overall representation of the theories but must include something like "perception" and the construction of representations carried out by the theories themselves. The second feature of the interaction mechanism will be tested as soon as the Kitcher agent is ready to proceed.

SiSiFOS is hoped to be instructive for empirical research in Science Studies by producing plausible hypotheses. Its processing may give new insights, where and how — with respect to special explanation demands and concrete research questions — theories of Science Studies can build interdisciplinary and cooperative explanation networks.

Bibliography

[1] Ahrweiler, Petra, and Nigel Gilbert (eds.), Computer Simulations in Science and Technology Studies. Springer (1997, forthcoming).

[2] Ahrweiler, Petra, and Rolf Wolkenhauer, Modeling a Theory Agency as a Toolkit for Science Studies. in: Gilbert, N., U. Mueller, R. Suleiman, and K.G. Troitzsch (eds.), Social Science Microsimulation: Tools for Modeling, Parameter Optimization, and Sensitivity Analysis. Physica (1997, forthcoming).

[3] Bond, Alan, and Les Gasser (eds.), Readings in Distributed Artificial Intelligence. Morgan Kaufmann (1988).

[4] Engelmore, R.S., and A.J. Morgan (eds.), Blackboard Systems. Addison-Wesley (1988).

[5] Kuhn, Thomas, The Structure of Scientific Revolutions. University of Chicago Press (1962).

[6] Luhmann, Niklas, Gesellschaftsstruktur und Semantik [Social Structure and Semantics], vol. 2. Suhrkamp (1981).

[7] Popper, Karl, The Logic of Scientific Discovery. Hutchinson (1959).

Figure 0.1: Converting Theories into Agents

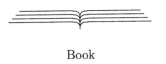

Book

... problems that, while the paradigm is taken for granted, can be assumed to have solutions. To a great extent these are the only problems that the community will admit as scientific or encourage its members to undertake (Kuhn 1962: 37).

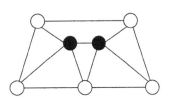

Notion System

A problem becomes relevant to a scientific community, if and only if it can be assumed that there is a solution inside the paradigm of the community.

Agent

evaluation_problem(T,Community,
Paradigm,
Problem,relevant)   :-
evaluation_problem(T,Community,
Paradigm,
Problem,solvable).

Table 0.1: The Algorithm

As long as there is data that has not been written on the blackboard:
Agent 0:
Write the next set of data on the front of the blackboard.
Agent 1,2,... (in parallel processing):
Read the new data off from the front of the blackboard.
For every theory-specific concept:
Establish to which groups of data it applies and to which groups it possibly applies.
Write these detections on the front of the blackboard.
Write them on the back of the blackboard, each detection together with the data units that are relevant to it.
Form compilations of groups of data to be explained and concepts that (possibly) apply to these groups.
For each of these compilations:
Formulate new hypotheses to explain the data taking into consideration only own hypotheses that have been formulated earlier.
For each of these new hypotheses:
Establish, if it is compatible with all theory-specific theses and all data.
Store it together with the just established result.
If it is compatible with the theses and the data: Write it on the front of the blackboard together with the data it explains.
In any case: Write it on the back of the blackboard together with the result concerning compatibility and (if compatible) together with the data it explains.
For each of the compilations mentioned above:
Formulate new hypotheses that might be able to explain more aspects of the data this time also taking into consideration hypotheses that have been formulated earlier by one of the other agents.
For each of these new hypotheses:
Establish, if it is compatible with all theory-specific theses and all data.
Store it together with the just established result.
If it is compatible with the theses and the data: Write it on the front of the blackboard together with the data it explains and the hypotheses of others that have been used.
In any case: Write it on the back of the blackboard together with the result concerning compatibility and (if compatible) together with the data it explains and the hypotheses of others that have been used.
Agent 1,2,... (in parallel processing):
Read the detections (concerning the theory-specific concepts) and the explanations (hypotheses and theses and the data they explain) off from the front of the blackboard.
For every new explanation proposed by one of the other agents:
Establish, if it follows from the own explanations.
If it does not follow: Establish, if it contradicts one of the own theses and explanations.
If it contradicts: Write the contradiction on the front of the blackboard.
If it does not contradict: Store it together with the detections that are relevant to it as a hypothesis of others.
In any case: Write the established result on the back of the blackboard.

Figure 0.2: The Interaction Mechanism

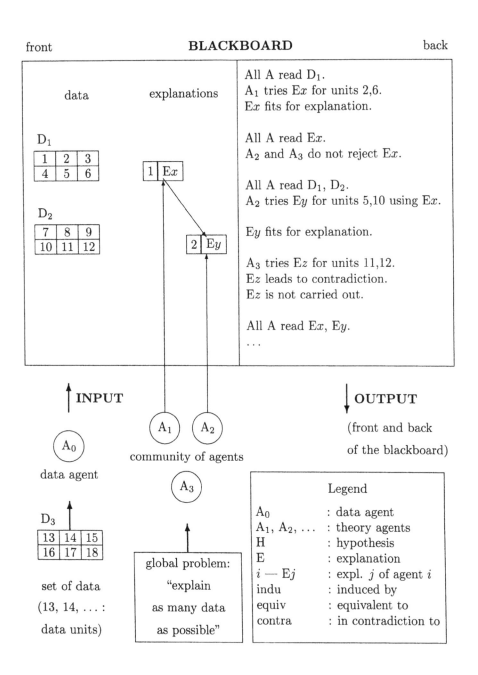

Figure 0.3: The Explanation Network

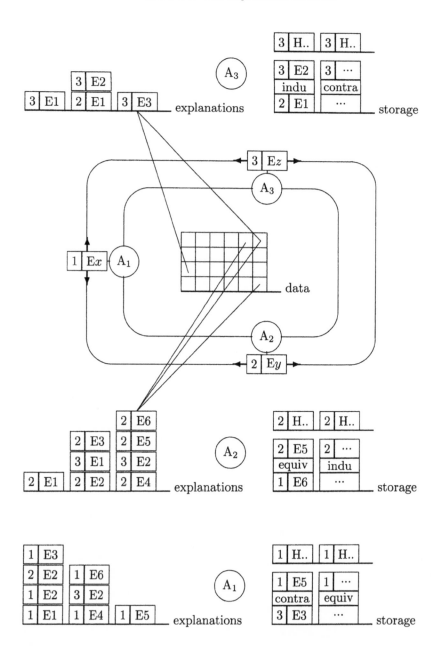

Figure 0.4: Possibilities of Interaction

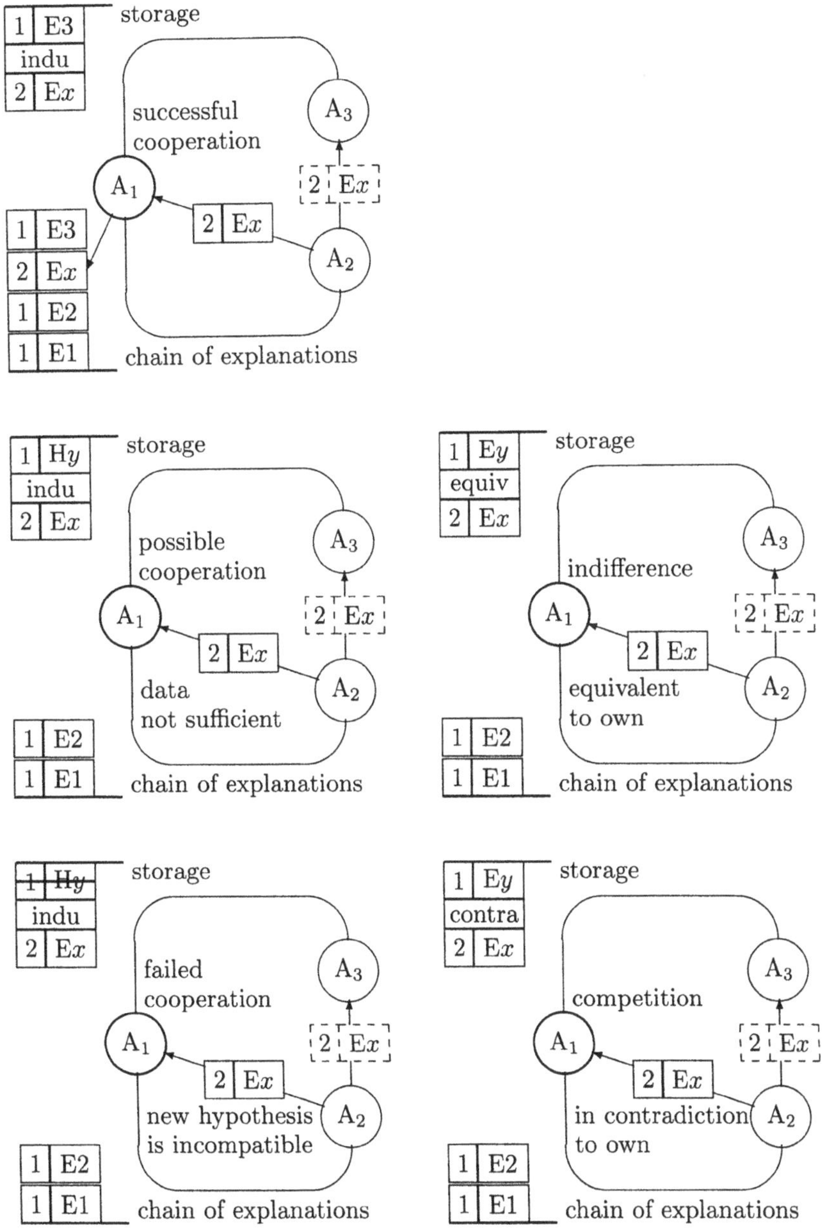

Modeling fractal patterns with Genetic Algorithm solutions to a variant of the inverse problem for Iterated Function Systems (IFS)

Olympia Lilly Bakalis[1]

Los Alamos National Laboratory

We investigate the use of Iterated Function Systems (IFS) for modeling 2 dimensional fractal structures by seeking solutions to a variant of the Inverse IFS Problem: Given a fractal pattern we are looking for parameters in 24 dimensions for a small set of contractive affine maps and their associated probabilities which constitute the IFS. Upon iteration the IFS solution produces an attractor with the characteristics which describe the image under consideration.

We define the "Mandelbrot set" for a 3-map IFS family, and demonstrate the complexity of the error hypersurface on a cross section within it. We therefore chose, for an automated search, a Genetic Algorithm (GA). We designed a general objective function, incorporating the specified errors, to accomodate two different classes of IFS attractors, namely "just touching" and "minimally overlapping". Solutions obtained with the GA, in the 24 dimensional parameter space, meet the desired specifications which characterize the given pattern to within the prescribed discretization.

The solutions are applicable to modeling spatial or temporal fractal structures, which exhibit scale invariance and power laws, such as those encountered during critical phenomena. An IFS model allows us to study the system under consideration which

---

[1]Supported in part by funds made available by DOE

belongs to the same universality class as the model.

1. Introduction

An abundance of natural systems exhibit complex behavior [1], [2] which often results in the formation of spatial or even temporal structures which cannot be described by Euclidean geometry. Examples are objects whose fragmentation is preserved at all scales such that a small piece of them is structurally similar to the whole. Trees, clouds, fractures, and coastlines are among objects depicting spatial scale-free structure. Mandelbrot named these objects, whose dimension is most often *fractional* rather than integral, *fractals*, and the geometry necessary for their description, *fractal geometry* [3]. As he stated [4], *fractal geometry is a workable new middle ground between the excessive geometric order of Euclid and the geometric chaos of general mathematics.* During critical phenomena, in addition to scale invariance, systems, however diverse from each other, also incorporate temporal scale-free behavior, while their macroscopic quantities exhibit power laws. Scale invariance and power laws are intrinsic to fractal geometry. The underlying mechanism responsible for such pattern formation is governed by nonlinear dynamics. Consequently, fractals are *attractors* of dynamical systems.

Due to the fact that there exist physically dissimilar systems which have similar properties, especially in critical phenomena, in order to characterize and study their phase space attractors, we are prompted to construct artificial models whose attractors belong in the same category. Such modeling may be applicable to *prediction* of continuous phase transitions. The model is therefore a mathematical tool which is a dynamical system in itself. An Iterated Function System (denoted as IFS) is an example of such a tool. As sources of deterministic fractals, IFSs were extensively studied by Michael Barnsley and his colleagues, and were used for image compression [5], [6].

An Iterated Function System consists of a specified number of appropriately chosen functions or transformations (linear or nonlinear) which as a set operate iteratively on their own output in a metric space. An *IFS with probabilities* may be constructed when a probability-weight is associated with each of the functions. The IFS produces an orbit which converges within an object of fractal dimension. The functions used in this work are contractive affine maps acting in the Euclidean plane. A probability is associated with each of them. The object produced, namely the *attractor* of the prescribed IFS, is an image whose fractal density is distributed on top of an underlying fractal shape, i.e., its *support.*

In order to use IFS for modeling, the solution to the *Inverse Problem for IFS* becomes the subject of attention, for the answer to the following question is given by it. Given some fractal pattern corresponding to a complex physical system, can we find a simple IFS whose attractor incorporates properties and characteristics of the pattern in question? For this purpose, the inverse problem for IFS may be defined as finding parameters for a small number of affine maps and their associated probabilities.

1.1. The IFS

All terms, definitions, and theorems concerning Iterated Function Systems, used herein, may be found in the book *Fractals Everywhere* by Barnsley [5]. As mentioned earlier, our IFS consists of N affine transformations operating on shapes in the Euclidean plane. Under affine transformations, several properties of shapes are preserved: straight lines remain straight lines; parallel lines remain parallel; the ratio of volumes, areas and line segments is preserved. Thus, ellipses transform into ellipses, parabolas into parabolas, etc. In $\mathbb{R}^2$, an affine map is representable by an 2×2 matrix together with a shift:

$$w \begin{pmatrix} x \\ y \end{pmatrix} = \begin{pmatrix} a & b \\ c & d \end{pmatrix} \begin{pmatrix} x \\ y \end{pmatrix} + \begin{pmatrix} e \\ f \end{pmatrix}. \tag{1.1}$$

In our test case, the image to be modeled is a discretized attractor μ^* of an IFS. μ^* represents the density of visited points on its support $\mathcal{A}^*$ (all the raised pixels). Using Barnsley's *Collage Theorem* [5] we assume that we can find at least one IFS of contractive affine maps whose attractor $\mathcal{A}$ is close in Hausdorff distance to $\mathcal{A}^*$, and whose density distribution μ (coded in color) is close in Hutchinson distance to μ^*. [2] The density of points on the attractor is distributed proportionally to the probability associated with each affine map [5]. During the search the probabilities are being adjusted implicitly to be proportional to the normalized areas of their associated transformation.

We explore solutions to the IFS inverse problem with the use of a Genetic Algorithm (GA), in the *forward iteration method* as defined in [5]. Forward iteration guarantees that every iteration on the attractor with the IFS which generated it, will map it onto itself with probability 1. A significant number of iterations (an average of 20 for images embedded in 128×128 grid) is necessary for determining whether a chosen parameter set actually causes the orbit of the IFS to converge on an attractor whose Hausdorff distance from $\mathcal{A}^*$ is sufficiently small.

We are searching for solutions in the parameter space of the IFS, the hypercube $\boldsymbol{P} = [-1, +1]^{6N}$, where each point represents an IFS attractor. We are interested in the neighborhood of the boundary of the Mandelbrot set (that is, the set of stable periodic points of the dynamical system) associated with our parametrized IFS family. The test images, embedded in the unit square, covered two categories of IFS attractors:

1. *Just touching*, with test image the "fern" attractor of Fig. 3.3(A).

2. *Minimally overlapping*, with test image the "leaf" attractor of Fig. 4.1(A).

1.2. The GA

We chose to use a GA in order to search for the $6N$ affine parameters of an IFS whose attractor is an acceptable solution, for the following reasons: First,

---

[2] The Hausdorff distance measures "nearness" between shapes of two images. The Hutchinson distance measures "nearness" between density distributions residing on top of those shapes.

it has been observed [7] that GAs consistently outperform other methods of stochastic search on problems involving discontinuous, noisy, high-dimensional, and multimodal objective functions, as is the case of the Inverse IFS problem. This is demonstrated in Fig. 1.1: A projection of the Mandelbrot set in the 18-D parameter space of the 3-map IFS family is plotted (top). The projection was purposefully chosen to partly cover the niche containing the point "fern" (white regions). In the bottom section, the error landscape within the inner rectangle is plotted. Here, however, we observed that this projectional slice misses the "fern" minimum (0.067, 0.51) (due to the introduced randomness) because the largest peak is not located exactly on it. We concluded that an automated search will be sensitive to perturbations in the parameters and for that we decided to represent the IFS with a binary string.

Genetic Algorithms were developed by Holland in the 1960s [8], [9] after the paradigm of natural evolution. Detailed description of GAs may be found in the books by Goldberg [10] and Michalewicz [11]. In brief, each generation of species is represented with a time step of an iterative procedure. The GA maintains a population P of n individuals s_i, $i = 1, \ldots, n$,

$$P(t) = \{s_1(t); \ s_2(t); \ \ldots; \ s_n(t)\}. \tag{1.2}$$

During each generation the members undergo genetic operations and are evaluated according to their fitness in their environment thus producing a new population of candidate solutions to an objective (fitness) function. Each individual s_i is formed by a chromosomal binary string of length l in which the vector of parameters of the fitness is encoded. A "Simple GA" (SGA) procedure is shown in Fig. 1.2. The genetic operator *crossover*, applied with a probability C_rate, produces individuals representing new points in the search space. Additional variation in the genotype is achieved with the operator of *mutation*. Under mutation, a gene's value (bit) is altered with a very small probability M_rate, thus simulating a random walk through the string space. The fittest individual of the last population will be considered as the acceptable approximate solution to the objective function.

2. Encoding the IFS on a GA

Each IFS-chromosome consists of a binary string. We assigned 10 bits per parameter, of which one bit was reserved for the sign, to ensure an accuracy of 0.005. Encoding 6 parameters per affine transformation and up to N affines per IFS ($N = 4$), yield a string length $l = 60 \cdot 4 = 240$ bits.

2.1. Initializing the population

In the real world the only available data is a discretized image which is not necessarily an IFS attractor. We therefore do not know of any particular niche in the parameter space P where an IFS attractor (represented by a point in P) might resemble its characteristics. For an automated search, we must randomly

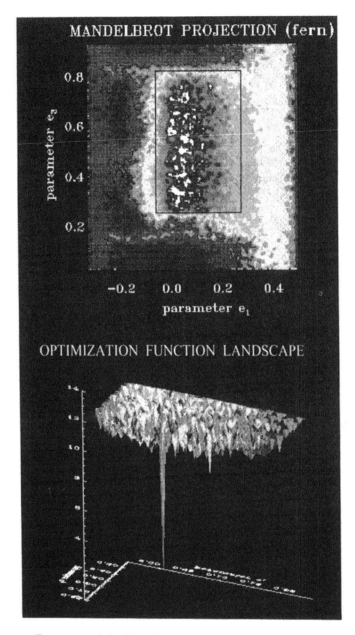

Figure 1.1: Projection of the Mandelbrot set (top) and error landscape (bottom) for the 3-affine IFS family near the fern minimum. The slice is generated by randomly deviating two of the 18 parameters within the region [−0.01, 0.01], while all other 16 parameters remain unaltered at their original fern values. Solutions exist only in the white regions (top).

```
t ← 0;
initialize P(t);
evaluate P(t);
while (not finished) do
        t ← t + 1;
        select P(t) from P(t - 1);
        operate on P(t);
        evaluate P(t);
end;
```

Figure 1.2: Sketch of a Simple GA procedure.

choose a vast number of points from the entire hypercube $P = [-1, +1]^{24}$! To bypass this problem, we select *vital* points as follows: While randomly scanning the parameter hypercube, we construct a large number of chromosomal strings which loosely satisfy constraints derived from the Collage Theorem. For that to hold, the area generated after one iteration on $\mathcal{A}^*$ should cover it almost exactly. We therefore require that each affine transformation should be contractive and map the image mostly within itself in one iteration. The individual affine maps should be arranged within $\mathcal{A}^*$ such, that there is only minimal overlap between them. The IFS chromosome, then, consists of N such maps stacked in a single string. To reduce initialization time, we approximated the area of convex attractors with the area enclosed by only eight extremities. We then imposed the constraint that after being mapped once, the eight extremities should lie within the approximated area. The "gene pool", described further below, is also formed by this method. Next, assuming that the area of each affine map is proportional to the determinant of the matrix [12], we required that each determinant is smaller than unity while their total sum nearly equals one, so that the following holds:

$$\text{Area of } \mathcal{A} = \sum_{k=1}^{N} \left(|\det \mathbf{a_k}| \cdot (\text{Area of } \mathcal{A}) \right). \tag{2.1}$$

where $\mathbf{a_k}$ is the linear part of the 2-D affine transformation $w_k(\mathbf{x}) = \mathbf{a_k x} + \mathbf{b}$ [see (1.1)].

The population size n represents the number of sampling points in parameter space and is usually dependent on the chromosomal length 1. For large 1, n can be of the order of thousands, resulting in slow evaluation. An immense population size would defeat the purpose of GAs who are meant to introduce new sampling points. A tiny one, on the other hand, could force the GA into premature convergence on the first "good enough" local minimum. The required size also depends on how diverse and fit the initial points are. To ensure a diverse selection of them, we introduced a *gene pool* [12]. That consists of a very large number of individuals (sampling points) which satisfy the aforementioned constraints. The first n members of the pool constitute the initial population. Unacceptable chromosomes on every third generation are replaced with randomly chosen ones from the gene pool. This method permits the use of small enough populations, typically of $100 - 500$ members, which minimizes computational time, while preventing premature convergence.

2.2. IFS-GA genetic operators

The fractal depends continuously on the IFS parameters. A flip of a sign on one of the parameters, however, which represents a jump in parameter space, may result in some minor alteration of the given image by causing, for example, a reflection on one of the transforms. Despite visual similarities of such attractors, the error between them may be quite large, while their surrounding neighborhoods may possibly constitute viable niches in the fitness landscape. The GA is capable of reaching such distant points in a single step with the crossover or with the mutation operators. The population size, and the crossover and mutation probabilities were adjusted empirically, and the values used for obtaining the minima of Fig. 3.3(B) and Fig. 4.1(B) are summarized in Table 2.1.

Table 2.1: Genetic Algorithm parameters

Image	n	l	C_rate	M_rate
fern 5(B)	500	240	0.75	0.00067
leaf 6(B)	500	240	0.75	0.00067

3. The GA search

The search aims towards recovering the fractal properties of the invariant support $\mathcal{A}^*$ and the p-balanced [3] measure μ^* of the IFS. The value of the invariant total measure is recovered in both categories of the test images. *Just touching* IFS attractors (as is the fern) are more sensitive to the discretization than the *minimally overlapping* IFS. It is therefore more difficult to recover the fractal dimension. For attractors of *minimally overlapping* IFS (as is the leaf) the GA has a much better performance in approximately identifying intersecting areas and optimizing the total density distribution [Fig. 3.3(B)].

---

[3] $\mathbf{p} = (p_1, p_2, \ldots, p_N)$ is the probability vector.

3.1. Definition of symbols

M: Total number of pixels on the grid. Usually $M = 128^2$.

N^*: Total number of raised pixels in the test image.

N : Total number of raised pixels in the constructed image.

N_0: Total number of raised pixels after 1 iteration on the test image with the IFS. N_0 is used to verify the Collage theorem.

μ^*: Total invariant measure of the test image.

μ : Total invariant measure of the constructed image.

D_f: Fractal dimension.

s : Contractivity factor.

not_raised: Number of pixels in the test image that are not raised.

plot_out: Number of raised pixels in the constructed image that do not belong to the test image.

coverage: Fraction of the test image that is covered by the constructed image.

overlap: Number of pixels in the intersection between the test and the constructed image.

3.2. Objective (fitness) function

The objective function both specifies the features to be looked for in terms of their errors, and guides the algorithm towards an acceptable solution. We formulated a general function to accomodate both classes of images within the unit square. The exact location of the image inside the square is to be found. Flipping the sign of one of the parameters moves the search in a separate niche, which in general implies the existence of discontinuities. To ensure diversity in the chromosomal structure we maintained the seemingly uninteresting regions of the error landscape, as high plateaus, by imposing penalties.

Our fitness function [12] consists of the sum of the following errors between the given image μ^* with support $\mathcal{A}$ and the GA IFS attractor μ with support $\mathcal{B}$:

1. V_1 is the the Hausdorff distance:

$$V_1 = \sup_{p^* \in \mathcal{A}} \inf_{p \in \mathcal{B}} d(p^*, p) \vee \sup_{p \in \mathcal{B}} \inf_{p^* \in \mathcal{A}} d(p^*, p), \qquad (3.1)$$

where $d(p^*, p)$ is the Euclidean distance between two pixel-points.

2. V_2 is the sum of the Hutchinson distance ($\mathcal{H}$) and the difference in the

total invariant measure:

$$V_2 = \mathcal{H}(\mu^*, \mu) + \frac{|\mu^* - \mu|}{\mu^*},$$

$$\mathcal{H}(\mu^*, \mu) = \sum_{1 < k \leq M} \left(\frac{\mu_k^*}{\mu^*} - \frac{\mu_k}{\mu} \right)^2. \tag{3.2}$$

3. V_3 is the pixel overlap together with the difference in the total number of raised pixels between the test image and the constructed image.

$$V_3 = \frac{1}{9} \left[\frac{|N^* - N|}{N^*} + \left(1 - \frac{2 \cdot \text{overlap}}{N^* + N} \right) \right]. \tag{3.3}$$

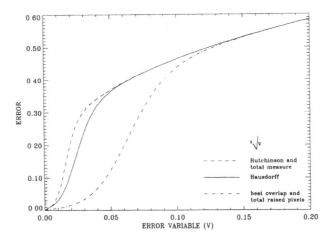

Figure 3.1: The three contributions to the objective function.

The error landscape is smoothed with sigmoidal filters (3.4) as shown in Fig. 3.1 [12], whose constants were empirically chosen in the first few (< 10) evaluations. The three errors adjust relative to each other during each evaluation, thus preventing dominance of one of them, a primary cause of premature convergence. The overall form of the objective function is:

$$\begin{aligned} \text{ERROR} &= \text{ERROR}_1 + \text{ERROR}_2 + \text{ERROR}_3 = \\ &= \sum_{i=1}^{3} \frac{\sqrt[3]{V_i}}{1 + \exp[\delta_i \cdot (V_{0i} - V_i)]}. \end{aligned} \tag{3.4}$$

The function's robustness is verified in the vicinity of the 3-map fern (no stem) parameters with the gradient descent (**powell** [13]) method. The initial condition is shown in Fig. 3.2(A). Smoothing the landscape had a tremendous

Table 3.1: The gradient descent algorithm `powell` performed remarkably well in minimizing our objective function, when starting from the initial condition of Fig. 3.2(A).

	initial condition (IC)	true fern 3 affines	powell on IC
N	534	1680	1695
N_0	1428	1680	1632
not_raised	1185	0	261
plot_out	39	0	276
coverage	44.72%	100%	84.09%
μ	13.26	20.54	20.54
D_f	1.629	1.638	1.638
s	0.75	0.85	0.85

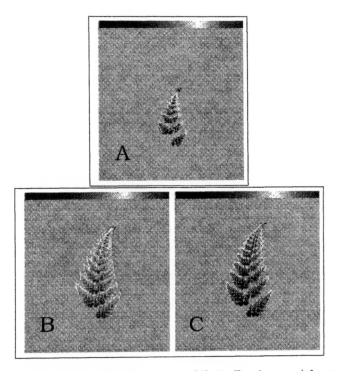

Figure 3.2: (A): Initial Condition for `powell`. (B): 3-affine (no stem) fern test image. (C): `powell` minimum.

effect, as shown in Fig. 3.2(C) which depicts the minimum where `powell` arrived and which is very close to the global [Fig. 3.2(B)], (see Table 3.1). [4]

3.3. The GA's search in the parameter hypercube $P = [-1.0, \; 1.0]^{24}$

Just touching IFS (fern)

For *just touching* IFS, a very fine grid should be used in order to preserve the intricate fine structure. Doubling the resolution increases the computation time 8-fold: each magnification is four times larger than the previous, and the iterations necessary for convergence double. Although parallel computing would be most appropriate, we performed all experiments serially. A grid of 128^2 pixels comprises the smallest size in which the most important characteristics of the image are still preserved. Many of the points of the fern in this size grid appear to be touching, which means that the GA will be looking for an overlapping IFS, and we do not expect the fractal dimension to be recovered in this class of images. In general, the fractal dimension is determined by the slope in the ratios

$$\frac{\log(\text{number of raised pixels})}{\log(1/\text{size of a pixel})}$$

calculated at various magnifications. This ratio on the 128^2 grid, however, is found exactly:

$$\frac{\log(\mathrm{N}^*)}{\log(1/\texttt{pixel\_size})} \; = \; \frac{\log(1680)}{\log(128)} \; = \; 1.530$$

$$\frac{\log(\mathrm{N})}{\log(1/\texttt{pixel\_size})} \; = \; \frac{\log(1663)}{\log(128)} \; = \; 1.528 \tag{3.5}$$

From the above numbers and from the summarized results of Table 3.2 we conclude that the GA performed quite well on the 128×128 grid for $N = 4$ affines. But we also conclude that coarse grids may convey misleading information for *just touching* IFS attractors. The fittest IFS chromosome, found by the Genetic Algorithm, produced the attractor shown in Fig. 3.3(B).

Minimally overlapping IFS (leaf)

It is easier to find solutions for the *minimally overlapping* leaf than for the *just touching* fern attractor, because the overlap of its maps is preserved throughout various magnifications, thus allowing for a larger selection of potential solutions. The minimum found by the Genetic Algorithm is shown in Fig. 4.1(B).

The contractivity factor for this IFS ($N = 4$) is $s = 0.61$, which is very close to the original $s = 0.60$. Its fractal dimension $D_f = 1.942$, is almost exactly the same as the original $D_f = 1.947$. We note that the GA approximately recovers the density distribution. The summarized results for the leaf attractor are presented in Table 4.1.

---

[4]The IFS parameters for all the images may be found in [12].

Table 3.2: Summarized results for the fern attractor: The GA solution found on the hypercube $[-1.0, \ 1.0]^{24}$ of the parameter space P, overall recovers some fractal characteristics of the test image more accurately than the minimum in the neighborhood of the true fern, found with Powell's method (despite the apparent similarity of the patterns of Fig. 3.2(B),(C)).

	true fern 4 affines	GA + powell on hypercube $[-1.0, \ 1.0]^{24}$
N	1723	1723
N_0	1723	1663
not_raised	0	255
plot_out	0	195
coverage	100%	86.71%
μ	20.70	20.70
D_f	1.638	1.920
s	0.85	0.65

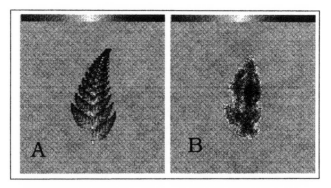

Figure 3.3: *Just touching* case: (A): 4-affine fern test image. (B): GA minimum (initial points are randomly selected). These images, apparently dissimilar, share similar characteristics (Table 3.2) and coordinates.

Final convergence on all the GA minima is achieved with the use of a local optimizer, namely either the *gradient descent* algorithm `powell` or the *simplex method* called `amoeba` subroutine (due to Nelder and Mead) [13]. From the two, `powell` is sensitively dependent on the form of the fitness function but is more accurate than `amoeba`.

4. Applications

4.1. Critical phenomena

The fitness function is designed such that the GA solutions are suitable for modeling fractal patterns associated with complex physical systems, whose structural complexity makes them difficult to study. Many natural phenomena undergo phase transitions where fractal features appear. Existence of universality classes for different and unrelated phenomena allow us to study their behavior by analyzing a corresponding artificial model. Solutions to the variant IFS inverse problem identify those necessary features for the construction of the model.

Table 4.1: Summarized results for the leaf attractor: The IFS found by the GA on the hypercube $[-1, 1]^{24}$ of the parameter space P, satisfies the criteria for solution, and also approximates the mass distribution.

	true leaf	GA + powell on hypercube $[-1.0, 1.0]^{24}$
N	3563	3588
Collage N_0	3563	3569
not_raised	0	350
plot_out	0	375
coverage	100%	89.86%
μ	980.07	979.72
D_f	1.947	1.942
s	0.60	0.61

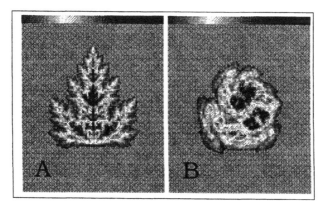

Figure 4.1: *Minimally overlapping* case: (A): 4-affine leaf test image. (B): GA minimum (initial points are randomly selected). These images, apparently dissimilar, share similar characteristics (Table 4.1) and coordinates.

In particular, the IFS-GA method can correlate fractal spatio-temporal structures, as those observed during critical phenomena, to simple hierarchical lattices for which the renormalization group (RG) transformation can be formulated. From the RG a nonlinear IFS is constructed whose attractor is the Julia set of the critical manifold. The Julia set resides on the basin boundary of the fixed points of the renormalization transformation. The critical exponents are extracted from the Julia set thus revealing the behavior of the macroscopic quantities near the critical region. This information is useful for prediction purposes.

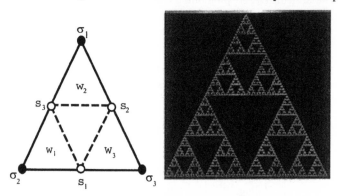

Figure 4.2: The Sierpinski gasket ($D_f = \log 3/\log 2 = 1.585$) is the attractor of an IFS with 3 affine transformations (w_1, w_2, w_3). A hierarchical lattice of spins is formed by placing spins ($\sigma_1, \sigma_2, \sigma_3$) on its vertices and the mid-points (s_1, s_2, s_3) and thus subdividing *ad infinitum*.

Fig. 4.3, for example, illustrates the basin of attraction for the renormalization transformation (4.1) derived for the Potts model on a Sierpinski lattice (Fig. 4.2) of spins which can assume one of two allowable states (0 or 1). We define $\omega \equiv e^{-J/kT}$ with T being the control parameter. [5]

The RG transformation is derived [12] by zooming into successive levels of magnification of the lattice and renormalizing the partition function. In ω it is:

$$\left(\omega'\right)^2 = \frac{\omega^2(1 + 3\omega^2)}{1 - \omega^2 + 4\omega^4}. \tag{4.1}$$

In this manner the Sierpinski lattice becomes a simple IFS model for patterns belonging to its universality class.

A subset of IFS attractors in the Euclidean plane are graphs of fractal functions. Interesting time series data are fractal. IFS-GA solutions can model and classify blocks of the series, which in combination with the renormalization group can be used for prediction of criticalities, a very desirable result for financial markets of our era.

---

[5]T is normally the temperature, k is the Boltzmann constant and J is the coupling strength. The interaction energy between neighboring sites is zero if the corresponding values of the spin are different, and equal to the negative of the coupling strength J if they are identical.

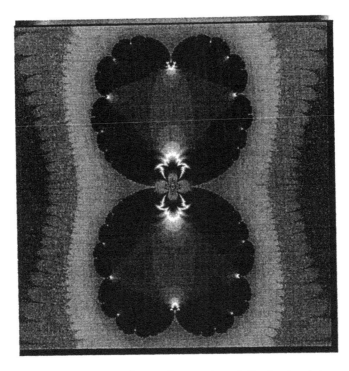

Figure 4.3: Sierpinski lattice: Basin of attraction of the fixed point $\omega = 0$, corresponding to $T = 0$ in the complex ω plane for the two state Potts model. We note that the Julia set (basin boundary) touches the real axis only at $\omega = 0 \Rightarrow T = 0$, indicating that this lattice does not undergo phase transitions at finite temperatures.

IFS models become, in essence, classifiers of universality classes for spatio-temporal fractal patterns encountered in criticality.

4.2. Other applications

IFS attractors incorporate locations and density of points. The densities are related to the probability of future occurrences of the corresponding events, as is the case of earthquake epicenters.

Finally, when taking into account the multi-dimensionality of the IFS parameter space, very high-dimensional data may be represented with 2-D IFS attractors. Thus, such data sets could be visually compared, a desirable attribute for data mining.

5. Conclusions

Many interesting natural features are fractal. Despite the intrinsic complexity of the fractal objects, the rules under which they were formed may be very simple.

IFS attractors are, in general, fractal. Although the IFS itself is nonlinear, a simple IFS consists of (but is not limited to) a set of linear maps. The rules are simple and the functions are simple: a desirable attribute for the construction of models.

In order to extract the essential properties which characterize the fractal pattern to be modeled, we used a GA to explicitly find parameters for 4 affine maps and implicitly adjust their associated probabilities. The nature of the IFS parameter space led to the choice of the algorithm (GA) for the automated search, which has to be stochastic, dynamic, adaptive and robust.

The structure of the objective function is crucial for finding solutions which model the given image; the properties of the image are incorporated in the errors which constitute the function. The GA is found to be a very robust algorithm in the sense that it is easily adaptable to the IFS inverse problem and that it does find solutions as specified by the fitness function.

The fractal characteristics of the images to be modeled, especially in the *minimally overlapping IFS* case, were recovered to within 1% error of the total image overlap, to less than 1% error of the Hausdorff distance, and to less than 1% error of the total measure of the fractal attractor. The fitness function was modified with sigmoids so that the landscape of the error hypersurface is smoothed. This resulted in a better performance of the algorithm than the "raw" distance functions used in the literature on both GAs and SAs (Simulated Annealing algorithm) [14], [15].

The major intent is to use such "simplistic" models in order to classify and model systems undergoing second-order phase transitions. The hierarchical nature of the model allows us to construct the renormalization transformation whose corresponding Julia sets reveal the critical points.

Bibliography

[1] COWAN, S. C., D. PINES, and D. MELTZER, eds., *Complexity. Metaphors, Models, and Reality*, Proceedings Volume XIX, Santa Fe Institute Studies in the Sciences of Complexity, Addison-Wesley (1994).

[2] STEIN, D. L., ed., *Lectures in the Sciences of Complexity*, Lectures Volume I, Santa Fe Institute Studies in the Sciences of Complexity, Addison-Wesley (1989).

[3] MANDELBROT, B. B., *The Fractal Geometry of Nature*, Freeman and Company (1983).

[4] FLEISCHMANN, M., D. J. TILDESLEY, and R. C. BALL, eds., *Fractals in the Natural Sciences*, Princeton University Press (1990).

[5] BARNSLEY, M. F., *Fractals Everywhere*, Academic (1993).

[6] BARNSLEY, M. F., and L. P. HURD, *Fractal Image Compression*, A. K. Peters (1993).

[7] SCHRAUDOLF, N., and J. GREFENSTETTE, "A User's Guide to GAUCSD 1.4", *Tech. Rep. no. CS92-249*, The University of California at San Diego (1992).

[8] HOLLAND, J. H., *Adaptation in Natural and Artificial Systems*, MIT Press (1992).

[9] HOLLAND, J. H., "Genetic Algorithms", *Scientific American* **7** (1992), 66–72.

[10] GOLDBERG, D. E., *Genetic Algorithms in Search, Optimization and Machine Learning*, Addison-Wesley (1989).

[11] MICHALEWICZ, Z., *Genetic Algorithms + Data Structures = Evolution Programs*, Springer-Verlag (1992).

[12] BAKALIS, O. L., "Encoding the Invariant Measure of Iterated Affine Transforms with a Genetic Algorithm", *Dissertation*, The University of New Mexico (1996).

[13] PRESS, W. H. , B. R. FLANNERY, S. A. TEUKOLSKY, and W. T. VETTERLING, *Numerical Recipes in C*, Cambridge University Press (1991).

[14] LÉVY-VÉHEL, J., and E. LUTTON, "Optimization of Fractal Functions Using Genetic Algorithm", *Fractals in the Natural and Applied Sciences* (M. M. NOVAK ed.) North-Holland (1994).

[15] NETTLETON, D. J., "Evolutionary Algorithms in Artificial Intelligence: A Comparative Study Through Applications", *Dissertation*, The University of Durham (1994).

Chapter 13

An artificial life model for investigating the evolution of modularity

Raffaele Calabretta
Yale University, EE Biology Dept.
Institute of Psychology, C.N.R., Rome
Stefano Nolfi
Institute of Psychology, C.N.R., Rome
Domenico Parisi
Institute of Psychology, C.N.R., Rome
Günter P. Wagner[1]
Yale University, EE Biology Dept.

To investigate the issue of how modularity emerges in nature, we present an Artificial Life model that allow us to reproduce on the computer both the organisms (i.e., robots that have a genotype, a nervous system, and sensory and motor organs) and the environment in which organisms live, behave and reproduce. In our simulations neural networks are evolutionarily trained to control a mobile robot designed to keep an arena clear by picking up trash objects and releasing them outside the arena. During the evolutionary process modular neural networks, which control the robot's behavior, emerge as a result of genetic duplications. Preliminary simulation results show that duplication-based modular architecture outperforms the nonmodular architecture, which represents the starting architecture in our simulations. Moreover, an interaction between mutation and duplication rate emerges from our results. Our future goal is to

[1]The financial support by NSF grant BIR 9400642 and the Yale Institute of Biospheric Studies is gratefully acknowledged.

use this model in order to explore the relationship between the evolutionary emergence of modularity and the phenomenon of gene duplication.

1. Introduction

In evolutionary biology, the concept of modularity is used to capture the fact that the bodies of higher organisms appear to be composed of semi-autonomous units ([9]; [11]; [1]). It has been argued that modularity is a prerequisite for the adaptation of complex organisms: modularity would allow the adaptation of different functions with little or no interference from other functions ([1]). However, this explanation raises two important questions. First, to say that modularity is a prerequisite for the adaptation of complex organisms seems to imply the need to explain the origin of a 'trait' (module) potentially useful for the species but not for the organism at the time of its emergence. This question is related to the more general issue of the evolution of the boundary conditions of evolution, in particular of the genotype-phenotype mapping (see [12]). Second, to say that modularity would allow the adaptation of different functions with little or no interference from other functions implies that we should be able to find a class of selective forces that can shape the genotype-phenotype mapping to allow for the existence of selective pleiotropic effects between genes, complexes of characters, and functions (see [11]). (Pleiotropy is "the influence of the same genes on different characters", [3], p. 429).

In more general terms, modularity requires that we look at a complex system from the point of view of three different and chronologically successive phases: the phase of emergence, the phase of actual functioning, and the phase of maintenance. In fact, for modularity to exist it is necessary for many different 'elements' to interact locally and nonlinearly at a number of different levels: genetic, phenotypic, physiological, and behavioral level. This complexity makes it more difficult to choose the right model for study and, therefore, to find the answers to important questions. As a consequence, even if the fact and the importance of modularity seems to be widely appreciated, there is little understanding of how modularity originates, works, and remains incorporated in the genome.

To evolve a neural controller for a mobile robot, Nolfi ([7]) used a modular neural network architecture that clearly outperformed other architectures in performing a task of garbage collecting (see below). To investigate the issue of how modularity can emerge in nature, we present a modification of Nolfi's model ([7]) in which gene duplication is also included as part of the evolutionary process and, therefore, modular neural networks can evolve starting from a population of non-modular ones as a result of gene duplication. Our future goal is to use the model to explore the relationship between the evolutionary emergence of modularity and the phenomenon of gene duplication.

Our preliminary simulation results show that duplication-based modular architecture outperforms non-modular architecture, which represents the starting architecture in our simulations. Moreover, an interaction between mutation and duplication rate emerges from our results.

2. The model

We ran a set of simulations in which neural networks ([10]) are evolutionarily trained to control a mobile robot designed to keep an arena clear by picking up trash objects and releasing them outside the arena. The robot has to look for 'garbage', somehow grasp it, and take it out of the arena (see [8]).

The organism is a miniature mobile robot called Khepera, developed at E.P.F.L. in Lausanne ([6]). The robot is supported by two wheels that allow it to move in various directions by regulating the speed of each wheel. In addition, the robot is provided with a gripper module with two degrees of freedom. The two arms of the gripper can move in parallel through any angle from vertical to horizontal while the gripper can assume only the open or closed position. The robot is also provided with eight infrared proximity sensors (six sensors are positioned on the front of the robot and two on the back) and an optical barrier sensor on the gripper capable of detecting the presence of an object between the two arms of the gripper. The infrared sensors allow the robot to detect obstacles to a distance of about 4 cm. The environment is a rectangular arena 60x35 cm surrounded by walls containing 5 target objects. The walls are 3 cm in height and target objects are cylindrical boxes with a diameter of 2.3 cm and a height of 3 cm. The targets are positioned randomly inside the arena. To speed-up the evolutionary process a simulator was used (see [8]).

In the present work we compare the results obtained with three different neural network architectures (see Fig. 2.1). In all cases the robot has 7 sensor neurons and 4 motor neurons. The first 6 sensory neurons are used to encode the activation level of the corresponding 6 frontal sensors of Khepera (the two back sensors are ignored) and the seventh sensory neuron is used to encode the light sensor on the gripper. On the motor side the 4 neurons respectively codify for the speed of the left and right wheels and for the triggering of the 'object pick up' and 'object release' procedures.

The activation values of the infrared sensors (which have 1024 different values ranging from 0 to 1023) and of the activation of the light-barrier sensor (which can have two values, 0 or 1023) were encoded in sensory neurons as floating point values between 0.0 and 1.0. The logistical function was used to determine the activation of the motor neurons. The activation of the first two motor neurons controlling the left and right wheels was transformed into 21 different integer values ranging from -10 to +10 (maximum speed backward and forward, respectively). The activation of the third and fourth motor neurons controlling the picking-up and releasing procedures, respectively, were thresholded into two values (1 = trigger the corresponding procedure, 0 = do not trigger the corresponding procedure).

The first architecture (a) is a simple feedforward network with 7 input units encoding the state of the 7 sensors and four output units encoding the state of the four effectors. The input units are directly connected to the output units through 28 connection weights (plus 4 biases). This architecture is not divided into modules.

106

The second architecture (b) is a modular one and it has been called emergent modular architecture ([7]) because it allows the required behavior to be broken down into sub-components controlled by different neural modules, although it does not require the designer to do such a partition in advance. (Notice that in this paper the emergent architecture is referred to as hardwired modular architecture). There are two modules for each of the four outputs (the two wheels, the object pick up procedure, and the object release procedure). In any particular input/output cycle only one of the two competing modules can control the corresponding output.

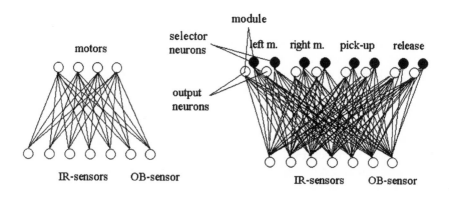

Figure 2.1: Architectures (a) and (b-c) are shown on the left and right side, respectively. Architectures (b) and (c) are structurally identical. However, in architecture (b) two modules compete to gain control of each of the four actuators. Individuals of the initial population with architecture (c) have only one module for each motor. However, a second competing module may be added in individuals of subsequent generations as a result of the duplication operator. Another difference is that in case (b) competing modules start with different random weights while in case (c), when a second competing module is introduced, the two competing modules have identical weights.

Each module includes two output units: a motor output unit and a selector unit. The motor output unit determines the speed of the corresponding wheel or whether or not the two procedures are executed. The selector unit determines the probability that the module will control the corresponding output. In other words, which of the two competing modules determines the output depends on which of the two competing selector units is more activated. Both the motor output unit and the selector unit of each module receives 7 connections (plus one bias) from the 7 sensory neurons.

The third architecture (c) is also modular and is denoted as duplication-based modular architecture because, in this case, the modules are not hardwired in the architecture from the beginning of evolution but they can be added during the evolutionary process. Each module, as in the case of architecture (b), consists of two output units (one motor output unit and one selector unit) which receive connections from the 7 sensors. At the beginning of the evolutionary process there is only one module for each of the four outputs, i.e., always the same module controls the corresponding output. However, at reproduction, modules may be duplicated (see below). Duplicated modules, which are exactly the same when duplication takes place, can differentiate across generations because of genetic mutations.

A genetic algorithm ([4]) was used to evolve the connection weights of our neural networks. Each connection weight or bias is encoded as a sequence of 8 bits in the genotype. We begin with 100 randomly generated genotypes each representing a network with the same architecture and a different set of random connection weights. Each individual is allowed to 'live' for 15 epochs, each epoch consisting of 200 input-output cycles or actions. At the beginning of each epoch the robot and the target objects are randomly positioned in the arena. An epoch is terminated either after 200 actions or after the first object had been correctly released. At the end of life, the best individuals are selected for reproduction. The 20 individuals that have accumulated the highest 'score' (i.e., performance measure; see below) during their lives generate 5 copies each of their neural networks. These 20x5=100 new robots constitute the next generation. The process is repeated for 1000 generations.

Reproduction consists in generating copies of an individual's genotype encoding the network's connection weights (we are assuming non-sexual reproduction in haploid populations) with the addition of random changes to some of the bits of the genotype sequence (genetic mutations) and, in the case of architecture (c), the duplication of a random selected neural module. The fitness formula is the way in which individuals are evaluated in order to decide who is allowed to reproduce. Individuals were scored for their ability to perform the complete sequence of correct behaviors, i.e., for their ability to release objects correctly outside the arena. However, in order to facilitate the emergence of this ability individuals were also scored (even if with a much lower reward) for their ability to pick up targets. More precisely, individuals were scored with 5 for each cycle they had an object in the gripper and with 10000 for each object correctly released outside the arena.

In the present preliminary model the maximum number of duplicated modules allowed in the case of architecture (c) is one for each motor output and no module-deletion operator was used. As a result, the architecture (b), already described in Nolfi ([7]), is the more complex architecture that can possibly evolve starting from architecture (a). However, the addition of competing modules during the course of evolution (instead than right from its beginning) that are initially identical to their competing module (instead of being completely unrelated) may produce qualitatively different results in the case of architecture (b) and (c), respectively.

3. Preliminary results

We present the results of several simulations in which we compare a simple feedforward neural network, the hardwired modular architecture, and the duplication-based modular architecture (see Fig. 2.1). In all simulations a mutation rate of 1% was used, i.e., 2% of the bits of the genotype randomly selected were replaced with a new randomly selected value. For the duplication-based modular architecture we investigated the performance obtained with a duplication rate of 0.02%, 0.03% and 0.04%, i.e., 0.02%, 0.03% and 0.04% of the modules were duplicated in each replication. We ran 10 simulations for each of the 3 different architectures described above. Each simulation started with a population of 100 networks with random connection weights and lasted 1000 generations.

Fig. 3.1 shows the average and peak performance for non-modular robots and for duplication-based modular robots with a duplication rate of 0.04%. In both conditions the performance level increases until a plateau is reached. However, modular robots achieve a higher terminal performance level and need less time (generations) to reach this level. This result confirms Nolfi's observation ([7]) that a modular architecture is useful in accomplishing a complex task.

Let us consider the individual results obtained in the 10 different repetitions of the simulation. In the case of duplication-based modular architecture, the ability to accomplish the desired task rapidly evolves in all replications. If we examine the genotypes of the best individuals across generations, we see that they incorporate at least one duplicated module at the time the performance level increases significantly (results not showed). The picture is different in the case of the non-modular architecture. In some of the replications it takes many generations for the performance level to reach the plateau; moreover, in one replication performance does not increase at all (results not showed).

Both populations with modules reach a higher fitness level than a population with only the basic architecture and no modules (see Fig. 3.1 and Fig. 3.2). However, the two populations with modules do not differ in terms of overall fitness except that fitness growth is slightly slower in the population with duplication-based modules (see Fig. 3.3).

In order to demonstrate that modularity plays a critical role, we varied the duplication rate in the population with duplication-based modules, with the result that both average and peak performance decreased linearly with a decreased duplication rate (0.04%, 0.03%, and 0.02%; results not shown). Fig. 3.4 shows the results obtained with a duplication rate of 0.02% and compares these results with those obtained with a non-modular architecture: the advantage of modular design is lost. This result underscores the importance of the interaction between mutation and duplication rate.

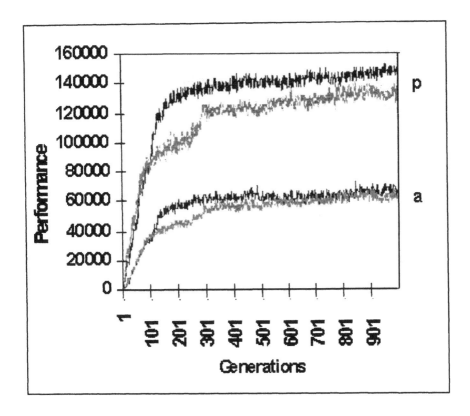

Figure 3.1: Average (a) and peak (p) performance of non-modular robots (grey curve) and duplication-based modular robots (black curve) with a duplication rate of 0.04%. Average of 10 different replications of the simulation.

4. Conclusions

In this paper we have described an Artificial Life model based on neural networks and genetic algorithms which can be used to understand the evolutionary mechanisms underlying the origin of modularity in nature, and we have presented some preliminary results obtained by comparing a nonmodular architecture, a hardwired modular architecture ([7]) and a duplication-based modular architecture. We used the same simulation scenario of Nolfi ([7]) but we added the genetic operator of gene duplication to explore the relationship between the evolutionary emergence of modularity and the phenomenon of gene duplication.

The cross-fertilization between Artificial Life and biology can take place since Artificial Life partially shares the theoretical apparatus and vocabulary of evolutionary biology and can offer additional methodological tools to biology. More specifically, our model allows us to reproduce in a computer both the organisms and the environment in which they live, behave and reproduce. An organism

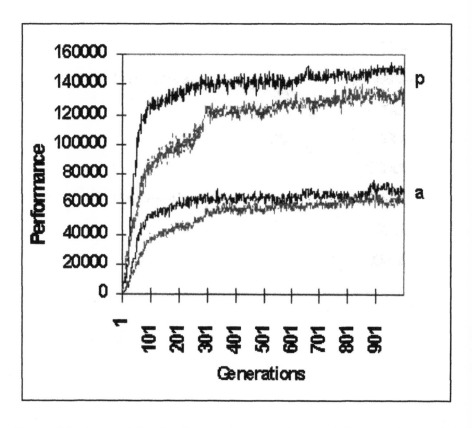

Figure 3.2: Average (a) and peak (p) performance of non-modular robots (grey curve) and hadwired-modular robots (black curve). Average of 10 different replications of the simulation.

is simulated as having a body with a specific size, external shape, sensory and motor organs, etc., and an internal structure made up of a genotype, the nervous system, and other organs. Artificial organisms can be analyzed at the genetic level, at the level of the mapping from genotype to phenotype (development), at the neural and behavioral level, at the level of the effects of the network's output on the environment, at the level of the reproductive success of each individual (fitness), and at the level of populations of individuals and of entire ecosystems.

Examining organisms at various levels could be crucial for understanding their behavior, because often an explanation of what happens at one level can be found at another level (see, for example, [5]; [2]). In particular, one could hypothesize that the evolution of modularity results from the interaction among processes at different levels. In future work we will focus on the evolutionarily emergence of functionally different modules at the neural-behavioral level from gene duplication. We will try to test the hypothesis that the different origin and evolutionary history of modules that arise out of genetic duplication instead

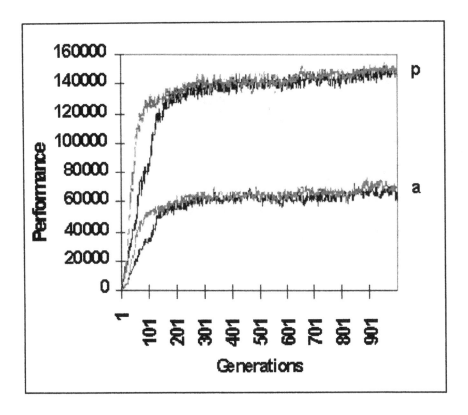

Figure 3.3: Average (a) and peak (p) performance of hadwired-modular robots (grey curve) and duplication-based modular robots (black curve) with a duplication rate of 0.04%. Average of 10 different replications of the simulation.

of being hardwired in the artificial organisms since the beginning of the evolutionary process results in modules endowed with a greater amount of functional meaning at the behavioral level.

Acknowledgments

This work has been made possible thanks to a fellowship from the Italian National Research Council (National Advisory Committee for Biological and Medical Sciences), which financially supported the one-year sabbatical stay of Raffaele Calabretta at Yale University in 1997. RC would also like to acknowledge the support of Aurelio Romeo, Jeffrey R. Powell, Valerio Sbordoni and Riccardo Galbiati, the assistance of Ginger Booth and Jan Taschner, and the useful discussions with the members of the GPW's lab at Yale University during weekly meetings.

112

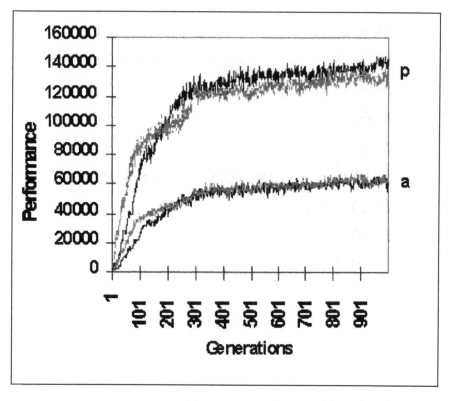

Figure 3.4: Average (a) and peak (p) performance of non-modular robots (grey curve) and duplication-based modular robots (black curve) with a duplication rate of 0.02%. Average of 10 different replications of the simulation.

Bibliography

[1] BONNER, John T., *The evolution of complexity*, Princeton University Press (1988).

[2] CALABRETTA, Raffaele, R. GALBIATI, S. NOLFI and D. PARISI, "Two is Better than One: a Diploid Genotype for Neural Networks", *Neural Processing Letters* 4 (1996), 149–155.

[3] FUTUYMA, Douglas, *Evolutionary biology*, Sinauer (1998).

[4] HOLLAND, John, *Adaptation in Natural and Artificial Systems*, University of Michigan Press (1975).

[5] MIGLINO, Orazio, S. NOLFI and D. PARISI, "Discontinuity in evolution: how different levels of organization imply pre-adaptation", *Adaptive Indi-*

viduals in Evolving Populations (Rick BELEW and Melanie MITCHELL eds.), Addison-Wesley (1996).

[6] MONDADA, Francesco, E. FRANZI and P. IENNE, "Mobile robot miniaturisation: a tool for investigation in control algorithms", *Experimental Robotics III, Lecture Notes in Control and Information Sciences* (T. YOSHIKAWA and F. MIYAZAKI eds.), Springer-Verlag (1994).

[7] NOLFI, Stefano, "Using emergent modularity to develop control systems for mobile robots", *Adaptive Behavior* **5** (1997), 343–363.

[8] NOLFI, Stefano, "Evolving non-trivial behaviors on real robots: a garbage collecting robot", *Robotics and Autonomous Systems* **22** (1997), 187–198.

[9] RAFF, Rudolph, *The shape of life. Genes, development, and the evolution of animal form*, University of Chicago Press (1996).

[10] RUMELHART, David and J. MCCLELLAND, *Parallel distributed processing: Explorations in the microstructure of cognition, Vol. 1.*, MIT Press (1986).

[11] WAGNER, Günter, "Homologues, natural kinds and the evolution of modularity", *American Zoologist* **36** (1996), 36–43.

[12] WAGNER, Günter and L. ALTENBERG, "Complex adaptations and the evolution of evolvability", *Evolution* **50** (1996), 967–976.

Chapter 14

From inductive inference to the fundamental equation of measurement

Bart De Boeck
Paul Scheunders
Dirk Van Dyck
Vision Lab, University of Antwerp, Department of Physics
Groenenborgerlaan 171, 2020 Antwerp, Belgium
deboeck@ruca.ua.ac.be

By considering inductive inference of the viewpoint of a gradual inclusion of information, instead of forecasting a given sequence, it will be shown that conditional algorithmic complexity decreases during learning. Based on a theorem of Levin, conditional algorithmic complexity and mutual algorithmic complexity are shown to be approximated by conditional entropy and mutual information, respectively. Furthermore, physical randomness and physical complexity are shown to be given by conditional algorithmic complexity and mutual algorithmic complexity, hence sum up to algorithmic complexity. A relation between computation and measurement will be suggested.

1. Introduction

In Inductive Inference an enumerable set of partial recursive functions represents the 'known' models. The optimal hypothesis is associated with the partial recursive function with the highest inferred probability [6]. In this paper we consider the influence of a gradual inclusion of information on the inferred probability, by rewriting Inductive Inference with conditional algorithmic complexity [8], [5], [3], [6] and consider its consequences for physical complexity and physical randomness.

We show that if the inferred probability of a model increases by the inclusion of information, conditional algorithmic complexity should monotonically decrease during learning. Furthermore, conditional algorithmic complexity and mutual algorithmic complexity sum up to the algorithmic complexity of the considered model. This shows that decreasing conditional algorithmic complexity during learning increases mutual algorithmic complexity.

As an immediate consequence, this property remains valid for the definition of physical randomness (conditional algorithmic complexity) and physical complexity (mutual algorithmic complexity). This shows that a clear relation exists between physical randomness and physical complexity (or regularity), contrary to previous ideas [1]. Furthermore, if the observer has no a priori knowledge, physical randomness and algorithmic complexity equal. Although this result seems 'counter intuitive' remind that randomness essentially means one has no logical explanation for a phenomenon, which has to be represented on an 'unknown' part of a Universal Turing Machine.

It was realized by Levin that the average algorithmic complexity is given by the stochastic entropy of the studied process [9]. As an application of this result we show that conditional algorithmic complexity and mutual algorithmic complexity are approximated by conditional entropy and mutual information, respectively. Furthermore since conditional algorithmic complexity or physical randomness and mutual algorithmic complexity or physical complexity sum up to algorithmic complexity, calculating the expectation gives us the fundamental equation of measurement. Stated otherwise : by calculating the average we pass from computation to measurement.

In §2, we reveal the interaction of reasoning with partial recursive functions. In §3, we calculate an approximation of conditional algorithmic complexity and mutual algorithmic complexity and explain the duality between physical complexity and physical randomness. Based on this result we suggest a relation between computation and measurement. In §4 we offer some concluding remarks.

2. The evolution of a model during learning

2.1. Algorithmic complexity

A physical system is considered understood when a model (or its associated partial recursive function) describing its properties is known [7]. This implies that the specification of the rules, input and matching properties describe the model and its relation to the physical system.

A measure indicating the accuracy of the observers description of the physical system, should in this context be restated as a 'distance' between the observers partial recursive function o and the unknown systems partial recursive function s. Notice that we are not satisfied with a correct prediction of the output only, the partial recursive functions should match somehow. We briefly review the definition of algorithmic complexity, for a complete treatment see [6].

Definition 1 *Fix a Universal Turing Machine U with binary input alphabet. The machine takes two inputs p and m. The algorithmic complexity of a binary string m given d is defined as*

$$K(m|d) = \min\{l(p) : U(p,d) = m\}$$

where l(p) denotes the number of bits of program p. Let $K(m) = K(m|\epsilon)$.

Definition 2 *The K−complexity of information in m about d is*

$$I(d:m) = K(m) - K(m|d)$$

The representation of models as partial recursive functions (or binary strings) does not explain how models evolve during knowledge retrieval, which is defined as the gradual inclusion of information about the physical system. To describe the evolution of models we need to include the mechanisms of reasoning and to reveal the interaction of this reasoning mechanism with models or the associated partial recursive functions.

2.2. Bayesian Inductive Inference

Let us assume reasoning is a generalized form of Aristotelian logic, essentially the idea that successive indications strengthen someone's belief for a certain proposition *relative to another*. The mathematical translation of this scheme leads to Bayesian Inductive Inference [4].

Recalling Bayes's rule

$$P(m|d) = \frac{P(m,d)}{P(d)} = \frac{P(d|m)P(m)}{P(d)}$$

m denotes a model or hypothesis and d the observed data.

On the assumption that m is admissible, or the Fundamental Inequality holds [6], we may write for the universal probability P_u :

$$\log P(d|m) = \log P_u(d|m) + O(1) \tag{2.1}$$
$$\log P_u(d|m) = -K(d|m) \pm O(1) \tag{2.2}$$

which gives us a relation between Bayesian Inductive Inference and algorithmic complexity. Throughout this article, log denotes the base-2 logarithm, which is appropriate for measuring information in bits.

2.3. Evolution of a model

Suppose now, as a matter of convenience, that an observer, after learning long enough, is capable of describing the systems model. Then assuming that models which explain the same phenomenon are one and the same (in the sense of conditional algorithmic complexity), learning essentially consists of the evolution of the observers partial recursive function to the systems partial recursive function. If the observer, for whatever reason, is not capable to explain the system accurately enough, his behavior should still be described by an evolution towards the systems partial recursive function (otherwise the observer is not acquiring new knowledge about the system), indicating that the conditional algorithmic complexity $K(s|o)$ or $K(o|s)$ should monotonically decrease during learning.

In order to describe the evolution of a model, the inclusion of information about the system s has to be taken into account. For the sake of generality, assume that the observer has, say n, different admissible models to explain the data with. Principally, for each model the associated probability, conditioned on the acquired information, can be calculated.

Consider a set of models or partial recursive functions $m_i : m_1, ... m_n$ and a set of data d_1, d_2. The conditional probabilities for the models given the data d_1 and d_1, d_2 respectively, are given by ((2.1)) and ((2.2)) :

$$P(m_i|d_1) = \frac{P(m_i, d_1)}{P(d_1)} = 2^{-K(m_i|d_1)-O(1)} \tag{2.3}$$

$$P(m_i|d_1, d_2) = \frac{P(m_i, d_1|d_2)}{P(d_1|d_2)} = 2^{-K(m_i|d_1,d_2)-O(1)} \tag{2.4}$$

where $O(1)$ is a computer dependent constant.

The relative probability r_i is given by

$$r_i = \frac{P(m_i|d_1, d_2)}{P(m_i|d_1)} = 2^{-K(m_i|d_1,d_2)+K(m_i|d_1)}$$

Suppose d_2 increases the probability for a certain model m_i. By rewriting

$$
\begin{aligned}
-K(m_i|d_1, d_2) + K(m_i|d_1) &\geq -K(m_i, d_1|d_2) + K(m_i|d_1) \\
&\quad +K(d_1|d_2) + O(1) \\
K(m_i, d_1|d_2) &= K(m_i, d_1) \\
&\quad -I(d_2 : m_i, d_1) \\
K(d_1|d_2) &= K(d_1) - I(d_2 : d_1)
\end{aligned}
$$

it is easy to see that by minimizing $I(d_2 : d_1)$ and maximizing $I(d_2 : m_i, d_1)$ the relative probability r_i will increase for a certain m_i.

Since for a certain model

$$
\begin{aligned}
-\log r_i &= K(m_i|d_1, d_2) - K(m_i|d_1) < 0 \\
&\text{iff}
\end{aligned}
$$

$$I(d_1, d_2 : m_i) > I(d_1 : m_i)$$

we state that conditional algorithmic complexity monotonically decreases during learning and that mutual algorithmic complexity monotonically increases.

Notice that for the special case $d_1 = \varepsilon$ the former reduces to

$$
\begin{aligned}
-\log r_i &= K\left(m_i|d_2\right) - K\left(m_i\right) < 0 \\
&\quad \text{iff} \\
I\left(d_2 : m_i\right) &> 0.
\end{aligned}
$$

Furthermore $K\left(m_i|d_1, d_2\right) + I\left(d_1, d_2 : m_i\right) = K\left(m_i\right)$, showing that decreasing conditional algorithmic complexity always increases mutual algorithmic complexity for fixed m_i. This property will remain valid for the definition of physical randomness and physical complexity. It is an indication of the idea that increasing the mutual algorithmic complexity of the description of a system reduces the randomness of the *observed* systems output. Stated otherwise : the regularity of the observed system increases ! This shows that a clear relation exists between physical randomness and physical complexity or regularity (see §3.2), contrary to the ideas in [1].

3. Shannon entropy

3.1. Conditional entropy and mutual information

Fix two binary strings r and q, it is known that $K\left(q|r\right) = K\left(q, r\right) - K\left(r\right)$ up to an $\Omega\left(\log K\left(q, r\right)\right)$ additive term [6], calculating the average over the ensemble gives, by the Theorem of Equality between Stochastic Entropy and Expected Algorithmic Complexity [9], [6] :

$$
\begin{aligned}
\langle K\left(q|r\right)\rangle &= \langle K\left(q, r\right)\rangle - \langle K\left(r\right)\rangle \pm \langle \Omega\left(\log K\left(q, r\right)\right)\rangle \\
&= -\sum_{q,r} p\left(q, r\right) \log p\left(q, r\right) + \sum_r p\left(r\right) \log p\left(r\right) \\
&\quad \pm \langle \Omega\left(\log K\left(q, r\right)\right)\rangle \\
&= -\sum_{q,r} p\left(q, r\right) \log \frac{p\left(q, r\right)}{p\left(r\right)} \pm \langle \Omega\left(\log K\left(q, r\right)\right)\rangle \\
&= -\sum_{q,r} p\left(q, r\right) \log p\left(q|r\right) \pm \langle \Omega\left(\log K\left(q, r\right)\right)\rangle \\
&= H\left(q|r\right) \pm \langle \Omega\left(\log K\left(q, r\right)\right)\rangle
\end{aligned}
\tag{3.1}
$$

whereby $p\left(x\right)$ denotes the probability of a random variable $X = x$. The first term in equation ((3.1)) is referred to as conditional entropy.

The reduction of r by the knowledge of q is easily calculated to be

$$
\begin{aligned}
\langle I\left(r : q\right)\rangle &= \langle K\left(q\right)\rangle - \langle K\left(q|r\right)\rangle \\
&= \sum_{q,r} p\left(r, q\right) \log \frac{p\left(r, q\right)}{p\left(r\right) p\left(q\right)} \mp \langle \Omega\left(\log K\left(q, r\right)\right)\rangle \\
&= H\left(q; r\right) \mp \langle \Omega\left(\log K\left(q, r\right)\right)\rangle
\end{aligned}
\tag{3.2}
$$

or mutual information. Notice the asymmetry in q and r of $K(r) - K(r|q)$ and the symmetry in q and r of $H(q;r)$, stating that the average reduction will become equal for observer and observed system (see ((3.3)) and ((3.4))).

The asymmetry in equation ((3.2)) can be eliminated by using $Kc(r|q) \equiv K(r|q^*)$ whereby q^* is the shortest program for q. Although this definition restores the symmetry between q and r, it does not eliminate the remaining term, previously $\Omega(\log K(q,r))$, now $O(1)$ [6]. We will see in equation ((3.3)) and ((3.4)) that these constants do not really matter.

For the special case $r = \varepsilon$, the conditional entropy reduces to entropy and since

$$
\begin{aligned}
K(q|r) &= K(q,r) - K(r) \pm \Omega(\log K(q,r)) \\
&= -\log m(q|r) - O(1)
\end{aligned}
$$

conditional entropy can be interpreted as the average uncertainty of the observer given a priori knowledge or as an average quantity of information given a priori knowledge.

3.2. Complexity and randomness of strings

Seen in the context of Universal Turing Machines (UTMs), Adami's environment ("the particular rules of mathematics of this "universe"" [2] validity of Church's Thesis, a UTM is capable of representing every logically deducible 'reasoning' process. This implies that the environment of Adami is necessarily imbedded on a UTM. The definition of a physical universe then consists of the definition of a subspace on a UTM. This view has the advantage that "the particular rules of mathematics" [2] can be interpreted as the knowledge one has about the physical universe, and are imbedded in a universal space, which allows the representation of any algorithm. Notice that we are obliged to consider a construction of this kind, since a UTM is capable of representing any algorithm, indicating that physical randomness cannot be defined as : not being representable on a UTM.

In order to calculate conditional algorithmic complexity, we conditioned on the prior knowledge of the observer (a subspace on a UTM). Notice that in this calculation conditional algorithmic complexity essentially represents the randomness of a string, or the amount of bits we cannot 'explain' with the given a priori knowledge.

Physical randomness becomes in this context the complexity of a string, say s, given the set of partial recursive functions induced by the observers a priori knowledge, say m_i : $K(s|m_1, ..., m_n)$ or $-\log P(s|m_1, ..., m_n)$.

Physical complexity is the representation of the possibility of the observer to explain the observed system or, formally, the mutual algorithmic complexity of a string, say s, given the set of partial recursive functions induced by the observers a priori knowledge, say m_i : $I(m_1, ..., m_n : s)$ or $-\log \frac{P(s)}{P(s|m_1,...,m_n)}$.

If $K(s|m_1, ..., m_n) = K(s)$, indicating that the observer has no a priori knowledge to describe s with, physical randomness and algorithmic complexity become equal. Obviously $K(s|m_1, ..., m_n) + I(m_1, ..., m_n : s) = K(s)$ or physical randomness plus physical complexity equals algorithmic complexity, the formal result for the duality between complexity and randomness indicated in §2.3.

The above results show that classifying regularity according to conditional algorithmic complexity is good sense, reducing to algorithmic complexity if no prior knowledge is present (in contradiction with "absence of an environment" [2]).

By the results of §3.1,

$$\langle K(s|m_1,...,m_n) \rangle + \langle I(m_1,...,m_n:s) \rangle = \langle K(s) \rangle \qquad (3.3)$$

$$H(s|m_1,...,m_n) + H(m_1,...,m_n;s) = H(s) \qquad (3.4)$$

which is the fundamental equation of measurement. Notice that this argument is derived by calculating the average of an equation expressing a relation between the length of algorithms, whereby the additive terms automatically cancel! Stated otherwise : by calculating the average we go over from computation to measurement.

The conceptual difficulties in [1] with algorithmic complexity emerge from the definition of algorithmic complexity on a UTM, which is capable of representing every partial recursive function, hence giving a 'counter intuitive' definition for 'physical randomness'. Remind that randomness essentially means one has no logical explanation for a phenomenon, which has to be represented on a 'unknown' part of the UTM (implicitly assumed a logical explanation exists).

4. Conclusion

By the interpretation of learning as an evolution of a probability distribution of partial recursive functions, induced by the inclusion of data, we were able to show that conditional entropy monotonically decreases during learning. Conditional and mutual algorithmic complexity are approximated by conditional entropy and mutual information, respectively.

Furthermore, physical randomness and physical complexity showed to be concepts which can be explained by conditional algorithmic complexity and mutual algorithmic complexity, respectively. Accordingly, increasing the $K-$complexity of information in the description of a system reduces the randomness of the *observed* output .

Acknowledgments

It is a pleasure to thank L. Lemmens for an Introduction on Bayesian Inductive Inference and Statistics.

Bibliography

[1] ADAMI, Chris, and Nicolas J. CERF, "Complexity, Computation, and Measurement", *Proceedings of the 4th Workshop on Physics and Computation*

122

(T. TOFFOLI, M. BIAFORE and J. LEAÕ ed.), PhysComp96 (22-24 November 1996), 7.

[2] ADAMI, Chris, and Nicolas J. CERF, "A lower bound on the Complexity of Symbolic Sequences", *Physica D* (1997), submitted.

[3] CHAITIN, Gregory J., "On the Length of Programs for Computing Finite Binary Sequences", *Journal of the ACM* **13** (1966), 547–569.

[4] JAYNES, E.T., *Probability Theory – The Logic of Science*, Washington University, St. Louis MO 63130 USA (Fragmentary Version Sept. 1993).

[5] KOLMOGOROV, Andrei N., "Three Approaches to the Quantitative Definition of Information", *Problems of Information Transmission* **1** (1965), 3–11.

[6] LI, Ming, Paul M.B. VITANYI, *An Introduction to Kolmogorov Complexity and its Applications*, Springer Verlag (1997).

[7] SHAPIRO, Stewart, "Understanding Church's Thesis", *Journal of Philosophical Logic* **10** (1981), 353–365.

[8] SOLOMONOFF, Ray J., "A Formal Theory of Inductive Inference", *Information and Control* **7** (1964), 1–22, 224–254.

[9] ZVONKIN, A.K., and L.A. LEVIN, "The Complexity of Finite Objects and the Development of the Concepts of Information and Randomness by Means of the Theory of Algorithms", *Russian Math. Surveys 25(6)*, (1970), 83–124.

Chapter 15

Controlling chaos in systems of coupled maps with long-range interactions

Michael Doebeli[1]

Zoology Department, University of Basel, Switzerland
doebeli@ubaclu.unibas.ch

We propose a simple method for controlling spatial chaos in an ecological model describing many local populations that are coupled by long–range dispersal. Control is achieved by making adjustments to the population sizes in a subset of the local populations. No detailed knowledge of the dynamic mechanisms underlying the chaotic population fluctuations is required. Every few generations, the populations in the censused subset of habitats are either decreased or increased by a fixed proportion of the current size. This reduces spatial chaos to regular cycles over a broad range of parameters. Similar results can be obtained with variants of the control technique. The models we use are of a very general nature, describing systems of coupled oscillators. The control of chaos in such systems has so far mostly been achieved in cases where coupling is restricted to nearest neighbours, and where chaos is controlled by perturbing system parameters according to a complicated scheme based on detailed knowledge about the system's dynamic behaviour. In contrast, our methods are very simple and consist of perturbing dynamic variables, which are generally amenable to measurement and manipulation. Therefore, our methods may prove useful in natural systems where underlying processes are imperfectly understood and long-range interactions are the norm.

[1]A longer version of this paper with slightly different contents was published in *Bull. Math. Biol.* **59**, 497–515, coauthored by G. D. Ruxton, whom I wish to thank for many fruitful discussions. This work was supported by the Swiss National Science Foundation.

1. Introduction

An important line of ecological research concerns the influence of spatial structure on the dynamics of ecosystems (Gilpin and Hansli 1991). Although spatial factors can simplify population dynamics in some circumstances (Hastings 1993, Gyllenberg et al. 1993, Doebeli 1995), the addition of space can lead to new levels of complexity as exemplified by the phenomenon of spatial chaos (Chow and Mallet-Paret 1995). Here we present a very simple recipe which can be used to tame chaotic fluctuations in a metapopulation model and replace them with a simple, repeating cycle.

Our model consists of a chain of oscillators (the local populations) that are coupled by long–range dispersal (Hastings and Higgins 1994), so that local populations interact over large distances. The control of chaos in systems of coupled oscillators has so far mostly been restricted to cases where the coupling is between nearest neighbours (Auerbach 1994, Astakhov et al. 1995, Brayman et al. 1995). A notable exception is a high–dimensional neural network model with non-local coupling (Sepulchre and Baboyantz 1993). In this model, the control is obtained by using the classical method of adjusting system parameters (Ott et al. 1990, Shinbrot et al. 1993). These methods require a detailed knowledge about the dynamical behaviour of the system, knowledge which is likely to be difficult to obtain for natural systems with many degrees of freedom and high–dimensional attractors.

In ecology, even apparently simple systems are not sufficiently understood or us to hope to use these control methods. A few years ago, Güémez and Matías (1993) have proposed a different control method which relies on modifying dynamic variables rather than system parameters. Their method works well in simple systems (Solé and Menéndez de la Prida 1995), and here we demonstrate that the technique can be generalised to high–dimensional systems of linked oscillators.

2. Model and results

We consider a 1–dimensional array of habitat patches $0, \ldots, n$ that might represent suitable fragments of a coastline, forest edge or lake shore. Each patch contains a population with density dependent reproduction alternating with dispersal. The population in patch i at the start of a given generation t is denoted by $N_t[i]$, $i = 0, \ldots, n$. After reproduction, but before dispersal, the population in patch i is given by the difference equation

$$M_t[i] = G(N_t[i]) = \frac{\lambda N_t[i]}{1 + (aN_t[i])^b}.$$

The demographic parameters λ, a, and b describe the intrinsic growth rate, the carrying capacity and the type of competition that leads to density–dependence (for more details on the ecological interpretation of these parameters see e.g. Bellows 1981). It is well known that, depending on these parameters, the difference equation (1) can exhibit a whole range of dynamic behaviours, including chaos (e.g. May and Oster 1976).

We assume that the parameters are the same in each patch. After reproduction, fixed proportions of the populations on each patch migrate to the other patches according to a Gaussian distribution function. The proportion leaving patch i for patch j is given by

$$p(i,j) = (\frac{\pi}{D})^{-1/2} \exp(-D(i-j)^2),$$

where D is a constant: the lower D, the further individuals disperse. The size of the local population in patch i at the beginning of the next generation is then given by

$$N_{t+1}[i] = \sum_{j=0}^{n} M_t[j] \cdot p(j,i).$$

This implies that individuals which disperse beyond the ends of the metapopulation are lost from the system. Eqs. (1)–(3) determine the dynamics of the metapopulation recursively. For our simulations, we scaled distances so that the distance across the lattice is unity, and we used absorbing boundary conditions.

For the control mechanism we choose a subset $K \subset \{0, \dots, n\}$ of all patches to which the control will be applied. If a patch is not a member of K then reproduction is given as before by eq. (1). For patches belonging to K, we modify the size of the local population before reproduction occurs. Let p be some positive integer; it is the period of the control. Then, in all generations that are multiples of p, we add (or subtract) a fixed proportion γ of the local populations before reproduction in each of the patches in K. Mathematically, if $i \in K$, and with $\gamma \in (-1, 1)$, we replace eq. (1) by

$$M_t[i] = G(N_t[i]) \quad \text{if } p \text{ does not divide } t$$
$$= G(N_t[i](1 + \gamma)) \quad \text{if } p \text{ divides } t.$$

Dispersal occurs as before. The uncontrolled model has been studied previously (Hastings and Higgins 1994, Ruxton and Doebeli 1996), and it is known to produce very long transients with varying dynamic regimes. For a wide range of parameters, however, the model shows short transients leading to spatial chaos, as illustrated in Fig. 2.1a. In Fig. 2.1b, the simulation of 1a is repeated, except that after 1000 generations, a period 2 control is activated on a randomly chosen subset K comprising 20% of all patches in the metapopulation. The control quickly acts to replace the violent chaotic fluctuations with a simple cycle of period 2.

Further numerical simulations suggest that such simplifications occur over a wide range of system parameters, provided that the control is strong enough, i.e. $|\gamma|$ is large enough. With longer control periods the system settles on cycles of longer periods, often (but not always) twice that of the control (Fig. 2.1c). If the control pulse $|\gamma|$ is weak, then control may "almost work": the system is almost 2–cyclic for intermittent periods of time which are abruptly broken up

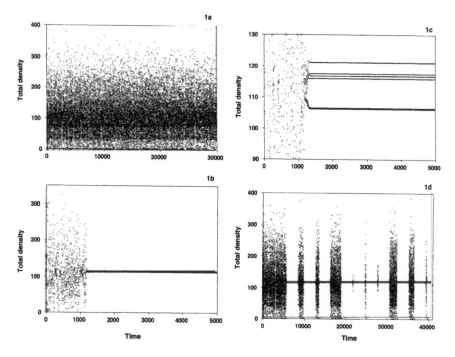

Figure 2.1: Control of chaotic metapopulations. The time series of the total density of a chaotic metapopulation consisting of 80 local populations is shown in 1a for 30000 generations. In the following panels, variants of the control mechanisms described in the text are applied to this metapopulation after 1000 generations. In 1b , the pulse control (4) is applied with a control period of 2. The control was applied to 17 randomly chosen patches ($\sim$ 20% of the 80 patches) and the magnitude of the control was $\gamma = -0.2$. For higher control periods larger pulses and a larger proportion of controlled patches are necessary for control. In 1c, $\gamma = -0.3$, and 23 randomly chosen patches ($\sim$ 30%) were used for a period 3 pulse control. The system settles not on 3–, but on a 6-cycle. Very similar results would be obtained if the values of γ were positive with the same modulus. 1d is the same as 1b, but the magnitude of the control pulse is lower, $\gamma = -0.09679$, resulting in intermittent chaos. For the metapopulation shown in 1a, the demographic parameters of the local populations in eq. (1) were $\lambda = 15$, $b = 7.5$ and $a = 1.4217$. The dispersal constant in eq. (2) was $D = 0.00111$, which was chosen so that the chance of dispersing over more than 25 patches is less than 0.5%.

by periods of highly chaotic fluctuations (Fig. 2.1d). This is a rare example of spatial intermittent chaos.

The effectiveness of the control depends of the distribution of the controlled patches in the metapopulation. The strength of the control can also be lessened by reducing the number of local populations to which the control is applied, and the control works least well if the selected sites are bunched together away from the periphery of the metapopulation. Surprisingly, the control also ceases to work if the set K of controlled patches is too large, e.g. if the control is applied to all patches in the metapopulation.

We have obtained similarly effective control in a variety of models using different density dependent reproduction functions and a variety of dispersal functions. Further we found that other control mechanisms also work. For example, if the pulse control (4) is replaced by the wave control

$$
M_t[i] = \begin{cases} G(N_t[i]) & \text{if } i \notin K \\ G(N_t[i]\{1 + \gamma \cdot \cos \frac{2\pi t}{p}\}) & \text{if } i \in K, \end{cases}
$$

in which the control is applied in every generation, but the strength and direction of the control changes sinusoidally with period p, then similar results as for pulse control are obtained.

In fact, simulations suggest that wave control is generally even more effective than pulse control. Fig. 2.2 and Fig. 2.3 summarize some of the effects of wave control. In Fig. 2.2 bifurcation diagrams are shown with the parameter D determining the dispersal distance as bifurcation parameter. Compared to the uncontrolled dynamics shown in Fig. 2.2a, the control mechanism induces windows of regular dynamics for intermediate dispersal distances. The control is least effective for very large and very low dispersal ranges. Similar results hold for the pulse control method. While for large dispersal distances control can often still be achieved if the control magnitude γ is big enough, control is impossible for very small dispersal distances, e.g. when dispersal occurs to nearest neighbours only. In these cases, our numerical simulations suggest that not even high values of γ are sufficient for control. Thus it seems that the control methods proposed here only work for systems that are sufficiently connected, i.e. in which interactions between local populations occur over a sufficiently wide range.

The windows of regular dynamics induced by the control are larger if the control is applied only to marginal patches (Fig. 2.2b) than when it is applied also to patches lying in the center of the metapopulation (Fig. 2.2c). This reiterates what was said earlier about the influence of the distribution of the controlled patches on the effectiveness of the control. The phenomenon is again apparent in Fig. 2.3, where the control magnitude $|\gamma|$ was used as bifurcation parameter. If only marginal patches are used for the control, it works for larger ranges and for lower values of γ than if the control is also applied to central patches. In addition, control altogether ceases to be effective in the scenarios considered in Fig. 2.2 and Fig. 2.3 if too many central patches are used for control. For example, even very high control magnitudes γ are ineffective if the control is applied to the central patches 8–32, and a corresponding bifurcation diagram for the dispersal distance would be very similar to the one shown in Fig. 2.2a for an uncontrolled metapopulation.

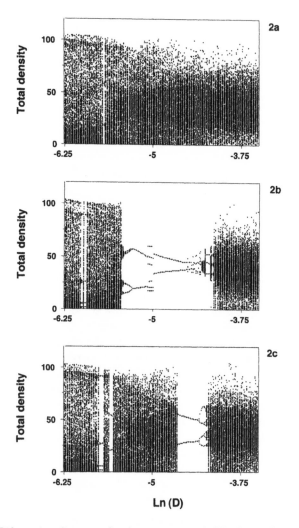

Figure 2.2: Bifurcation diagrams for the wave control. The logarithm of the dispersal parameter D was used as bifurcation parameter. Small values of $\ln(D)$ correspond to large dispersal distances. **2a** shows the uncontrolled dynamics of the total density of a metapopulation consisting of 41 local populations (i.e. $n = 40$). In **2b**, a period 2 wave control with $\gamma = -0.2$ was applied to the marginal patches 0–11 and 29–40 in the metapopulation, and in **2c** the same control was applied to the marginal patches 0–7 and 33–40 and to central patches 16–24. The windows of regular dynamics are larger in the former case. The demographic parameters of the local populations were $\lambda = 5$, $b = 8.75$ and $a = 1.172$. For 200 equally spaced values of $\ln(D)$ the metapopulation was first run without control for 500 time steps, then the control was applied for 700 time steps (no control in **2a**). During the last 200 steps the density was plotted.

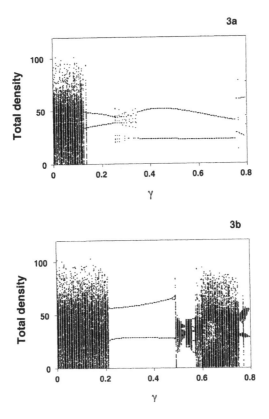

Figure 2.3: Bifurcation diagrams for the wave control. The control magnitude γ was used as bifurcation parameter for the same metapopulation as in Fig. 2.2, and for a dispersal parameter $D = 0.0136$. In **3a** the marginal patches 0–11 and 29–40 were controlled, while in **3b** the control was applied to the marginal patches 0–7 and 33–40 and to the central patches 16–24. The control works better when a larger proportion of marginal patches is controlled. The same procedure as for Fig. 2.2 was applied to obtain the diagrams.

3. Discussion

How feasible would it be to apply these control methods to real populations? We have already argued that it is more feasible than for other control mechanisms which perturb system parameters and require detailed knowledge about the dynamics of a system. Our methods only require that the population size can be measured in some of the local populations. Moreover, this measurement need not be very accurate, because the mechanisms are robust against noise in the proportion of the population which is added or subtracted in the controlled patches. This is illustrated in Fig. 3.1, which shows that the control still works in the presence of noise, even though the system does not move on an exact

2-cycle. In the presence of noise it typically takes longer until the control starts
to be effective (cf. Fig. 3.1 and Fig. 2.1b).

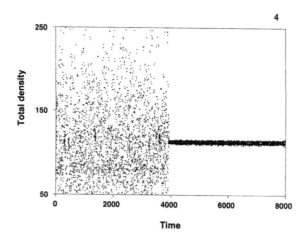

Figure 3.1: The same as 1b, except that there is noise in the control: instead of
adding γN in generations that are multiples of the control period 2, where N is the
population size in a controlled patch, we added $\gamma(N + \epsilon)$, where the error ϵ was chosen
for each control from a Gaussian distribution with mean 0 and variance equal to 20%
of the actual (local) population size. The control was started after 1000 generations
but is only effective after ca. generation 4000.

Moreover, we have found that our control methods also work when local
extinctions are included in the metapopulation model. For this we have assumed that every local population that falls below a certain threshold density
goes extinct. That the control also works in this situation indicates that the
method is robust not only against measurement error, but also against internal,
demographic stochasticity in the population dynamics.

We can also speculate on the question of whether natural populations are
likely to behave chaotically. Our results suggest that chaos should be less likely
under natural conditions, bearing in mind how much of the natural world is
subject to the kind of cyclical driving which we used for control, be it related
to day and night, tides, or the seasons of the year. However, we also suggest
that control will only work if applied to a fraction of the total metapopulation
which is neither too small not too big. Hence, we speculate that chaos is more
likely in metapopulations with a small longitudinal spread than in those whose
component populations vary considerably in the force with which they receive
seasonal fluctuations.

In this paper we present a simple and apparently robust method for controlling chaos in a system of coupled oscillators with long–range interactions. Controlling chaos in systems of coupled oscillators has generated interest in physics,
chemistry and neurology. The classical methods used to control chaos rely on
adjusting system parameters in a subtle way and requires detailed knowledge of

the dynamic behaviour of the system (Ott et al. 1990, Shinbrot et al. 1993). Typically, these methods have only been successfully applied to systems of coupled oscillators with nearest-neighbour interactions (Auerbach 1994, Astakhov et al. 1995, Brayman et al. 1995). In contrast, high–dimensional systems with long–range interactions have mostly eluded the traditional methods of chaos control.

For such systems, the methods proposed here seem more promising, because the perturbations are not based on information about the dynamic history of the system, and they are applied to the dynamic variables themselves, and not to system parameters. This is particularly important if the systems to be controlled are not models, but real systems, whose dynamics are typically only poorly understood. We note that, mathematically, the control perturbations we used can be viewed as perturbations to system parameters, but the big difference to the classical method is that the magnitude of the perturbations does not have to be calculated from the usually complicated dynamic history of the system, and instead can be determined in a simple and less error prone manner as a constant fraction of the current state of the system variables. Since the magnitude of the perturbations still depends on the current state of the system, the situation is quite different from those in which "periodic forcing" of parameters can promote complex dynamics in a system, which occurs for example in epidemiological models of childhoood diseases (Olsen and Schaffer 1990). In such cases, the periodic perturbations are not tuned to the system's dynamics and instead are independent of its current state, which leads to incommensurabilities that can induce chaos.

The simplicity and robustness of the methods is obtained at the expense of having to apply on average slightly larger perturbations than in the classical methods. However, our results show that already moderate perturbations of a magnitude of 10% to 20% of the current state of system variables are enough to obtain simple dynamics. Although explained here for ecological models, the proposed methods for controlling chaos in high-dimensional systems are expected to be useful in a number of other systems consisting of coupled oscillators, for example in chaotic neural networks (Sepulchre and Baboyantz 1993, Solé and Menéndez de la Prida 1995). In neural nets, the regular patterns induced by the controls could be viewed as a model for cognitive processes that are induced by external stimuli.

Bibliography

[1] Astakhov, V. V., Ansihchenko, V. S. and Shabunin, A. V. 1995. *IEEE Trans. Circ. Syst. I* **42**, 352–357 (1995).

[2] Auerbach, D. *Phys. Rev. Let.***72**, 1184–1187 (1994).

[3] Bellows, T. S. Jr. *J. Anim. Ecol.* **50**, 139–156 (1981).

[4] Brayman, Y., Lindner, J. F. and Ditto, W. L. *Nature* **378**, 465–467 (1995).

[5] Chow, S. N. and Mallet-Paret, J. *IEEE Trans. Circ. Syst. I* **42**, 746–751 (1995).

[6] Doebeli, M. *Theor. Pop. Biol.* **47**, 82–106 (1995).

[7] Gilpin, M. and Hanski, I. (eds.) *Metapopulation dynamics: empirical and theoretical investigations.* London: Academic Press (1991).

[8] Güémez, J. and Matías, M. A. *Phys. Lett. A* **181**, 29–32 (1993).

[9] Gyllenberg, M., Söderbacka, G. and Ericsson, S. *Math. Biosci.* **118**, 25–49 (1993).

[10] Hastings, A. *Ecology* **74**, 1362–1372 (1993).

[11] Hastings, A. and Higgins, K. *Science* **263**, 1133–1136 (1994).

[12] May, R. M. and Oster, G. F. *Am. Nat.* **110**, 573–599 (1976).

[13] Olsen L.F. and Schaffer W.M. *Science* **249**, 499-504 (1990).

[14] Ott, E., Grebogi, C. and Yorke, J. A. *Phys. Rev. Lett.* **64**, 1196–1199 (1990).

[15] Ruxton, G. D. and Doebeli, M. *Proc. R. Soc. Lond B* **263**, 1153–1158 (1996).

[16] Sepulchre, J. A. and Baboyantz, A. *Phys. Rev. E* **48**, 945–950 (1993).

[17] Shinbrot, T., Grebogi, C., Ott, E. and Yorke, J. A. *Nature* **363**, 411–417 (1993).

[18] Solé, R. V. and Menéndez de la Prida, L. *Phys. Lett. A* **199**, 65–69 (1995).

Chapter 16

Assessing software organizations from a complex systems perspective

Javier Dolado[1]
Dept. of Computer Languages and Systems
Univ. of the Basque Country, Spain
Alvaro Moreno
Dept. of Logic and Philosophy of Science
Univ. of the Basque Country, Spain

Nowadays, the words "software process improvement" are buzzwords in the software management activities. Assessments of the status of the software process in organizations are made by using models such as CMM, SPICE or ISO 9000. We see some uses of those models as an example of the technological (or mechanical) paradigm that is trying to pervade most activities of software construction.

We can consider some types of software products as the final outcome of human design processes. Thus, to use mechanical procedures as patterns to follow may not be the apropriate way for dealing with emergent structures. Looking at the biological area, a morphogenetic metaphor allows us to introduce the concepts of deducible emergence and hyperstructures. Thus, software processes could take many appearances, and improvement could be evaluated with respect to what extent processes allow intermediate structures to evolve and reach the final form.

[1] Work supported by UPV-EHU under project 141.226 - EA 114/96

1. Introduction

Software is an element that can be considered complex from different perspectives, and the complexity that arises in software development has been acknowledged for some time. The phrase "software crisis" tries to reflect many of the problems that software development organizations face in their projects. One of the factors that has been blamed for the source of the obstacles in achieving a controlled project is the software process. However, the ideal normative process for development activities is still unknown. In this work we aim to link some of the concepts of the complex systems area to the management of software.

Basically the question that is posed here is whether it is sound to have a mechanistic view of software development given that the social and human aspects are so significant. The view of software creation as a sort of form generation conveys an 'organic perception' of software houses. This suggests that the evaluation of the excellence of an organization has to take into consideration not only the 'manufacturing' processes, but also other less structured, but effective, processes.

2. The software process and its evaluation

A *software process* can be defined as a set of activites, methods, practices and transformations that people employ to develop and maintain software ant the associated products. Software process maturity is the extent to which a specific process is explicitly defined, managed, measured, controlled, and made effective. There are some methods that aim to assess the state of a software production organization, one of them being the well-known SEI's Capability Maturity Model – CMM– [16]. Other trends are the ISO 9000 [12] guidelines and the new SPICE model.

2.1. Different evaluations of the process

Every evaluation of a software process is made under a theoretical framework, whether or not it is explicitly defined. The framework provides the basis for the specific rules to be followed in the assessment. These may be one those previously listed or one of other more general guidelines like Total Quality Management, etc. From the practical point of view, the evaluation takes the general form of a review of the extent to which the areas, aspects or issues are fulfilled in a verification list. The two best known methods (CMM and ISO 9000) provide separate approaches to the evaluation, that of CMM being more flexible. As an equivalence can be established between those models, and since the purpose of this paper is not to dig into the technicalities of each model, we will refer to some aspects of the CMM as it is carried out in the actual practice.

Schematically, this model establishes a 5 level scale
1. Initial, sometimes called "chaotic" (the process is "ad hoc")
2. Repeatable, in which the process depends on individuals

3. Defined, the process is defined and institutionalized
4. Managed, the process is quantitatively measured
5. Optimizing, in which the improvement is fed back to process.

Thus, a software organization is better rated as a higher point is attained. Each of the levels is composed of several *key process areas*. Improvement is achieved by taking action in the activities involved in the key process areas. Reaching the highest level implies that the organization is mature with respect to the software engineering process and that it is able to manage effectively many aspects of the software construction.

While trying not to be simplistic, and acknowledging that technology is an important part in software, we view this model as rooted in the technological paradigm of improvement by standardizing repeatable procedures.

Several facts can contradict this way of assessment and could lead us to think of that scale as a very artificial evaluation. There are software houses that could be rated at level 1 and, despite this, they are able to produce good software. Although it seems that some economic benefits are obtained as higher point in the scale is attained [8], it is not clear that a level 3 software house (or labeled as ISO 9000) could guarantee a good product.

2.2. Some facts: Feedback and emergent patterns

One thing that has to be made clear here is that software is not a unique type of engineered object, but that there are many types of software. Thus Lehman [10] establishes two broad categories:
a) S-type software is required to be correct in the full mathematical sense with respect to a fixed specification;
b) E-type programs model real world applications and are frequently subjected to system evolution. The development of these systems constitutes a complex feedback learning system. Data collected in projects such as the IBM OS-360 shows trends in evolution that support the existence of feedback and evolution in the software process.

Also, confronting the idea of a pure technological view we can hear categorical phrases such as "*Software development is a predominantly social activity*" [4]. Cain and Coplien take a sociological view of the software process and empirically study the underlying (self–organizing) structures that are created among the members of the organizations of software development.

Several of their discoveries are worth noting here:
a) there was a strong focus on process documentation in the organization, but what was documented was often distant from what was seen in actual practice;
b) in one case 80% of the developers were working on different processes from the official one, because the project's process standard did not capture the stable structure of the process;
c) communication is a key element for obtaining good organizational behavior. Highly productive organizations are cohesive; but more importantly: "Good organizations nurture roles that work directly with product artifacts".

They suggest that the patterns generated within the organization can be used for improving the software process. Another author, King [9], uses CASE tools to support systems emergence in information systems development, differentiating it from software engineering development. The link that is still missing is the relationship of the organizational structures with the series of intermediate structures that they are creating through the development process.

In this section we have observed two important views of software development: the first one embodies the characteristics of complex feedback systems and the second one highlights the communication aspects of an organization. The next section introduces a paradigm that can easily accomodate these facts.

2.3. Design: Moving away from the technological paradigm

In a recent book Blum [2] addresses the software construction problem as an ecological design problem. Blum does not deal with software as a mechanic artifact, but as a product of the design process, the design resultant of an activity. But the design process he proposes is not a technological design, but an *adaptive design*, wherein the design process takes into account changes in the requirements. Design is not a phase of the development of a software system, but the process itself.

The metaphor that Blum proposes is the 'sculpting process', in such a way that the sculptor interacts with the materials, adjusting the goals while the process goes on. The modifications of a part of the design extend inmediately to the other design objects influenced. He presents his evaluations of the ecological design paradigm as a way of moving away from the machine paradigm, but does not assume that it is the unique solution.

It is the concept of *design* that can accommodate the facts stated in §2.2 and that links to the concepts of the following section §3. Design is a synonym of form, and a design is a transition from concepts and ideas to concrete descriptions [3]. But as Braha and Maimon state, "the concept of form is elusive, abstract, and complex". However, the design process can be characterised and situated within a paradigm. The five major paradigms of design are: analysis–synthesis–evaluation, case–based, cognitive, algorithmic and artificial intelligence. So, human design activities are developed within these frameworks. Design is also an ubiquitous process in nature as we state in the next section.

3. A metaphor for the software process: Morphogenesis

Nature can provide us with examples of what are the results of effective processes that guarantee good products. Looking at the biological area for models we can initially pay attention to insect colonies. Ant colonies, for instance, are able to make some effective and reliable constructions. From the point of view of the process, nature has positioned those systems at the higher level of the scale, but that does not say much about the organization of the elements. The plan and the product is achieved with only a very effective communication among the agents.

But, if we focus on the aspects of design, the model to follow is that of *morphogenesis*. Morphogenesis in nature is also design. The metaphor of morphogenesis is taken here as a framework that allows us to view the process of software development as form generation. This is one of the biological similes that is more appropriate for comparison with software development.

Form is one of the properties that organisms show [7]. The word *phenotype* represents what organisms reveal to the senses and, according to Rosen [14], it is a uniquely biological concept. The generation of a phenotype starts from an initial set of genetic (explicit, informational) instructions that only specify the construction of low level building blocks. The interactions (implicit, dynamical) of those blocks unfold the next organizational levels in the system. As result of the internal interactions among the processes and with the environment a form emerges.

The distinction between phenotype and genotype resembles in some ways the distinctions between specifications and final code in software systems (E-type or real world systems). If in biology little can be said of the genotype by the study of the phenotype, in software systems it is very questionable that it would be possible to recover exactly the original specifications from the final product. More difficult still is the issue with the software process, since the product -code- does not entail much knowledge of the process used to built it.

Taking this perspective, the final product of the software process can be seen as an emergent structure. This emergent outcome is the result of a group of dynamic processes that are induced by a set of initial specifications. These specifications, in turn, are handled by the next level of procedures. Nature through evolution discards those forms that are not adapted to the environment. But there are also other reasons for limiting the variety of forms, like structural or "architectural coherence". If we take this analogy, then the evaluation of an organization not only has to take into account how the process is implemented, but also what the software products (final and intermediate) it manufactures are like. This can help us to understand why software houses that could be rated as inferior from the normative assessments can still survive in the market.

3.1. Emergence of software systems

The biological concept of *emergent property* is important since it can be made operational by using some formalized theories. There are several approaches to emergence as explained by Emmeche et al. [6], but almost all of them share the general definition of "creation of new properties", regardless of the substance involved. With the inclusion of the concept of *levels*, more power is introduced in the description of phenomena. The elements of a level give shape to the entities of the emerging level. Several constraining conditions act on the low level to form the next level. Different levels and entities can arise as the combination of primary and secondary emergence. This idea of emergence among levels can be applied to multiple fields.

The concept of emergence that is useful here is the one named *deductional emergence* defined by Baas [1]. Emergence is the creation of new properties in a structure that is the result of the interactions among the agents involved. The conceptual element that allows emergence to occur is the concept of *hyperstructure* which is a multi-level emergent structure. Hyperstructures apart from being the organizative and design principle, allow to incorporate the concept of *evolution*. A hyperstructure is a cumulative structure, not necessarily recursive. Hyperstructures are higher order structures (structures of order n), that are the representation of complexity and of complex designs. Interactions among structures can take the form of communication or information processing.

Baas establishes a basic distinction between deducible emergence and observational emergence. In the first case there is a deductional process that allows us to deduce the new properties. This type of emergence is applied to engineering constructions and to some nonlinear systems and it seems adequate for representing software development. Observational emergence, on the contrary, can not be deduced. The causes for emergence are attributed to two basic circumstances: 1) the existence or nonlinear interactions and 2) the existence of a large number of structures. The second cause is ubiquitous in software, and the first is the object of the work of [10]. So, as occurs in morphogenesis, emergence implies not only upward processes (from the bottom level to the upper), but also downward ones.

It seems that hyperstructures could be the way to represent the organization of hierarchies in various fields. It is argued also that the formation principle for the hyperstructures is the nonlinear interaction among structures, objects and systems, with emergence of new properties at each level. Hyperstructures allow the combination of the processes of evolution, adaptation and recursion. *Category theory* is the vehicle for manipulating those structures. Category theory has already been used in computer science for manipulating some object constructs, and it is a valid tool for analyzing transformational processes.

The previous paragraphs suggest that in the creation of an artifact several distinct types of processes concur, and the interaction among them leads as result to the generation of new properties in the resultant product. It is not difficult to envisage that in the creation of software products a combination of multiple types of processes cooperating until the final product is built. Many unknown relationships are established among the intermediate products and the designers of the software system that are invisible to the normative process (if existing).

3.2. The human element as the substratum for the creation of processes

The substratum that allows the creation of E-type software is a human organization. So, the relationships between the groups and the intermediate structures they create should be made explicit in order to analyze the software process and its improvement. The main problem is that the view that each developer has of the system at each time step is not represented or documented, so that the traceability of the combination of representations is laborious. Only part of the emergent structures are recognized externally.

But it is possible to have less "mechanical" software processes. Zhuge et al. [17] have proposed a software process that is a special case of the general human problem-solving process. This perspective involves using analogy and abstraction as essential components of software process models. Curiously, these authors use the category theory as the representation for the objects and its operators within the software process. The cognitive-based problem-solving model is an example that deviates from other classical models and includes the possibility of constructing designs in a cognitive space. This clearly assumes (although not explicitly stated) that emergence of intermediate representations is a mechanism that allows the creation of an engineered product.

These cognitive-based problem-solving models have also other advantages, such as the ability to improve the software process by enhancing the group and developer's cognition. The inclusion of the cognitive-space for the group of developers, is an example of how a mechanism can be developed for group learning which bypasses some other structured processes for developing software. This supports the hypothesis that it is possible to develop software without having a "defined" software process, since abstraction and analogy are working at the cognitive level, interacting with the intermediate products that are being refined until the final solution is reached.

4. Conclusion

The thread of this paper is that there are some facts which speak against the consideration of software development as a purely mechanical matter. Thus, assessment from that point of view may be flawed. The concept of design has a biological interpretation that is the process of morphogenesis. This process is characterized by the creation of emergent structures, and emergence can be theoretically formalized by using hyperstructures. The construction of E-type software systems can be seen as the generation of forms by social bodies, so processes of adaptation, evolution and recursion are pertinent to the evaluation of those systems. More research is needed for the formalization of the software processes using, for instance, category theory.

Some of the recommendations of the technological paradigm that are suggested for improving the software process may achieve progress in the technological part of the process. But as products and human resources are so intertwined, establishing a comparison with a pure mechanical paradigm is difficult to maintain. Some well-known software companies do not follow that path [5]. Having strictly defined processes may go against the self-organization processes that help to improve those very processes. Some specific uses of models such as those mentioned in §2.1 can convey this view, although other interpretations

140

may overcome the limitations of the technological approach. The evaluation of the processes should be made taking final and intermediate products into account, while considering the emergent intermediate system representations that the set of developers have when making decisions.

Finally, there remains the question of what is the exact relationship between process and product. This depends on the paradigm that is used for software developement. In fact, there is a way of building programs –genetic programming– in which the final product does not tell anything about the steps of the process (survival of the fittest). Thus, taking an organic perspective, many of the concepts of the complex system area could be useful for software development.

Bibliography

[1] BAAS, N.A., "Emergence, Hierarchies and Hyperstructures", *Artificial Life III* (C.G. LANGTON ed.), Addison-Wesley (1994), 515–537.

[2] BLUM, B., *Beyond Programming: To a New Era of Design*, Oxford University Press (1995).

[3] BRAHA, D., and O. MAIMON, " The Design Process: Properties, Paradigms and Structure", *IEEE Trans. on Systems, Man and Cybernetics, Part A: Systems and Humans* **27(2)** (1997), 146–167.

[4] CAIN, B.G., and J.O. COPLIEN, "Social patterns in productive software development organizations", *Annals of Software Engineering* **2** (1996), 259–286.

[5] CUSUMANO, Michael A., and Richard W. SELBY, "How Microsoft Builds Software", *Comm. of the ACM* **40 (6)** (1997), 53–62.

[6] EMMECHE, C., S. KOPPE, and F. STJERNFELT, *Emergence and the Ontology of Levels. In Search of the Unexplainable*, Arbejdspapir 11, ISSN 0908-0589, Kobenhavn, (1993).

[7] GOODWIN, Brian, *How the leopard changed its spots* , Phoenix Giant (1994).

[8] HERBSLEB, James, et al., "Software Quality and the Capability Maturity Model", *Comm. of the ACM* **40, 6** (1997), 30–40.

[9] KING, S., "Tool support for systems emergence: a multimedia CASE tool", *Information and Software Technology* **39** (1997), 323–330.

[10] LEHMAN, M.M., "Feedback in the Software Evolution Process", *Information and Software Technology* **38** (1996), 681–686.

[11] LUMLEY, T., "Complexity and the 'Learning Organization", *Complexity* **2 (5)** (1997), 14–22.

[12] PAULK, M., "How ISO 9001 Compares with the CMM", *IEEE Software* (January, 1995), 74–83.

[13] RACOON, L.B.S., "The Chaos Model and the Chaos Life Cycle", *ACM Software Engineering Notes* **20 5** (1995), 40–47.

[14] ROSEN, Robert, "Mind as Phenotype", *Matter Matters? On the Material Basis of the Cognitive Activity of Mind* (P. ARHEM, H. LILJENSTRÖM, and U. SVEDIN eds.), Springer (1997), 39–56.

[15] SANDERS, J., "Product, Not Process: A Parable ", *IEEE Software* (March, 1997), 6–8.

[16] SOFTWARE ENGINEERING INSTITUTE, *The Capability Maturity model. Guidelines for Improving the Software Process*, Addison-Wesley (1995).

[17] ZHUGE, Hai, Jian MA, and Xiaoqing SHI, "Abstraction and analogy in cognitive space: A software process model", *Information and Software Technology* **39** (1997), 463–468.

Chapter 17

Hazards, self-organization, and risk compensation: A view of life at the edge

Keith Donnelly

Ontario Hydro, Risk Analysis & Decision Support
1549 Victoria St East, Whitby, Ontario, Canada L1N 9E3

The response of a human system or organization to hazards may often be viewed in terms of self-organized criticality. That is, the operating point of the system, in terms of resources, scheduling, work methods, safety barriers, etc, evolves until a balance is reached between competing pressures such as the simultaneous desires for high work productivity and a low probability of accidents. From a safety viewpoint, the system may be viewed as having migrated away from high safety to a point where it is "on the edge of chaos". Chaos in this case may be defined as a situation where a small perturbation may or may not be absorbed by the system; the outcome is often tragic. At this operating point "on the edge", the frequency of small incidents already gives an indication that more severe incidents are distinctly possible, even though much less likely to occur. Empirically, a power law relates the frequencies of different severity events.

Examples drawn from balanced systems or systems with power laws are: air traffic control, powerline maintenance work, the Great Fire of London, and the King's Cross Underground Fire.

In the safety literature, Risk Compensation is a theory claiming that most safety improvements will be nullified because humans will increase their exposure to or level of hazard and claim the benefit in some other currency such as time or performance, while keeping the net risk unchanged. Driving faster and following more closely after the introduction of seat belts, anti-lock brakes and airbags are often cited as examples. This is often described as being governed by a "risk thermostat" whereby a target level of acceptable risk is maintained. The insight provided by self-organization is that the feedback provided by small incidents, and the protection against severe events afforded by a power law, allows the operating point to be set "on the edge."

Ultimately, it is hoped that theories of self-organized systems will lead to better management indicators of safety risk and lead to insights when organizational or economic changes may be inadvertently moving a system into an unacceptable regime.

1. Introduction

When faced with hazards, those charged with managing or regulating them typically approach this as a straightforward process. Objectives may be set, policies, strategies and plans developed, and implementation and monitoring carried out. Will the objectives be met? Will the organization, the human components in a society, or nature itself, comply with this linear and deterministic view of the world?

This note adopts a somewhat different viewpoint, one that has been stimulated by developments in dynamical systems theory within the past decade or so. Simply put, systems which have the capability of evolving or of making complex decisions will find themselves positioned between a number of competing forces. Furthermore, the "operating point" of such systems may be approaching instability in order to take best advantage of opportunities, or to stretch out precious resources as much as possible.

Such systems are said to have adapted; their properties and behaviour often display unexpected characteristics. In evolving, for perfectly good reasons, to a state that is marginally stable, their response may no longer be that of a simple system responding linearly to its inputs. These systems are sometimes described as being on "the edge of chaos".

"The edge of chaos" is a recognizable state though perhaps not rigorously defined; it is indicated by the number of undesired events increasing rapidly as "the edge" is approached. It is tempting to postulate that many personal and societal situations are of this nature. In fact, the situation in personal life may be so widespread as to be viewed as a generalization of Parkinson's Law: "the number of commitments requiring time or attention increases to fill the capacity available." The "edge of chaos" metaphor gains strength by noting that the number of larger unwanted events roughly follows a power law, characteristic of self-organized systems.

Our intent in this note is to begin a discussion of whether man-made systems which have adapted to some set of goals will respond to unusual stresses in a simple, predictable manner, or will their response be unpredictable, in the manner of a chaotic system. This addresses the crucial question of whether many of our engineered and social systems have left us unexpectedly and tragically vulnerable to the effects of change. Does this explain the tragedy of the heat-related deaths in Chicago during the summer of 1995?

As an example of engineered systems, we mention the 1995 European floods. It has been suggested [1] that increased urbanization and modern farming practices in this century have reduced the land's capacity to absorb rainwater. Simultaneously, rivers have been straightened to increase their capacity for barge traffic, reducing their role in delaying and stretching out the arrival of floodwaters downstream. No doubt these changes have occurred gradually, with confidence growing that even extremes of weather could be absorbed. In reality, the overall system may have moved into a regime where a relatively small change in climate produced unpredictable results.

We will briefly lay out a description of self-organized systems. This will include a discussion of adapted systems, some so trivial and commonplace that they may have escaped previous attention, illustrating that adapted systems may be more widespread than recognized.

2. Self-organized criticality

2.1. Sandpiles

The term "self-organized criticality" has been used by the physicist Per Bak [2][3] and others in studying physical systems such as sandpiles, having certain characteristics in common. These systems are observed to evolve out of an initial state (addition of sand to the top of a pile), accompanied by discrete events which change the configuration of the system (avalanches of varying quantity of sand), finally reaching a critical state where a balance has been reached and the dynamic events redistributing energy through the system (avalanches carrying gravitational energy away from sand added to the top of the pile) become stable *in a statistical sense* over time.

These avalanches of sand are described only in a statistical sense, in that there is a distribution of probabilities of an avalanche having a certain size following the addition of a grain of sand. This distribution follows a power law whose exponent depends on the number of dimensions in which the system is modeled; sandpiles and their analogies also show self-similarities in both time and space, leading to the use of the term "critical" in reference to the behaviour of many other physical systems.

2.2. Other self-organized systems

To illustrate the far-reaching applicability of the concept of self-organization, let us ask what the following have in common:

1. A Beethoven piano sonata,

2. First base on a baseball diamond,

3. A jet fighter aircraft,

4. An electrical utility powerline maintainer?

These items all are elements of highly evolved systems. These are systems in a general sense of being comprised of interacting procedural, technological and human components, and the overall system being designed to achieve some goal.

These systems, it may be argued, are also self-organized in the sense that the evolution to their final properties is driven by internal, not external, forces.

Let us explore these systems in more detail to focus on their common aspect of internal self-organization.

1. A piano composition by Beethoven (perhaps Franz Liszt might be a better example, although his music is somewhat less familiar) is part of a system which contains the elements of a physical part (a musical instrument, in this case the pianoforte circa 1810) and a human performer who is called upon to play the music as prescribed by the composer (a procedure). Beethoven was a virtuoso keyboard performer; not surprisingly, many of his compositions call for a level of performance that approaches the ultimate that a human is capable of. When attempted by a performer of only average skill, the result will be a presentation containing a vast array of mistakes and "fluffs", ranging from mis-hit notes in running passages, wrong notes in large chords, and perhaps entire sections that cannot be played at anything like the correct tempo. When attempted by a virtuoso, the performance is likely to be one where virtuosity is pushed only to the point where mistakes are possible but infrequent.

2. The distance from home plate to first base, ninety feet, is crucial in the variety of plays that are seen after the batter hits. How did this distance come to be selected? If the distance were increased by only a couple of feet, hitters would be thrown out so regularly that the game would cease to be interesting. If, on the other hand, the distance were decreased by a couple of feet, they would seldom be thrown out and the game would lose much of its challenge. In addition, the double play would no longer be possible, and the game would lose one of its most exciting plays[4].

3. Modern jet fighter aircraft have evolved to the point where their pilots are called on to carry out manoeuvres at the limits of their capabilities of rapid perception, decision making, and motor control over the plane's responses. The plane's purely technical capabilities such as speed and manoeverability have advanced under pressure to remain superior in hostile wartime encounters, but are now limited by the skills of a human operator, and the challenge is to provide information and control systems to overcome these limitations.

4. An electrical utility powerline maintainer works in a physically dangerous environment; the principal risks are from high-voltage electrical contacts and from falls. The work system of procedures, hardware, trade practices, and supervision has evolved to maintain a balance between high work productivity and a low probability of accidents. This work is heavily reliant on human skills to avoid accident-producing mistakes: key aspects are visual perception, short- and long-term memory, and the ability to focus attention despite difficult weather and other distractions. Procedures and hardware have evolved to provide a margin of safety against errors. In hazardous work, it is often found that accidents result when a human error occurs in conjunction with a procedural violation [5].

Readers are invited to suggest further examples from their own experience in sports or other endeavour. Self-organization in these examples appears as a self-consistency between tasks undertaken and capabilities of the performers.

It is perhaps a tautology to observe that the hardest mathematical problems are just at the limit of the best mathematicians' ability. However, this simple observation contains a great deal of insight into the structure underlying these situations.

Along with this unique structure, we suspect there may be a peculiarity in any statistics describing self-organized systems, in contrast to situations that are created arbitrarily. One imagines, for example, computer generated musical scores, that might range from trivially playable through devilish to completely unplayable, depending on the randomly chosen values of a few parameters. Would the spectra of errors have the same shape as for a virtuoso's first attempt at an unfamiliar piece of music by a demanding composer?

It is not clear which self-organized systems also exhibit criticality, in the sense of having global properties of scaling and so on[6]. Most of the hazard situations we are concerned with appear to have power law behaviour in the likelihood versus consequence curve, in common with critical systems. Further modeling of these situations will be required to answer this question. We have begun constructing event tree models of near miss incidents; it is hoped these can be used to simulate the chains of probabilistic events that lead to accidents or varying degrees of near miss.

3. Risk compensation

3.1. Risk compensation theories

Theories of adaptive behaviour in hazardous situations already exist, though not entirely without controversy. These are known as "risk compensation" theories[7][8][9], and typically take the following form: a safety improvement such as seat belts is introduced, and it is expected that fatality and serious injury rates will decline. However, it is also argued that individuals may drive faster or more dangerously, leaving their risk of injury roughly unchanged. Statistical evidence to support a decline in fatalities and injuries is notoriously difficult to obtain, as there are often concurrent changes in the driving, social and legislative environments. In fact, the case that there is no evidence has been vigorously argued by Adams[10].

Risk compensation may be intuitively interpreted in several ways. We may think of individuals as having a "risk thermostat", which determines the level of risk regardless of changes in the safety defenses or the hazardous environment. Such feedback mechanisms exert a powerful control over the behaviour of systems, and in fact can cause a diametric shift from a deterministic, clockwork system to one governed completely by the feedback parameters.

Risk compensation may also be thought of as executing a tradeoff between safety improvements and performance, perhaps in the nature of a driver's opportunity to exhibit skill or to work more productively.

Another side of risk compensation could be mentioned here: it often operates to transfer risk between parties, such as when faster or less careful driving endangers pedestrians and cyclists. We will not, however, embark on the thorough exploration that risk compensation deserves.

3.2. Risk compensation, adaptation and power laws

Risk compensation behaviour may also be understood as an example of an adaptive system. We will consider an example in detail to see the points at which risk compensation theory is extended by this paradigm.

Consider a driver rounding a sharp curve. Knowing the car's handling characteristics from past experience, the driver reduces speed until a certain "comfort level" is reached; this implies that he or she has a reasonable degree of confidence that the vehicle will stay under control.

What are the consequences and likelihood of the occasional slight loss of control? We can imagine the car's path veering slightly into the oncoming lane or onto the shoulder, or a slight slippage of the rear tires. We could also imagine other less likely consequences, ranging in severity from a major but recoverable skid to the vehicle going out of control and perhaps overturning. Due to the handling characteristics of most automobiles, these severe consequences are much less likely to occur.

Most drivers' experience would be that the frequencies of minor, moderate and serious slips are given by a decreasing function. It is very difficult to quantify these consequences or obtain data on their rate of occurrence. However, it would appear that a coarse-grained ordering of severity could be established, with frequency of occurrence diminishing with increasing severity. In very rough terms, an order of magnitude decrease in severity for our driving example corresponds to an order of magnitude decrease in frequency - we could argue that a power law underlies this situation.

Interpreted as risk compensation, the driver has, over time, adjusted his or her curve-taking to incorporate the latest advances in tires, suspensions and road construction, and has traded these improvements for an increase in speed or "performance" while maintaining the same level of risk.

Interpreted as adaptive behaviour, the driver has taken advantage of the spectrum of small events experienced to provide a calibration or gauge of how near "the edge" is, while depending on a power law for protection against severe consequences. In this way, small events are employed as a predictor of the likelihood of larger ones, while also serving as an indicator of a potential change in regime if the small events become too frequent. Self-correction and near-misses are known to be features of human performance[11], suggesting that this self-calibration may be a common strategy.

4. Hazards and the balancing act

In this section, we will discuss several real-world hazard situations which illustrate the balance that often occurs between competing objectives or pressures. As these are hazard examples, one of the pressures is always a desire for an "acceptably" low rate of accidents[1]. Some of these examples are tragic, and this is the aspect that brings them to our attention, in comparison to similar situations leading to only minor consequences. However, by bringing a viewpoint of adaptation and self-organization to bear, we may be able to gain a broader perspective on the underlying mechanisms and dynamics.

4.1. The Great Fire of London

The Great Fire of London, 1666, is known to have started in a baker's shop in Pudding Lane, and before it eventually burnt itself out, destroyed 13,000 houses. An enormous tragedy, but was it totally unexpected? Although our limited research has not uncovered any statistics on fires in that period, it is well known that fires consuming a few houses were a familiar occurrence, and fear of a large conflagration was a common preoccupation[12].

What else can we surmise of the conditions around the Great Fire? Fire was both extremely beneficial to society, and extremely hazardous. Society as it then existed was dependent on fire for cooking and heating. At that time, fires were open and in close proximity to daily activities, and carried out in surroundings that were highly flammable. Practices for safe use of fire, such as separating fuel and fire, building common walls between dwellings out of masonry, and damping out open fires before retiring at night, were commonplace. Perhaps there were fire control measures within the social system, such as collection of supplies of sand or water and formation of informal fire brigades. In some self-organizing process, the extent of fire use, the urban density and housing conditions, and the daily risk of fire of various consequences adapted to a balance believed to be satisfactory.

What can we say of the statistics of fires in London at that time? The number of fires of different sizes appears to follow a decreasing function, though without more information we cannot say if it follows a power law. Is the Great Fire consistent with the predictions of whatever that function was? Detailed research might show that urban conditions were changing more rapidly than the adaptive process of fire safety practices, leading to a risk spectrum that passed some critical point and changed its form dramatically, losing the protection of a power law against very large events.

4.2. Deaths of four Army Ranger candidates

On March 3, 1995, The New York Times carried a report on the deaths of four Army Ranger candidates who had died of hypothermia during the last stages of a grueling nine-week course. The men died after being urged by their instructors to carry on an exercise in an icy cold, rain-swollen river, in fog that hampered eventual rescue attempts.

---

[1]A difficult concept. As a pragmatic definition, we may say that a risk is acceptable if it is comparable to or less than the risk of similar activities the participant might normally undertake or be exposed to, in recreation or occupation or life in general, and having similar benefit to the participant and being of a similar degree of voluntariness.

The objective of the training course is to simulate, as nearly as possible, the true conditions of warfare. As The Times reports: "The Ranger course, the Army says, is designed to induce fatigue and hunger, mental and emotional stress, to take men to the breaking point. It is not supposed to break them."

Clearly this training program was deliberately positioned "on the edge" for compelling reasons. Nonetheless, the backup and monitoring systems that normally intervene to prevent such tragedies failed in this case.

4.3. Air traffic control

Air traffic control, in its simplest terms, is a system that allows aircraft to land and take off from airports in an efficient manner while maintaining adequate safety margins against collisions in mid-air or on the runways. A balance is struck between the pressures to increase airport traffic while minimizing takeoff and landing delays, and the need to minimize safety risks dependent on the number and spacing of aircraft in the airport vicinity. According to a lay report[13], "The number of near collisions has steadily decreased since the mid-1980s. The FAA has kept things running by instituting conservative spacing margins between aircraft and by retaining extensive backup systems." In this case, a system that was moving uncomfortably close to the edge was backed away when the situation was recognized by the number of near misses. These low consequence occurrences act as a predictor of the likelihood of high consequence events.

5. Statistics and indicators

Certainly an industry records its major injuries and fatalities. However, from the safety point of view, statistics on a wider spectrum of unwanted events ranging from violations of procedures through near misses and minor injuries should also be kept in order to observe trends that may be the result of changing work practices or organizational structures. This would allow intervention, control or reduction of exposure before the increased risks of serious injury have been translated into events.

There is some belief[14] that in many classes of industry, the severity of accidents, measured in days of work lost, follows a power law of slope around -1. There is also some belief[2] that as the absolute frequency of accidents increases, the slope becomes flatter and severe accidents relatively more frequent.

When a major disaster has occurred, it is perhaps a matter of taste whether one interprets previous moderately severe events as precursors or as indicators of a heightened risk for all events, with a power law or similar decreasing function acting as a predictive link.

A string of moderately severe events often leads to false complacency, with the belief that these were the worst that could occur, or that control measures would always be effective. This belief can be disastrous. In the case of the 1987 King's Cross Underground Fire[15], resulting in the loss of 31 lives, at least

---

[2]The author was unable to locate the original reference for this claim.

six serious fires resulting in evacuations and minor injuries had occurred in the London Underground system within a few years previous. Whether one believes that such events are related by a power law, or that similar initial conditions have a statistically related range of outcomes a la chaos, it is foolhardy to believe that there is more than an order of magnitude difference in the probabilities between near miss and tragedy.

The reader can imagine the practical difficulties of quantifying unwanted events in order to determine a power law. This is especially true for events which require speculation, such as near misses, violations of procedure and human error.

6. Multifactor disasters

Many disasters, when carefully scrutinized, are found to have resulted from the confluence of a number of human and organizational failures. Most systems have built-in error-identifying and correcting mechanisms, leading to interdependence. This number of simultaneous failures is typically 4 or 5. It is tempting to model these events with percolation theory in 4 and 5 dimensions, with the failure probabilities related proportionally to simulate an entire organization degrading under change, and noting how they might approach some edge of chaos value.

7. Risk compensation and progress

We have described risk compensation in somewhat pessimistic terms. Alternatively, if one has an optimistic view of technological progress, then the increase of some measure of production or performance, following some safety improvement while keeping risk unchanged, will be of benefit to society. In a dynamic sense, it may be that the rate of progress in many spheres of endeavour is limited by the rate of development of safety improvements that allow change without increase of risk.

8. Summary

We have outlined several areas in safety science and management that may be approachable with the new insights of Complex Systems. Some of the aspects that may be considered are severity-frequency curves that are similar to power laws, and a structure governed by self-organization under competing pressures. The "edge of chaos" metaphor is appealing in systems where production pressures move risk levels to the edge of unacceptability. Ultimately, it is hoped that theories of self-organized systems will lead to better management indicators of safety risk and lead to insights when organizational or economic changes may be inadvertently moving a system into an unacceptable regime.

Bibliography

[1] WALLACE, Bruce, "Deluge: Critics say massive flooding was manmade", *Maclean's* (Feb. 13, 1995), 33.

[2] BAK, Per, and Kan CHEN, "Self-Organized Criticality", *Scientific American* (January 1991), 46-53.

[3] BAK, Per, C. TANG, and K. sc Wiesenfeld, "Self-Organized Criticality", *Phys Rev A* **38** (1988), 364–374.

[4] WARD, G.C., and K. BURNS, *Baseball: An Illustrated History*, Knopf (1994).

[5] LAWTON, Rebecca, and Dianne PARKER, "Individual differences in Accident Liability: A Review and Integrative Approach", *Human Factors* **40**(1998), 655-671.

[6] BAK, Per, *How Nature Works*, Copernicus Springer Verlag (1996).

[7] WILDE, Gerald, *Target Risk*, PDE Publications: Toronto (1994).

[8] ADAMS, John G.U., *Risk*, UCL Press: London (1995).

[9] GLENDON, Ian, and Neville STANTON, eds., "Risk Homeostasis and Risk Assessment", special issue of *Safety Science* **22**(1-3) (1996) 1–262.

[10] ADAMS, John G.U., "Seat Belt Legislation: the Evidence Revisited", *Safety Science* **18** (1994), 135–152.

[11] REASON, James, *Human Error*, Cambridge University Press (1990).

[12] BELL, W.G., *The Great Fire of London*, Lane: London (1920).

[13] STIX, G., "Aging Airways", *Scientific American* (May 1994), 96–104.

[14] BIRD, F.E. Jr., and G.L. GERMAIN, *Practical Loss Control Leadership* 2d Ed., International Loss Control Institute: Loganville GA (1992).

[15] FENNELL, Desmond, *Investigation into the King's Cross Underground Fire*, HMSO (1988).

Chapter 18

Structure formation by Active Brownian particles with nonlinear friction

Udo Erdmann
Humboldt Universität zu Berlin, Institut für Physik
Invalidenstr. 110, 10115 Berlin

Active Brownian particles are one particle systems described by stochastic differential equations (Langevin equations) which are driven by a nonlinear deterministic force. The interaction of the particles is described by a self consistent field. An ensemble of these particles far from equilibrium shows a complex behavior and pattern formation. The self consistent field is described by a reaction diffusion equation. These equations analyzed numerically. The stationary distribution of the velocities of the particles is derived by the corresponding Fokker Planck equation. For a macroscopic description we use a hydrodynamic approach to interpret the numerical results and to understand the behavior of nonlinear systems far from equilibrium.

1. Introduction

There are three conditions to be fulfilled for a dynamical system to show structure formation in non-equilibrium.

- The process supposed to form structures has to take place over-critically far from equilibrium.

- The equations of motion must be nonlinear. This implies interaction. An ensemble of N objects with linear equations of motion is equivalent to N one particle systems what is useless to describe interaction processes.

Nonlinear partial differential equations as opposed to linear partial differential equations have more than one fixed point, limit cycles or chaotic attractors, hence permanent motion of the particle.

- Finally we need stochastic forces. These forces give the particles the opportunity to switch between the stable manifolds if the stationary case has already been reached and can originate new structure.

The system presented in this article has been investigated to understand the aspects of the mechanism of structure formation. Furthermore it serves as a simple model to describe the behavior of ensembles of self-moving objects like ant looking for food or colonies of bacteria.

2. Self-moving particles

So far the work of SCHWEITZER, SCHIMANSKY-GEIER et al. [16, 17, 15] concerned with particles moving in a viscous medium. In the stationary case the particles did not move anymore, except of thermal motion, because of this assumption. The mean velocity was equal to zero because the influence of friction was a linear one.

In biological systems objects can be observed which move constantly [14]. In the stationary case, the absolute value of the velocity does not vanish. Equations whose stationary solution shows non-vanishing finite velocities have to be nonlinear. In this paper we want to investigate the case with nonlinear friction.

2.1. Langevin equations

The dynamic system introduced here obeys all of the upper requirements needed to form structures. The Langevin equations are:

$$\dot{\mathbf{r}} = \mathbf{v} \tag{2.1}$$
$$\dot{\mathbf{v}} = (\alpha - \mathbf{v}^2)\mathbf{v} + F(\mathbf{r}) + \sqrt{2\epsilon}\boldsymbol{\xi}(t).$$

For $F(\mathbf{r}) = -\omega_0^2\mathbf{r}$ and vanishing noise ($\epsilon = 0$) the equations represent the well-known van-der-Pol oscillator. The system shows different behavior according to different values of the parameter α due to its stability (Fig. 2.1). The pump term causes a acceleration into the direction the particle is already going to.

Because of the existence of a deterministic pump with strength α the particles do more than just random motion. Therefore we want to call them Active Brownian particles [16, 17, 15]. As has been shown therein these particles are very useful to simulate active media like chemical reactions in mesoscopic systems.

2.2. Fokker-Planck equation

To understand a system of N Brownian particles taking into account the behavior of every single particle and the interaction with its environment would be a

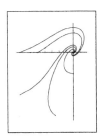  

Figure 2.1: For the 1-dimensional case, the bifurcation of the van-der-Pol oscillator is shown above. For $\alpha = -1.0$ (left) the behavior is equivalent to a system with linear friction. The system has one fixed point at $v_{\text{fix}} = 0$. At $\alpha = 0.0$ (middle) we can observe a Hopf bifurcation and at $\alpha = 1.0$ (right) a limit cycle is formed. All trajectories have a stable solution which is on the limit cycle. Note that even in the stationary case the observed objects still move.

difficult problem to tackle. It is enough to consider the equation of motion of the probability distribution of the particles $P(\mathbf{r}, \mathbf{v}, t)$, the so called Fokker-Planck equation, to get an understanding of the probabilistic behavior of the whole system. With the help of the Kramers-Moyal series we can find a Fokker-Planck equation belonging to a system of Langevin equations.

For the case of a conservative force $F(\mathbf{r}) = -\nabla_r U$ the Fokker-Planck equation corresponding to (2.1) is:

$$\frac{\partial P}{\partial t} = -\mathbf{v}\nabla_r P - \text{div}_v \left\{ \left[\left(\alpha - \mathbf{v}^2 \right) \mathbf{v} + \nabla_r U \right] P - \epsilon \nabla_v P \right\} . \tag{2.2}$$

With $\nabla_r U = \text{const.}$ the stationary solution of (2.2) is easy to derive:

$$P(\mathbf{v}) = N \exp\left(\frac{\alpha}{2\epsilon} \mathbf{v}^2 - \frac{1}{4\epsilon} \mathbf{v}^4 \right) . \tag{2.3}$$

As we can see in Fig. 2.2 particles are not in a equilibrium state. Computer simulations verify the assumptions above (see Fig. 2.3).

The Fokker-Planck equation is a partial differential equation while the Langevin equations are (ordinary) stochastic differential equations. These equations are much easier to handle on a computer than partial differential equations due to the stability of the discretization scheme [15]. All the computer simulations are done with the stochastic equations.

Using Brownian particles for simulating partial differential equations like reaction diffusion systems etc. because of the equivalence between Langevin equation and Fokker-Planck equation reminds one to the simulating algorithm "Dissipative Particle Dynamics". Right here, one uses stochastic differential equations to simulate hydrodynamic systems described by the Navier-Stokes equations. Several instructive papers were published about this algorithm e.g. [8, 5, 3, 6, 12, 13]. One needs to point out the differences to the system described

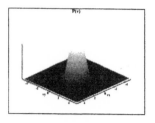 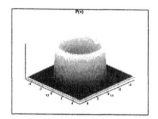

Figure 2.2: Stationary solution (2.3) of (2.2) $\alpha = 0$ (left picture) $\alpha = 10$ (right picture). For $\alpha < 0$ the distribution of velocities has its maximum at the mean of zero with a deviation depending on the intensity ϵ of the noise. If $\alpha > 0$ the mean of the velocity distribution is still zero, but the maxima are different. All particles move with a constant absolute value of velocity $\sqrt{\alpha}$. Therefore the distribution has the form of a hat.

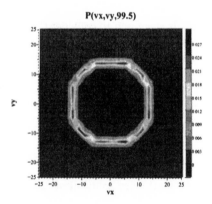

Figure 2.3: For different values of ω_0 and $\alpha > 0$ the velocity distribution shows similar behavior. As expected it has the form of a hat.

here. "Dissipative Particle Dynamics" is used for mesoscopic systems where the basic description is given by partial differential equations (e.g Navier-Stokes). In the opposite to that the basic equations are the Langevin equations here. The Fokker-Planck equation is just to figure out what are the differences between equilibrium and non-equilibrium systems. While the stationary solution of the Fokker-Planck equation of "Dissipative Particle Dynamics" is the Gibbs ensemble [5] it is a non-equilibrium solution (Fig. 2.2, Fig. 2.3) for positive pumping here.

Due to [14] we can derive a fluctuation-dissipation theorem leading to a effective diffusion constant for the system. Therefore we look at the Fokker Planck equation in polar coordinates ($v_x = v \cos \phi$, $v_y = v \sin \phi$ and $v = |\mathbf{v}|$). In this case equation (2.2) leads to:

$$\frac{\partial}{\partial t} P(v, \phi, t) = -\frac{\partial}{\partial v} \cdot \left[(\alpha - v^2) v P \right] \tag{2.4}$$

$$+ \epsilon \frac{1}{v} \left(v \frac{\partial}{\partial v} \frac{\partial P}{\partial v} \right) + \frac{\epsilon}{v^2} \frac{\partial^2 P}{\partial \phi^2} .$$

If we suppose that the velocity fluctuations are very small we can neglect the velocity dependent terms and replace v by constant velocity v_0. Equation (2.4) is reduced to a diffusion equation. After an integration over the v-space we get:

$$\frac{\partial}{\partial t} P(\phi, t) = \frac{\epsilon}{v_0^2} \frac{\partial^2 P}{\partial \phi^2} . \tag{2.5}$$

With the help of the well-known solution of equation (2.5) we are able to calculate $\langle \phi^2 \rangle = \frac{\epsilon t}{v_0^2}$. Due to [14] it is possible to derive the mean square displacement of the particle according to its coordinates

$$\langle \mathbf{r}^2 \rangle = \frac{2 v_0^4 t}{\epsilon} + \frac{v_0^6}{\epsilon^2} \left[\exp \left(-\frac{2 \epsilon t}{v_0^2} \right) - 1 \right] \tag{2.6}$$

and with the approximation $t \gg \frac{v_0^2}{\epsilon}$ we get a expression for the effective diffusion constant D_{eff}[1]:

$$D_{\text{eff}} = \frac{2 v_0^4}{\epsilon} . \tag{2.7}$$

Another opportunity to formulate a fluctuation-dissipation relation is given by KLIMONTOVIC in [9]. He introduces a different presentation of the stochastic integral which is used to calculate the corresponding Fokker-Planck equation of a certain Langevin system. For Langevin systems with velocity-dependant friction the so called "kinetic" approach of handling the stochastic integral gives the only opportunity to derive a fluctuation-dissipation theorem which has a similar structure compared with the Einstein relation of the classical linear system.

2.3. Macroscopic approach

For experimental reasons the distribution $P(\mathbf{r}, \mathbf{v}, t)$ is not very useful. The macroscopic variables which can be experimentally verified are variables like density, mean velocity and temperature. With the help of the Fokker-Planck equation we can derive [4] the dynamics of every moment of the distribution P:

$$\frac{\partial}{\partial t} (M_0 M_n) = -\frac{\partial}{\partial x} (M_0 M_{n+1}) \tag{2.8}$$

$$+ n[\alpha M_0 M_n - \gamma M_0 M_{n+2}$$

$$+ F(x) M_0 M_{n-1}]$$

$$+ n(n-1) \epsilon M_0 M_{n-2} ,$$

---

[1] According to the "Dissipative Particle Dynamics" you cannot derive a Einstein relation as in [5, 13]. This is because of the velocity dependant friction here [9].

where the moment of the power n is calculated as:

$$M_n(x,t) = \langle v^n \rangle = \frac{1}{\rho} \int v^n P(x,v,t)\, dv \;. \tag{2.9}$$

This leads to *hydrodynamic* equations for the ensemble of Brownian particles:

$$\frac{\partial \rho}{\partial t} = -\frac{\partial u}{\partial x}\rho - u\frac{\partial \rho}{\partial x} \tag{2.10}$$

$$\frac{\partial u}{\partial t} + u\frac{\partial u}{\partial x} = \left(\alpha - u^2 - 3T\right)u - \frac{\partial T}{\partial x} \tag{2.11}$$

$$+ F(x) - \frac{T}{\rho}\frac{\partial \rho}{\partial x}$$

$$\frac{1}{2}\left(\frac{\partial T}{\partial t} + u\frac{\partial T}{\partial x}\right) = \left(-3u^2 - \frac{\partial u}{\partial x} + \alpha\right)T \tag{2.12}$$

$$- T^2 + \epsilon \;.$$

The normalization factor $\rho(x,t) = M_0 = \int P(x,v,t)\, dv$ is the density of the particles, $u(x,t)$ is the mean velocity and $T(x,t)$ is the temperature. Consider that $T(x,t)$ is not the temperature of the bath which is equivalent to the strength of the noise in the Langevin equations.

Because of the stationary non-equilibrium we must take into account higher cumulants of the generating function of the probability function $P(\mathbf{r}, \mathbf{v}, t)$ than just the second one [11]. Our consideration was made up to the fourth order of accuracy [4].

(2.10) is equivalent to the continuity equation with $u\rho = j$ and j is the flux. Assuming that $T\rho = p$, calling p pressure, (2.12) leads to an Euler-like equation:

$$\frac{\partial u}{\partial t} + u\frac{\partial u}{\partial x} = -\frac{1}{\rho}\frac{\partial p}{\partial x} + F(x) + \alpha u - \frac{u}{\rho}\left(\rho u^2 + 3p\right) \;. \tag{2.13}$$

In contrast to the classical Euler equation of hydrodynamics the particles form a "fluid" which velocity field varies with the velocity and the pressure itself in time. $\left(\rho u^2 + p\right)$ one can call momentum flux density [10]. (2.13) leads to:

$$\frac{\partial p}{\partial t} + u\frac{\partial p}{\partial x} = -\frac{2}{\rho}p^2 + \left(2\alpha + 6u^2 + 3\frac{\partial u}{\partial x}\right) + 2\epsilon\rho \;. \tag{2.14}$$

One can observe that the hydrodynamic equations of the system described here distinguish from the Navier-Stokes equations which can be simulated with the help of "Dissipative Particle Dynamics". The "fluid" formed by Active Brownian particles is not a viscous one.

2.4. Interaction by a self-consistent field

If we assume generation of a self-consistent field $h(r,t)$ by the Brownian particles the force $F(\mathbf{r},t)$ looks like:

$$F(\mathbf{r},t) = \beta\frac{\partial h(r,t)}{\partial \mathbf{r}}\bigg|_{\mathbf{r}=\mathbf{r}_i(t)} \tag{2.15}$$

With the idea of self-consistency one could imagine that the particles produce a kind of chemical which can be attractive or repulsive[2]:

$$\dot{\mathbf{r}}_i = \mathbf{v}_i \tag{2.16}$$

$$\dot{\mathbf{v}}_i = \left(\alpha - \mathbf{v}^2\right)\mathbf{v}_i + \beta \frac{\partial h}{\partial \mathbf{r}}\bigg|_{\mathbf{r}=\mathbf{r}_i(t)} + \sqrt{2D}\boldsymbol{\xi}_i(t)$$

$$\dot{h} = q\sum_{i=1}^{N(t)} \delta\left(r - \mathbf{R}_i(t)\right) - \kappa h(r,t) + D_h\Delta h(r,t). \tag{2.17}$$

This assumption is due to [16]. $\mathbf{R}_i$ means the coordinate the particle with number i $(i = 1,\ldots,N)$.

If the self-consistent field is attractive we can find processes in nature which could be described by this approach. Ants looking for food are said to spread a chemical at the place where they have already been [2]. This substrate smells attractive to the rest of the ants standing around. The rest of these small animals follow the trail the first ant left. Furthermore a attractive self-consistent field could describe pattern formation by bacteria colonies. Dependant on the oxidative stress the bacteria produce a chemical sensed by the aspartate receptor which becomes attractive [1].

With attractive field coupling, the stationary case for the system of Brownian particles are clusters which do not move anymore. The probability distribution of the velocity of the particles is similar to Fig. 2.3 what proofs that the system is in the stationary case. The number of the clusters depends on the ratio $\frac{\alpha}{\beta}$. If the ratio is to big the influence of the field coupling vanishes. The pumping force driven by parameter α rules the behavior of the whole system of particles at that time. For ratios becoming smaller the number of clusters increases. This is shown in Fig. 2.4.

Another example for active media are activator inhibitor systems. We have a fast moving activator spot surrounded by a slower inhibitor cloud. Due to SCHIMANSKY-GEIER in this case the Langevin equations (2.16) could describe the motion of the center of mass of an activator spot. The cloud of inhibitor substance defines an interaction field from one spot, which is surrounded by the cloud, to the rest of the activator. The effective repulsion, created by the inhibitor field, led to clusters of spots. These spots dissolve with a constant frequency and are built up at position with lower inhibitor concentration. Similar effects are described in [7].

Acknowledgments

I appreciate the fruitful discussions with H. Rosé and T. Aßelmeyer. Furthermore I thank Ch. Koch for being a private referee.

---

[2]One can recognize another difference to "Dissipative Particle Dynamics". The space-dependant force has not to be a conservative one. The interaction between the single particles are not directed kicks. It is an indirect interaction here.

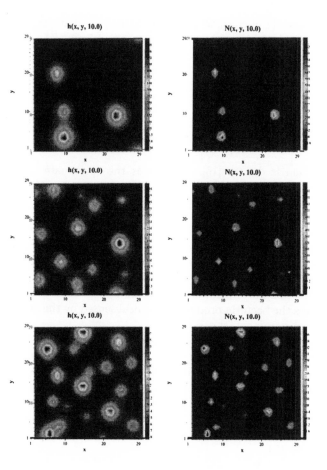

Figure 2.4: For increasing values of an field strength the number of clusters increases as well. Positive β means attractive field in all of the three cases. As can be seen it looks like effects Simulated Annealing shows too.

Bibliography

[1] BUDRENE, Elena O., and Howard C. BERG, "Complex patterns formed by motile cells of Escherichia coli", *Nature* **349** (1991), 630–633.

[2] CALENBUHR, V., and J.-L. DENEUBOURG, *Biological Motion*, (W. ALT AND G. HOFFMANN eds.) vol. 89 of *Lecture Notes in Biomathematics*. Springer (1990), pp. 453–469.

[3] COVENEY, Peter V., and Keir E. NOVIK, "Computer simulations of domain growth and phase separation in two-dimensional binary imiscible fluids using dissipative particle dynamics", *Physical Review E* **54**, 5 (1996), 5134–5141, Erratum in Phys. Rev. E 55, 4831.

[4] ERDMANN, Udo, "Ensembles von van-der-Pol-Oszillatoren" (1997).

[5] ESPAÑOL, P., and P. WARREN, "Statistical mechanics of dissipative particle dynamics", *Europhysics Letters* **30**, 4 (1995), 191–196.

[6] GROOT, Robert D., and Patrick B. WARREN, "Dissipative particle dynamics: Bridging the gap between atomistic and mesoscopic simulation", *Journal of Chemical Phyics* **107**, 11 (1997), 4423–4435.

[7] HEMPEL, Harald, Ilse SCHEBESCH, and Lutz SCHIMANSKY-GEIER, "Travelling pulses under global constraints", (1997).

[8] HOOGERBRUGGE, P. J., and J. M. V. A. KOELMAN, "Simulating microscopic hydrodynamic phenomena with dissipative particle dynamics", *Europhysics Letters* **19**, 3 (1992), 155–160.

[9] KLIMONTOVICH, Yu. L., "Nonlinear brownian motion", *Physics-Uspekhi* **37**, 8 (1994), 737–766.

[10] LANDAU, L. D., and E. M. LIFSCHITZ, *Lehrbuch der theoretischen Physik* vol. VI (Hydrodynamik), Akademie-Verlag, Berlin.

[11] MARCIENKIEWICZ, J., *Math. Z.* **44** (1939), 612.

[12] MARSH, C. A., Backx G., and Ernst M.H., "Fokker-planck-boltzmann equation for dissipative particle dynamics", *Europhysics Letters* **38**, 6 (1997), 411–415.

[13] MARSH, C. A., Backx G., and Ernst M.H., "Static and dynamic properties of dissipative particle dynamics", *Physical Review E* **56**, 2 (1997), 1676–1691.

[14] MIKHAILOV, Alexander, and D. MEINKÖHN, "Self-motion in physicochemical systems far from thermal equlibrium", *Stochastic Dynamics*, (L. SCHIMANSKY-GEIER AND T. PÖSCHEL eds.) vol. 484 of *Lecture Notes in Physics*. Springer (1997), pp. 334–345.

[15] SCHIMANSKY-GEIER, Lutz, Michaela MIETH, Helge ROSÉ, and Horst MALCHOW, "Structur formation by active brownian particles", *Physics Letters A* **207** (1995), 140–146.

[16] SCHWEITZER, Frank, and Lutz SCHIMANSKY-GEIER, "Clustering of "active" walkers in a two-component system", *Physica A* **206** (1994), 359–379.

[17] SCHWEITZER, Frank, and Lutz SCHIMANSKY-GEIER, "Clustering of active walkers", *Fluctuations and Order*, (M. MILLONAS ed.). Springer (1996), ch. 18, pp. 293–305.

Chapter 19

Systems properties of metabolic networks

David A. Fell[1]
School of Biological & Molecular Sciences
Oxford Brookes University

Metabolism is the collection of chemical interconversions carried out by living organisms as they feed, grow and reproduce. It is the overall sum of several thousand individual chemical reactions brought about by catalytic proteins called enzymes. This set of reactions forms a highly branched network in which the cross-connections are increased because the majority of reactions involve at least two reactants. Although each enzyme reaction consists of a number of elementary steps obeying linear chemical kinetics, its overall kinetic behaviour is nonlinear in the concentration of the chemical intermediates of the metabolic net. Thus a possible description of the dynamics of the network is as a system of nonlinear ordinary differential equations, though severe practical problems confront attempts to model even small parts of the metabolic net.

In spite of this complexity, the structure of the network imposes linear constraints on its attainable steady states, and these constraints can be analysed independently of the nonlinear kinetics, for example to give maximum conversion efficiencies.

Further, even though algebraic solution of the nonlinear equation system is intractable even for short sequences of steps, it has proved possible to establish some general principles governing the sensitivity of steady state properties to the catalytic activities of individual enzymes. These studies in *metabolic control analysis* have shown that the total amount of control that can be exerted by enzymes on the steady state fluxes through the network is limited and distributed between the enzymes. This distribution of control between the enzymes of the network can be related to their kinetic responses in the vicinity of the steady state even without having a global description of each enzyme's kinetics. By such means, it is them possible to derive results governing

---

[1]I thank the Wellcome Trust for a Sir Henry Wellcome Award for Innovative Research (048728).

the control, genetics and evolution of metabolic pathways, and also to determine the patterns of interactions within the network that maintain the best homoeostasis.

Within the totality of metabolism, there is a hierarchical organisation in which relatively isolated modules have partially separable time scales. It has been shown that in principle this partitioning of the network can be reflected in partitioning of the control properties.

1. Introduction

Metabolism is the collection of chemical interconversions carried out by living prganisms as they feed, grow and reproduce. Characteristically it involves the synthesis and degradation of complex organic molecules, corresponding to its two major branches: anabolism and catabolism respectively. The schematic representation in Fig. 1.1 applies equally to nutrient cycling at the level of whole ecosystems (with plants as producers and animals as consumers), or to whole organisms, or even to single cells. The central core of metabolic biochemistry is essentially similar in all extant living organisms, so there is a single biosphere, not a number of parallel but incompatible ones.

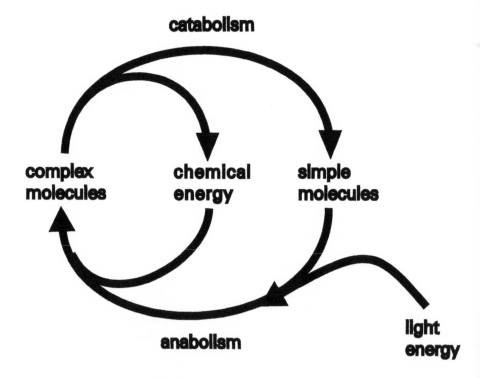

Figure 1.1: An overview of metabolism.

Metabolism is therefore chemical conversion, but chemistry at ambient temperatures, mostly in water at near neutral pH. It achieves extensive and complex chemical conversion in series of very small steps, each catalyzed by special protein molecules called enzymes. The enzymes are themselves metabolic products, though as catalysts they need only be present in small amounts.

1.1. Enzymes

There are over 3500 known basic enzymic reactions, and a greater, though ill-defined, number of actual reactions because many of these enzymes catalyze reactions on sets of chemically similar metabolic intermediates (or metabolites). In addition, there is a greater number of enzymes because even in a single organism there can be more than one enzyme catalyzing the same reaction (isoenzymes). On the other hand, no single organism performs the full range of reactions, so cells might typically contain several hundred to a few thousand enzymes.

Thus metabolism is formed of an interconnected mesh of reactions organised into short 'pathways' of 3 to 10 reactions. Catabolism converges on a central core of about a dozen metabolites that form the starting points for the synthesis of all cellular components. Described in this way, it appears that the metabolic network might bear resemblance to an electrical circuit, a pipeline network, or a digraph. Although some analyses of metabolism have used approaches borrowed from these disciplines, the analogies are limited by a crucial difference. About 85% of reactions are bimolecular or more in either substrates, products or both. Consider the most typical case, a two reactant (substrate), two product reaction, such as:

$$\text{glucose} + \text{ATP} \longrightarrow \text{glucose 6–phosphate} + \text{ADP}$$

Most biochemical maps will emphasize the role of this reaction in processing the carbon atoms of the universal nutrient, glucose. The product, glucose 6–phosphate goes on to be further metabolised. However, the reaction mechanism ensures an obligatory coupling of the conversion of ATP to ADP at exactly the same rate, and such stoichiometrically coupled flows have no counterpart in electrical networks.

In fact, it is equally valid to concentrate on metabolism as a system for recycling ATP and ADP, because it is these molecules that transfer chemical energy between catabolism and anabolism. Similar pairs are the two reduction-oxidation couples NAD^+ and NADH, and $NADP^+$ and NADPH. Around 25% of the multireactant enzymes catalyze reactions involving one of these three pairs, so an alternative way of representing the metabolic network would show it tightly clustered around these 6 molecules.

Even at this level of viewing metabolism, we have already ascended a level in the hierarchy of similar elements, since each enzyme–catalyzed reaction is in effect a miniature metabolic pathway composed of elementary chemical reactions. An example is given in Fig. 1.2. Since the time scale of most of the steps is much shorter than that of a metabolic pathway, for many purposes it

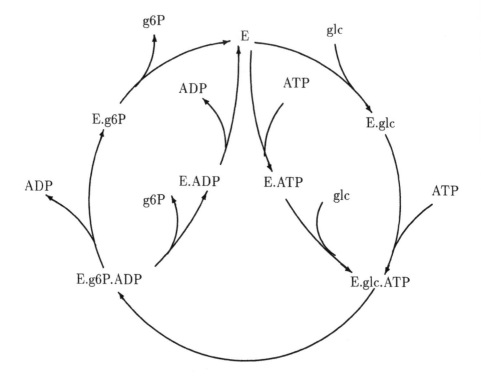

Figure 1.2: A feasible mechanism for the catalysis of the conversion of glucose (glc) to glucose 6–phosphate (g6P) by an enzyme (E).

is not necessary to describe metabolism at this level of detail. The kinetics of each elementary step in an enzyme mechanism follow ordinary chemical kinetics, where the rate depends linearly on the concentrations of the reactants. (It must be noted though that the details of the mechanism and the values of the rate constants are not easily accessible and have been determined for a relatively small fraction of enzymes.) At the level of the overall enzymic reaction, however, the rate of conversion is nonlinear in the metabolites such as glucose and ATP, though it is generally still linear in the total concentration of the enzyme.

1.2. Problems in describing metabolism from the bottom up

Biochemistry and molecular biology have a strongly reductionist ethos and seek explanations based on the properties of biological molecules. The ultimate expression of this is the genome–sequencing programs currently under way. How feasible is it to start from DNA sequences and arrive at the metabolic behaviour of an organism? Currently there are some formidable obstacles to the processes outlined in Fig. 1.3.

There is often little problem in deriving the amino acid sequences of putative transcribed proteins from the base sequence in the DNA, but in the genomes currently sequenced, about 50% of these proteins have not had their functions determined, and cannot be identified by means of sequence comparison with proteins of known function [2], posing the problem of how 'reverse genetics' is to be performed [15]. At the moment, linking a gene sequence to a known enzyme activity is the only way to arrive at the kinetic characteristics of the process catalyzed by the gene product, and this assumes that the kinetics have been measured. Most work on measuring kinetic properties of enzymes has ceased for lack of funding, and there is only one, large but incomplete, database [23] of the results obtained so far. Work on structural interpretation of protein sequences only functions as an aid to identifying genes that cannot be identified on the basis of sequence comparisons and remains very far from being a tool to predict kinetic properties.

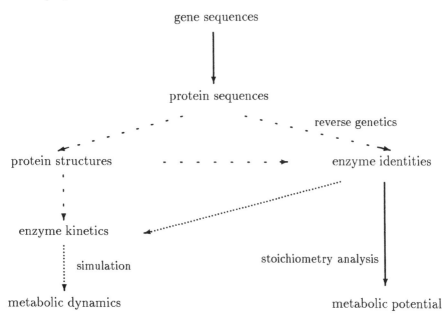

Figure 1.3: Relationships between sequence and structural data and functional interpretation in terms of metabolism. Solid lines indicate sound practical or methodological links; dashed and dotted lines indicate decreasing degrees of feasibility at present.

Thus although we can envisage receiving some supplementary information from genome sequencing projects, at present our knowledge of the sets of enzyme reactions available in particular cells is still based on conventional biochemical assay methods. Interpretation of this information in terms of metabolic capabilities can be aided by route–finding algorithms [20, 21] that are adapted to the particular characteristics of metabolic nets, and techniques such as linear programming (to be discussed below). However, these only establish the possibilities inherent in the network and cannot predict whether particular conversions are kinetically feasible.

In principle, computer simulation can be used to make the step to prediction of metabolic dynamics and my group, along with others, have developed simulation packages specifically adapted to this problem domain, e.g. [14, 17]. A key problem remains the availability of kinetic data, which as mentioned above, is not currently being addressed. A more fundamental worry, however, is whether it is in principle feasible to collect sufficiently detailed kinetic information about all the enzymes in a cell (or even in a particular metabolic pathway). Most enzyme kinetic information was collected to elucidate the mechanisms of catalysis, and not to be useful in simulating metabolism. Consequently vital information is not available. A simple example is that enzymes in cells function in the presence of their reaction products; for technical reasons, most kinetic measurements are made with essentially no product present. Further, for some of the enzymes with more complex kinetics, a complete characterisation may not in any case be practically feasible. One estimate is that the kinetics of glutamine synthase would require a minimum of 10^9 measurements for a minimal determination [19].

Generally most biochemists have not attempted quantitative modelling of metabolic system properties, but have relied on qualitative, verbal reasoning assisted by various rules of thumb that have not withstood close scrutiny (such as the control of metabolic flux by 'rate–limiting enzymes' located near the start of a pathway). Nevertheless, progress has been made in the last 30 years in characterising some general features of metabolic systems without requiring highly detailed knowledge of lower level behaviour. Within this article I shall concentrate on some of the general properties of metabolic networks in asymptotically stable steady states, particularly some of the conclusions obtained on how the steady state fluxes in networks respond to changes in the quantities of enzymes present, according to metabolic control analysis [3]. Metabolic networks can exhibit other interesting behaviours such as oscillations and hysteretic transitions between alternate steady states, the former of which is described elsewhere in this volume (see Goldbeter).

2. Steady state of a metabolic network

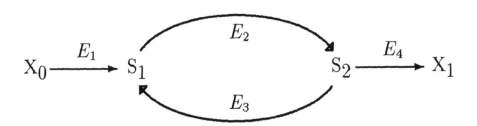

Figure 2.1: A simple specimen metabolic pathway.

The analysis of the steady state of a metabolic pathway begins with the total mass balance equation, which states that for each internal metabolite, its rate of formation is equal to its rate of consumption. A compact expression of this equation is obtained by use of the stoichiometry matrix, $\mathbf{N}$, whose elements n_{ij} represent the number of molecules of metabolite i formed (or, if negative, consumed) in a reaction step j. For this purpose, we are only concerned with the internal metabolites of the network, and not the external source and sink metabolites, for which there is net consumption and production respectively. Given a vector of metabolite concentrations $\mathbf{S}$ and a vector of reaction rates $\mathbf{v}$, the equation is:

$$\frac{d\mathbf{S}}{dt} = \mathbf{N}.\mathbf{v} = \mathbf{0} \tag{2.1}$$

As an example, the pathway shown in Fig. 2.1 gives the equation

$$\begin{bmatrix} 1 & -1 & 1 & 0 \\ 0 & 1 & -1 & -1 \end{bmatrix} \cdot \begin{bmatrix} v_1 \\ v_2 \\ v_3 \\ v_4 \end{bmatrix} = \begin{bmatrix} 0 \\ 0 \end{bmatrix} \tag{2.2}$$

The nonlinearity of the enzyme kinetics functions is present in the velocity vector: each element v_i of $\mathbf{v}$ is a nonlinear function of:

- some of the variable metabolites, $S_1 \ldots$;

- perhaps some of the external concentrations, $X_0 \ldots$, and

- its kinetic parameters, $K_{i,1} \ldots$,

and a linear function of

- its enzyme concentration, E_i.

i.e.:

$$v_i = f(\mathbf{S}, \mathbf{X}, \mathbf{K}, E_i) \tag{2.3}$$

2.1. Linear analysis

For all except very simple cases, it is not possible to obtain analytical solutions of Eqn. 2.1 for the rates and metabolite concentrations, but it can be solved numerically with appropriate nonlinear solvers. If we consider only the reaction rates, then Eqn. 2.1 is linear in them, but is under–determined. That is, it cannot be solved for unique values of the rates, but it places some constraints on valid solutions (such as a requirement that all the rates are the same in a linear sequence of unimolecular reactions).

In other words, any observed set of velocities at steady state is spanned by a set of basis vectors, $\mathbf{B}$, for the null space of the stoichiometry matrix. Thus for the pathway shown in Fig. 2.1:

$$\mathbf{B} = \begin{bmatrix} 1 & 0 \\ 1 & 1 \\ 0 & 1 \\ 1 & 0 \end{bmatrix} \tag{2.4}$$

so that two parameters a and b suffice to describe any observable set of reaction rates:

$$\begin{bmatrix} 1 & 0 \\ 1 & 1 \\ 0 & 1 \\ 1 & 0 \end{bmatrix} \cdot \begin{bmatrix} a & b \end{bmatrix} = \begin{bmatrix} a \\ a+b \\ b \\ a \end{bmatrix} = \begin{bmatrix} v_1 \\ v_2 \\ v_3 \\ v_4 \end{bmatrix} \tag{2.5}$$

Unfortunately, an arbitrary basis for the null space of a stoichiometry matrix does not necessarily correspond to a set of physically realisable routes through the network. By using modified algorithms [20, 21], it is possible to determine sets of feasible routes through the network, and use them as the components of a description of an observed set of fluxes through the pathways.

If additional linear constraints are placed on the acceptable values of the reaction rates, it is possible to solve for sets of rates satisfying the constraints using linear programming [6]. For example, it may be possible to calculate the set of rates that gives the most efficient conversion of a nutrient into a given end–product. To apply linear programming, it is necessary to use a full stoichiometry matrix with additional rows for the external metabolites that are not at steady state. In the case of the specimen pathway in Fig. 2.1, the equation is:

$$\begin{bmatrix} -1 & 0 & 0 & 0 \\ 1 & -1 & 1 & 0 \\ 0 & 1 & -1 & -1 \\ 0 & 0 & 0 & 1 \end{bmatrix} \cdot \begin{bmatrix} v_1 \\ v_2 \\ v_3 \\ v_4 \end{bmatrix} = \begin{bmatrix} -x \\ 0 \\ 0 \\ \geq 0 \end{bmatrix} \tag{2.6}$$

Since the cyclic flow involving v_2 and v_3 is clearly futile, if we seek a solution minimising v_3, the (trivial) solution is

$$v_1 = v_2 = v_4 = x; \qquad v_3 = 0.$$

The same principles can be applied to larger networks where the solution is far from obvious.

3. Metabolic control analysis

Even though it is not possible to obtain general solutions for the metabolite concentrations and velocities satisfying Eqn. 2.1, it turns out that there are some interesting general properties of the sensitivity of the solutions to changes in the amounts of the enzymes in the network. The application of this particular form of sensitivity analysis to metabolism was inedependently initiated by Kacser and Burns in Edinburgh [11] and Heinrich and Rapoport [8] in Berlin, though both were drawing on earlier work by Higgins.

The particular focus of interest is the dependence of the steady state propeorties on variation of enzyme amounts because in many instances where metabolic fluxes are changed (for example by environmental, nutritional or hormonal stimuli), the amounts of one or more enzymes are changed, or else the activities, which in many cases can be equated to changes in enzyme amounts.

Metabolic control analysis considers scaled sensitivity coefficients for the effects of individual enzymes on system variables. If the variable is a flux J, which is a linear function of one or more of the reaction rates, then the measure is called the flux control coefficient, where the coefficient for the effect of enzyme xase on the flux is the fractional (or percentage) change in flux caused by a fractional (or percentage) change in the enzyme:

$$ C_{xase}^{J} = \frac{\partial J}{\partial E_{xase}} \cdot \frac{E_{xase}}{J} = \frac{\partial \ln J}{\partial \ln E_{xase}} \tag{3.1} $$

Even though we have no explicit solution for flux as a function, $J(E_1 \ldots E_n)$, of the amount of each enzyme, we can derive some properties of the set of flux control coefficients. We start from the observation that for some factor t:

$$ J(tE_1 \ldots tE_n) = tJ(E_1 \ldots E_n) \tag{3.2} $$

This follows from Eqn. 2.1 since if $\mathbf{N}.\mathbf{v} = \mathbf{0}$, $\mathbf{N}.(t\mathbf{v}) = \mathbf{0}$, and the v_i are linear functions of the E_i. Eqn. 3.2 shows that J is homogeneous of degree 1 in $\mathbf{E}$, and therefore by Euler's theorem:

$$ E_1 \frac{\partial J}{\partial E_1} + \ldots + E_n \frac{\partial J}{\partial E_n} = J \tag{3.3} $$

Substituting the definition of the flux control coefficient from Eqn. 3.1 gives what is known as the summation theorem for flux control coefficients [11]:

$$ \sum_{i=1\ldots n} C_{E_i}^{J} = 1 \tag{3.4} $$

A number of conclusions have been derived from this result and the observation that flux–enzyme curves are often convex curves tending asymptotically to a limiting flux value at high enzyme levels:

- The total amount of control available is limited.

- The amount of control of one enzyme depends on the amount exerted by others [11].

- Measurements of enzyme control coefficients confirm control is shared over a number of enzymes — not the result biochemists expected [3].

- Changing flux by action on a single enzyme will be difficult [24].

- Distributed control means the genetic phenomenon of *dominance* (where a 50% loss of enzyme activity makes no observable difference to the organism's phenotype) arises from the systems properties of biochemical networks [12].

- The summation theorem constraint (and other similar ones) on sensitivity of the system to these key parameters originate from mass conservation in the system equation [16].

Control analysis can also relate the control coefficients to the molecular properties of the network. Briefly, differentiation of the steady state equation (Eqn. 2.1) with respect to the enzymes gives relations between their control coefficients and their *elasticities*, or local kinetic responses at the steady state, to the metabolites [8, 11, 16]. Together with the summation theorems (and related structural constraints), this gives enough equations to express the control coefficients in terms of the elasticities [5, 16]. In this way, the flux control coefficient of any one enzyme is seen to depend on the kinetics of all the other enzymes.

The advantages of this approach over attempting to determine control coefficients by sensitivity analysis of a detailed computer model of metabolism based on *in vitro* enzyme kinetics data include:

- Elasticities require much less detailed kinetic information than a full kinetic characterisation.

- There are in any case techniques for measuring elasticities *in vivo*, which avoids problems of the uncertain validity of laboratory data in biological conditions.

- It is also possible to define the elasticities of sequences/blocks of enzymes and carry out the analysis on modules of metabolism without needing to know all the internal molecular details of the pathway. The experimental implementation of this using *in vivo* measurements of elasticities is known as 'top–down' control analysis [1].

4. Feedback regulation

The interpretation of feedback inhibition illustrates the divergence between the traditional qualitative notions of control and the quantitative systems view afforded by metabolic control analysis. It is a common feature of metabolic pathways that an enzyme near the beginning, or just after a branch point, is inhibited by the product of the pathway, or by a metabolite produced several steps further along. In the traditional view, the inhibited enzyme controlled the flux through the pathway, and the function of feedback inhibition was flux control. On the other hand, Kacser & Burns [11] showed that feedback inhibition reduced the flux control coefficient of the inhibited enzyme and increased the flux

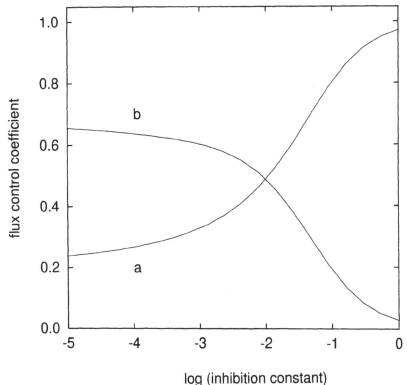

Figure 4.1: The effect of feedback inhibition on flux control coefficients. The abscissa shows the strength of the inhibition, with strong inhibition to the left. In this simulated example, curve a) is the flux control coefficient of the inhibited enzyme at the beginning of the pathway; curve b) is the control coefficient for the enzyme 4 steps further along that uses the feedback metabolite.

control coefficients of enzymes further down the pathway that use the inhibiting metabolite, as shown in Fig. 4.1. Now that molecular genetics allows manipulation of the amount or type of an enzyme at will, there are several examples where overexpression of a feedback–inhibited enzyme has not caused a significant change in pathway flux, and also cases where replacement of an inhibited enzyme by a feedback–resistant form has little effect either.

Quantitative analysis of feedback loops in metabolism, first by Savageau [18] and then Hofmeyr & Cornish–Bowden [10] showed that a major effect of feedback inhibition is homoeostasis of metabolite concentrations. The degree of metabolite homoeostasis often achieved by living organisms in the face of large changes in metabolic flux is striking [25], but generally ignored by biochemists. Using metabolic control analysis, it is possible to show that it is generally easier to achieve good metabolite homoeostasis by controlling a pathway from the bottom, at the demand steps using the feedback metabolite, when the control has been passed there by a feedback loop, than by exercising control at the start of the pathway, on the supply side for the feedback metabolite [25].

5. Large changes in metabolic rate

There is an apparent paradox revealed by this new analysis of the control of metabolism. The theory and experimental evidence suggest that that the flux in metabolic pathways is relatively insensitive to variations in the activities of individual enzymes. In many cases, there is no one enzyme that has a flux control coefficient close to 1. How then, can the metabolic flux be controlled? The answer seems to be that large flux changes can be achieved by changing the activity of several, maybe most, of the enzymes along a pathway in parallel. Biochemists know many mechanisms that work in this way (e.g. signal transduction mechanisms working through covalent modification of enzymes; coordinated induction and repression of enzyme synthesis), but have always tended to focus attention on a single enzyme because of the influence of the 'rate–limiting step' dogma. Simon Thomas have I have assembled evidence in favour of the necessity of coordinated changes in enzyme activity, which we termed multisite modulation [4, 7]. The advantage of this mechanism is that it also accounts for metabolite homoeostasis during flux changes.

6. Hierarchical organisation of metabolism

So far, we have dealt with enzymes as the agents that catalyze metabolism, but the relationship between enzymes and metabolites is more complex than this, because enzymes are themselves metabolites, in that:

- they are made from other metabolites (amino acids), and
- they can be reversibly, catalytically modified by other enzymes (in signal transduction pathways that mediate hormonal and environmental effects on metabolism).

The justification for regarding enzymes separately from other metabolites is that their production and degradation tends to occur on a slower timescale than that of the main metabolic pathways themselves, and the manufacture of enzymes often involves metabolic flows that are of much smaller magnitude than those through the central pathways of metabolism.

Because the metabolism of an organism is coded in its genes, there is an information flow, with much weaker associated mass–flows, leading down to metabolism. Genes on the DNA sequence specify the synthesis of messenger RNA molecules that in turn catalytically direct synthesis of enzymes from amino acids. Once again though, DNA and RNA are themselves metabolic products because they are assembled from metabolites, in this case nucleotides. Even neglecting this, The information hierarchy is still not completely dictatorial, because the amounts of certain metabolites present in the cell affect which of the genes are transcribed and translated to make enzymes.

When a stoichiometry matrix is written for the whole system of nucleic acids, enzymes and metabolites, the information, catalytic, and effector, connections are not recorded, solely the sequences of mass flows. It is then possible to determine the independent modules that make up the stoichiometry matrix by determining the blocks of reactions that have no direct connections [9]. In effect, this will lead to a partitioning and diagonalisation of the stoichiometry matrix (which would not be possible when the level of analysis is just a single metabolic pathway). Though the modules are not stoichiometrically coupled, there are still the information, catalytic and regulatory effects by which the metabolites in one module affect the rate of metabolism in another. It has been shown that it is possible in principle to generalize metabolic control analysis to such modular systems [13, 22], so that the analytical tools can be applied to help understand the control properties of these more complex systems. Currently, the experimental application of modular control analysis has not reached the same degree of development as in metabolic control analysis, but it has great potential.

Bibliography

[1] BROWN, G. C., R. P. HAFNER, and M. D. BRAND, "A 'top–down' approach to the determination of control coefficients in metabolic control theory", *Eur. J. Biochem.* **188** (1990), 321–325.

[2] CLAYTON, R. A., O. WHITE, K. A. KETCHUM, and J. C. VENTER, "The first genome from the third domain of life.", *Nature* **388** (1997), 539–547.

[3] FELL, D. A., "Metabolic Control Analysis: a survey of its theoretical and experimental developments.", *Biochem. J.* **286** (1992), 313–330.

[4] FELL, D. A., *Understanding the Control of Metabolism.*, Portland Press, London (1996).

[5] FELL, D. A., and H. M. SAURO, "Metabolic Control Analysis: Additional relationships between elasticities and control coefficients", *Eur. J. Biochem.* **148** (1985), 555–561.

[6] FELL, D. A., and J. R. SMALL, "Fat synthesis in adipose tissue: an examination of stoichiometric constraints.", *Biochem. J.* **238** (1986), 781–786.

[7] FELL, D. A., and S. THOMAS, "Physiological control of flux: the requirement for multisite modulation.", *Biochem. J.* **311** (1995), 35–39.

[8] HEINRICH, R., and T. A. RAPOPORT, "A linear steady–state treatment of enzymatic chains; general properties, control and effector strength", *Eur. J. Biochem.* **42** (1974), 89–95.

176

[9] HEINRICH, R., and S. SCHUSTER, *The Regulation of Cellular Systems.*, Chapman & Hall New York (1996).

[10] HOFMEYR, J.-H. S., and A. CORNISH-BOWDEN, "Quantitative assessment of regulation in metabolic systems.", *Eur. J. Biochem.* **200** (1991), 223–236.

[11] KACSER, H., and J. A. BURNS, "The control of flux", *Symp. Soc. Exp. Biol.* **27** (1973), 65–104, Reprinted in Biochem. Soc. Trans. 23, 341–366, 1995.

[12] KACSER, H., and J. A. BURNS, "The molecular basis of dominance", *Genetics* **97** (1981), 639–666.

[13] KAHN, D., and H. V. WESTERHOFF, "Control theory of regulatory cascades.", *J. Theor. Biol.* **153** (1991), 255–285.

[14] MENDES, P., "Gepasi — a software package for modeling the dynamics, steady–states and control of biochemical and other systems.", *Comp. Appl. Biosci.* **9**, 5 (1993), 563–571.

[15] OLIVER, S. G., "From DNA–sequence to biological function.", *Nature* **379** (1996), 597–600.

[16] REDER, C., "Metabolic control theory: a structural approach.", *J. Theor. Biol.* **135** (1988), 175–201.

[17] SAURO, H. M., "SCAMP: a general–purpose simulator and metabolic control analysis program.", *Comput. Applic. Biosci.* **9** (1993), 441–450.

[18] SAVAGEAU, M. A., "Optimal design of feedback control by inhibition: Steady state considerations.", *J. Mol. Evol.* **4** (1974), 139–156.

[19] SAVAGEAU, M. A., *Biochemical Systems Analysis: a Study of Function and Design in Molecular Biology*, Addison–Wesley, Reading, Mass. (1976).

[20] SCHUSTER, S., and C. HILGETAG, "On elementary flux modes in biochemical reaction systems at steady state.", *J. Biol. Syst.* **2** (1994), 165–182.

[21] SCHUSTER, S., C. HILGETAG, J. H. WOODS, and D. A. FELL, "Elementary modes of functioning in biochemical networks.", *Computation in Cellular and Molecular Biological Systems* (Singapore,) (R. CUTHBERTSON, M. HOLCOMBE, AND R. PATON eds.), World Scientific (1996), 151–165.

[22] SCHUSTER, S., D. KAHN, and H. V. WESTERHOFF, "Modular analysis of the control of complex metabolic pathways", *Biophys. Chem.* **48**, 1 (1993), 1–17.

[23] SELKOV, E., S. BASMANOVA, T. GAASTERLAND, I GORYANIN, Y. GRETCHKIN, N. MALTSEV, V. NENASHEV, R. OVERBEEK, E. PANYUSHKINA, L. PRONEVITCH, E. SELKOV, and I. YUNIS, "The metabolic pathway collection from EMP — the enzymes and metabolic pathways database.", *Nucleic Acids Res.* **24** (1996), 26–28.

[24] SMALL, J. R., and H. KACSER, "Responses of metabolic sytems to large changes in enzyme activities and effectors. 1. the linear treatment of unbranched chains.", *Eur. J. Biochem.* **213** (1993), 613–624.

[25] THOMAS, S., and D. A. FELL, "Design of metabolic control for large flux changes.", *J. Theor. Biol.* **182** (1996), 285–298.

Chapter 20

Complex dynamics of molecular evolutionary processes

Christian V. Forst[1]
Inst. of Molecular Biotechnology
Beutenbergstr. 11, 07745 Jena, Germany

Evolution is an extreme complex dynamical phenomenon which has to be partitioned into simpler systems for an adequate description. Peter Schuster suggested a partition of evolutionary dynamics into three such subsystems

Population Dynamics
Population Support Dynamics
Genotype–Phenotype Mapping

In this paper we present an adequate description of all three systems and a structural and dynamical characterisation of their combination. Describing genotype–phenotype relations by RNA sequence–structure maps is the most delicate part in this approach. Such a relation plays a central role in evolution, because genetic mutation acts upon genotype but selection operates on phenotypes. The set of all sequences which map into a particular structure is modelled as random graph in sequence-space. Population-support dynamics by means of a catalytic network is realized as a (large) random digraph and secondary structures as vertex set. Studying a population of catalytic RNA-molecules exploring and *utilising* a large catalytic network shows significantly different behavior compared to a deterministic description: hypercycles are able to coexist and survive resp. a parasite with superior catalytic support. A *switching* between different dynamical organisations of the network can be reported.

Evolutionary dynamics in its full capability of describing interacting evolving species provides a powerful tool for molecular evolutionary biology. Not only evo-

---

[1]now at: University of Illinois, Beckman Institute, 405 N. Mathews Ave., Urbana, IL 61801

lutionary experiments yield better models but also these models are capable for the design of optimal experimental strategies. This synthesis between theoretical model and experimental setup is the key of a comprehensive and complete description of molecular evolution.

1. Introduction

Dynamics of molecular-biological, such as cellular metabolisms, are easily interpreted as complex and non-linear bio-chemical system. Any comprehensive understanding of such biological phenomena requires an interpretation in evolutionary terms as Theodosius Dobzhansky [6] formulated: *"Nothing in biology makes sense expect in the light of evolution"*. This sentence, rephrasing Galilei's famous quote [18, 28], is much stronger since it postulates the existence of a formal language to describe and explain observations in nature.

Providing the ignition spark by outstanding discoveries made by Francis Crick and James Watson in the year 1953 the research field of *molecular biology* was born [44]. Initial steps in the direction of a *molecular evolutionary biology* have already been performed in the late sixties by the pioneer work of Sol Spiegelman [40] who developed *serial transfer experiments* as a new method of molecular evolution in the test tube. Manfred Eigen [7] at about the same time formulated a kinetic theory of molecular evolution. Since then studying evolution of molecules in laboratory systems has become a research area of its own. This approach simplifies evolutionary systems as much as possible and makes them accessible to an analysis by the conventional methods of physics and chemistry.

Evolutionary dynamics itself is a highly complex process. Therefore we omit additional difficulties in considering spatiotemporal patterns and introduce a comprehensive model which tries to account for most of the relevant features of molecular evolution. Peter Schuster proposed an interaction of three processes described in three different abstract metric spaces [35] as essential building blocks of *evolutionary dynamics*:

- the *sequence space* of genotypes being DNA or RNA sequences,

- the *shape space* of phenotypes, and

- the *concentration space* of biochemical reaction kinetics.

Both a description of each subsystem and the characterisation of properties emerged by combination of all three systems is the main goal of this paper. We first give an introduction about graph-topology in sequence-space induced by genotype–phenotype maps (§2). By combining these genotype–phenotype relationships with a Darwinian "hill-climbing" scenario in §3 we observe new characteristics of the optimisation process. But formulating molecular evolution in terms of evolutionary dynamics has to take into consideration not only approaching a steady state but also nonlinear dynamical phenomena like oscillations or chaotic behavior in space and time. In §4 we construct both small "deterministic" and large random catalytic networks and report stability of hypercycles in competition with a superior parasite and emergence and optimisation of hypercycles in a large system.

2. Biomolecules

The function of biomolecules, especially peptides and nucleic acids is predetermined significantly by their tertiary structure in space. Active residues of these molecules are kept in precise position by a huge spatially organised framework of interacting residues and backbone. As conserved active residues in e.g. catalytically active sites are as flexible is the structural framework. Here complete motives can be omitted maintaining (almost) unperturbed functionality. Thus a relevant structure of biopolymers in a given context is seldom described with atomic resolution. In order to detect phylogenetic relations, e.g., structures of proteins are often considered to be similar when polypeptide backbones coincide roughly. A large fraction of amino-acid residues can be exchanged without changing these coarse-grained structures that are apparently relevant in an evolutionary context. Rost and Sander investigated that 25% pairwise sequence identity of residues are sufficient for folding into the same structure [33]. I.e. 75% sequence dissimilarity (in best cases) is compatible with conserved structures.

Similar results are known for RNA structures. Here an adequate coarsegraining is represented by the secondary structure. They commonly are understood as list of Watson-Crick ($\mathbf{A}=\mathbf{U}$ and $\mathbf{G}\equiv\mathbf{C}$) and Wobble ($\mathbf{G}-\mathbf{U}$) base-pairs which are compatible with unknotted and pseudo-knot-free two-dimensional graphs (for a precise formal definition we refer to Waterman [43]). The relevance of RNA secondary-structures for biomolecular function is significantly reflected in viral life cycles [46]. Replication of RNA molecules in the Qβ-system[2] depends exclusively on the structural feature of a hairpin at the 5'-end [3]. Especially kinetics of RNA replication by Qβ-replicase and the dependence of structural features have been studied [4]. A different system — which is extensively examined — is the internal initiation of translation for specific +-strand RNA-viruses. This so called IRES-region (*Internal Ribosomal Entry Site*) — a highly structured region close to the 5'-end of the virus genome — is responsible for the success of the genome translation in the host cell [29].

2.1. RNA secondary structures and compatible sets

RNA secondary structures and the induced sequence – structure relationship are a suitable and generic description for genotype – phenotype mapping which is important in molecular evolutionary biology. One great advantage of RNA molecules are their capability of relatively simple evolutionary experiments in

---

[2]Qβ-phage is a virus which infects bacteria

the test-tube. Here genotype (the sequence) and phenotype (spatial structure) are two features of the same molecule [40].

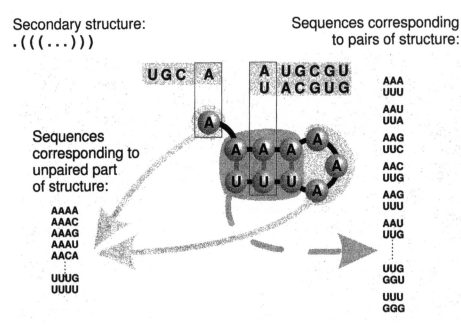

Secondary structure:

`.((( ... )))`

Sequences corresponding to pairs of structure:

Sequences corresponding to unpaired part of structure:

Figure 2.1: Compatible sequences with respect to a fixed secondary structure. A sequence is called *compatible* with a given secondary structure if for all base pairs in the structure there are pairs of matching bases in the sequence. Sequences compatible to a structure *do not* fold in general into this structure. However the structure will always be found as result of suboptimal folding.

Defining secondary structures independently of chemical or physical restrictions yields a general description based on contacts with respect to arbitrary alphabets $\mathcal{A}$ with arbitrary pairing rules. A *pairing rule* on $\mathcal{A}$ is given as a set of pairs of letters from the given alphabet. As an extension of secondary structures a general *contact structure c* is determined by a *set of contacts* of c omitting the trivial contacts due to adjacent letters in the succession of the sequence [26].

A relevant concept in studying sequence – structure relation is how sequences has to be composed to fulfil necessary conditions for folding into a desired structure. In the following we define *compatibility* of a sequence to a given structure: A sequence x is said to be *compatible* to a structure s if all base-pairs required by s can be provided by x_i and $x_j \in x$ with respect to the pairing rule for each base pair (Fig. 2.1). $\mathbf{C}(s)$ is the set of all sequences which are compatible to structure s. The number of compatible sequences is readily computed for secondary structures (with n_u unpaired bases and n_p base pairs this evaluates to $4^{n_u} \cdot 6^{n_p}$).

2.2. Covering sequence-space

The relation between RNA sequences and secondary structures is understood as a surjection f_n from sequence space $\mathcal{Q}_\alpha^n$ into shape space $\mathcal{S}_n$. Essential insights of a graph-theoretical approach characterising generic properties of such maps are presented here. A mathematical framework with proofs can be found elsewhere [32].

The set of all sequences folding into a given structure is denoted as *neutral network* $\Gamma_n(s)$ with respect to s. $\mathcal{Q}_\alpha^n$ denotes the generalised hypercube of dimension n over an alphabet $\mathcal{A}$ of size α (i.e. the number of letters in $\mathcal{A}$ is α), and $s \in \mathcal{S}_n$ is a fixed secondary structure. Mathematically $\Gamma_n(s)$ refers to the induced subgraph of $f_n^{-1}(s)$ in $\mathbf{C}(s)$ ($f_n^{-1}(s)$ indicates the *preimage* of a fixed structure s w.r.t the mapping f_n). A sketch of these embeddings is shown in Fig. 2.2.

Remark 1 *The graph of compatible sequences $\mathcal{C}(s)$ to a fixed secondary structure s is*

$$\mathcal{C}(s) = \mathcal{Q}_\alpha^{n_u} \times \mathcal{Q}_\beta^{n_p} \tag{2.1}$$

α is the number of different nucleotides, and β is the number of different types of base pairs that can be formed by α different nucleotides.

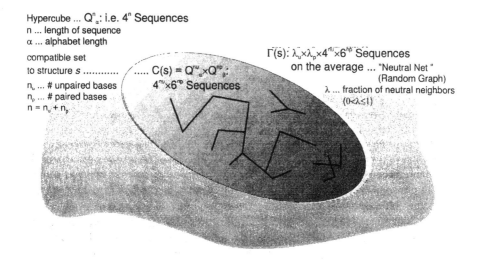

Figure 2.2: Sketch of a neutral network $\Gamma(s)$ (shown as solid line graph) embedded in the set of sequences compatible to structure s (i.e. $\mathbf{C}(s)$ — indicated as oval) which itself is embedded in sequence space $\mathcal{Q}_\alpha^n$ (realized as shaded background).

Sequences folding into the same structure are thus represented by a subgraph that is *randomly* induced on the graph of compatible sequences corresponding to this structure [32]. Accordingly this model is of probabilistic nature and properties of random subgraphs are studied as functions of a single parameter — a probability measure over all possible induced subgraphs in a given sub-hypercube with a *choosing parameter* λ. This parameter represents the mean fraction of neighbors that are neutral with respect to the structure. – the *fraction of neutrality* with $0 \leq \lambda \leq 1$. Considering paired and unpaired regions we are dealing with two independent assignments for each corresponding choosing probability λ_p and λ_u resp. are used. Now vertices in both sub-cubes $\mathcal{Q}_\alpha^{n_u}$ and $\mathcal{Q}_\beta^{n_p}$ are chosen independently with these probabilities λ_u and λ_p resp. This is equivalent to choosing pairs of vertices in $\mathcal{C}(s)$ and yields exactly the desired neutral network $\Gamma_n(s)$ as probability space with its corresponding probability measure.

Two remarkable results are assertions about *density* and *connectivity* of subgraphs. In analogy to percolation theory these properties are fulfilled almost sure if λ exceeds the threshold value

$$\lambda^* = 1 - {}^{1-\alpha}\!\sqrt{\alpha}.$$

Once we know how to construct a neutral network $\Gamma_n(s)$ for a single structure we extend the description of a "folding landscape" towards many structures. Given an ordered set of secondary structures $\mathcal{S}_n$ and define a complete mapping by iterating the construction process of the corresponding neutral network w.r.t. the ordering. The preimage for the structure with highest rank s_1 is assigned independently. For all other structures s_i, $i > 1$ the mapping depends on all previous assignments. Note that the given ranking is arbitrary. The actual ordering into common and rare structures is essentially dependant on $|\mathbf{C}(s)|$, the rank and the choosing probabilities for accepting a sequence as element of the preimage to a given structure.

3. Evolutionary dynamics

A canonical approach studying molecular evolution is a combination of a genotype–phenotype mapping with a reaction scheme. Thus we consider such a (bio)chemical reaction system and study induced dynamics of a population of individuals with genotypes and phenotypes (with a distinct relationship characteristic for biopolymers) living in an artificial world. Error-prone autocatalytic and catalyzed replication, unspecific dilution due to limited recourses, specific dilution due to predation, alteration due to reaction between individuals happen and change the composition of the population in time. As an example a possible scenario is shown in Fig. 3.1: On the phenotypic level a reaction system describes a cycle of interactions between three phenotypes. "Interactions" can be understood as catalysis, predation, transformation into a different phenotype, … The underlying genotype–phenotype mapping implies a partly neutral

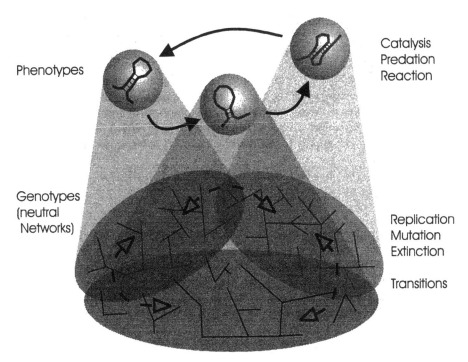

Phenotypes

Catalysis
Predation
Reaction

Genotypes
(neutral
Networks)

Replication
Mutation
Extinction

Transitions

Figure 3.1: Driving forces of Evolutionary Dynamics: three molecular species are connected in a catalytic cycle. Directed catalysis of replication takes place on the level of phenotypes. On the contrary on the level of genotypes pairs of neutral networks are almost always in close vicinity to each other. By stochastic flow transitions between two neutral networks happen.

landscape. Due to error-prone replication mutants emerge and explore new regions in sequence-space. The close vicinity between pairs of neutral networks enables populations to transit between them. The reaction system itself and the corresponding reaction rates judge the moves of the population by benefits or treatment.

Let us start with a well known representative of evolutionary dynamics — *Darwinian Evolution*. A Darwinian scenario of optimising species of an population by evolutionary processes is easily realized as hill-climbing in a high-dimensional fitness landscape. The underlying dynamics of this process is linear (in most cases) and approaches a stationary state. One quite important class of dynamics is the so called *quasispecies dynamics* which describes a population of replicating individuals under mutational forces in an constant environment. In terms of biochemical reactions the corresponding reaction-equation are as follows [8, 27]:

$$I_i \xrightarrow{f[s(I_i)] \cdot W_{ij}} I_i + I_j \qquad (3.1)$$

$$I_i \xrightarrow{\Phi} \emptyset$$

We denote I_i, $i = 1, \ldots, n$ as reacting species with a fixed phenotype $s(I_i)$, W_{ij} as stochastic matrix indicating the probability of reproducing I_j by replicating I_i. $s(I_i)$ corresponds to the structure (or phenotype) of the individual, and $f(s)$ is the fitness of phenotype s. Thus the top reaction-equation of Equ. 3.1 describes an autocatalytic, error-prone replication of I_i. the bottom equation refers to an unspecific dilution flux maintaining the total numbers of individuals constant in the system. Applying chemical reaction-kinetics yields following selection-mutation equation originally formulated by Eigen [7] in a simpler representation:

$$\dot{x}_i = x_i \left[f(s(I_i))W_{ii} - \Phi(\mathbf{x}) \right] + \sum_{j \neq i} k_j W_{ij} x_i \qquad (3.2)$$

$$i = 1, \ldots, n.$$

Eigen in his original paper described evolution in molecular-biological systems without explicit usage of a genotype–phenotype mapping. Thus he assigned constant reaction rates $k_i \equiv f(s(I_i))$ for each genotype. He focused especially on dynamics on the so called *single-peak landscape* as mean field approach, where a single genotype has superior fitness (high reaction rate k_i implying fast replication) upon all other genotypes with equal but lower fitness.

3.1. Neutral evolution

Parallel to the investigation of evolution in Darwinian scenarios neutral evolution on model landscapes has been studied extensively [5, 22]. More recently these studies where extended to neutral evolution on RNA folding landscapes by simulations [24]. In case of selective neutrality populations drift randomly on the corresponding neutral networks by a diffusion-like mechanism. The error-rate p of the underlying replication-deletion process acts as a temperature. By means of stochastic processes one can derive analytical expressions of pair-distance distribution in the neutrally evolving population and of the diffusion constant [16, 31]. Using quasispecies-dynamics, transitions between two different neutral nets with equal fitness-values assigned have been studied [45]. Here essential dependencies between chosen pair of secondary structure (which induce two neutral networks), the topology of the corresponding neutral network — especially for the intersection region, and of the dynamics can be reported. It has been shown (ibidem) that these transitions take place at intersection regions or in close vicinity. Three distinct scenarios for the dynamic behavior of the population have been observed:

- long time fixation on one (or the other) neutral network,

- sharp transitions between the neutral networks,

- long term fuzzy transitions.

An essential parameter for these different classes of transitions is the coupling rate between the networks (dependent on the pair of chosen secondary structures) and the distribution of the intersection in sequence space. Fig. 3.2a shows the time behavior of an evolving population in a two network scenario. In Fig. 3.2b the corresponding probability distribution of the population on two networks is shown (c.f. [45]).

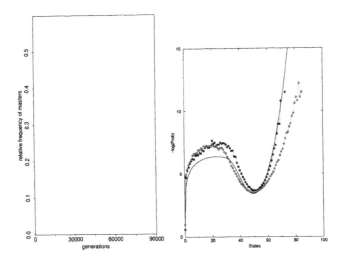

Figure 3.2: Transition in a two-network scenarios are shown: a) Relative frequencies of masters on network 1 (dark) and on network 2 (light) are shown. Two secondary structures with chain-length 30 are chosen. The alphabet of the underlying sequences is binary (**G**, **C**), neutral network are formed under minimum free energy conditions (RNA-folding). The initial population size is set to $N = 1000$. b) Stationary probability distributions are shown.

3.2. Adaption and error thresholds

Let us return to adaption on partly neutral landscapes. Introducing genotype–phenotype mapping motivated by sequence–structure relations of RNA molecules yields an expansion of the quasispecies dynamics with new properties. First studies of quasispecies-dynamics on RNA secondary-structure folding-landscape have been done by Fontana and Schuster, and by Fontana et al.[12, 13]. Forst et al. performed evolutionary dynamics on *single-shape land-scapes* where shape refers to a fixed secondary structure [16]. Analogue to the approach of a single-peak landscape Forst et al.classified the sequence-space in fit sequences (sequences which are mapped in the distinct structure) and non-fit sequences (all other sequences). This implies a classification of an evolving population in masters (fit individuals) and non-masters (non-fit individuals). Starting with low error rates the population is localised around a non-moving master as the quasi-species in the single-peak landscape. At a distinct error-rate (the error-threshold for a single-peak landscape) the population starts moving and drifts on a neutral network analogue to a diffusion-process. for even higher error rate the population breaks into small clusters with lifetime obeying a power law (large clusters have long lifetime, small clusters have short lifetime) [24]. Exceeding a sufficiently high error-rate a so called *phenotypic error-threshold* can be observed. In this case the population is no longer able to conserve the information of the phenotype but diffuses randomly all over sequence-space. Analytical expressions of stationary distributions of a population in this landscape and of diffusion constants have been reported [17]. Derivations and proofs of these formulas can be found at Reidys [31].

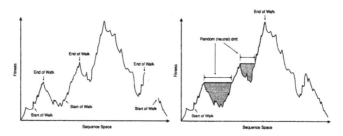

Figure 3.3: Optimisation on Fitness Landscapes: The difference of optimisation in landscapes without and with selective neutrality is shown. On landscapes without partly neutral properties optimisation reaches a local optimum and stays there. Having selective neutrality optimisation don't have to stop at local optima but is able to move neutrally in sequence space finding new opportunities to optimise.

As a consequence of neutral networks, a population seeking the global optimum is likely to find its goal [36]. Fig. 3.3 shows different dynamical behavior of an optimising population for landscapes without and with selective neutrality. Landscapes without partly neutral properties lead an evolving population to the nearest local optimum. In landscapes with selective neutrality a population reaches a local optimum but is not doomed to stay there forever. Instead the population drifts neutrally in genotype space, changes neutral networks by transitions occasionally and looks for a better place.

This behavior is not restricted to RNA landscapes. Studies performed in protein space by computers indicate neutral networks [1] which would induce similar dynamics.

4. Catalytic reaction networks

As a canonical step towards a description of interaction between species we extend Equ. 3.1 by a second order reaction yielding following reaction scheme:

$$
\begin{aligned}
I_i \quad &\xrightarrow{\;f[s(I_i)]\cdot W_{ij}\;} \quad I_i + I_j \\
I_i + I_k \quad &\xrightarrow{\;g[s(I_i),s(I_k)]\cdot W_{ij}\;} \quad I_i + I_j + I_k \\
I_i \quad &\xrightarrow{\;\Phi\;} \quad \emptyset
\end{aligned}
\tag{4.1}
$$

The new function $g[s(I_i),\ s(I_k)]$'s denotes the rate of the catalyzed reaction. First studies of such dynamical systems (Equ. 4.1 with $f \equiv 0$) as homogeneous ordinary differential equations (referred to as *Replicator Equations* [37] without mutation and distinct genotype–phenotype mapping) have already performed by Hofbauer *et al.*[23] in the early eighties:

$$\dot{x}_i = x_i \left[\sum_k A_{ik} x_k - \sum_{jk} A_{jk} x_j x_k \right], \qquad i = 1, \ldots, n \qquad (4.2)$$

Similar to Equ. 3.2 the reaction rates $g[s(I_i), s(I_k)] \equiv A_{ik}$ are kept constant in these studies. Due to the nonlinear property even for small system size (four individual are sufficient) chaotic behavior can be observed [14, 34]. In the last 17 years studying Replicator Equations and derivates (which can be described as perturbations of Replicator Equations by perturbation theory [41]) have become a huge research field of its own. As a recent example may deal the work upon *Autocatalytic networks with intermediates* by Hecht et al.[21].

Remark 2 *In this paper we restrict ourselves to an arbitrary assignment of values to functions f and g. It is straightforward to introduce reaction rates derived from specific properties of the involved phenotypes (e.g. structure, structural motives, combination between structure and sequence composition, structure of transition-states, ...).*

Important for the dynamical characteristics of the system are the reaction-graphs of phenotypes and the error-prone replications on the level of genotypes. The topologies of the underlying neutral nets assure that there are couplings between each two of them.

Thus parts of the population can switch from one net to the other and thereby cause a stabilising effect for the hypercycle. This stabilising effect is not restricted to hypercyclic organisations an can be extended to (all) reaction networks with cooperative behavior.

4.1. Hypercycles and parasites

Pioneer work in studying natural self-organisation have been performed by Eigen and Schuster in the late 70's [9, 10]: their seminal work about *hypercycles* reports about emergence of hypercycles, their abstract description and their occurrence in nature. Thus it is straightforward that hypercycles are a canonic starting point in studying complex reaction systems.

As an extention to earlier investigation on structural stability of hypercyclic organisations [38, 39], we study a five-member hypercycle with a shortcut to a three-member hypercycle. Attaching a single parasite to a vertex of the hypercyclic reaction graph yields five distinct scenarios (Fig. 4.1). Studying the stability of these networks itself and in competition with other entities like parasites reveals deep insights in dependencies between topology of the network, stability of the dynamics and dynamical behavior. In terms of ODE's upper reaction graph (Fig. 4.1) is not *permanent*. A formal definition about permanence in general [39] and for *sparse catalytic networks* [42] has been given elsewhere.

Expected scenarios by the theory for the deterministic case are therefor:

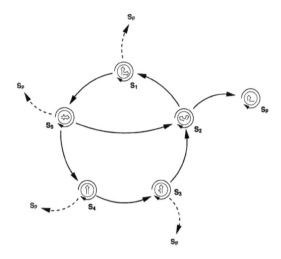

Figure 4.1: A sketch of a five-member hypercycle is shown here with a single detour between member 5 and member 2. Additionally a single parasite is catalyzed by one member of the hypercycle (i.e. S_2) yielding five possible scenarios (one single parasite attached to member 1 to 5).

- dominance of the five-member hypercycle (periodic behavior)

- dominance of the three-member hypercycle (fixed point)

- dominance of the parasite (c_{rel}(Parasite) $\longrightarrow 1$)

In the stochastic model a more relaxed standpoint is necessary. Different quasi-stationary distributions change in time. By altering selection parameters distinct scenarios have been observed:

- dominant parasite

- dominant three-member hypercycle

- dominant five-member hypercycle for distinct parameters

- irregular (stochastic) behavior

An *ad hoc*-conclusion for above scenario is: large member hypercycles are hard to develop and to maintain if their support in sequence-space is insufficiently populated. Previous studies show [15] that without alternatives in terms of shortcuts a five-member hypercycle can successfully compete with a parasite. But introducing alternative catalytic pathways to shorter hypercycles often the systems prefers latter for small population size. Here undirected transition phenomena between neutral networks are a dominating driving force of the dynamics. Increasing the number of individuals in the population shows a different scenario. Preliminary studies indicate that for a distinct topology of the catalytic network

Hypercycles with Parasite and Shortcut

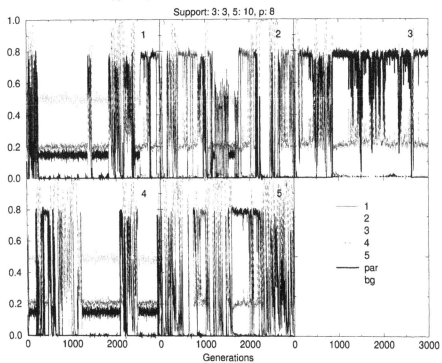

Support: 3: 3, 5: 10, p: 8

Figure 4.2: A five-member hypercycle with a parasite and a shortcut. Depending on the member which supports the parasite the dynamic changes. The parasite dominates part-time the hypercycle by a support from member 3. Support of the parasite by members 1 and 4 favours the dynamics of a three-member hypercycle, support by member 2 induces dynamical behavior of a five-member hypercycle. No regular dynamics at all shows the support by member 5. Here transition phenomena dominate catalytic support. Kinetic parameters are: 10 for catalytic support along the five-member hypercycle, 3 for the shortcut and 8 for the parasite. The error-rate in the system is fixed to $p = 0.01$, the population-size N is 1000.

the five-member hypercycle is preferred against the three-member one (parasite attaches to member #1 — data not shown). Additionally in more than one scenario hypercycles are able to out-compete the parasite due to higher support-capacity (reaction rates are $hypercycle : shortcut : parasite = 10 : 10 : 12$).

4.2. Voyaging large catalytic nets

A natural extension of the system outlined in the previous section are network topologies towards large random catalytic networks. Interesting in this context are questions about self- and re-organisation of dynamical systems where evolving biomolecules are involved. Early studies about autocatalytic sets and

192

emergence of metabolisms have been performed by various people [2, 11, 25, 30].

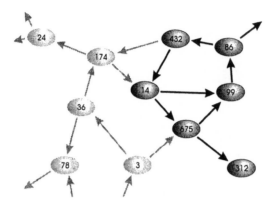

Figure 4.3: Given a large catalytic network with phenotypes (structures) as vertices and randomly chosen edges to an maximal out-degree of 2. A population will establish a catalytic unit (light region). Depending on alternative catalytic pathways on the level of phenotypes or stochastic effects on the level of genotypes (neutral drift of the support or transitions between different neutral nets) new phenotypes are "built" and new dynamic organisations can emerge.

An essential feature which all these systems are lacking, is an adequate genotype–phenotype mapping. Again we use our model of evolutionary dynamics and setup a large catalytic network which has to be explored by an evolving population. This *exploration* of reaction networks can be interpreted as reorganisation of (bio)chemical reaction-units under selection-pressure. How such "voyages" happen is sketched in Fig. 4.3. Part of the network has already been explored by the population. Due to undirected and directed moves new regions — thus new dynamical organisations — are discovered.

Using an ordered set of 1000 secondary structures as phenotypes, a catalytic network of roughly 420 random hypercycles of size 3 to 7, and about 18 additional random reaction paths is constructed. This yields a total of about 2000 possible reaction-paths. By an initial population of $N = 1000$ individuals with phenotype of rank 1 the catalytic network is explored. The reaction-rates g for the catalyzed reaction are randomly distributed with distinct mean and variance. The population explores the catalytic network in search for better support (Fig. 4.4). The following scenario can be observed: although long hypercycles are predefined in the given reaction-networks the population prefers short hypercycles of length three or four resp. This is due to small population-size relative to the capacity of the catalytic units. Only for short hypercycles the catalytic organisation can overwhelm stochastic effect due to transitions and stabilise.

This still quite simple model gives deep insights how new catalytically active units can be formed, how they can reorganise by neutral drift and how optimisation to more efficient units take place.

5. Conclusion

Nonlinear dynamics are an inherent property of biological systems. Not only due to the complex matter of biology itself but also due to complex feedback loops and nonlinear relationship between different subunits. By todays biotechnological tools these molecular-biological systems are quite easily designed and evolved [19, 20]. Theories of molecular evolution can be proven and give novel insights for new experiments. An essential property of the dynamics of such systems are partly-neutral mappings between genotype and phenotype. Neutral networks provide a powerful medium through which evolution can become really efficient. Adaptive walks of populations, usually ending in one of the nearby minor peaks of the fitness landscape, are supplemented by random drift on neutral networks. Periods of neutral diffusion end when the population reaches areas of higher fitness values. Series of adaptive walks interrupted by phases of neutral random walk allow to approach the global fitness-maximum provided the neutral networks are sufficiently large. For more complex organised dynamical units, such as hypercycles, similar behavior can be observed. Hypercycles are able to out-compete parasites by avoiding parasitic regions in sequence-space due to neutral drift off such infected areas. By same mechanism new dynamical organisations emerge.

Acknowledgements

I thank Prof. Peter Schuster for fruitful discussions and for the support with historic quotes. Part of this work presented here is joint research with Drs. Michael Gebinoga, Ulrike Göbel, Janos Palinkas, Christian Reidys, Peter Stadler and Jacqueline Weber which is published elsewhere. I also would like to thank the community of the Santa Fe Institute where part of this publication has been done.

Bibliography

[1] BABAJIDE, Aderonke, Ivo L. HOFACKER, Manfred J. SIPPL, and Peter F. STADLER, "Neutral networks in protein space: A computational study based on knowledge-based potentials of mean force", *Folding & Design* (1996), in press, SFI Preprint 96-12-085.

[2] BAGLEY, Richard J., and J. Doyne FARMER, "Spontaneous emergence of a metabolism", *Artificial Life II*, (C. G. LANGTON, C. TAYLOR, J. D. FARMER, AND S. RASMUSSEN eds.) vol. X of *Santa Fe Institute Studies in the Science of Complexity*. Addison Wesley Redwood City (1992), ch. Origin/Self-Organization, pp. 93–141.

[3] BIEBRICHER, C. K., "Darwinian selection of self-replicating RNA molecules", *Evolutionary Biology* **16** (1983), 1–52.

194

[4] BIEBRICHER, Christop K., and Manfred EIGEN, "Kinetics of RNA replication by $Q\beta$ replicase", *RNA Directed Virus Replication* (Boca Raton, Fl,) (E. DOMINGO, J. HOLLAND, AND P. AHLQUIST eds.), vol. Vol.I of *RNA Genetics*, CRC Press (1988), 1–21.

[5] DERRIDA, Bernard, "Random-energy model: An exactly solvable model of disorderes systems", *Phys. Rev. B* **24**, 5 (1981), 2613–2626.

[6] DOBZHANSKY, T., "Nothing in biology makes sense expect in the light of evolution", *Am. Bio. Teacher* **35** (1973), 125–129.

[7] EIGEN, M., "Selforganization of matter and the evolution of biological macromolecules", *Die Naturwissenschaften* **10** (1971), 465–523.

[8] EIGEN, M., J. MCCASKILL, and P. SCHUSTER, "Molecular Quasi-Species", *Journal of Physical Chemistry* **92** (1988), 6881–6891.

[9] EIGEN, M., and P. SCHUSTER, "The Hypercycle A: A principle of natural self-organization: Emergence of the hypercycle", *Naturwissenschaften* **64** (1977), 541–565.

[10] EIGEN, M., and P. SCHUSTER, "The Hypercycle B: The abstract hypercycle", *Naturwissenschaften* **65** (1978), 7–41.

[11] FONTANA, Walter, and Leo W. BUSS, "The arrival of the fittest: toward a theory of biological organizazion", *Bull. Math. Biol* **56** (1994), 1 – 64.

[12] FONTANA, Walter, Wolfgang SCHNABL, and Peter SCHUSTER, "Physical aspects of evolutionary optimization and adaption", *Physical Review A* **40**, 6 (1989), 3301–3321.

[13] FONTANA, W., and P. SCHUSTER, "A computer model of evolutionary optimization", *Biophysical Chemistry* **26** (1987), 123–147.

[14] FORST, Christian V., "Chaotic interactions of selfreplicating RNA", *SFI Preprint 95-10-093, Computers Chem.* **20**, 1 (3 1996), 69–84.

[15] FORST, Christian V., and Christian REIDYS, "On evolutionary dynamics", *Artificial Life V* (Cambridge, Massachusetts,) (C. G. LANGTON AND K. SHIMOHARA eds.), Complex Adaptive Systems, Nara-Ken New Public Hall, Nara, Japan, May 16–18, 1996, MIT Press (1997), 453–461.

[16] FORST, Christian V., Christian REIDYS, and Jacqueline WEBER, "Evolutionary dynamics and optimization: Neutral Networks as model-landscape for RNA secondary-structure folding-landscapes", *Advances in Artificial Life* (Berlin, Heidelberg, New York,) (F. MORÁN, A. MORENO, J. MERELO, AND P. CHACÓN eds.), vol. 929 of *Lecture Notes in Artificial Intelligence*, ECAL '95, Springer (1995), 128–147, Santa Fe Preprint 95-10-094.

[17] FORST, Christian V., Jacqueline WEBER, Christian REIDYS, and Peter SCHUSTER, "Transitions and evolutive optimization in Multi Shape landscapes", in prep. (1995).

[18] GALILEI, Galileo, *Opere* A. Favaro ed., Babera, Firenze, Italy (1968), p. 232, The original quotation reads: *"... It (the great book of the universe) is written in the language of mathematics and its symbols are triangles, circles, ... ".*

[19] GEBINOGA, Michael, "Hypercycles in biological systems", *J. Endocyt.* (1995), submitted.

[20] GEBINOGA, Michael, and Frank OEHLENSCHLÄGER, "Comparison of self-sustained sequence replication reaction systems", *Eur. J. Biochem.* **235** (1996), 256–261.

[21] HECHT, Robert, Robert HAPPEL, Peter SCHUSTER, and Peter F. STADLER, "Autocatalytic networks with intermediates I: Irreversible reactions", *Math. Biosc.* **140** (1997), 33–74, Santa Fe Institute preprint 96-05-024.

[22] HIGGS, Paul G., and Bernard DERRIDA, "Genetic distance and species formation in evolving populations", *J. Mol. Evol.* **35** (1992), 454–465.

[23] HOFBAUER, Josef, Peter SCHUSTER, K.SIGMUND, and R. WOLFF, "Dynamical systems under constant organisation II: Homogeneous growth functions of degree $p=2$.", *Siam. J. Appl. Math.* **38**, 2 (1980), 282–304.

[24] HUYNEN, Martijn A., Peter F. STADLER, and Walter FONTANA, "Smoothness withing ruggedness: The role of neutrality in adaption", *Proc. Natl. Acad. Sci.* **93** (January 1996), 397–401.

[25] KAUFFMAN, Stuart A., "Autocatalytic sets of proteins", *J. Theor. Biol.* **119** (1986), 1–24.

[26] KOPP, Stephan, Christian M. REIDYS, and Peter SCHUSTER, "Exploration of artificial landscapes based on random graphs", *Self-Organization of Complex Structures: From Individual to Collective Behavior* (London, UK,) (F. SCHWEITZER ed.), Gordon and Breach (1996).

[27] MCCASKILL, John S., "A localization threshold for macromolecular quasispecies from continuously distributed replication rates", *J. Chem. Phys.* **80** (May 1984), 5194–5202.

[28] PARK, D., *The How and Why. An essay on the Origins and Development of Physical Theory*, Princeton University Press Princeton, NJ (1988).

[29] PELLETIER, J., and N. SONENBERG, "Internal initiatino of translation of eukaryotic mRNA directed by a sequence derived from poliovirus RNA", *Nature* **334** (1988), 320–325.

[30] RASMUSSEN, Steen, Carsten KNUDSEN, and Rasmus FELDBERG, "Dynamics of programmable matter", *Artificial Life II* (Redwood City,) (C. G. LANGTON, C. TAYLOR, J. D. FARMER, AND S. RASMUSSEN eds.), vol. X of *Santa Fe Institute Studies in the Sciences of Complexity*, Addison Wesley (1991), 211–254.

[31] REIDYS, Christian, *Neutral Networks of RNA Secondary Structures*, PhD thesis Friedrich Schiller Universität, Jena (May 1995).

[32] REIDYS, Christian, Peter F. STADLER, and Peter SCHUSTER, "Generic properties of combinatory maps and neutral networks of RNA secondary structures", *Bull. Math. Biol.* **59**, 2 (1997), 339–397.

[33] ROST, Burkhard, and Chris SANDER, "Bridging the protein sequence-structure gap by structure predictions", *Ann. Rev. Biophys.* **25** (1996), 113–136.

[34] SCHNABL, Wolfgang, Peter F. STADLER, Christian FORST, and Peter SCHUSTER, "Full characterization of a strange attractor: Chaotic dynamics in low dimensional replicator systems", *Physica D* **48** (1991), 65 – 90.

[35] SCHUSTER, Peter, "Artificial life and molecular evolutionary biology", *Advances in Artificial Life* (Berlin, Heidelberg, New York,) (F. MORÁN, A. MORENO, J. MERELO, AND P. CHACÓN eds.), Lecture Notes in Artificial Intelligence, ECAL '95, Springer (1995).

[36] SCHUSTER, Peter, "Landscapes and molecular evolution", *Physica D* (1996), in press, SFI Preprint 96-07-047.

[37] SCHUSTER, Peter, and Karl SIGMUND, "Replicator dynamics", *J. theor. Biol.* **100** (1983), 533–538.

[38] SCHUSTER, Peter, Karl SIGMUND, and R. WOLFF, "Dynamical systems under constant organisation I.topologigal analysis of a family of non-linear differential equations - a model for catalytic hypercycles", *Bull. Math. Biophys.* **40** (1978), 743–769.

[39] SCHUSTER, Peter, Karl SIGMUND, and R. WOLFF, "Dynamical systems under constant organisation III.cooperative ond competitive behaviour of hypercycles.", *J. Diff. Equ.* **32** (1979), 357–368.

[40] SPIEGELMAN, S., "An approach to the experimental analysis of precellular evolution", *Quart. Rev. Biophys.* **17** (1971), 213.

[41] STADLER, Peter F., and Peter SCHUSTER, "Mutation in autocatalytic networks - an analysis based on perturbation theory", *J. Math. Biol.* **30** (1992), 597–631.

[42] STADLER, Peter F., and Peter SCHUSTER, "Permanence of sparse catalytic networks", *Math.Biosc.* **131** (1996), 111–133.

[43] WATERMAN, M. S., "Secondary structure of single - stranded nucleic acids", *Studies on foundations and combinatorics, Advances in mathematics supplementary studies, Academic Press N.Y.* **1** (1978), 167 - 212.

[44] WATSON, J. D., and F. H. C. CRICK, "Molecular structure of nucleic acid. A structure for Desoxyribo Nucleic Acid.", *Nature* **171** (1953), 964–969.

[45] WEBER, Jacqueline, *Dynamics of Neutral Evolution – A case study on RNA secondary structures*, PhD thesis Friedrich Schiller Universität Jena (1997).

[46] WEISSMANN, C., "The making of a phage", *FEBS Letters (Suppl.)* **40** (1974), S10–S12.

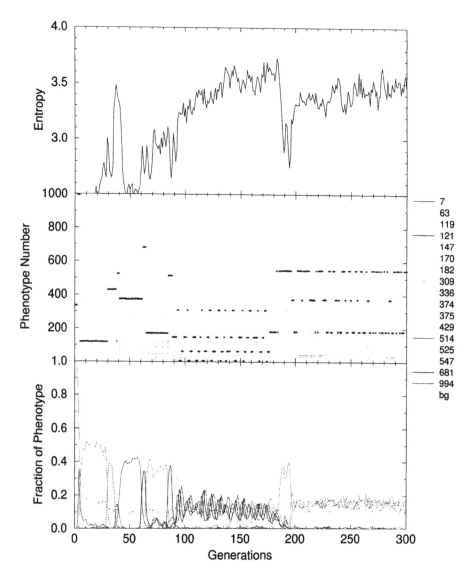

Figure 4.4: Voyaging large catalytic networks: embedded in an artificial secondary-structure landscape of 1000 phenotypes 420 randomly chosen hypercycles of length 3 to 7 are assigned. In addition 18 reaction paths between randomly chosen phenotypes are defined. In this figure an optimisation from one catalytic organisation (four-member hypercycle) to a three-member hypercycle is shown. In the bottom plot concentrations of only these phenotypes are shown whichever existed with maximal concentration at least once in the simulation. The middle plot shows all phenotypes in time. Stars (large dots) denote phenotypes with maximal concentration. In the upper graph the Shannon entropy of the population is plotted. Reaction rates for the catalyzed replication are randomly distributed between $2f$ and $65f$ ($f = const.$).

Chapter 21

Genetic network inference

Stefanie Fuhrman
Xiling Wen
Molecular Physiology of CNS Development, LNP/NINDS/NIH
36/2C02, Bethesda, MD 20892
George S. Michaels
Institute for Computational Sciences and Informatics
George Mason University, Fairfax, VA 22030
Roland Somogyi
Molecular Physiology of CNS Development, LNP/NINDS/NIH
36/2C02, Bethesda, MD 20892

Development may be conceptualized as the dynamic output of a genetic program. We have applied this analogy experimentally by using reverse transcription-coupled polymerase chain reaction (RT-PCR) to measure the expression (mRNA) of large numbers of genes over the course of central nervous system (CNS) development. We have determined mRNA expression at nine developmental time points for 65 genes in rat cervical spinal cord and hippocampus.

Clustering these temporal gene expression patterns using the euclidean distance measure revealed groups of genes which may share inputs. We found that these 65 genes expressed in spinal cord and hippocampus—two distantly related CNS regions— cluster into a small number of basic temporal expression patterns, and that the majority of genes assayed are at least moderately similar in their individual temporal expression patterns when the two regions are compared. This finding suggests the existence of a basic neural genetic program, common to all CNS regions.

The collection of temporal gene expression patterns and development of computational methods, such as clustering and reverse engineering algorithms for analyzing these data, may provide new hypotheses concerning the study of development, tissue regeneration and cancer.

1. Introduction

Given the explosion of data in molecular biology, we are now faced with the challenge of a functional genomics. We must now consider experimental and computational approaches for determining the self-organizing principles that govern parallel interactions among large numbers of genes. Such an approach will be necessary for the generation of a comprehensive theory of biological development.

The advent of completely sequenced genomes and the ongoing human and animal genome projects have provided biologists with an opportunity to study the global dynamics of genetic information flow. In order to determine the principles governing these dynamics, we must study not only individual genes as isolated entities, but also the nature of the parallel interactions among large numbers of genes. This strategy should provide new hypotheses for the study of the phenotypic changes which characterize normal and abnormal development, cancer, and disease.

We have taken a systems approach to development, in which the tissue is regarded as a whole. In this view, differentiation and maintenance of a cell phenotype occur within a set of complex interactions, with some cell types or tissue regions influencing the development of others. *In vivo* gene expression patterns characteristic of stem cells, pluripotent progenitor cells, and mature neurons and glia should be reflected in the patterns of gene expression at different developmental time points.

The developmental process may be conceptualized as the dynamic output of a genetic program in which genes interact as a parallel distributed processing network. This concept is related to the Boolean network idealization[2]. We have applied this idea experimentally by measuring the gene expression (mRNA) levels of large numbers of genes in the rat central nervous system (CNS) over the course of development. Because of its reliability, sensitivity, and dynamic range, we used reverse transcription-coupled polymerase chain reaction (RT-PCR) to assay mRNA levels. In order to determine the degree of order discernible in our whole-tissue samples, we clustered the gene expression patterns for both hippocampus and spinal cord using the euclidean distance measure. A comparison of the temporal gene expression patterns for spinal cord[8] and hippocampus revealed a high degree of similarity between these two anatomically and functionally distinct CNS regions, suggesting the existence of common genetic control processes. We have conceptualized these shared patterns as a generalized genetic network of neural development, which may be common to all CNS regions. To our knowledge, this is the first study to use large-scale temporal gene expression mapping to compare two different tissue regions in an animal.

2. Methods

We used RT-PCR[8] to assay the expression levels of 65 genes in rat cervical spinal cord at nine developmental time points (embryonic days 11, 13, 15, 18, 21, and postnatal days 0, 7, 14, and 90). The same sixty-five genes expressed in rat hippocampus were assayed in the same manner at embryonic days 15 and 18, and postnatal days 0, 3, 7, 10, 13, 25, and 60. Each time series begins with the developmental stage at which its respective CNS region becomes identifiable. Animal care and use was in accordance with NIH guidelines. RNA isolated from cervical spinal cord tissue (Sprague-Dawley albino rats) using RNAstat 60 (Tel-Test, Inc., Friendswood, TX) was adjusted to 200 ng/μl according to absorption at 260 nm, prior to reverse transcription and PCR (Perkin Elmer GeneAmp RNA PCR kit, Applied Biosystems); PCR involved preheating a mixture of Taq antibody (TaqStart, Clontech, Palo Alto, CA), primers, cDNA, and PCR components to 97°C for 90 sec before amplification. PCR cycle: 30 sec at 95°C (dissociation), 45 sec at 60°C (annealing), and 60 sec at 72°C (extension) [35 PCR cycles]. PCR reactions were within the exponential range. Control RNA (transcribed from PAW 108 plasmid DNA; PAW 108 forward and reverse primers; Applied Biosystems, Foster City, CA), in all reverse transcription and PCR amplification reactions allowed for a ratiometric method for determining gene expression levels, and permitted detection of inefficient PCR reactions. PCR products were visualized using PAGE and ethidium bromide staining.

We clustered genes according to their patterns of expression over the nine developmental time points using the hierarchical clustering functions of the Splus statistical software. We chose to include slopes as well as expression values to take into account offset but parallel temporal patterns. The pair-wise distances were entered into a 65 gene × 65 gene distance matrix for spinal cord and a 65 × 65 matrix for hippocampus, which served as the inputs for the clustering program.

3. Results

Gene expression levels among independently run triplicate samples were generally uniform[8, 1], with occasional differences arising perhaps from actual variations in expression levels or from differences in maturational levels among animals of the same age. Previously, two of the primers sets used here were tested in calibration reactions under similar conditions[7]. In that study, the PCR products were well within the exponential range. In the present study, the vast majority of PCR products have sample to control ratios close to 1. This suggests that our PCR reactions were well below the plateau phase. Therefore, because of the large number of genes assayed and the linearity of the control ratios, we determined it was not necessary to perform calibration reactions for every set of PCR primers.

Because multicellular organisms depend on intercellular signals for differentiation, we assayed major gene families known for their involvement in intercellular signaling in hippocampus and spinal cord, and a smaller number known to be important in intracellular functions: neurotransmitter synthesizing and metabolizing enzymes (relating to GABA, acetylcholine, catecholamines), ionotropic neurotransmitter receptors (GABA$_A$ receptors, nicotinic acetylcholine receptors, 5HT3 receptor), metabotropic neurotransmitter receptors (metabotropic glutamate receptors, muscarinic acetylcholine receptors, 5HT receptors), neurotrophins and their receptors, heparin-binding growth factors and their receptors, insulin and IGF family and their receptors, intracellular IP$_3$ receptors, cell cycle proteins, transcriptional regulatory factors, and housekeeping genes. We also included genes for phenotypic markers to correlate expression time series with differentiation. Data were normalized to maximal expression for each gene[8].

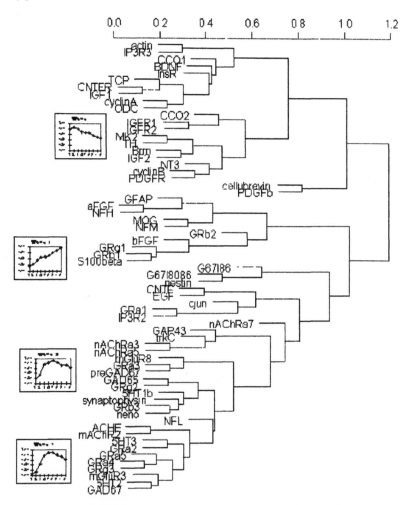

Figure 3.1: Euclidean distance clusters for spinal cord. The average gene expression pattern for each cluster or "wave" is shown as an inset. Common branch points indicate similarities in expression patterns.

Fig. 3.1 and Fig. 3.2 show the clustering of gene expression patterns according to euclidean distance for spinal cord (Fig. 3.1) and hippocampus (Fig. 3.2). The genes assayed for spinal cord cluster into four basic temporal patterns, or waves: high early, two mid-developmental high expression waves, and high late (waves 1 through 4, respectively). Similarly, genes expressed in hippocampus cluster into five groups, similar to those of spinal cord. One of these groups, hippocampal wave 2, is similar to spinal cord waves 2 and 3 in expression pattern shape, but also in content: these waves contain mostly neurotransmitter-related genes. Further, hippocampal waves 1 and 5 and spinal cord wave 1 both contain a high proportion of marker and peptide-related genes, and have similar high-early-decreasing-toward-adult expression patterns.

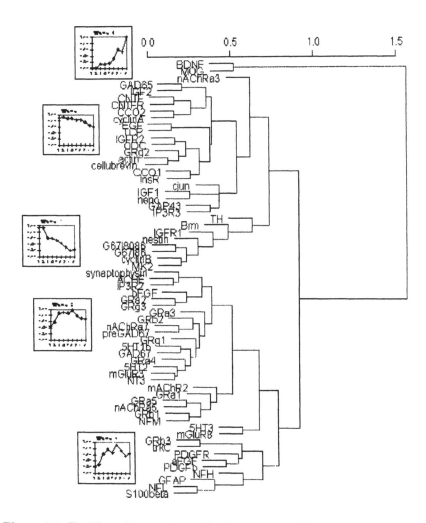

Figure 3.2: Euclidean distance clusters for hippocampus. The average gene expression pattern for each cluster or "wave" is shown as an inset. Common branch points indicate similarities in expression patterns.

We assayed sixty-five genes in common for spinal cord and hippocampus. A visual comparison of the expression maps revealed a similarity between spinal cord and hippocampus in the expression patterns of 60%, or thirty-nine of these genes. Of those, ten patterns are highly similar, as shown in Fig. 3.3; these include genes for four neurotransmitter-related proteins, GABA A receptors alpha 4 and 5, pre-glutamic acid decarboxylase 67, and metabotropic glutamate receptor 8; insulin-like growth factor; the amino acid metabolizing enzyme, ornithine decarboxylase; the marker, nestin; two subunits of the mitochondrial enzyme cytochrome C oxidase; and the cell cycle protein, cyclin B.

Many gene expression patterns appear to begin leveling off late in development. This is reflected in hippocampal waves 1 and 2, and in spinal cord wave 1 at the P25 and P14 time points, respectively. It appears that gene expression has begun to reach its adult level at that stage.

4. Discussion

Large-scale temporal gene expression patterns are essentially the output patterns of genetic networks. Despite the multivalued, continuous nature of gene expression, a genetic network is analogous to a Boolean network in a number of respects. A genetic network self-organizes into cell types (attractors) and exhibits cycling, such as cell cycles (dynamic attractors); as in a Boolean net, this is accomplished via rules of interaction which generate state transitions[2, 6]. Our 65-gene RT-PCR data on spinal cord and hippocampus represent only a fraction of the thousands of genes acting in parallel to generate the fully developed CNS. Nevertheless, these data provide important information concerning the architecture of the genetic program of neural development.

Although most gene-gene interactions involve proteins, there is currently no protein expression assay comparable to RT-PCR in reliability. However, it is reasonable to assume that most protein expression patterns are closely correlated with their respective gene expression patterns. In addition, the use of a semi-quantitative assay provided relative gene expression levels, sufficient for the detection of temporal patterns. Absolute quantitation, although possible with PCR, would not have yielded further information of direct relevance to the questions being addressed in this study.

Both spinal cord and hippocampus exhibit a high degree of order, apparently organizing their gene expression into a small number of simple patterns. These patterns do not exhibit large, sporadic fluctuations, although lower amplitude fluctuations might have been noticeable with smaller sampling intervals more suitable for the study of the underlying biochemical network. The intervals used here, however, are appropriate for the important events of differentiation, which in the rat, occur over the course of days. The apparent order exhibited by these two CNS regions suggests that the cells are well-coordinated in the timing of their gene expression. A random timing among cells would have resulted in a relatively invariant temporal expression pattern for each gene, when averaged over whole tissue. This suggests that cells of the CNS do not run their genetic programs independently of one another. Interestingly, of the four basic temporal expression patterns, or "waves" in spinal cord, and the five in hippocampus, none is "U-shaped"—high-early, low mid-development, and high-late. Further studies will be necessary to determine whether these patterns reflect all the basic types of temporal gene expression patterns generated during CNS development, or whether other types exist.

The apparent leveling off of many gene expression patterns by P14 in spinal cord and P25 in hippocampus suggests that large developmental changes have ceased by this stage. One interpretation of this is that the genetic network has reached its attractor at that point in development.

These waves of expression may reflect common genetic control processes for all members of a wave. In that context, it is notable that, of the different waves, those which have their highest expression levels at mid-development contain a very high proportion of neurotransmitter-related genes. This suggests a link between a general class of gene function and a control process. It may therefore be possible, using large-scale temporal expression mapping, to generate hypotheses concerning the functions of newly discovered genes. This may be particularly relevant to the new genes discovered recently during the sequencing of complete microbial genomes, as a large proportion of these genes have no known homologs, and therefore, no hypothesized functions.

It should be noted that our use of tissue homogenate precluded any determination of cell type contributions to these gene expression patterns. For example, declining expression may be caused by apoptosis in some cases. The question of why the expression patterns occur as they do will require studies involving sorted cells and in situ hybridization.

A comparison of the euclidean distance trees for spinal cord and hippocampus reveals a high degree of similarity in their basic temporal expression patterns. Essentially the same patterns appear in both CNS regions. At the level of individual genes, 60% of the genes appear to exhibit at least moderately similar expression patterns when each hippocampal gene is compared with its counterpart in spinal cord, and ten of these are highly similar. This is surprising given the anatomical differences between spinal cord and hippocampus; the fact that spinal cord begins its development approximately four days ahead of hippocampus in the rat; and the fact that spinal cord appears much earlier in vertebrate evolution. Such a high degree of similarity between hippocampus and spinal cord in the expression patterns of specific genes suggests the existence of a generalized genetic program of neural development, common to all CNS regions. The assumption that this finding can be extrapolated to other CNS structures is not far-fetched given the evolutionary distance between hippocampus, a structure derived from cerebral cortex, and spinal cord. The gene expression patterns shared in common undoubtedly reflect the development of the essential features of nervous tissue, such as neurons, glia, and synapses. Consistent with this hypothesis are the inter-regional similarities in the expression patterns of "marker genes": glial fibrillary acidic protein (GFAP), a marker for astrocytes; myelin-oligodendrocyte glycoprotein (MOG); synaptophysin, a synaptic vesicle-associated protein; and the neuronal marker, neurofilament light (NFL) (individual gene expression patterns can be found in Wen et al., 1998, and Fuhrman et al., in preparation).

Large-scale temporal gene expression data, such as these, can be clustered using different methods. Michaels et al.[4] used mutual information[5] as an alternative cluster analysis method for the spinal cord data described here, and compared the results with those of euclidean distance. Given the assumption that genes whose temporal expression patterns cluster together share genetic inputs, euclidean distance clusters genes which respond in the same manner to those inputs. Mutual information, on the other hand, captures both positive and negative, linear and nonlinear correlations, clustering genes which may respond differently to their shared inputs. With much larger temporal gene expression maps and laboratory studies involving targeted perturbations, it should be possible to go beyond clustering, and determine actual interconnections between genes, as we are now beginning to do with hippocampus[1]. The results of perturbation studies will provide constraints for reverse engineering algorithm[3] capable of producing putative "wiring diagrams" of genetic networks. Such diagrams, complete with combinatorial rules for multigenic regulation, could be used to form new hypotheses for the study of cancer, regeneration, and developmental disorders.

5. Abbreviations

GAP43, growth-associated protein 43; NFL, NFM, NFH, neurofilament light, medium, and heavy; neno, neuron-specific enolase; GFAP, glial fibrillary acidic protein; MOG, myelin-oligodendrocyte glycoprotein; GAD65, glutamate decarboxylase 65; GAD 67, G67I80/86, and G67I86, glutamate decarboxylase 67 splice variants; AChE, acetylcholinesterase; ODC, ornithine decarboxylase; TH, tyrosine hydroxylase; GR, gamma-aminobutyric acid (GABA) receptor; mGluR, metabotropic glutamate receptor; nAChR, nicotinic acetylcholine receptor; mAChR, muscarinic acetylcholine receptor; 5HT, serotonin (5-hydroxytryptamine) receptor; NT3, neurotrophin 3; BDNF, brain-derived neurotrophic factor; CNTF, ciliary neurotrophic factor (CNTFR, receptor); trk, NGF (nerve growth factor) receptor; MK2, midkine 2; EGF epidermal growth factor; bFGF, aFGF, basic and acidic fibroblast growth factor; PDGF and PDGFR, platelet-derived growth factor and its receptor; InsR, insulin receptor; IGF, insulin-like growth factor; IGFR, IGF receptor; IP3R, inositol 1,4,5 trisphosphate receptor; Brm, brahma; TCP, t-complex protein; CCO1 and CCO2, cytochrome c oxidase, subunits 1 and 2.

Bibliography

[1] Fuhrman, Stefanie, in preparation.

[2] Kauffman, Stuart A., The Origins of Order: Self-Organization and Selection in Evolution. Oxford University Press, New York (1993).

[3] Liang, Shoudan, Stefanie Fuhrman, and Roland Somogyi, REVEAL, A general reverse engineering algorithm for inference of genetic network architectures. Proceedings of the Pacific Symposium on Biocomputing, 18-29 (1998).

[4] Michaels, George S., Daniel B. Carr, Manor Askenazi, Stefanie Fuhrman, Xiling Wen, and Roland Somogyi., Cluster analysis and data visualization of large-scale gene expression data. Proceedings of the Pacific Symposium on Biocomputing, 42-53 (1998).

[5] Shannon, Claude E. and Weaver W., The Mathematical Theory of Communication. University of Illinois Press (1963).

[6] Somogyi, Roland, and Carol A. Sniegoski, Modeling the complexity of genetic networks: understanding multigenic and pleiotropic regulation. Complexity, 1(6):45-63 (1996).

[7] Somogyi, Roland, Xiling Wen, Wu Ma, and Jeffery L. Barker, Developmental kinetics of GAD family mRNAs parallel neurogenesis in the rat spinal cord. J. Neurosci. 15: 2575-2591 (1995).

[8] Wen, Xiling, Stefanie Fuhrman, George S. Michaels, Daniel B. Carr, Susan Smith, Jeffery L. Barker, and Roland Somogyi. Large-scale temporal gene expression mapping of central nervous system development. Proc. Natl. Acad. Sci. USA, 95: 334-339 (1998).

208

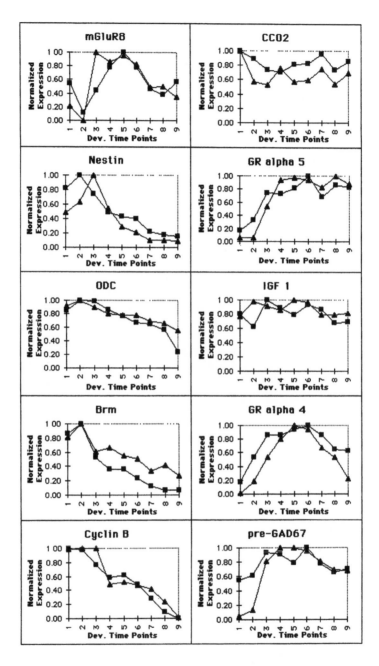

Figure 3.3: Ten genes with highly similar expression patterns between cervical spinal cord and hippocampus. Triangles, spinal cord; squares, hippocampus. Developmental time points range from embryonic day 11 to postnatal day 90 for spinal cord, and embryonic day 15 to postnatal day 60 for hippocampus. See Fig. 3.1 and Fig. 3.2.

Chapter 22

Socioeconomic systems as complex self-organizing adaptive holarchies: The dynamic exergy budget

M. Giampietro
Istituto Nazionale della Nutrizione, Rome, Italy
K. Mayumi
The University of Tokushima, Tokushima Japan
G. Pastore
Istituto Nazionale della Nutrizione, Rome, Italy

This paper presents a model that describes socioeconomic systems as complex, self-organizing, adaptive holarchies (SAH for short) stabilized by informed autocatalytic cycles. SAH is based on a resonance in terms of "computational capability of society" and "process of exergy degradation". In our model, the profile of "computational capability of society" corresponds to the profile of allocation of human time on different human activities. On the other hand, the size of the different processes of exergy degradation corresponds to the amount of energy input consumed in the different sectors of the economy. Central to the model is the dynamic exergy budget of society, which is approached as a dynamic equilibrium between both supply and demand of useful energy per unit of working time. A set of equations for dynamic equilibrium is described for three different hierarchical levels: individuals, the socioeconomic system as a whole, and the environment within which society operates. Because of the complex nature of socioeconomic systems, two distinct perspectives have to be considered when analyzing societal development. One is the quasi-steady-state perspective, which analyzes the system at a particular point in time and space. It provides a definition and assessment of improvements in the "efficiency" of the system in terms of a given set of present boundary conditions, technological levels and system goals. The other perspective is evolutionary, which allows for definition and assessment of "adaptability" of the system, that is, its ability to perform well according to unknown future boundary conditions and different goals. These two contrasting perspectives cannot be blended into a single description of the system. Technological changes imply a trade-off between efficiency and adaptability, and never represent "absolute improvements". Biophysical analyses can provide indicators referring to both perspectives to analyze the nature and the effects of trade-offs but cannot provide a single denominator to define costs and benefits.

1. Introduction

This series of two papers presents a model of analysis that describes socioeconomic systems as complex, Self-organizing Adaptive Holarchies (SAH) stabilized by "informed autocatalytic cycles" (using the concept proposed by Brooks and Wiley, 1988; Brooks et al. 1989) or by "a mechanism of reciprocal entailment between two systems of entailments across hierarchical levels" (using the concept proposed by Rosen, 1991). The model can be used to analyze pre-industrial and post- industrial societies in their interaction with the environment - both the model and possible applications have been presented in detail elsewhere (Giampietro, 1997a; 1997b; Pastore et al. 1996; Giampietro and Mayumi, 1996; 1997).

According to the metaphor suggested by Simon (1962 pp. 477-482), adaptive dissipative systems are based on a resonance that can be described as: "recipes inducing processes which in turn makes better recipes". Following this rationale the model describes such a resonance in terms of "computational capability of society" controlling the existing set of "processes of exergy degradation" which in turn stabilize and improve the original computational capability (Fig. 1.1).

The proxy used to assess the profile of allocation of "computational capability" - defined at the hierarchical level of society - is the profile of allocation of human time on different activities. The assumption is that - at the hierarchical level of society - a given level of technology (assessed by the ratio exosomatic/endosomatic energy metabolized by society) translates into a certain level of amplification of human activity. Due to the fact that "a human being is more nearly a serial than a parallel information-processing system" (Simon 1962, p. 476), human societies face an internal constraint to their expansion when the interactive capacity of lower-level holons (individuals) is saturated, (e.g., because it takes time to care for a friend, one can not have an infinite number of them). Hence, the limited ability of humans to exert control over flows of useful energy allocated to various societal activities is a fundamental factor in shaping the process of self-organization of human society. Redundancy in the requirement of controls on the established set of activities needed to stabilize the current steady-state reduces the ability to expand the set of activities performed by social systems.

The proxy used to assess the size of the different processes of exergy degra-

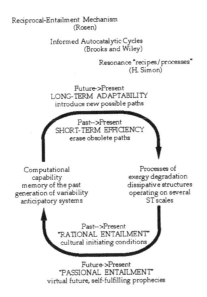

Reciprocal-Entailment Mechanism
(Rosen)

Informed Autocatalytic Cycles
(Brooks and Wiley)

Resonance "recipes/processes"
(H. Simon)

Future->Present
LONG-TERM ADAPTABILITY
introduce new possible paths

Past-->Present
SHORT-TERM EFFICIENCY
erase obsolete paths

Computational
capability
memory of the past
generation of variability
anticipatory systems

Processes of
exergy degradation
dissipative structures
operating on several
ST scales

Past-->Present
"RATIONAL ENTAILMENT"
cultural initiating conditions

Future->Present
"PASSIONAL ENTAILMENT"
virtual future, self-fulfilling prophecies

Figure 1.1:

dation is the amount of energy input consumed in the different sectors of the economy. The assumption is that the unavoidable difference between "useful energy" and "energy input consumed" (generated by differences in the efficiency of conversions, different qualities of energy inputs used by society) can be assumed to be at a particular point in time and space as given (determined by the average technological level, the mix of energy sources, and the mix of end uses) for a society. Therefore, assessments of the amounts of energy consumed in different activities can be used as indicators of the investment (in terms of exergy degradation) allocated to stabilize structures and functions.

Central to the model is the dynamic exergy budget of society, which is approached as a dynamic equilibrium between: (i) supply of useful energy per unit of working time; and (ii) demand of useful energy per unit of working time. Such a dynamic equilibrium can be described by a set of equations that refer to three different hierarchical levels: individuals, the socioeconomic system as a whole, and the environment within which society operates.

Recognition of the complex nature of socioeconomic systems implies that when analyzing societal development two distinct perspectives have to be considered. The quasi-steady-state perspective, which analyzes the system at a particular point in time and space, provides a definition and assessment of improvements in the "efficiency" of the system, that is doing better of what the system already does according to the given set of goals and the present picture of boundary conditions. On the other hand, the evolutionary perspective allows for a definition and assessment of "adaptability" of the system, that is the ability to perform well according to unknown future boundary conditions and different goals. Improvements in "efficiency" can only be defined at a particular point in time and space (quasi-steady-state view), whereas improvements in "adaptability" can only be defined from an evolutionary perspective (on a different and much larger space-time scale). Therefore, these two contrasting perspectives can not be blend into a single description of the system (Giampietro, 1994). Indeed, the model shows that technological changes imply a trade-off between efficiency and adaptability and never represent "absolute improvements." Biophysical analyses can provide indicators referring to both perspectives to analyze the nature and the effects of trade-offs, but can not provide a single denominator to define costs and benefits.

2. Efficiency and adaptability (hypercyclic and purely dissipative compartment)

To remain away from thermodynamic equilibrium -an essential requisite to have any process of self-organization- complex systems have to stabilize in time the process of exergy degradation (Morowitz, 1979; Jørgensen 1992, Schneider and Kay 1994). As pointed out by H.T. Odum (1983, 1996), in economic and ecological systems - both based on a dynamic equilibrium of energy flows - an energetic investment (an activity) in order to be stable in time must pay back its cost. In other words, an activity (structure performing a function) must be able to take advantage of favorable boundary conditions to sustain the process of energy dissipation required for its own survival. On the other hand, the process of exergy degradation needed to guarantee the metabolism of a dissipative structure has the effect of destroying the very same gradients of free energy on which the dissipative process depends. Especially when dissipative systems can amplify their own activity through replication - when systems are affected by "Malthusian instability" (Layzer 1988) - boundary conditions will change eventually. This implies that self-replicating dissipative systems have to become, sooner or later, something else (Prigogine, 1978). Therefore, the process of self-organization of society requires two types of activities, those related to *efficiency* and those related to *adaptability*.

Activities related to efficiency are those that stabilize the current process of energy dissipation according to existing stored information about boundary conditions in order to match in time and space the flows of materials required by societal metabolism. Hence, by increasing its efficiency society tries to stabilize and strengthen its current pattern of interaction with the environment (stabilize the existing quasi-steady-state).

Activities related to adaptability are those that expand the possible solutions that will stabilize the process of energy dissipation according to future changes in boundary conditions. In other words, activities that will be able to match in the future the (unknown) needs and aspirations of the socioeconomic process during its continuous process of evolution (in the long run) when facing different (unknown) boundary conditions. By improving its adaptability, the system increases the probability to reach new situations of quasi-steady-state in thermodynamic non-equilibrium in the future when boundary conditions and internally generated goals will have changed. Thus, adaptability refers to the ability to find new ways of stabilizing processes of energy degradation that will keep alive the resonance between processes and recipes (the reciprocal entailment between two systems of entailments - Rosen, 1991) in spite of a changing environment and a changing identity of the system - Fig. 1.1.

Analyzing ecosystem structure, Ulanowicz (1986) finds that the network of matter and energy flows making up what we call an ecosystem can be divided in two parts. One part that generates a hypercycle, that is, a part that is a net energy producer for the rest of the system. Since some dissipation is always "necessary to build and maintain structures at sub-compartment level" (ibid., p. 119) this part comprises activities that, taking advantage of sources of free energy outside the system (e.g., solar energy, stocks of energy inputs), generate a positive feed-back by introducing degradable exergy into the system at a higher rate than it is consumed. The role of this part is to drive and keep the whole system away from thermodynamic equilibrium. The other part has a purely dissipative nature. This part contains activities that are net energy degraders. However, this second part is not useless for the system: it has the role of providing control over the entire process of energy degradation and stabilizing the whole system. An ecosystem made of a hypercyclic part alone could not be stable in time. Without the stabilizing effect of the dissipative part a positive feedback "will be reflected upon itself without attenuation, and eventually the upward spiral will exceed any conceivable bounds" (ibid., p. 57). Hypercycles alone cannot survive, they just blow up.

Using this analogy we can describe the society as made of two compartments, one of which is hypercyclic (a net producer of useful energy for the rest of society) and the other purely dissipative (a net consumer of useful energy). This model assumes that those processes of exergy degradation and the fraction of computational capability invested in CI activities (to run and maintain the existing quasi-steady-state) are allocated to "efficiency", whereas processes of exergy degradation and computational capability invested in FI activities (in expanding the possible paths of exergy degradation and adding functions to the information systems of society) are allocated to "adaptability". This approach is illustrated in Fig. 2.1, where the energy consumed by society is allocated to three sectors involving distinct types of activities:

- Activities in the household sector (HH) are purely dissipative in nature as they consume net energy in the short term. They are not strictly encoded in the form of defined roles or protocols. These activities include sleeping, personal care, leisure time, and all activities performed by the non-economically active population.

- Activities in the service sector (SS) are also dissipative in nature, but are 'encoded' in the form of defined social roles (e.g., job positions). These activities include all services such as police, army, health care, education, insurance, etc..

- Activities in the primary economic sectors (CI) have a positive return in terms of energy flows and are 'encoded' in the form of defined social roles (e.g., job positions). This includes the energy and mining sector, the manufacturing sector in modern economies, food security, and environmental security. CI activities generate the hypercycle by stabilizing the autocatalytic loops of endosomatic and exosomatic energy. They must be able to generate the energy surplus consumed by the dissipative sectors HH and SS.

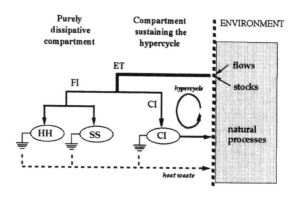

Figure 2.1:

3. The dynamic exergy budget

The energy throughput at which society's exergy budget can be stabilized (when demand is equal to supply) is defined by the socioeconomic characteristics of society that generate the demand as well as the characteristics of the exosomatic autocatalytic loop that generates the supply.

Several variables can be used to characterize the socioeconomic organization of society (demand side) and the nature of the exosomatic autocatalytic loop of energy (supply side). The dynamic equilibrium between demand and supply can then be studied by using the existing relationship between groups of parameters determining demand and supply.

DEMAND SIDE

The flow of exosomatic energy consumed by society (ET) can be expressed as:

$$ET = (MF \times ABM) \times \frac{Exo}{Endo} \times THT \qquad (3.1)$$

where

- ET = Energy throughput or flow of exosomatic energy ($Joules/yr$) in society.

- MF = Metabolic flow = flow of metabolic energy per kg of body mass ($Jkg^{-1}hr^{-1}$).

- ABM = Average body mass = Total mass (kg) of population divided by the number of individuals in society.

- $Exo/Endo$ = Ratio between exosomatic and endosomatic energy flows in society.

- THT = Total human time = number of individuals (population) × hours in a year (hy).

The ratio working time/total human time can be expressed as:

$$\frac{WS}{THT} = \frac{B+C}{\text{population size} \times \text{hours in a year}} \qquad (3.2)$$

where

- WS = Work supply = $B+C$ = amount of time (hours) that the economically active population allocates to work on a year basis (as opposed to sleeping, leisure, etc.).

- C = Hours of work delivered in the primary sectors (CI) of the economy (these sectors include: energy and mining; manufacturing; food security; environmental security, and industrial transportation).

- B = Hours of work delivered in the service sector (SS) of the economy.

By combining relations 3.1 and 3.2 we obtain a definition of the bioeconomic pressure (BEP) that measures the exosomatic energy throughput consumed at the level of society per unit of labor time delivered in the primary sectors:

$$BEP = \frac{ET}{C} = (ABM \times MF) \times \frac{Exo}{Endo} \times \frac{THT}{C} \qquad (3.3)$$

BEP is an *intensive* variable, which means that to assess the size of the energy demand of the socioeconomic system we have to couple this parameter to an *extensive* variable, such as the size of the work supply in the primary sectors of the economy.

SUPPLY SIDE

After defining the following parameters,

- CI/C = exosomatic energy throughput per hour of labor in the CI sectors of the economy (MJ/hour),

- C/THT = fraction of total human time (THT) which is allocated to CI activities (in the primary sector of modern economies),

- $hy = 8760$ = hours in a year,

- ET/CI = return for the socioeconomic system of the energetic investment in CI sectors,

we can write the relation:

$$SEH = \frac{ET}{C} = \frac{ET}{CI} \times \frac{CI}{C} \tag{3.4}$$

where

- SEH = Strength of the hypercycle that is defined as the exosomatic energy throughput (societal power) generated at the level of society per unit of work delivered in CI activities (in those sectors stabilizing the exosomatic energy autocatalytic loop = primary sectors).

Therefore, SEH on the supply side is the analog of BEP on the demand side. Clearly, also SEH is an intensive parameter (ET supply per hour of labor in CI sector) and to assess the total energy supply to the socioeconomic system this parameter must be coupled to an extensive variable, such as work supply in the CI sector.

SEH depends on two characteristics of the exosomatic compartment:

(i) ET/CI, that measures how much of the exosomatic energy throughput of society is "eaten" by the hypercycle, which is related to the output/input energy ratio of the processes that make available resources to the economy. In other words, a certain fraction of ET (CI) must be invested to guarantee (1) the building and maintenance of the exosomatic devices themselves (manufacturing), (2) the procurement of energy and matter inputs as well as waste disposal, and (3) food and environmental security for humans.

(ii) CI/C, that measures the power level per worker in the primary sectors. In developed societies, where the ratio WS/THT decreases (aging of population, longer education, smaller work load) and where the service sector absorbs a large fraction of the available work supply (the ratio $B/(B + C)$ increases), a continuous decrease of the ratio C/THT is feasible only if there is a concomitant increase in the ratio CI/C.

Linking the demand and supply side

An overview of the parallel allocation of exosomatic energy and human time to different compartments of the economy is provided in Fig. 3.1. Recalling relations 3.3 and 3.4 , a constraint on the energy budget of society can be formulated that refers to intensive variables: the demand of exosomatic energy per unit of work supply in CI sectors (BEP) must equal the supply of exosomatic energy per unit of work supply in the CI sectors (SEH). That is:

$$(ABM \times MF) \times \frac{Exo}{Endo} \times \frac{THT}{C} = \frac{ET}{CI} \times \frac{CI}{C} \qquad (3.5)$$

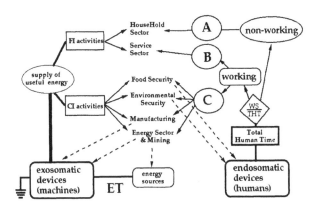

Figure 3.1:

The dynamic exergy budget, therefore, establishes a link between the set of socioeconomic variables indicated in Fig. 3.2 and the set of technical coefficients describing the performance of different economic sectors of the society indicated in Fig. 3.3. The analysis of technical coefficients (input/output values) indicated in Fig. 3.3 has to include: labor among inputs, and the household sector among the economic sectors.

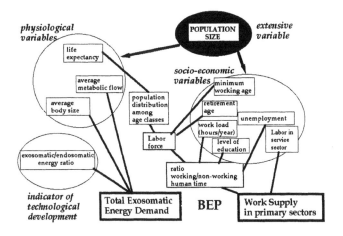

Figure 3.2:

218

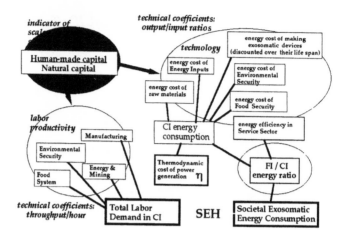

Figure 3.3:

4. The scale issue : Environmental loading and need for adaptability

The larger is the energy throughput (ET) in the system, the larger the size of the process of energy degradation that is based on the destruction of available, favorable biophysical conditions. In practical terms, this means faster depletion of stocks and faster filling of sinks. The resulting stress on boundary conditions is determined by the difference in speed at which favorable conditions (e.g., gradients of free energy) are generated by biophysical processes and that at which favorable conditions are destroyed by the self-organization process of society. After a certain threshold has been reached, the larger the energy throughput of society within a defined ecosystem, the greater will be the stress on boundary conditions and the more pronounced the need for adaptability in the socioeconomic system. Hence, a larger energy throughput requires a higher ratio FI/CI, which is reflected in changes in the pattern of human time allocation (THT/C). This process is confirmed by the current trend of development of human society - post-industrialization of modern economies (see examples given in next paper of this series).

An analysis of the stability of boundary conditions requires a comparison of the scale of the self- organization process of human society, which is proportional to ET, with that of natural ecosystems, which can be assumed proportional to the solar energy used in sustaining their self-organization process. H. T. Odum (1996) has developed a methodology that can be used to make this comparison. Briefly, the flow of useful work in society (loops of energy that amplify back under control of humans in the economic process) is compared (after correction for a quality factor) to the flow of useful work in ecosystems (loops of energy that amplify back under control of biophysical processes). The ratio between the indicator of scale of human activity and the indicator of scale of biophysical activity of ecosystems on which society depends then provides an indicator of *environmental loading*. H.T. Odum applies this rationale through a methodology that estimates indicators in terms of EMergy assessments. This operationalization is, however, still controversial.

Nevertheless, approximations can be made of the requirement of an economy for ecological services in terms of requirement of space-time of biophysical activity needed to make available the inputs of matter and energy to the economic process and to absorb the related outputs. Possible approaches, following H.T. Odum's methodology, include: defining the amount of inputs and/or wastes, assessing the space-time requirement per unit of input and waste, and checking whether sufficient biophysical activity is actually available within the area occupied by society to sustain its economy without degrading the environment. Examples are the "BIOSTA budget" (Giampietro and Pimentel 1991) and "Ecological Footprint" (Rees and Wackernagel 1994).

When society's requirements exceed available ecological services, the gap can be filled through importation by using embodied space-time activity of other ecosystems that are distant in time (e.g., oil stocks) or in space (e.g., imported commodities). Another temporary solution is overdrafting of the ecological services available, which will steadily erode the resilience of the local ecosystems and eventually destroy the complexity of natural systems.

In conclusion, using the rationale proposed by H.T. Odum, the intensity of the activity of self- organization of human society, defined as the ratio ET/exploited area can be used as an indicator of environmental loading. Such an indicator can then be related to the insurgence of negative effects caused by human activity on the environment. Based on cases where this has happened, a critical environmental loading ratio (ELR_{max}) can be defined that represents the threshold value for ecological compatibility.

The different parameters proposed in this model can be used to put in relation the characteristics of societal exergy budget with levels of environmental loading by using the scheme given in Fig. 4.1.

5. Conclusion

The model of analysis presented in the first paper of this series:
1. describes human societies by using only measurable biophysical variables for which data are available;
2. can assess - by using intensive variables describing the feasibility domain for the exergy budget (where $BEP = SEH$)-the level of development of a society according to a socioeconomic perspective;
3. can assess - by using extensive variables - the degree of environmental loading induced by the stabilization of the exergy budget of a society. That is, it can put the scale of the process of self- organization of human societies in relation to the scale of biophysical processes of self-organization occurring in the biosphere stabilizing favorable boundary conditions. This can be used to study the feasibility domain for the exergy budget according to an ecological perspective;

220

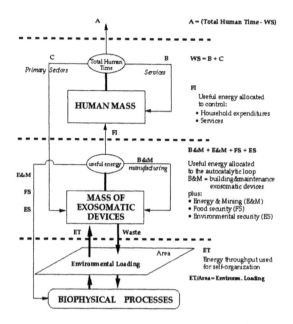

Figure 4.1:

4. addresses explicitly the hierarchical nature of human societies as self-organizing adaptive holarchies. In fact, it provides different sets of variables describing changes on different hierarchical levels and equations of congruence linking changes occurring on different levels.

A validation of the model and numerical examples are given in the next paper of this series.

Bibliography

[1] BROOKS, D. R., J. COLLIER, B. A. MAURER, J. D. H. SMITH and E. O. WILEY, *Entropy and information in evolving biological systems*, Biology and Philosophy 4 (1989), 407–432.

[2] BROOKS, D. R. and E. O. WILEY, *Evolution as Entropy*, Chicago: The University of Chicago Press (1988).

[3] GIAMPIETRO, M., *Using hierarchy theory to explore the concept of sustainable development*, Futures 26 (1994), 616–625.

[4] GIAMPIETRO, M., *The link between resources, technology and standard of living: a theoretical model*, Advances in Human Ecology Vol. 6 L. Freese (Editor), JAI Press Greenwich, CT (1997a), 73–128.

[5] GIAMPIETRO, M., *Energy budget and demographic changes in socioeconomic systems*, S. Dwyer, U. Ganslosser and M. O' Connor (Editors), Ecology Society Economy: Life Sciences Dimensions - Proceedings from the Inaugural Conference of the European Society for Ecological Economics - Filander Press - in press (1997b).

[6] GIAMPIETRO, M., Bukkens S.G.F., and D. PIMENTEL, *The link between resources, technology and standard of living: examples and applications*, Advances in Human Ecology Vol. 6 L. Freese(Editor), JAI Press Greenwich, CT (1997), 129–199.

[7] GIAMPIETRO, M., and D. PIMENTEL, *Energy analysis models to study the biophysical limits for human exploitation of natural processes.*, C. Rossi and E. Tiezzi (Editors), Ecological Physical Chemistry, Amsterdam: Elsevier Science Publishers B.V. (1991), 139–184.

[8] GIAMPIETRO, M., and K. MAYUMI, *Carrying Capacity and Optimum Population theoretical definitions and simulations*, Paper for the Fourth Biennial Meeting of The international Society for Ecological Economics "Designing Sustainability" August 4–7, Boston University, Boston, MA (1996).

[9] GIAMPIETRO, M., and K. MAYUMI, *A Dynamic Model of Socioeconomic Systems based on Hierarchy Theory and its application to Sustainability*, Structural Change and Economic Dynamics in press (1997).

[10] JØRGENSEN, S. E., *Integration of Ecosystem Theories: A Pattern.*, Dordrecht: Kluwer Academic Publishers (1992).

[11] LAYZER, D., *Growth of order in the universe.*, B.H. Weber, D.J. Depew and J.D. Smith, eds., Entropy, Information, and Evolution, Cambridge, MA: MIT Press (1988), 23–40.

[12] MOROWITZ, H.J., *Energy Flow in Biology.*, Woodbridge, CT: Ox Bow Press (1979).

[13] ODUM, H.T., *Systems Ecology.*, New York: John Wiley (1983).

[14] ODUM, H.T., *Environmental Accounting: Emergy and Decision Making.*, New York: John Wiley (1996).

[15] PASTORE, G., Giampietro, M., and K. MAYUMI, *Bio-Economic Pressure as indicator of material standard of living*, Paper for the Fourth Biennial Meeting of The international Society for Ecological Economics "Designing Sustainability" August 4–7, Boston University, Boston, MA (1996).

[16] PRIGOGINE, I., *From Being to Becoming.*, San Francisco: W.H. Freeman and company (1978).

[17] ROSEN, R., *Life Itself: A Comprehensive Inquiry into Nature, Origin and Fabrication of Life.*, New York: Columbia University Press (1991).

[18] REES, W.E., and M. WACKERNAGEL, *Ecological Footprints and Appropriated Carrying Capacity: Measuring the Natural Capital Requirements of the Human Economy.*, A. M. Jansson, M. Hammer, C. Folke and R. Costanza(Editors), Investing in Natural Capital, Washington D.C.: Island Press (1994), 362–390.

[19] SIMON, H. A., *The architecture of complexity.*, Proc. Amer. Philos. Soc. 106 (1962), 467–482.

[20] SCHNEIDER, E. D., *Life as a manifestation of the second law of thermodynamics.*, Mathl. Comput. Modelling 19 (1994), 25–48.

[21] ULANOWICZ, R.E., *Growth and Development: Ecosystem Phenomenology*, Springer-Verlag, New York (1986).

Chapter 23

Socioeconomic systems as nested dissipative adaptive systems (holarchies) and their dynamic energy budget: Validation of the approach

M. Giampietro
Istituto Nazionale della Nutrizione, Rome, Italy
K. Mayumi
The University of Tokushima, Tokushima Japan
G. Pastore
Istituto Nazionale della Nutrizione, Rome, Italy

The text of this paper is divided into 5 sections: the first section describes the procedure used to set up a database referring to 107 countries of the world comprising more than 90present 4 applications of the approach which confirm the 4 hypotheses presented in the previous paper: (1) the indicator of development (BEP) suggested according to theoretical consideration, which is obtained combining only biophysical variables for which assessments are available, correlates better than GNP with a set of more than 20 traditional indicators of development used by the World Bank; (2) it is possible to visualize a common trajectory of development for all countries comprised in the sample, when we describe their evolution in an appropriate state space. This seems to confirm that the need of congruence among flows of energy, human time, and money across levels imposes constraints on the path of development; (3) equations of congruence across levels can be used to establish a direct link between demographic variables, level of development, existing technology and the availability of natural resources. This makes possible to discuss the feasibility of different scenarios by checking the perspective about the effects generated by the very same change on different hierarchical levels; (4) our dataset seems to indicate that the so-called phenomenon of "demographic transition" can be studied, by adopting our approach, in terms of a shift from one metastable equilibrium of the dynamic societal energy budget to another. This finding is particularly relevant since, at the moment, the only two points of zero-growth equilibrium accessible for human societies are: (i) the short life span-poor material standard of living typical of pre-industrial societies and (ii) the high-waste-based typical of developed post-industrial societies. To increase the sustainability of our civilization we have to look for a different combination among the relevant variables.

1. Setting up the data base

The database started with 187 world countries, from which 55 counties with less then 2.000.000 inhabitants were excluded because of a too small size (this excluded 0.6% of the total world population). For 25 of the remaining 132 countries (countries formerly included in ex USSR, ex Yugoslavia, ex Czechoslovakia, plus South Africa, Libya, Algeria, Cambodia - which comprise 9% of the total world population) data are not available. Thus, our database is now based on 107 countries, comprising more than 90% of world population. The database has been created using official data of UN, FAO and World Bank statistics (more details later).

1.1. Assessments of parameters needed to calculate BEP

(1) the term "ABM × MF"

- ABM has been calculated by pondering the average weights (by age and sex classes) and the structure of population as reported by James & Schofield (1990) on the total population of 1992 as reported by the World Tables published for the World Bank (1995).

- MF has been computed separately for each sex and age class (of each country) following the database and protocols given by James and Schofield (1990) and merged into national averages.

(2) the term "Exo × Endo"

- The annual flow of exosomatic energy was evaluated according UN energy statistics (1995) for commercial and traditional biomass consumption in 1992 using a conversion factor of 29.3076 terajoules per thousand metric tons of coal. However a minimum value of 5/1 has been adopted for countries with a resulting value of exo/endo ¡5, since for rural communities official statistics tend to underestimate the contribution of animal power, biomass for cooking and building shelters (see Giampietro et al. 1993)

- The annual flow of endosomatic energy has been computed starting with the value of "ABM × MF" and the population size of 1992 as reported by the World Tables published for the World Bank (1995).

(3) the term "THT/C"

- the fraction of economically active population and the distribution of labor force in different sectors of economy derive from UN statistics and regard the latest available data in the period 1990-93.

- In this model, productive sectors of economy include: agriculture, hunting, forestry and fishing; mining, quarrying; manufacturing; electricity, gas, water; construction and a fraction of transport. Transport was in fact divided between productive sectors and service sector, proportionally, for each country, according to the working time spent in the productive sectors and the working time spent in service sectors (which include trade, restaurants, hotels; financing, insurance, real estate, business; community, social and personal services).

- work-load was estimated at a "lat" value of 1800 hours/year including vacations, absences and strikes.

1.2. Conventional indicators of material standard of living and socio-economic development.

Such a set of indicators was basically taken from World Tables. The 24 indicators used in the analysis can be divided in three groups:

(i) 8 indicators of nutritional status and physiological well being: (# 1) life expectancy; (# 2) energy intake as food; (# 3) fat intake; (# 4) protein intake; (# 5) average BMI adult; (# 6) prevalence of children malnutrition ($Wt/Ht < 2$ z-score of NCHS reference growth curve); (# 7) infant mortality; (# 8) % low birth weight.

(ii) 7 indicators of economic and technological development: (# 9) GNP per capita; (# 10) % GDP from agriculture; (# 11) ARL$_\$$ (Average Return of Labor in terms of added value = GNP/WS); (# 12) % of labor force in agriculture; (# 13) % of labor force in services; (# 14) Energy consumption per capita; (# 15) % of GDP expended for food.

(iii) 9 indicators of social development : (# 16) television/1000 people; (# 17) cars/1000 people; (# 18) Newspaper/1000 people; (# 19) Phones/100 people; (# 20) Population/physician ratio; (# 21) Population/hospital bed ratio; (# 22) Pupil/teacher ratio; (# 23) Illiteracy rate; (# 24) Access to safe water (% of population).

All data about (or used to calculate) these 24 indicators come from FAO (FAO Yearbook 1995); UN (Statistical Yearbook 1995) and World Bank (Social Indicators of Development 1995) and each one of them refers to the latest available year between 1991 and 1993. Data on prevalence of malnutrition in children come from ACC/SCN (1993).

2. BEP as an indicator of development for socioeconomic systems

The model of analysis of socio-economic systems presented in Paper 1 indicates that the process of development can be seen as a "decoupling" between: (i) the profile of allocation of investements of human time on different societal activities (CI and FI activities) and (ii) the profile of allocation of investements of exosomatic energy on the same set of activities. The more developed is the society the smaller is the fraction of total human time which is used to run the productive sector of the economy (food security, energy and mining, manufacturing), whereas the energy throughputs within these sectors is dramatically increased (using the encoding proposed in the first paper we can say that BEP tends to increase).

This trend can be explained in terms of balancing returns in the long- term and short term (adaptability versus efficiency). The higher is the rate of energy dissipation (the faster the consumption of natural resources), the more probable becomes that sooner or later the society will have to face changes in boundary conditions. That is, systems consuming more must invest more in developing adaptability.

We used our database to check whether or not BEP could be used as an indicator of development within our model of analysis.

Our data base shows that BEP is strongly correlated with all classic economic indicators of development considered - see Table 1 for correlation coefficients:

(i) for the indicators of group 1 (nutritional status and physiological well being) the average is r = 0.88 [ranging from 0.77 to 0.92]; Group 1 includes eight indicators reading of the socioeconomic system at the lower hierarchical level: (# 1) life expectancy; (# 2) energy intake as food; (# 3) fat intake; (# 4) protein intake; (# 5) average BMI adult; (# 6) prevalence of children malnutrition ($Wt/Ht < 2$ z-score of NCHS reference growth curve); (# 7) infant mortality; (# 8) % low birth weight.

(ii) for the indicators of group 2 (economic and technological development) the average is r = 0.78 [ranging from 0.65 to 0.87]; Group 2 includes seven indicators of referring to a reading of the economic system: (# 9) GNP per capita; (# 10) % GDP from agriculture; (# 11) ARL (Average Return of Labor in terms of added value = GNP/WS); (# 12) labor force in agriculture; (# 13) % of labor force in services; (# 14) Energy consumption per capita; (# 15) % of GDP expended for food.

(iii) for the indicators of group 3 (social development) the average is r = 0.76 [ranging from 0.44 to 0.89]; Group 3 includes nine indicators referring to the societal dimension of development: (# 16) television/1000 people; (# 17) cars/1000 people; (# 18) Newspaper/1000 people; (# 19) Phones/100 people; (# 20) Population/physician ratio; (# 21) Population/hospital bed ratio; (#

22) Pupil/teacher ratio; (# 23) Illiteracy rate; (# 24) Access to safe water (population).

According to these findings, BEP could subsitute GNP as an indicator of development according to a socio-economic perspective.

The novelty of this model of analysis is that BEP, in addition of being a good indicator of material standard of living according to the conventional economic perspective, does establish a direct link between different perspectives (readings) of the process of development referring to different hierarchical levels of analysis (different non-equivalent descriptions on different scales).

Recalling the definition of BEP given in the first paper:

$$BEP = (ABM \times MF) \times \frac{Exo}{Endo} \times \frac{THT}{C} \qquad (2.1)$$

The three terms on the right side of equation (1) reflect three different views of the material standard of living in society obtained by looking at this concept from three different hierarchical levels of analysis:

'ABM $\times$ MF' = physiological hierarchical level - endosomatic metabolism per capita (MJ/hour); the higher this value, the better are physiological conditions of humans living in the society. According to our data base, the feasibility domain of 'ABM $\times$ MF' is within a minimum of 0.33 (= short life expectancy at birth, small average body mass) and a maximum value of 0.43 (plateau reached in developed countries).

'Exo/Endo energy ratio' = socioeconomic hierarchical level; short-term efficiency - exosomatic metabolism per capita expressed as a multiple of (ABM $\times$ MF). According to our data base and previous studies on pre- industrial societies, the feasibility domain of 'Exo/Endo energy ratio' is within a minimum value of 5 (when exosomatic energy is basically in the form of traditional biomass fuels and animal power) and a maximum value of 90 (when exosomatic energy is basically in the form of machine power, electricity obtained by relying on fossil energy stocks). Exo/Endo is a good indicator of economic activity and it is strongly correlated to GNP p.c.. The higher the Exo/Endo, the more goods and services are produced and consumed per capita.

'THT/C' = socioeconomic hierarchical level; long-term adaptability - Total Human Time available in the society / Working Time allocated in productive sectors of the economy. According to our data base, the feasibility domain of 'THT/C' is within a minimum value of 10 (subsistence socioeconomic systems in which agriculture absorbs a large fraction of work force) and a maximum value of 45 (post-industrial societies with a large fraction of elderly and a work force mainly absorbed by services). This indicator reflects social implication of development (longer education, larger fraction of elderly, more leisure time for workers), since it assesses the allocation of human controls on "long-term returns" rather than on "short-term returns" (adaptability versus efficiency)..

'BEP' = is the combination of previous three indicators; it provides an overall assessment of these different aspects of the material standard of living assessed on three hierarchical levels (1. physiological/endosomatic autocatalytic loop for

228

individuals; 2. economic/exosomatic autocatalytic loop for the whole society; 3. level of social development/updating of the system of control regulating social behaviour). According to our data base, the feasibility domain of 'BEP' is within a minimum value of 18 MJ/hr, and a maximum value of 1500 MJ/hr. An increase in the value taken by this parameter can be seen as reflecting the ability of increasing the fraction of resources (computational capability and exergy dissipation) that a socio- economic system is actually investing in adaptability

3. Existence of an internal set of constraints on the evolutionary pattern of socio-economic systems

In the theoretical discussion given in the first paper of this series we hypothesized that the trajectories of development of different societies should show similarities, since the need of congruence of flows of energy and human time across hierarchical levels imposes constraints on the shape of possible paths. Put it in another way, the different parameters considered in our model, even if referring to non-equivalent descriptions and assessed on different hierarchical levels cannot take whatever value since they are affected, when aggregated, by forcing functions (due to the nested hierarchical nature of the system).

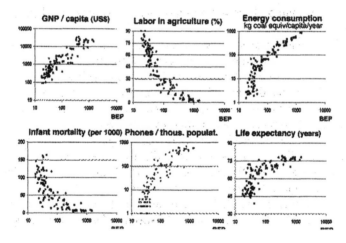

Figure 3.1: Comparison between *BEP* energy ratio and major indicators of standard of living

Actually, a simple look at Fig. 3.1 clearly indicate that when the path of development respect of each one of the selected indicator is graphed against BEP we can see that the 107 countries of our sample are clustered around a given trajectory rather than being scattered over the plane - for reason of space we are showing here only 6 graphs out of the 24 available, but the picture does not change in the others - (more graphs are available on the literature in reference).

Even more striking is the analysis of the same trajectory if we put on the x-axis one of the 3 factors determining BEP at the time. An example of this is presented in Fig. 3.2, where the values taken by 6 indicators of development are graphed against the value of "exo/endo". In this case, (due to the linear scale) it is quite easy to identify a threshold value for "exo/endo" - which is about 25/1 - above which the trajectory of development seems to reach a plateau. A similar observation could be seen by using the other two factors making up BEP: (i) "ABM × MF" (the pleateau is reached around 9 MJ/day or 0.4 MJ/hour), and (ii) "THT/C" (the threshold value is 30/1). As discussed later, we believe that the existence and the convergence over these threshold values of all the countries considered in the sample can be put in relation to a shift from one metastable point of equilibrium of the dynamic energy budget to another.

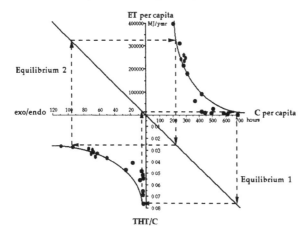

Figure 3.2:

4. Establishing links across levels to check the feasibility of future scenarios

The approach of analysis of socioeconomic systems interacting in biophysical terms with their environment can be applied to the discussion of the feasibility of future scenarios (tutoring the envisioning of more sustainable societies). In fact, as soon as we define for a given socio- economic system a set of characteristics describing its demographic structure and material standard of living (following the list of parameters described in paper 1) we can calculate: (1) the technical coefficients that would be required in specific economic sectors (e.g. agricultural sector and energy sector) to match the SEH demand which would be generated by such an "envisioned society" and (2) the correspondent environmental loading (or in alternative what type of technological achievement would be required to keep the environmental loading of such a society below a critical value). We want to briefly touch upon three applications of this type of analysis (for more details published papers are available):

1. it is actually possible to establish a direct link between technological changes occurring at the farm level, macroeconomic changes ocurring at the level of society and levels of environmental loadings occuring in the agro- ecosystems. This can be usefully applied to the discussion of scenarios of technical development of agriculture (Giampietro, 1997a; 1997b). In this way it is possible to show that techniques of agricultural production are not only affected by a growing demographic pressure (= the need of increasing the throughput per hectare due to the shrinking of available land per capita) but also by a continuously increasing socio-economic pressure (= the need of increasing the productivity of labor of farmers and the economic return on the economic investement) - Giampietro, 1997a. Also this hypothesis has been confirmed by a statistical analysis over a database comprising 73 countries (Conforti and Giampietro, 1997). Actually, socio-economic pressure can result more crucial than demographic pressure in the abandonement of ecologically friendly techniques of production (Giampietro, 1997b).

2. it is possible to analyze the performance of farming systems using different set of indicators reflecting different perceptions of "improvements" on different hierarchical levels, and establish a link between effects generated by changes within the hierarchical systems on its various levels. This applications has been developed in a 4-year field project in China to study the mechanism of intensification of rural areas in relation to the issue of sustainability (Giampietro et al. 1998; Pastore et al. 1998). The application of the approach, illustrated in Fig. 4.1, uses a radar-type graph on which different sets of indicators are divided into four quadrants. Each quadrant contains a set of indicators related to the perspective of a particular holon considered (e.g. farmer household, national economy, agroecosystem). On each axis there is a "viability domain" (the range of values within which the holon can be considered stable). Equations of congruence are used to establish a link between changes in the values of indicators within a quadrant and the induced changes in the values of indicators within another quadrant. For example, lowering the price of rice by imposing a "fixed price" policy will result into a worsening of the situation for the "holon" farmers (reduction of income) and an improvement for the "holon" government of China (reduction of cost of food for the cities). According to the distance of the indicators from the edges of the viability domain (minimum acceptable values) we can evaluate the implications of the correspondent trade-off. The mixing of socio-economic and biophysical readings makes also possible to discuss trade-offs between economic and ecological considerations (e.g. an increase of the yield per hectare obtained by increasing the nutrient input - a change that results to be good for the holons: "farmers" and "China government" - implies, within the quadrant of ecosystem perspective, an increase in the value of environmental loading - a change perceived as "bad" by the set of indicators selected for the holon "agro-ecosystem").

3. it is possible to discuss whether or not large-scale biofuel production

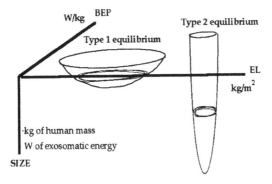

Figure 4.1:

is a feasible option to fuel developed society currently heavily dependent on fossil energy stocks (Giampietro et al. 1997a). An energy sector running on biofuels must be compatible with the socioeconomic characteristics of society: it must provide a SEH of the same order of magnitude of the existing BEP, and be compatible with ecological constraints: it must have a demand of natural resources (e.g. arable land and fresh water) compatible with current supply. Data on modern biofuel systems found in the literature can be used to estimate the biophysical requirements per unit of net energy supply to society. Depending on the production system, these requirements per gigajoule (1 GJ = 109 joules) of net energy are: 0.015-0.100 ha of arable land, 200-400 ton of fresh water, and 0.6-5.5 hours of labor (Giampietro et al. 1997a). Because in developed society (BEP > 400 MJ/hour) work supply in the energy sector is only a small fraction of the work force in the productive sectors (in general less than 5%) it is evident that in order to achieve such a result the throughput per hour of labor in the energy sector must be in the order of 10,000 MJ/hour of labor.

For example, in Italy, with a population of 57 million, only 7.3% of the total of 499 billion hours of human time available in a year were spent in paid work in 1991. Of this yearly labor supply, 60% was absorbed by the service sector, 30% by the industrial sector, and 9% by agriculture, fishery and forestry, leaving a tiny 1% or 360 million labor hours to run the entire energy sector (ISTAT, 1992). Total energy consumption in Italy that year was 6,500,000 terajoules, implying that in 1991 the Italian energy sector delivered almost 18,000 MJ of energy throughput per hour of labor in that sector. This throughput was achieved while using mainly fossil energy (about 90%). Thus, the characteristics of a developed society require that the energy throughput per hour of labor in the energy sector is in the range of 10,000 to 20,000 MJ/hr. These levels are well beyond the range of values achievable with biofuel, that is 250-1,600 MJ/hr (Giampietro et al. 1997b). Energy resources - as biofuels - operating with a much smaller throughput per hour of labor would require, in a developed society with very high levels of energy consumption per capita an energy sector absorbing between 20 and 40% of the labor force of society. Such a scenario would be completely incompatible with the current profile of labor allocation to the various economic sectors.

5. The demographic transition as a shift between two metastable equilibrium points of the dynamic energy budget

The value of BEP - which is referring to collective properties of socioeconomic systems - is the result of the product of three terms (two of which are affecting or are affected by demographic changes). Also the value of SEH depends on two parameters linked to technological development and availability of natural resources, which are both scale-dependent and, therefore, affected by population size - (more details in the first paper and in reference). Therefore, the constraint of balance BEP = SEH (that apparently involves only intensive variables) is in reality affected by demographic changes. In order to focus on this fact it is necessary to add extensive variables to such a constraint. That is we can impose the congruence of total ET supply as a function of SEH and population size as well as total ET demand as a function of BEP and population size. Some useful relations are (using the same variables used in the first paper of the series):

$$\frac{THT}{C} \times \frac{C}{CI} \times CI = pop \times \text{hy} \tag{5.1}$$

$$\text{ET supply} = SEH \times C = \frac{ET}{CI} \times \frac{CI}{C} \times C \tag{5.2}$$

$$\text{ET demand} = BEP \times C = BEP \times \frac{C}{THT} \times pop \times \text{hy} \tag{5.3}$$

ET at a particular point i in time can be expressed as:

$$ET_i = \left(\frac{Exo}{Endo}\right)_i \times pop_i \times (ABM \times MF)_i \times \text{hy} \tag{5.4}$$

When SEH > BEP (supply > demand) socioeconomic systems will adjust their parameters so as to increase the ability to use surplus of useful energy and therefore to expand the scale of their activity (assessed by a larger ET). At a point in time $i + 1$, we will then have:

$$ET_{i+1} > ET_i \tag{5.5}$$

which can be formulated as:

$$\left(\frac{Exo}{Endo}\right)_{i+1} \times pop_{i+1} \times (ABM \times MF)_{i+1}$$
$$> \left(\frac{Exo}{Endo}\right)_i \times pop_i \times (ABM \times MF)_i \tag{5.6}$$

An expansion of the process of self-organization of a socioeconomic system means an increase in its energy dissipation ($ET_{i+1} > ET_i$) which can be obtained through either (i) an increase in the metabolism of individual human beings ($ABM \times MF$), (ii) an increase in population size (pop) and/or (iii) an expansion of exosomatic devices operating in the socioeconomic system (Exo/Endo).

Relation (5.6) can also be written as:

$$(\frac{ET}{C})_{i+1} \times (\frac{C}{THT})_{i+1} \times pop_{i+1}$$
$$> (\frac{ET}{C})_i \times (\frac{C}{THT})_i \times pop_i \tag{5.7}$$

Relation (5.7) gives a different view on the possible ways of expansion of ET. When the surplus is absorbed by an increase in the Exo/Endo ratio rather than an increase in population, the solution implies an increase in ET/C (which is BEP) linked to a reduction of C/THT (the fraction of working time in productive sectors, which is also an indicator of development). Put it in another way, depending on the path of expansion followed by the system, increases in ET can be transformed either into (i) increased population size or (ii) improved material standard of living (a combined change of ET/C and C/THT). Changes in ET/C and in THT/C are subject to: (1) lag time in building up the technical infrastructures needed to expand exo/endo (jobs positions available in different sectors of the economy), and (2) speed of demographic changes determining the dependency ratio within the existing households; (3) cultural constraints to change the profile of allocation of human time on different sets of activities (moving to a different job or to a different housing type). These three factors create a certain 'intrinsic' lag- time in the adjustment of the values of ET/C and C/THT to new conditions. The difference in lag-time between ET/C and THT/C will determine how much of the increase in ET will end up with population growth rather than in improvement in BEP.

The path of expansion of ET

Starting from equation (8) we can study possible paths of expansion in ET by examining relative changes among these three parameters. An increase in size of the system (ET) can imply changes in (i) ET/C; (ii) ET per capita (Exo/Endo); and (iii) C/THT. The population term in equation (8) becomes stable when the increase in ET/C on the supply side is matched by a combination of changes in C/THT and Exo/Endo on the supply side. Note that due to the limited range of possible values of the term (ABM × MF) (the maximum increase amounts to about 20in ET per capita almost directly translates into an increase of the Exo/Endo ratio. Let us imagine a process of expansion of ET starting from a hypothetical socioeconomic system in phase 1 of the demographic transition (stable population due to high fertility and mortality). Assume that the parameter population is subject to fluctuations (in accordance with reality) and that the initial population size is small. Assume also that improvements in the exosomatic autocatalytic loop (e.g., new technologies, use of higher- quality resources, better knowledge and management of socioeconomic activities) make available to the socioeconomic system a surplus of energy (SEH > BEP). The surplus can be absorbed by an increase in population and/or by an increase in energy consumption per capita both of which will be reflected in a changing pattern of human activity (changes in THT/C). In the short term, population growth generates an increase in THT/C because of the increasing number of children in the system. On a longer time scale, the increase in size of the socioeconomic system due to population growth causes a further increase in THT/C because of an increasing ratio $B/(B + C)$. Indeed, with the expanding size of the system, the fraction of total work supply allocated to administration and other services gradually increases. However, in pre-industrial societies the possibility to increase the ratio THT/C is limited due to lack of devices that can amplify the power of workers in the productive sectors (CI/C in pre-industrial societies is small). History shows that animal, wind and water power were decisive factors in determining local spots of development in pre-industrial societies. Nevertheless, the very nature of these 'power devices'–they are location- specific and do not provide a continuous (reliable) flow of power in time– prevented pre-industrial socioeconomic systems to reach a type-2 equilibrium on a large scale. The only solution to have large-scale complex societies was the division of the population in widely different social classes and a strong taxation of farmers by the central administration. However, this could not avoid a strong instability of these systems (for a more detailed analysis see Giampietro et al. 1997b based on Tainter, 1988).

In industrial times the picture is totally different and population growth no longer implies a reduction in SEH because: 1. The supply of energy input is no longer related to availability of land; 2. the surplus of ET can be easily absorbed by an increase in the ratio exo/endo.

Using the encoding proposed earlier, it is possible to illustrate the establishment of two possible equilibrium points as shown in Fig. 5.1. All variables indicated in this figures are intensive to avoid a complicated three- dimensional representation. However, an extensive variable, such as population size, can be easily introduced on a third axis (perpendicular to the plane of the page) on which changes in the size of the system (ET or population) are indicated. This would make clear that type-2 equilibrium is achieved only at a much larger size of the system (higher on the third axis) than type-1 equilibrium. Data reported in Fig. 5.1 refer to a synchronic analysis of 107 countries - white dots - grouped in clusters (based on the database presented before), and a diachronic analysis of average values for OECD countries over 20 years (1970-1990) - black dots (based on a database similar to the one presented before but calculated over historic series of OECD countries - Pastore et al. 1996). It is worthwhile to note here that values coming from both the horizontal and longitudinal analysis lie on the same curve.

This can be taken as an indication that the reciprocal relation among set of variables generates two basins of attraction for the energy budget of a socioeconomic system. Indeed, the trajectory of development of socioeconomic systems

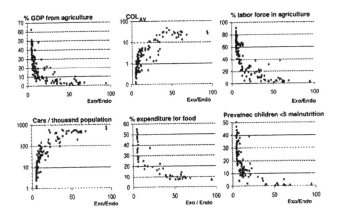

Figure 5.1: Comparison between $Exo/Endo$ energy ratio and major indicators of standard of living

can be described in a three-dimensional phase space: Axis 1: energy dissipation per unit of human control (ET per capita), Axis 2: size of the system (population size), Axis 3: environmental loading, an indication of the intensity of the activity of self-organization of the socioeconomic system relative to that of the natural processes that guarantee the stability of boundary conditions. As noted earlier this ratio affects ET/C, and therefore THT/C, and is affected by the population size. The type-1 and type-2 equilibria (presented in Fig. 5.1) can be imagined in this three-dimensional phase space as in Fig. 5.2.

Reconsidering the classic representation of demographic transition The traditional graphic representation of the demographic transition is a sigmoid curve that connects two values of stabilized population sizes. The horizontal axis is generally represented as an axis measuring time. However, available data for societies that have completed the demographic transition (e.g., France and Sweden) and societies that are in different stages of this process (e.g., Burundi and Singapore) suggest that the timing (speed) of the transition process is dramatically different for these countries (Chesnais, 1992). This would imply that the variable "time" in itself is not directly linked to the changes taking place during the transition from one metastable equilibrium to another.

Indeed, the analysis presented in this paper suggests that the "Exo/Endo ratio" may be a better variable for the horizontal axis. With this choice we get the following explanation for the process. At the beginning of the transition (stage 1), socioeconomic systems are not able to expand their size in terms of larger values of ET. They can only replicate themselves through "seedlings," that is through colonization of other ecosystems. Growth in this case generates redundancy: more of the same thing (replication), but no development (no qualitative

236

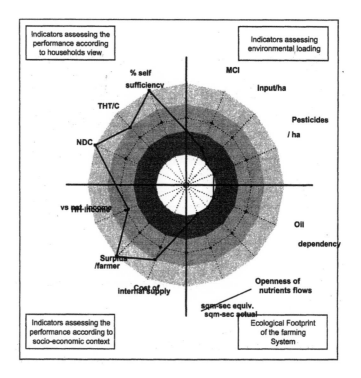

changes).

When socioeconomic systems become able to expand their autocatalytic loop of exosomatic energy they enter into a transitional phase in which they are able to keep SEH > BEP. In this phase they can expand their size (ET) by increasing both their population and their Exo/Endo ratio. These two forms of expansion have different intrinsic lag-times and therefore the two processes proceed at different speeds.

When changes in the structure of the exosomatic autocatalytic loop (industrialization) enable the socioeconomic system to absorb the entire surplus of ET by increasing the Exo/Endo ratio while maintaining a fixed population size, the socioeconomic system has completed the demographic transition. This final stage is coupled to an increase in the material standard of living and in changes of the profile of human time allocation over different activities (increases in THT/C). This transformation is linked to a dramatic change in social patterns of organization.

As already discussed in §2 when the 24 indicators of development present there are plotted against the Exo/Endo ratio for different countries (Fig. 3.2) we can actually see a "picture" of the trajectory that countries are following in the transition between the two metastable equilibria of the dynamic energy budget. In this context, the description of the demographic transition in terms of indicators of fertility and mortality (done by demographers) should be seen as just one of the possible descriptions of such a phenomenon. In the same way, the energetic description of the process provided in this paper is just another of the possible descriptions. Any type of description simply depends on the window of observation chosen by the scientist performing the analysis.

Bibliography

[1] ACC/SCN., *Second report on the world nutritional situation*. FAO Rome (1993).

[2] CHESNAIS, J.C., *The Demographic Transition.*, Clarendon Press, New York (1992).

[3] CONFORTI, P., and M. GIAMPIETRO, *Fossil energy use in agriculture: an international comparison.*, Agriculture, Ecosystems and Environment. 65 (1997), 231-243.

[4] FAO., *Production Yearbook*, 1994. FAO Statistic Series No. 125. Rome: FAO (1995).

[5] GIAMPIETRO, M., *Socioeconomic pressure, demographic pressure, environmental loading and technological changes in agriculture.*, Agriculture, Ecosystems and Environment. 65 (1997a), 201-229.

[6] GIAMPIETRO, M., *Socioeconomic constraints to farming with biodiversity.*, Agriculture, Ecosystems and Environment 62 (1997b), 145-167.

[7] GIAMPIETRO, M., Bukkens S.G.F., and D. PIMENTEL, *Labor productivity: a biophysical definition and assessment*, Human Ecology 21 (3) (1993), 229-260.

[8] GIAMPIETRO, M., Bukkens S.G.F., and D. PIMENTEL, *The link between resources, technology and standard of living: examples and applications*, Advances in Human Ecology Vol. 6 L. Freese(Editor), JAI Press Greenwich, CT (1997), 129–199.

[9] GIAMPIETRO, M., G. PASTORE, 1998 (in press) *A model of analysis to study the dynamics of rural intensification in China*, Paoletti et al (editors) Special Issue of Critical Reviews in Plant Sciences, CRC Press Boca Raton, Fl in press (1998).

[10] GIAMPIETRO, M., S. ULGIATI, and D. PIMENTEL, *Feasibility of large-scale biofuel production: Does an enlargement of scale change the picture?*, Bio-Science 47 (9) (1997a), 587-600.

[11] ISTAT., *Annuario Statistico Italiano.*, Rome. Istituto Centrale di Statistica (1992).

[12] JAMES, W.P.T., E.C. SCHOFIELD, *Human energy requirement*, Oxford: Oxford University press (1990).

[13] PASTORE, G., M. GIAMPIETRO, and Ji LI, *Understanding the dynamics of rural intensification in China: Land-time crossed budget of five villages in Hubei province*, Paoletti et al (editors) Special Issue of Critical Reviews in Plant Sciences - CRC Press Boca Raton, Fl in press (1998).

[14] PASTORE, G., M. GIAMPIETRO, K. MAYUMI, *Bio-Economic Pressure as indicator of material standard of living*, Paper presented at the Fourth Biennial Meeting of the International Society for Ecological Economics: Designing Sustainability. Boston University. August 4-7, 1996.

[15] TAINTER, J.A, *The Collapse of Complex Societies*, Cambridge University Press, Cambridge, UK (1988).

[16] United Nations, *Statistical Yearbook. 1993*, New York: U.N.Department for Economic and Social Information and Policy Analysis; Statistical Division (1995).

[17] World Bank, *Social Indicators of Development 1995*, The Johns Hopkins University press. Baltimore. U.S.A (1995).

[18] World Bank, *World Tables 1995*, The Johns Hopkins University press. Baltimore. U.S.A (1995).

Chapter 24

Psychology and corporations: A complex systems perspective

Jeffrey Goldstein
Adelphi University

Industrial/Organizational Psychology is now adding to its repertoire methods and constructs from the study of complex systems. Yet, complex systems theory can also benefit from the study of businesses and institutions because the latter abound in connectivities and scaling phenomena, are easily accessible and observable, and have established metrics of effectiveness. Research findings from Industrial/Organizational regarding group cohesion and conformity as well as emergent leadership are shown to have relevance to the study of complex systems.

1. Introduction

Industrial/Organizational Psychology conducts research with the aim of improving workplace productivity, quality, and job satisfaction. Industrial/Organizational is now being influenced by many new methods and constructs from complex systems theory [6, 7, 20, 24]. Yet, there can also be a reverse direction of influence: the study of organizational systems can aid complex systems researchers by providing real life laboratories of complex systems in action. Actually, this direction of influence is not all that new since in the nineteenth century physics and chemistry appropriated statistical tools for dealing with aggregates that were developed in the social sciences [5, 8]. Organizations can be a particularly apt place to ferret out various dynamics of complex systems because of the following reasons: businesses and institutions consist of networks of connectivities; they abound in scaling phenomena; they are extremely accessible in being the sites of our work as well as personal lives; they are easily observable not requiring any special type of observational technology; they easily track what is predicable and what is unpredictable (in fact, a great deal of a corporation's or institution's activity is devoted to such trackings); they have established metrics as criteria for successful performance such as profit, market share (and a host of other financial measures), quality of service or product, employee job satisfaction levels, employee stress levels, absentee rates, turnover rates, and so on; and they are rife with self-organizing, emergent phenomena (no matter how much official bureaucracies dislike this fact). In this paper findings from Industrial/Organizational research will be presented that can provide suggestions for complex system modellers.

2. Organizations as currently organized

Before we get to these findings, however, we list several reasonable postulates about the functioning of typical organizations:

1A. When organizations succeed, it is mostly in spite of, not because of the way they are organized.

1B. When organizations succeed, it is mostly in spite of, not because of the way leadership is exercised.

1C. The manner in which most organizational working units are organized, set-up and managed serves more to stifle than encourage the creativity and productivity of its members.

There seem to be two possibilities for understanding 1A, 1B, and 1C. First, organizational success may be a function of how effectively officially sanctioned rules of functioning are transgressed. Or, it may be the case that it is precisely the friction between the transgressions and the organizational orthodoxy that is the necessary ingredient for success [20].

2. Organizations are nonlinear, nonequilibrium, complex, interacting, self-organizing, adaptive systems but are managed as linear, equilibrium-seeking, simple, isolating, "other"-organized, nonadaptive systems.

Indeed, most managers seem to be experts at creating and maintaining policies, structures, procedures, and practices that dampen the complexity, self-organizing capability, and adaptability of organizations. For example, one of the four traditional management functions – controlling – refers to activities whose objective is to diminish the effect of unanticipated variances. But while such an objective seems crucial for organizational functioning, it needs to be pointed out that it also serves to diminish the degree of variance that may be exactly what is required for the adaptability of an organization to an unpredictable environment.

3. Adaptive organizational change is more the consequence of the serendipitous taking advantage of unplanned events than careful planning and implementation of plans.

The point here is not that planning is useless, rather, it is, instead, that, as traditionally conceived, corporate planning is based on a view of organizations as linear, predictable, and simple systems, whereas studies of successful corporate transformations (similar to many scientific discoveries) result from an attitude summed up by Pasteur as "Chance favors the prepared mind" [6].

4. For too many people, going to work everyday is not dissimilar to lock-up in a prison.

This unfortunate state of affairs is, in large part, the result of the bureaucratization of organizations. A century ago the German sociologist Max Weber [26] studied the process of bureaucratization taking place in government, industry, and the military (e.g., what Frederick the Great of Prussia did with the military) and identified the following key principles of bureaucracy: rational-legal authority for a hierarchical chain of command where each lower office is under the control and supervision of a higher office.; clearly specified task and decision-making responsibilities; reliance on written rules, standard operating procedures, and norms to control the behavior and the relationship between roles in an organization. Indeed, this manner of organizing businesses and institutions was only buttressed by the later emergence of so-called Scientific Management advocated by Frederick Taylor who proffered job specialization, the manager, not the worker, as the expert of a job, that there was only one best way to efficiently perform a task, and the pervasive use of time and motion studies as a guide to management (By-the-way, Taylor himself seems to have suffered from a bad case of obsessive compulsive disorder, so that in an important sense, the principles of modern management were founded on pathology [14]). We see in these foundations of modern management an attempt to simplify the complex, linearize the nonlinear, and predict the unpredictable.

3. Using organizations to study complex systems

3.1. Cohesion and conformity in work group dynamics

A major line of inquiry into the social-psychological dynamics of work groups concerns how group composition, structure, and process effects team performance. This research examinese such factors as homogeneity versus heterogeneity, the influence of minority sectors, the mechanics of group decision-making, and group cohesiveness. For example, Industrial/Organizational research suggests that as "team" cohesiveness increases, so does the tendency for social conformity. Sinclair [19] can thus point to the "tyranny of team ideology," which although camouflaged by overt consultation and cohesion, often leads to what Irving Janis called "group think" or a "deterioriation of mental efficiency, reality testing, and moral judgement that results from in-group pressure" [11]. Certainly a high degree of group cohesiveness may expedite productivity by increasing both motivation and coordination among members [3], yet it has also been shown that group cohesiveness engenders a sense of in-group favoritism that can distort cognitive processing among group members [22]. For example, in-group favoritism has been shown to skew statistical reasoning, attributions of causality, and various subtle cognitive processes that underlie judgements made about in-group versus out-group members.

We can call this tendency toward social conformity the "Borg Effect" after those alien villains, the dreaded "Borg," in the Star Trek Voyager television series. As a collective organism, the "Borg" aim at "assimilating" all individual autonomy into a collective group mind. Unfortunately, besides finding this "Borg Effect" in our businesses and institutions such a connotation of totalizing cohesiveness can also be found in certain descriptions of the emergent coherence demonstrated in complex systems research. Hermann Haken [9], for example, has described the coherence of self-organized patterns as "enslavement," and Stuart Kauffman's [12] explanation of coherent structures in his boolean networks resorts to "canalyzing" boolean rules that propagate conformity in the form of redundancy. Indeed, much of the theorizing in complex systems discusses coherent structures in terms of being "trapped" within attractors or order parameters. Such a way of conceiving emergent structures leaves little room for the potentially constructive nature of group diversity found in Industrial/Organizational research.

Complex systems theory needs to take heed of such studies – for example, creativity in terms of divergent thinking and originality increases when a marked minority influence is allowed in a work group [24]. Contrary to expectations, conflict in these groups did not increase although there was an increase in personal stress on the part of the minorities. Furthermore, Jackson, et.al [10] found that group heterogeneity in terms of a mix of personalities, gender, attitudes, and background of experience was positively related to the creativity and the decision-making effectiveness of teams. And, Bantel and Jackson [1] found that innovation in the banking industry was positively associated with a diversity in the experience of top management teams. Moreover, research has shown that choosing members to be part of a team solely on the basis of a match between each member's abilities and knowledge and the requirements of the task is not enough to ensure optimal team effectiveness [22]. Indeed, this demonstrates that in organizations, because they are complex, nonlinear sysems, the whole is decidedly greater than the sum of the parts.

Furthermore, research on the effectiveness of techniques in group idea generation and decision-making suggest that tendencies to group conformity can interfere with the creative and quality of decision-making processes. Processes arising in such groups tend to inhibit decision-making performance. Thus, creative idea generation is highest when individuals are first placed in a brainstorming session and then allowed to independently work at generating ideas [21]. To be sure, there is a pay-off between, on the one hand, having individuals generate ideas on their own but thereby warding-off a tendency toward conformity, and, on the other hand, having them work together which may serve to increase consensus and thereby acceptance and acting-upon the group's decision. Thus, Stumpf, Zand, and Freedman [21] have proposed a matrix for structuring groups depending on the kind of output needed: e.g., does a decision need to be of high quality but not accepted, or broad-based and accepted or original and not accepted and so on.

These findings about group process suggest that complex systems modellers need to take into consideration how coherent structures in complex systems have both an up and down side for creative system adaptability. Research in Industrial/Organizational suggests that the way coordinated emergent structures may "enslave" system behavior can reduce the creative efficacy of such structures. Therefore, organizations may provide an ideal site for investigating specifically what kinds of connectivities, rules, and process variables lead to a higher degree of adaptability and so on.

4. Leadership as an emergent phenomenon

Traditionally, leadership has been considered an imposed, bureaucratic, hierarchical chain of command enabling a large number of persons working together to accomplish their work goals. The presence of leadership phenomena outside of this officially sanctioned edifice has been acknowledged as either "informal" leadership with little real impact on the organization or as some kind of "grass roots" activity found, e.g., in the "charismatic" type of leadership often seen in social reform or religious movements. Pillai [17] for instance found that "charismatic" leadership tends to emerge in leaderless groups during crisis situations (and the ratings of these charismatic leaders were higher in leadership ability than leaders emerging during noncrisis situations). But it need not take a crisis for leadership to emerge and play a crucial role in an organization's success. Indeed, studies indicate that the nature of leadership effectiveness may be more the result of emergent rather than imposed processes.

An example of emergent leadership is found in Murnigan's and Conlon's [15] study of professional British string quartets. Composed of a first violinist, second violinist, viola player, and cellist, a string quartet's playing is an intensively reciprocal interdependent process where there is a concomitant need to both listen accutely and respond accordingly to each other. In this way, the string quartet qualifies as a highly nonlinear, complex, and self-organizing system. The music played by string quartets is usually part of a standard repertoire, therefore, the members try to stamp each performance with their own unique style and interpretation in terms of speed, emphasis, rhythm, balance and phrasing. In fact, an effective quartet actually plays each piece differently each time to surprise the audience with new interpretations.

Murrigan and Conlon looked at leadership patterns emerging in successful versus not-so-successful quartets as measured by: the number of concerts booked plus the reputation of the concert house; the number of cd's produced as well as sales of cd's; and the quantity and quality of reviews. Murnigan and Conlon identified three paradoxes of leadership/group faced by string quartets which generate emergent leadership patterns that can be lead to greater or lesser success:

1. The first violinist as leader versus the democratic values of the group: the first violinist is the acknowledged musical leader, yet the members join because

they will theoretically have input in musical and business decisions as well as sharing equally in concert income. Moreover, the first violinists also are typically spokespersons for the quartet.

2. The role of the second violinist in relation to the first violinist: this leadership paradox arises since the second violin may actually be the best player but the first needs the best musical "sense" The "second fiddle" accordingly receives less attention and recognition, is commonly thought to have less to do, and is often given more business responsibilities than other group members.

3. Conflict management through confrontation versus compromise: whereas conflict can disrupt, distract, and even injures the performance of a quartet, yet it may be necessary for change, creativity, and individual freedom. The leadership challenge is how to encourage different, even conflicting, views and input while creating some kind of shared framing of issues large enough to encompass those differences.

Murnigan and Conlon found that in successful quartets there was an emergent leadership pattern that enabled them to understand and implicitly manage their leadership/group paradoxes. Thus, in terms of the first paradox above, the top quartets acknowledged both sides of the paradox– between the first violinist's leadership role and the democratic ideals of the group. Whereas, in the less successful quartets the paradox was "resolved" by either having the first violinist taking control without the other's consent, or by insisting on participative democracy in all decisions. Paradoxically, the first violinist in the successful quartets did not need to advertise their leadership and they consistently and strongly advocated democracy as the ruling principle. Indeed this is an emergent pattern of leadership, a creative kind of leadership within democracy that transcends the traditional dichotomies between authoritarian leadership and distributed democratic-based decision-making. Moreover, it doesn't appear to be a synthesis but an actual emergent process [4].

Regarding the second paradox, in most successful quartets the first violinists attributed their leadership position to their personality and less importantly to their ability. The second violinists were content or resigned to their position although they could voice a desire to be first but did not seem to feel resentment about it; indeed, other membes were quite complimentary of the second fiddle. However, in the less successful quartets, there was more tension between the first and second violinists, the latter often feeling they were not being appreciated. A key question is what kind of "rules" of leadership and group interaction are more likely to lead to the successful resolution of this second paradox. Is it purely a result of "personality" congruence or clashing or does it have more to do with the implict "rules" governing the group interaction?

Finally, in respect to the third paradox – confrontation versus compromise – in the successful quartets conflicts about how to play a piece were often resolved by agreeing to play it one way in one concert and another way in another concert. However, this alternate playing seldom needed to take place because the alternative interpretations were somehow incorporated into the first play. As one members stated, "When you play, what is right and what is wrong emerges."

This emergent resolution of the paradox can be seen as a kind of adaptation in a complex system. In the more successful quartets members didn't concede when they had opposing views, but they played more than talked about their conflicts: "either we play or we fight." But in the less successful quartets conflict was either avoided or heightened: they talked more than played and thereby didn't practice enough or they avoided conflict altogether inviting frustration, simmering resentments, or passive-aggressive behavior. Even the supposed capacity of group members'committment to superordinate goals as a means of managing conflict was not enough since superordinate goals neither specify particular items over which members might disagree, nor do they constrain different means for implementing the group members' goals.

Another finding from Industrial/Organizational concerning leadership can be tied to what determined effective leadership in these string quartets. In spite of the pervasive notion that successful leadership is primarily a function of "vision" or the leader's articulation of the goals of the organization, Randall Peterson [16] found instead that a leader's directiveness of task group process rather than goal setting was a more potent predictor of the quality of both group process and outcome, whereas outcome directiveness was associated with a much smaller and less coherent array of group outcomes. Indeed, this may be an emergent effect as leader and followers interact in an ongoing process whereas the "vision" is achieved only at the end of the process. A key issue is to identify the "rules" of leader/follower interactions that lead to successful work outcomes. In this manner, studying the role of hiearchies in organizations can provide insight into the role of hiearchies in complex systems in general.

5. Conclusion: The need for a sufficiently rich complex systems perspective

In 1883, the American economist and sociologist Henry Carey espoused a theory of social dynamics based on Newton's theory of gravitation,

Man tends of necessity to gravitate toward his fellow-man...the greater the number [of men] collected in a given space the greater is the attractive force exerted...Gravitation is here, as everywhere else in the material world, in the direct ratio of the mass, and in the inverse of the distance [2].

The historian of science I. Bernard Cohen [2] identifies Carey's gravitational theory of sociology as an example of a "mismatched homology" between one domain of science and another, in part because Carey left out Newton's little "square of the distance." From our present-day perspective, of course, Carey's ideas may seem idiotic, yet we do talk about people tending to "gravitate" into groups although noone seriously today entertains Carey's gravitational theory of social dynamics. But, could it have been necessarily recognized during Carey's time whether or not gravitation was going to stick as a good theory of sociology? It couldn't have been that gravity dealt with physical systems whereas sociology is about the human dimension since in the nineteenth century, both August Comte and and Vilfredo Pareto applied the laws of mechanics to social systems

[5] with great and lasting success. But, what is there about complex systems theory that gives us the confidence it will prove to have greater worth and staying power than Carey's gravitation?

The point here is not a criticism of the application of complexity science to social dynamics. Rather, it is to call attention to what was probably the major problem with Carey's gravity theory: it was far too simple a theory to account for the complexity of social processes and structures. To be sure, complex systems theory is founded on notions of complexity, nonlinearity, and nonequilibrium but it also is subject to the temptation of a too simplistic model such as gravitation offered Carey. For example, the kind of research in N/K models [12] puts a great deal of the onus for emergent phenomena on rather simple constructs as the number of inputs in a connection and the type of rule (which themselves are very simple, after all). Now, of course, one of the things so fascinating about complex systems theory is that complex behaviors can emerge from such simple elements.

But such simplicity hardly does justice to the kind of complexity found in the way organizational networks are connected. Indeed, besides the N/K model, other modellers of complex systems, notably research conducted in Russia and Canada by I. Trofimova, A. Potapov, and W. Sulis [23], have suggested more enriched constructs of connectivities in terms of what they call "sociability" and "adaptability" parameters which seem a better fit to to the human connections in work groups. By using constructs for their models that have a closer correspondence to actual connectivities in human systems, these more enriched complex systems approaches provide a more fruitful groundwork for the intersection of Industrial/Organizational and complex systems theory. In support of their models, Industrial/Organizational research has shown that turnover rates depend on the number as well as strength of social network ties of individuals to each other within a group [13].

Bibliography

[1] BANTEL, K., and S. JACKSON, "Top management and innovations in banking: Does composition of the top teams make a difference?", *Strategic Management Journal 10* (1989), 107–124.

[2] COHEN, I. Bernard, "Analogy, homology, and metaphor in the interactions between the natural sciences and social sciences, especially economics", *Non-natural Social Science; Reflecting on the Enterprise of More Heat Than Light*, Duke University Press (1993), 7–44.

[3] DELBECQ, A., A. VAN DE VEN, and D. GUSTAFSON, *Group Techniques for Program Planning: A Guide to Nominal Group and Delphi Processes*, Scott Foresman (1975).

[4] GOLDSTEIN, Jeffrey, "Creative Processes in Emergence and Selforganization" (unpublished manuscript).

[5] GOLDSTEIN, Jeffrey, "Simple or Complex Dynamics: Connotations of 'Equilibrium' in Chaos and Complexity Theories", *International Conference on Chaos, Fractals, & Models '96 Proceedings*. Universita Degli Studi di Pavia & Rome and La Societa Italiana Caos & Complessità (In press).

[6] GOLDSTEIN, Jeffrey, *The Unshackled Organization: Facing the Challenge of Unpredictability Through Spontaneous Reorganization*, Productivity Press (1994).

[7] GUASTELLO, Stephen, *Chaos, Catastrophe, and Human Affairs: Applications of Nonlinear Dynamics to Work, Organizations, and Social Evolution*, Lawrence Erlbaum Associates (1995).

[8] HACKING, Ian, *The Taming of Chance*, Cambridge University Press (1990).

[9] HAKEN, Hermann, *Synergetics*, Springer (1977).

[10] JACKSON, S., K. MAY, and K. WHITNEY, "Understanding the dynamics of diversity in decision-making teams", *Team Effectiveness and Decision Making in Organizations* (R. Guzzo and E. Salas (ed.), Jossey-Bass (1995), 204–261.

[11] JANIS, Irving, *Victims of Groupthink: A Psychological Study of Foreign-policy Decisions and Fiascocos*, Houghton-Mifflin (1972).

[12] KAUFFMAN, Stuart, *At Home in the Universe: The Search for the Laws of Self-organization and Complexity*, Oxford University Press (1995).

[13] MCPHERSON, J., A. POPIELARZ, and S. DROBNIC "Social networks and organizational dynamics", *American Sociological Review 57* (1992)

[14] MORGAN, Gareth, *Images of Organization (Second Edition)*, Sage (1997).

[15] MURNIGAN, J. Keith, and Donald CONLON, "The dynamics of intense work groups: A study of British string quartets", *Administrative Science Quarterly 36* (1991), 165–186.

[16] PETERSON, Randall, "A directive leadership style in group decision making can be both virtue and vice: Evidence from elite and experimental groups", *American Journal of Personality and Social Psychology 72* (1997), 1107–1121.

[17] PILLAI, Rajnandini, "Crisis and the emergence of charismatic leadership in groups", *Journal of Applied Social Psychology 26* (1996), 543–562.

[18] SCHALLER, Mark, "In-group favoritism and statistical reasoning in social inference: Implications for formation and maintenance of group stereotypes", *Journal of Personality and Social Psychology 63* (1992), 61–74.

[19] SINCLAIR, Amanda, "The tyranny of team ideology", *Organization Studies 13* (1992), 611–626.

[20] STACEY, Ralph, *Complexity and Creativity in Organizations*, Berrett-Koehler (1996).

[21] STUMPF, S. D. Zand, and R. FREEDMAN, "Designing groups for judgemental decisions", *Academy of Management Review 4* (1979), 589–600.

[22] THORNBURG, Thomas, "Group size and member diversity influence on creative performance", *The Journal of Creative Behavior 25* (1991), 324–333.

[23] TROFIMOVA, I., A. POTAPOV, and W. SULIS, "Collective effects on individual behavior: Three questions in the search for universality", (Submitted for publication).

[24] VAN DYNE, Linn, and Richard SAAVEDRA, "A naturalistic minority influence experiment: Effects on divergent thinking, conflict, and originality in work groups", *British Journal of Social Psychology 35* (1996), 151–167.

[25] VALLACHER, Robin, Andrzej NOWAK, "The emergence of dynamical social psychology", *Psychological Inquiry 8* (1997), 73–99.

[26] WEBER, Max, *From Max Weber: Essays in Sociology* , (H. Gerth and C.W. Mills ed.), Oxford University Press, (1946).

Chapter 25

Symmetry breaking and the origin of life

Hyman Hartman
Institute for Advanced Studies in Biology
28 Banks street, Cambridge, MA 02138

Symmetry breaking in this paper is first discussed in the case of reaction-diffusion systems which have been considered as models for self-organization and the Origin of Life. The lack of a clear definition of an order parameter suggests the statistical mechanics of spin glasses and the breaking of replica symmetry might be a better model for the Origin of Life. The lack of a clear definition of a cost function for systems far from equilibrium further suggests that cellular automata and in particular probabilistic cellular automata might be the theoretical model of choice for the Origin of Life. The clay model for the Origin of Life is used to evaluate these various abstract models presented.

1. Thermodynamics and dissipative systems

Symmetry and symmetry breaking are central to modern theoretical physics. From phase transitions to the quantum field theory of elementary particles, symmetry and symmetry breaking are key concepts which govern these fundamental areas of mathematical physics. Is it possible that these concepts can be applied to the problem of the Origin of Life? In this paper we review various approaches to this question and ask whether these concepts can be applied to the Origin of Life.

In biology, symmetry breaking is the norm. Turing in his paper, "The Chemical Basis of Morphogenesis" demonstrated that a system of chemical reactions coupled with diffusion would lead to solutions which would break the symmetry of the initial state of the system. "An embryo in its spherical blastula stage has spherical symmetry ... One may take it therefore that there is perfect spherical symmetry. But a system which has spherical symmetry, and whose state is changing because of chemical reactions and diffusion, will remain spherically symmetrical for ever. (The same would hold true if the state were changing according to the laws of electricity and magnetism, or of quantum mechanics.) It certainly cannot result in an organism such as a horse, which is not spherically symmetrical.

There is a fallacy in this argument. It is assumed that deviations from spherical symmetry in the blastula could be ignored because it makes no particular difference what form of asymmetry there is. It is, however, important that there are some deviations, for the system may reach a state of instability in which these irregularities, or certain components of them, grow. If this happens a new and stable equilibrium is usually reached, with the symmetry entirely gone."(1)

What Turing had shown was that a set of chemical reactions especially autocatalytic chemical reactions coupled by diffusion could give rise to spatial and temporal patterns. The experimental discovery of the Belousov-Zhabotinskii reaction which involved the oxidation of malonic acid by bromate ion catalyzed by cerium, led to a widespread experimental and theoretical analysis of reaction and diffusion models. The role of autocatalysis in these systems was emphasized.

Prigogine and his group generalized the Turing analysis of reaction and diffusion systems to what they called dissipative systems far from equilibrium. Symmetry breaking in these systems led to spatial and temporal order. Self organization by means of symmetry breaking, due to fluctuations in systems of chemical reactions and diffusion, are seen as a clue to the Origin of Life. "Broadly speaking destruction of structures is the situation which occurs in the neighborhood of thermodynamic equilibrium. On the contrary, creation of structures may occur, with specific non-linear laws beyond the stability limit of the thermodynamic branch. This remark justifies Spencer's point of view: 'Evolution is integration of matter and concomitant dissipation of motion' ". (2)

My first experience with symmetry breaking came when I studied Bioconvection in 1976. Dense cultures of Tetrahymena form swimming patterns which are strikingly like the polygonal patterns formed when a liquid is heated from below (Benard convection). We used a set of equations which were similar to Turing's to explain these swimming patterns. The initial state was a homogeneous collection of cells which rapidly formed polygonal swimming patterns in which the cells swam up the middle of the polygon, swam to the sides of the polygon and fell down the sides. The explanation depended upon the Tetrahymena cell having a geotaxis which directed it to swim up and a chemotaxis which directed it to swim towards higher concentrations of carbon dioxide.

The pattern could then be explained by 1) the cells swimming up to the surface directed by their geotaxis 2) at the surface the cells respond to the carbon dioxide gradients formed by the autocatalytic collection of cells to each other (as they both form carbon dioxide and they also swim towards higher concentrations of carbon dioxide) 3) They then fall to the bottom of the container by means of a vertical density current (as the cells are denser than the liquid medium). It was clear that the carbon dioxide secreted by a single cell was instrumental in ordering the swimming pattern of a population of these cells. (3)

Haken has studied self-organizing systems by using analogies from phase transitions. He has coined the term synergetics to cover this field of self organizing systems. Examples of such systems are the laser and the Benard instability in a fluid heated from below. The major idea involved is that of an order parameter.(4)

Landau introduced this term in order to deal with the nature of thermodynamic phase transitions. "The order parameter is a quantitative measure of the loss of symmetry. The canonical one is M in a ferromagnet: the mean moment $\langle mi \rangle$ on a given atom ... The loss of symmetry requires a new thermodynamic parameter whose value is zero in the symmetric phase—for instance the magnetization of a ferromagnet ... The magnitude of the order parameter measures the degree of broken symmetry."(5).

It is the case that the phase transition takes place as the temperature is lowered. What Haken has attempted to do is expand this definition of order parameter to systems which are far from equilibrium. Unlike the orderly structures formed in phase transitions which occur with the lowering of temperature, self organization of living systems occur at room temperature. "Life processes obviously must be based on completely different principles, which have nothing to do with the reactions of super conductivity, in crystal formation or of ferromagnetism" (6). Nevertheless Haken still wants to take over the idea of order parameter from phase transitions and apply it to dissipative systems far from equilibrium.

Anderson later phrases the question:" Is there a theory of dissipative structures comparable to that of equilibrium structures, explaining the existence of new, stable properties and entities in such systems?"

The answer he gives is that "Contrary to statements in a number of books and articles in this field, we believe there is no such theory and it may even be that there are no such structures as they are implied to exist by Prigogine, Haken and their collaborators. What does exist in this field is rather different from Prigogine's speculations and is the subject of intense experimental and theoretical investigation at this time."

Finally, Anderson asks; "Can we see our way clear to a physical theory of the origin of life which follows these general lines?". The answer is "No, because there exists no theory of dissipative structures." (7). In systems which are in equilibrium we have an order parameter which measures the amount of symmetry-breaking. However in systems far from equilibrium there is no well-defined function which behaves as an order parameter. Thus there is as yet no well defined theory of dissipative systems far from equilibrium. The structures found in the reaction and diffusion systems of Turing and Belousov-Zhabotinskii are all related to a set of non-linear partial differential equations. This is an area which still does not have a unified theory of symmetry breaking as does the thermodynamics and statistical mechanics of phase transitions.

2. Statistical mechanics

It was very clear to the American philosopher C.S. Peirce that what Darwin had accomplished in his theory of Natural Selection was similar to the statistical mechanics of Maxwell and Boltzmann. " The Darwinian controversy is, in large part, a question of logic. Mr Darwin proposed to apply the statistical method to biology. The same thing has been done in a widely different branch of science, the theory of gases."(8)

The statistical mechanics of ideal gases was compared by R. Fisher to the theory of population genetics; "the particulate theory of inheritance resembles the kinetic theory of gases with its perfectly elastic collisions." (9) Statistical mechanics of gases studies systems with a large population of particles. The major conclusions of the theory deal with statistical averages of these populations. The first step in statistical mechanics is to define a hamiltonian of the particles (molecules or atoms). In mechanics, the hamiltonian $H = T + V$; where T is the kinetic energy of a particle and V is the potential energy describing the interaction between the particles. In the case of the ideal gas $H = T$ and the particles only interact by colliding with each other. If a potential function is defined and added to the hamiltonian of the ideal gas then the gas can be shown to undergo a phase transition to a liquid state. Aside from the ideal gas, the Ising model (in two dimensions) for ferromagnetism is one of few solvable statistical mechanical systems. In the case of the Ising model, we have a lattice on whose vertices are located spins capable of two values +1 (spin up) and –1 (spin down). Here the ferromagnetic interaction $-J$ between nearest neighbor spins on the lattice favors parallel alignment of the spins. This system in the case of two dimensions shows a phase transition at a non zero temperature whereas the one dimensional Ising system does not. There is an order parameter m which is the number of spins pointing in the same direction (e.g., up). The symmetry of the system is broken, as the critical temperature is reached, by a fluctuation in the conjugate field H. The free energy of the Ising model in the ground state has two minima: one with all the spins pointing up and the other with all the spins pointing down. The two dimensional Ising model is an infinite automaton with a one bit memory: It remembers whether the conjugate field was pointing up or down when it passed through the critical temperature.

The Ising model can be generalized to a spin glass if instead of just a ferromagnetic $-J$ interaction there is also a $+J$ antiferromagnetic interaction. These interactions are distributed at random on the lattice and quenched. This gives rise to frustration and a very complicated ground state. The free energy of the Ising spin glass has many valleys which gives rise to broken ergodicity (as well as broken replica symmetry) caused by the physics of frustration.

A variant of the Ising spin glass was used by P. Anderson to model the Origin of Life. The model was based on an "analogy with the equilibrium statistics of spin glass" . The spins were the bases guanine and cytosine which were distributed along a polynucleotide. The polynucleotides interacted, conjugated and replicated. The polynucleotides which result were then subjected to selection and this selection depended upon a long range frustrated random interaction between the bases and the environment. " Hartman [following Cairns-Smith] has proposed a logically consistent scheme in which the primitive replicating stage is not chains of RNA having a linear array of different bases but layers of silicate minerals having a two-dimensional array of different ions. From the "universality class" point of view, this proposal is totally equivalent: the crucial stage requires conjugation symmetry and random interaction with the environment."(10) The model of Hartman and Cairns-Smith which Anderson referred to was the Clay hypothesis for the Origin of Life.

The clay theory for the Origin of Life begins with the necessity of liquid water for life. A major effect of liquid water on a planet such as the earth is to weather rocks to clays. This would have occurred on the primitive earth. Clays are fundamentally "two-dimensional" aperiodic crystals. This is of extreme importance if one is going to postulate a replication of information stored in an aperiodic crystal. Since we live in a three-dimensional world, it would be impossible to replicate the information stored in a crystal (aperiodic) by a simple template mechanism unless the crystals were one-dimensional or two-dimensional. Thus we can model RNA as a one-dimensional aperiodic crystal whose sequence of the four bases (uracil, guanine, adenine, cytosine) encode the information stored in the RNA. The information in a clay is stored in the distribution of ions like silicon, aluminum, iron, and magnesium in the "two-dimensional" clay lattice. It was on the surface of replicating clays that the amino acids and nucleotides were synthesized which ultimately led to the form of life as we now know it.

With spin glasses, the concepts of symmetry and symmetry breaking has led to the theoretical formulation of systems which are complex enough to model some of the problems arising in the fields of nerve nets, protein folding and the Origin of Life. The question which can be asked is whether or not these models are rich enough to resolve the deep problems in the Origin of Life. In particular there is the issue of whether or not the fitness surfaces are well characterized. In other words what is the nature of the conjugate field.

The problem with spin glasses as models for the origin of life is that we must define a cost function analogous to free energy in a system which is not at equilibrium. In a sense the problem is to use a term in population genetics is; what is the fitness function?. The question arises as to how to generalize the Ising spin glass to non equilibrium systems. One answer would be to study the cellular automata, invented in 1948 by J. von Neumann in order to model self reproducing automata.

3. Cellular automata

Cellular automata consist of a discrete set of "spins" arranged in a regular lattice. The state of each spin is a discrete variable. Time advances in discrete steps. A local function is defined to compute the new state as a function of the previous spin state and the state of the neighborhood. When iterated over the array, it generates a global map and the dynamical evolution of the system. These systems have been used to model hydrodynamics and reaction-diffusion systems.

My first experience with cellular automata was to model the surface of a reactive clay particle as a two-dimensional reaction-diffusion system. Clays are well-known catalysts. In the parlance of chemical kinetics a clay particle is a heterogenous catalyst. Molecules are absorbed on to the clay surface where they are activated at certain sites to react and where they then diffuse to other sites and react further and finally the products are desorbed. These reactions and diffusions on the clay surfaces were a motivation to study cellular automata models of reaction-diffusion systems.

The first such cellular automata which we studied was the Greenberg-Hastings model for the Belousov-Zhabotinskii reaction. This was a two-dimensional lattice and on each vertex was a spin with three states. This was a three state spin model which were labelled [0] quiescent, [1] active, and [2] refractory. There were four neighbors (north, east, south, west). The basic local function was: if the spin was in the quiescent state it will remain in this state unless there is a active state in its neighborhood then it will enter the active state at the next time step. Once it has entered the active state it will enter the refractory state at the next time step independent of what is happening in its neighborhood. At the next time step it will return to the quiescent state again independent of its neighborhood.The space-time patterns which are generated from a random initial state of active and refractory spins are dominated by singularities which are characterized by a "winding" or topological number.(11)

We have also studied a "reversible" Greenberg-Hastings cellular automata model which leads to chemical turbulence from very simple initial states. These two-dimensional reversible cellular automata are more easily studied in one dimension where they also lead from simple initial states to a turbulent condition.(12)

In recent years an attempt has been made to classify one-dimensional (two states per spin) cellular automata. A local function is specified and the states of the spins on the one-dimensional lattice (chain) are specified at random with 0's and 1's.The evolution in time is then observed. The rules fall into 4 classes:

1. the chain becomes homogeneous (say all 0's)

2. Appearance of simple localized time periodic structures

3. Evolution leads to chaotic behavior

4. Evolution leads to complex localized structures.

The first three classes behave as limit points, limit cycles and chaotic attractors respectively. The fourth class is probably capable of universal computation, so that its infinite time behavior are undecidable. These cellular automata are all deterministic. The probabilistic cellular automata is a system where each spin, at each discrete time step, undergoes a transition depending probabilistically on the states of its neighbors. This is now a field of computational statistical mechanics which is not at equilibrium.

Domany and Kinzel mapped the time development of cellular automata in d dimensions onto equilibrium statistical mechanics of Ising models in $d + 1$ dimensions. "The class of CA (cellular automata) studied by Wolfram constitute the deterministic limit of a more general class of stochastic CA. d-dimensional stochastic CA were studied under the name of crystal-growth models. Their time development is equivalent to the equilibrium statistical mechanics of $(d+1)$-dimensional Ising models." Directed percolation was shown to be equivalent to a cellular automaton, and thus to an Ising model.(13)

We mapped a one dimensional cellular automaton of Domany and Kinzel into an inhomogeneous cellular automaton with the Boolean functions XOR and AND as transition rules. Wolfram's classification is recovered by varying the frequency of these two simple rules and by quenching or annealing the inhomogeneity. In particular, Class IV is related to critical behavior in directed percolation. The critical slowing down of second-order phase transitions is related to a stochastic version of the classical " halting problem" of computation theory.(14)

The problem of how to program such probabilistic cellular automata is of course not been solved.(15)

One way, of course, would be to program these systems by evolving them by means of replication, mutation and natural selection. One such system is the clay model for the origin of life.

Schrodinger had suggested that genes were aperiodic crystals: "compared with the aperiodic crystal, They [homogeneous crystals] are rather plain and dull. The difference in structure is one of the same kind as that between an ordinary wallpaper in which the same pattern is repeated again and again in regular periodicity and a masterpiece of embroidery, say a Raphael tapestry, which shows no dull repetition but an elaborate, coherent, meaningful design by the great master"(16). In a sense, the clay theory proposed to evolve a two-dimensional chemical tapestry by natural selection. The evolution of such systems resulted in the biochemistry and molecular biology of cells. In an evolving system, in which you are going from a simpler state to a more complex state, it is assumed that the simpler state is retained and the more complex state is built on top of it. The final system is very much like an onion in that layer is built upon layer. It thus becomes possible to reconstruct the history of the system by peeling the outer layers back to the origin. One can then reconstructed the evolution of metabolism (17), photosynthesis (18) and the genetic code (19).

The evolution of such systems would record history by amplifying unique events (i.e. mutations) by means of replication. The breaking of symmetry would now depend upon which mutation would survive. An example, would be the well known breaking of chiral symmetry by living systems. The clay ancestor, which survived, made L-amino acids and D-sugars on its surface rather than D-amino acids and L-sugars. It is a frozen accident rather than some deep physical principle which is involved.

In a deep sense the Origin of Life is related to the most important symmetry breaking of all, the arrow of time.

This points out that in biology it is the ancestor rather than the archetype which explains the uniformities or non-uniformities found in biology. It is the ability to record history rather than the laws of physics or chemistry which is the ultimate explanation for the origin of life. Symmetry breaking phenomena are one way of recording historical events.

In this paper we have dealt with first self-organization due to the breaking of symmetry by non-linear chemical reaction and diffusion equations (non-equilibrium systems). These equations suggested that we look into the symmetry breaking of the Ising spin glass. However the need for a complex free energy surface and its restrictions led us to look a cellular automata and their dynamics and finally to the clay model for the origin of life.

Bibliography

[1] Turing, A.M. (1953), "The Chemical Basis of Morphogenesis" *Phil. Trans.R.Soc.* ser.B., **237**, 5-72.

[2] Glansdorff, and Prigogine,I.,(1971), *Thermodynamic Theory of Structure, Stability and Fluctuations*, Wiley Interscience: New York, 288.

[3] Hartman,H., and High,R.(1976), "Bioconvection and Morphogenesis" General Systems Yearbook, Vol.21, 47.

[4] Haken,H. (1983), *Synergetics*, Springer-Verlag: New York.

[5] Anderson, P. (1984), Basic Notions of Condensed Matter Physics. Benjamin-Cummings:Menlo Park. 266

[6] Haken, H.(1981) *The Science of Structure: Synergetics*, Van Nostrand Reinhold Company: New York. 42.

[7] Anderson,P.(1984) *Basic Notions of Condensed Matter Physics*, Benjamin-Cummings: Menlo Park. 264

[8] Peirce, C.S.(1955) *Philosophical Writings of Peirce*, ed. by J.Buchler: Dover: New York. 7 .

[9] Fisher,R.A. (1958) *The Genetical Theory of Natural Selection*, Dover: New York. 11.

[10] Anderson,P. (1983) "Suggested model for prebiotic evolution: The use of chaos" *Proc.Natl.Acad.Sci.* vol. **80**, 3386-3390.

[11] Tamayo,P.and Hartman, H.(1989) "Cellular-Automata, Reaction-Diffusion Systems and the Origin of Life, in *Artificial Life*, ed. C.G. Langton, Addison-Wesley publishing Co:Menlo Park 105-124.

[12] Hartman,H. and Tamayo,P. (1990) "Reversible Cellular Automata and Chemical Turbulence", *Physica D* **45** 293.

[13] Domany,E. and Kinzel,W,(1984) " Equivalence of cellular automata to Ising models and directed percolation", *Phys. Rev. Lett.* **53**, 311.

[14] Vichniac, G.Y., Tamayo,P. and Hartman,H.(1986) "Annealed and Quenched Inhomogeneous Cellular Automata (INCA)" *Journal of Statistical Physics*, **45**, 875.

[15] Hartman,H.,Tamayo,P. and Klein,W. (1987) "Inhomogenous Cellular Automata and Statistical Mechanics", *Complex Systems* **1**, 245.

[16] Schrodinger,E. (1944). *What is Life*, Cambridge University Press . Cambridge.

[17] Hartman,H. (1975) Speculations on the Evolution of the GeneticCode *Origins of Life* **6**, 423.

[18] Hartman,H. (1975) Speculations on the Origin and Evolution of Metabolism *J. Molecular Evolution* **4**, 359.

[19] Hartman,H. (1992) Conjectures and Reveries, *Photosynthesis Research* **33**, 171.

Chapter 26

Complexity and functionality: A search for the where, the when, and the how

Piet Hut
Institute for Advanced Study, Princeton, NJ 08540, U.S.A.
Brian Goodwin
Schumacher College
The Old Postern, Dartington, Devon TQ9 6EA, U.K.
Stuart Kauffman[1]
Santa Fe Institute
1399 Hyde Park Road, Santa Fe, NM 87501, U.S.A.

We discuss three approaches to a study of complexity: the reductionist stance; a search for new laws; and a search for a new aspect of reality, besides space and time. We focus on the latter, introducing the term 'sense' as a candidate for such a third aspect. We point out some of the ramifications of such a move for the subject/object relationship in physics and in biology.

1. Complexity with an attitude – but which one?

How do complex phenomena such as life and especially consciousness fit into our scientific world view, based on physics as the most fundamental of the natural

[1]S.K. acknowledges grants from NIH and NASA

sciences? Is biology more than a complex form of applied physics? In general, what is the character of 'emergent properties'? There are three fundamentally different attitudes that we can take with respect to these questions.

1) the reductionist stance. We can deny that there is any problem, remaining satisfied with the 'explanation' that ultimately the most complex phenomena are, after all, layered upon some physical substratum, a dance of matter and energy in space and time. Whatever it is that is thus layered on top is seen as mere icing on the cake, nothing 'substantial', and hence nothing special, from a basic point of view.

2) a search for new laws. Accepting that physics in its current state is unable to capture phenomena such as life and consciousness, although it may suffice to describe the behavior of the physical substrata, it is natural to search for something else, something with which to augment physics. A natural move is then a search for new laws of physics, additional principles that may help explain properties such as autonomous agency and adaptive behavior.

3) a search for a new aspect of reality, besides space and time. This is the move that we are exploring in this paper. At first, this move may seem bewildering. What could there possibly be, in addition to space and time, and equiprimordial with space and time, irreducible to either or both? The answer may lie along the lines of intention, agency, cognition/feedback, relationship, and functionality.

In a nutshell, the move from space and time to a third aspect of reality can be motivated as follows. A movie shows motion in time, but does so by freezing temporality into a series of purely spatial snapshots. Similarly, a biological treatment of a living cell shows that cell's functionality, through a series of snapshots at different 'levels' of complexity, from that of atoms to that of molecules to that of organelles, etc. Where a movie freezes and carves up the time dimension, an analysis of a complex biological system freezes and carves up the different levels of description of emergent properties.

A single snapshot can be described spatially, but such an analysis fails to capture fully the temporal coherence of the series of snapshots. Similarly, a single level of description of a cell, on the molecular level, say, can be performed using physics, based purely on spatial and temporal concepts. However, such an analysis fails to capture in full the 'dimension' of functionality of a living cells, the intrinsic coherence between the different levels of description, that which gives the cell its unity, deserving the single name 'cell'.

2. Reductionism

Physics describes a world of complex reactions between a handful of simple entities (particles, fields, ...), situated in space and time.

On the one hand, physics is very successful: everything we see around us seems to have a place within the physics fold, if we restrict ourselves to objective (inter-subjective) phenomena. Even so-called secondary quantities such as the color of an object, can now be calculated from first principles, given the material composition: we can derive the fact that the sky is blue and the grass is green. That is to say, that their light has a wave length distribution corresponding to blue and green.

On the other hand, there a few aspects in which physics is not (yet) successful:

- the subjective experience of a color, the 'quale', may correlate with an objective description in terms of wavelength, but is clearly something else. In the near future, we may find a more and more accurate correlation with an intermediary phenomenon, between the wavelength and the quale of a color, in the form of a detailed description of the firing pattern of neurons corresponding to that color. But even if we would have the complete wiring diagram of the brain, and what is more, a full understanding of the data processing involved, we would still not have reached the quale.

- physics is an abstraction of reality, or better: actuality. It is the result of pushing actuality through the filter of physics, taking only what is objectifiable and repeatable. Ethics and esthetics are thus lost. The best that physics can hope for, it seems, is to reconstruct the objective counterparts to esthetic and ethical experiences in terms of a detailed understanding of the corresponding neurological processes. As is the case with color, there seems to be a rift between an understanding of such processes and 'the real thing'. Even more than with color, the internal logic of ethics and of esthetics seems to be lost here.

- even on the level of biological processes, anything with a function, aim, need, or intention already fails outside the purview of physics, based on a purely causal explanation. The very term self-organization that is used to describe the processes of maintenance and reproduction in organisms implies a self that perpetuates a distinctive type of organization, the living condition. This goes beyond current physics in two ways: 1) life involves an emergent organizational property which has yet to be precisely defined in physical terms and is not reducible to causal molecular interactions (including the template and coding properties of DNA)(Goodwin 1994; Kauffman 1995); 2) the self of an organism can be described as an autonomous agent that acts on its own behalf and in so doing it invokes a world which it knows and in which it can function successfully (see Kauffman, Investigations; Maturana and Varela 1992). Clearly the terms 'know', 'act on its own behalf', 'evoke', and 'function' take us into a territory of internal agency or subjectivity that transcends the terms of reference of causal explanation in physics, even allowing for an observer.

Let's take a closer look at those three objections. The root cause for all three to arise is the opening move of reductionism: to want to start with (what appears to be at a given moment in the history of physics) the most fundamental building blocks. A while ago, atoms and molecules seemed to be the most fundamental. We then descended to wave functions and relativistic quantum fields, and we may soon arrive at an even deeper level, perhaps given by string theory. Reductionism considers those 'deeper' levels to provide a foundation for the 'higher' levels of analysis of complex systems: physics founding biology founding chemistry.

This notion of the natural sciences as stacked on top of each other, with logic in the basement, math on the ground floor, physics on the second floor, and so on, provides a rather curious metaphor. Time and again, the 'underlying principles' of physics have changed dramatically, and yet the building never collapsed as a result. The switch from classical to relativistic mechanics replaced some 'fundamental' assumptions of physics, and yet most chemistry and biology went on with business as usual, on the higher floor. The switch to quantum mechanics provided an even greater change in 'basic' assumptions, but this, too, did not mean that we had to rebuild chemistry from scratch. The vast body of knowledge built up so far in chemistry did remain valid: verified phenomena remained verified, and so did their empirical relationships.

What type of building is it, this grand structure of the natural sciences, that you can cheerfully replace the foundations or first floor without affecting the higher floors? It sure would be a convenient building to work on, for architects and construction workers who had trouble making up their mind! The conclusion we draw from this picture is that the whole notion of 'foundations' is greatly flawed. Rather, the building seems to be held up in the middle, or better, all over the place. The real support for any scientific theory ultimately comes from experience: from experimentation in the lab, or from observations in nature.

What science provides is a divide-and-conquer strategy. Starting with the full buzzing and confusing diversity of phenomena in daily life, a severe filtering operation leads us to isolate the principles of mathematics first, then of physics and then of the more complex branches of natural science. If there is any real 'grounding' in this whole operation, the ground is provided by 'what happens', and by the relationships between these happenings, which can be described by carefully specifying where, when, and how each happening happens. Everything else, no matter how elegant and simple it may seem at any given time, is derivative (Nishida 1911, James 1912, Husserl 1913, van Fraassen 1994, Hut and Shepard 1996).

The where, when, and how of these happenings are always described by a human agent - one of us - busy doing, acting and constructing, as well as describing. As we climb the building through levels of complexity, we reach the level where we turn our probings on ourselves and describe the happenings that we can observe - as outside observers.

Here we encounter the agent that sets up the experiments defined by specific relationships between the observer and the observed happenings. This agent is also part of reality, and there is no way it is going to disappear in the dance of happenings that it itself has choreographed. The doings of a causal agent cannot be reduced simply to a set of happenings. Why not? Because an agent has a point of view, a framework of action, a relationship to actuality that gives direction to its doings. Happenings just happen - they do not serve any agent's purpose. Doings involve both the effects of outside influences on an agent and the point of view, the framework of action that defines the agent as an actor in the world.

To make these ideas clearer, consider a concrete biological example - a bacterium swimming towards a source of sugar that it has detected. Here is an example of an autonomous agent able to act on its own behalf in an environment defined by its selected relationship to the world. Going to get dinner! Good. A bacterium appears to be "just" a physical system. However, when we examine this agent in more detail we realize that it has a very interesting property: it is able to perform work cycles that perpetuate its own distinctive organization. There is a logical closure within such systems that allows the agent to make more of itself. Of course it can only do this within an environment that satisfies its needs - such as sources of sugar for energy to drive the work cycles that generate more of the structures that constitute the organization of the agent itself. The very language of description here is self-reflexive, revealing the logical cycle that defines closure. Such as agent can act on its world - swim towards a source of dinner, eat it, and in so doing change its world. These are "doings": they serve the specific purposes of the agent in perpetuating and propagating its distinctive organization.

When the bacterium does work in swimming to get dinner, its swimming slightly warms the liquid medium and jiggles other creatures. The warming and jiggling are happenings, not doings, for they do not serve the bacterium's purposes.

So we see the distinction between doings and happenings. The agent has a point of view underlying its actions, and its actions literally change the world. The agent has embodied knowhow - the knowhow to make its way in the world. That knowhow of doings that tend to sustain and propagate the bacterium are its mode of navigation in its world. None of this is standard physics. All of it is under our noses. All of it is true. But where does it take us in relation to new laws or new basic postulates of science?

3. In search for new laws

Let us look more closely at that question of what makes biology more than physics. Whence intention? We'll leave the first two questions of §1 (reductionism) for later, but keep them in mind as part of our motivation to grope for something beyond physics.

A living cell, a living organism, a living ecosphere: all seem to be composed out of known building blocks. On the atomic level, we have an intricate network of a handful of molecules, mostly H, C, O, N, and a few others. This may seem to suggest that we just have a complicated calculation at our hands, an advanced exercise in applied physics, nothing more.

Well, let's pause for a moment. In physics itself, what counts as the understanding and deriving of results is quite different in different fields. In particle physics, the drive is towards more and more fundamental levels of insight, in terms of finding more and more primitive building blocks. But in solid state physics, say, the drive is to find effective laws, emergent properties, laws that are layered on the underlying properties of the next layer of more primitive building blocks, but that cannot be understood purely in terms of the building blocks. The latter are needed, and show up when things break down, but fade out of view as long as we focus on the higher-level machinery in full action.

So, perhaps biology, too, has its own logic, its own laws, equally 'real' or 'fundamental' as the laws of, say, quantum field theory.

But wait, there is something funny going on here. If the most fundamental laws of physics already 'cover' everything, then where is there room for more laws? How can there be more 'fundamental laws' than an already complete set? And don't those extra laws hold sway over a substratum that perfectly runs by the existing laws as such? Isn't this picture of 'extra laws' not a bit like that of a child in the passenger seat, holding a toy steering wheel, turning it whenever the car turns, in great delight, but without adding anything to the dynamics of the car.

4. Where and when and how

How about an even bolder move. Perhaps biology can point the way to something extra, not on the level of laws, but on the level of the composition of reality itself. Instead of viewing space and time as fixed, and all of physics and biology as a game played within space and time according to given rules, we can question the notion of space and time as catching the whole picture. In other words, instead of searching for new rules of the game, we could search for ways to widen the playing ground itself (Hut 1996, Hut and Shepard 1996, Hut and van Fraassen 1997; cf. Tarthang Tulku 1977).

So let us introduce a third aspect of reality. Besides the 'where' of geometry and physics in the limit of statics, and the additional 'when' of physics in general, including dynamics, we want to ask a third question, characteristic for biology. What sets apart biology from physics is the fact that any living system has a functional structure. In physics, once you have specified the dynamics and the initial conditions, the system evolves in time. How it evolves is specified in the universal laws that are obeyed by the system under consideration. In biology, however, any particular system has its own 'how'. The bacterium mentioned above has an efficient 'how' for the question of how to get dinner: swimming in the direction of increase of sugar concentration.

In physics, the 'when' of dynamics does not negate the fact that we can still ask about the 'where' of moving objects. Time does not replace space as a mysterious alternative type of fluid or ether, spread out 'in' space; rather, time is fully complementary to space, and a complete description of dynamics needs a specification in both space and time. Similarly, the 'how' of biology, we propose, does not negate the fact that a living cells still partakes in the 'where' and 'when' of physics. We can still describe the cell on a molecular level, as an exceedingly complex dynamic system in terms of physics. The third aspect (meta-dimension?) is simply complementary to the first two, space and time. Just as time and space don't bite each other, so the 'how' of biology does not interfere with the 'where' and 'when'.

We propose to use the simple term 'sense' to give a name to this third aspect of reality. The answer to how? is then: through sense (in its aspect of 'meaning', not that of 'sense experience'). In order to interpret what is happening, we have to 'make sense' out of the situation. This is already implicitly the case in physics, where any form of experiment or observation entails particular choices on the side of the experimenter or observer, who plays the role of subject. Any description, no matter how objective, is a description made by a subject trying to make sense. Even though the relationships between subject and object were rarely discussed in classical physics, relativity theory and quantum theory have forced us to make them more explicit. Especially in biology, the polarity between an agent and its environment is prominent. It makes sense for a bacterium to look for a source of its dinner. Already in order to stay alive, organisms explore the 'everyhow' of relational possibilities side by side with the everywhere of space and the everywhen of time.

This provides an alternative direction, complementary to the search for 'laws of complexity' and 'emergent properties', mentioned in the previous section. Of course, the choice between 'extra laws' and 'an extra aspect of reality' is not mutually exclusive. On the contrary. If it is reasonable to talk about an extra 'dimension' of reality, then that dimension has been there already from the outset. Just as time is still there when we analyze a static configuration, sense has been there all along, when doing physics. So biology simply uses more of what is already there, and physics less, and geometry even less. Seen in that light, physics could be viewed as a specialization of biology, rather than the other way around.

5. From where to when

An analogy may help us here. The first time we hear about the possibility of a fourth dimension, it is hard to imagine what a four-dimensional world would look like. It is easier to go in the other direction, from three to two dimensions, and imagine how the world would appear for two-dimensional beings. Specifically, we can then ask ourselves how such beings could try to imagine the hypothetical existence of a third dimension. After thus getting some experience in flexing our 'dimension imagination muscles', we are then ready to move up, to contemplate the move from three to four dimensions.

In our case, too, rather than trying to figure out what our third aspect of reality may look like, let us descend to an understanding of the world, based purely on space. Imagine that we would encounter a group of scientists who would look at the world as a geometric landscape, aware of the subtleties of space, but more or less oblivious of time. Of course, they would know motion, change, origination and decay, but imagine that they never made the jump to postulating a background time, a single something that can act as the condition of possibility for *any* type of change or motion or occurrence to occur.

How could we possibly go about trying to convince this tribe of geometers that there is something else besides their beloved space, something called time that is really on a par with space, with neither of the two being reducible to the other? Let us imagine a dialogue between one of the geometers (G) and a physicist (P) who is trying to point to the existence of time.

G: So you are saying that there is something very important, something called time, but which is invisible, and in general, unmeasurable as such?

P: Yes and no; time as such cannot be measured, but what we can measure is the progress of time, reflected in all motions around us.

G: Do you mean that time is some sort of field, that is especially strong and concentrated around fast moving objects?

P: No, time is everywhere, equally present for static bodies as for bodies in motion.

G: So you mean that time is like space? Is it equal to space, or is it a type of subtle ether, something that is filling space everywhere equally?

P: Neither of the above. You're searching in quite the wrong direction. Hmm. How can I explain this. In a way, space and time are such basic concepts, you can almost feel them. If you wave your hand, you are waving through space, but at the same time, it takes time to wave your hands, so you are waving through time as well. Each breath you take, your chest moves rhythmically through space and time; it is exactly the balance between the spatial and temporal motions that defines the presence of rhythm.

G: Now you are really mystifying things. Are you really asking me to believe that there is anything more to motion that what can be analyzed in a series of snapshots? What more can there possibly be, over and above snapshots? I bail out. This is getting just too ridiculous.

6. From where and when to how

After this warm-up exercise, let us now switch to the topic of our present paper. Let us imagine a similar dialogue between a physicist (P) and a biologist (B). For definiteness, let us call the third aspect of reality 'sense.'

P: So you are saying that there is something very important, something called sense, but which is invisible, and in general, unmeasurable as such?

B: Yes and no; sense as such cannot be measured, but what we can measure are the many types of relational behavior, reflected in any and all biological processes around us.

P: Do you mean that sense is some sort of field, that is especially strong and concentrated around living objects, A type of vital spirit?

B: No, sense is everywhere, equally present for inanimate and animate bodies, and already implied by the existence of mass-energy in space.

P: So you mean that sense is like space? Is it an aspect of space, built into it, perhaps on the level of vacuum fluctuations or the like?

B: No, you're searching in quite the wrong direction. Hmm. How can I explain this? In a way, space and time and sense are all such basic concepts, you can almost feel them. If you wave your hand, you are waving through space, but at the same time, it takes time to wave your hands, so you are waving through time as well, and while doing all this, you are in some way 'making sense' within space and time. Each moment, you are making sense of your world in a different way, and so you can be said to move through a dimension of sense. When we talk about exploring the 'depth' of meaning, we use just one of many metaphors that point to meaning or sense as having a type of geometric interpretation. Similarly, each breath you take is an act, something that makes sense for you as an organism, whether you are conscious of it or not. And the most relaxed way of breathing, in fact, occurs when space, time, and sense are all in perfect balance, expressing and revealing the appropriate ratios that define the inherent know-how or embodied rationality of the action. Indeed, that way of breathing is most sensible.

P: Now you are really mystifying things. Are you really asking me to believe that there is anything more to life that what can be analyzed by physical processes occurring in space and time? What more can there possibly be, over and above the dynamics of matter and energy? I bail out. This is getting just too ridiculous.

7. Conclusion and outlook

In retrospect, adding time to space to get motion and dynamics is such an obvious move that you wonder why it wasn't done before Newton and Leibnitz. However, inventing an effective analytical structure such as the differential calculus, and an appropriate conceptual framework for space and time, was difficult and continues to challenge the scientific imagination, since time remains an enigma.

Similarly, the addition of sense to space and time seems obvious and, at first sight, trivial. However, its consequences grow rapidly in depth and significance as the implications are pursued. They lead directly to the dilemma of the observer and the subject, which is put to one side by physics but must be faced in biology where the intentional actions of agents, their purposes and functions, are ever-present aspects of reality. Developing an appropriate conceptual structure for space, time, and sense, and effective methodologies of investigation that allow us to make progress in understanding and explaining these properties of organisms and of nature in general, is the hard work that faces us on this path. We are convinced that something along the lines suggested is the move that is now required, and we invite anyone interested to indicate how they believe that we can proceed further.

Bibliography

[1] GOODWIN, Brian, *How The Leopard Changed its Spots*, New York: Simon and Schuster (1994).

[2] HUSSERL, Edmund, *Ideen zu einer reinen Phaenomenologie and phaenome-nologischen Philosophie I, Erstes Buch* (1913), translated as *Ideas Pertaining to a Pure Phenomenology and to a Phenomenological Philosophy, first book*, Dordrecht: Kluwer (1982).

[3] HUT, Piet, "Structuring Reality: the Role of Limits", in *Boundaries and Barriers*, eds. J. Casti and A. Karlqvist, Reading, MA: Addison-Wesley (1996), pp. 148-187.

[4] HUT, Piet, and Roger SHEPARD, "Turning 'The Hard Problem' Upside Down & Sideways", *Journal of Consciousness Studies*, **3** (1996), 313-329.

[5] HUT, Piet, and Bas VAN FRAASSEN, "Elements of Reality: A Dialogue", *Journal of Consciousness Studies*, **4** (1997), 167-180.

[6] JAMES, William, *Essays in Radical Empiricism* (1912), reprinted in *Essays in Radical Empiricism & A Pluralistic Universe*, Gloucester, MA: Peter Smith (1967).

[7] KAUFFMAN, Stuart, *At Home in the Universe*, New York: Oxford University press (1995).

[8] KAUFFMAN, Stuart, *Investigations*, in press (1997).

[9] MATURANA, Humberto R, and Francisco J. VARELA, *The Tree of Knowledge*, Boston: Shambala (1992).

[10] NISHIDA, Kitaro, *Zen no Kenkyu* (1911), translated as *An Inquiry into the Good*, New Haven: Yale Univ. Pr. (1990).

[11] TARTHANG TULKU, *Time, Space, and Knowledge*, Berkeley: Dharma Publ. (1977).

[12] VAN FRAASSEN, Bas, "The World of Empiricism", in *Physics and our View of the World*, ed. Jan Hilgevoord, Cambridge: Cambridge Univ. Pr. (1994), pp. 114-134.

Chapter 27

Biological design principles that guide self-organization, emergence, and hierarchical assembly: From complexity to tensegrity

Donald E. Ingber

Harvard Medical School, Pathology Dept.

A living organism represents the ultimate complex adaptive system. Our work focuses on the question of how groups of molecules self-organize to create living cells and tissues with emergent properties, such as the ability to change shape, move, and grow. Most complexity-based approaches focus on nodes, connections, and resultant pattern formation. We have extended this approach by taking into account the importance of architecture, mechanics and structure in the evolution of biological form. This work has led to the discovery of fundamental design principles that guide self-assembly in natural systems, from the simplest inorganic compounds to the most complex living cells and tissues. These building rules are based on the use of a particular form of geodesic architecture, known as tensegrity, which causes hierarchical collections of different interacting components to self-organize and mechanically stabilize in three dimensions. Shape and pattern stability emerge through establishment of a force balance between globally acting attractive (tensile) forces and locally acting repulsive (compressive) forces or, in simplest terms, through continuous tension and local compression (tensional integrity or "tensegrity"). Recent development of a mathematical explanation for the mechanical behavior of living cells and tissues based on tensegrity may provide a useful computational tool for analysis in other complex adaptive systems ranging from protein folding to cosmology.

1. Introduction

Complexity theory is based on the understanding that the behavior of a complex adaptive system depends on the connections between its components and that the whole is literally greater than the sum of the parts. Any insights into the functioning of these systems therefore will come from elucidating how the relationships between different components are stabilized, rather than through analysis of their substance. Complexity has provided a conceptual and mathematical handle to approach systems that contain multiple interdependent subcomponents. This approach has been found to be effective at predicting phase transitions (e.g., abrupt changes from chaos to pattern) and general evolutionary trends (e.g., speciation; stock market fluctuations) within large populations in a wide variety of disciplines. However, current approaches can not explain how or why specific patterns come into existence. Many other critical questions also still remain unanswered: How do emergent systems self-organize? How do they stabilize themselves? How do systems at different size scales assemble into even more complex hierarchical networks? In simplest terms: What is the driving force behind the evolution of natural forms?

Living organisms represent the most sophisticated examples of complexity at work. Every creature represents a complex adaptive system that exhibits self organization, emergence, hierarchical integration and self-propagation. If we could decipher the basic building rules that guide biological organization, then in essence we would understand complexity. Many people have the impression that DNA is the source of all pattern-generating information in biology. Well, we are getting awfully close to sequencing the entire human genome, however, we are still far from understanding how cells function, let alone self-organize.

In this chapter, I will describe recent advances we have made in terms of understanding fundamental design principles that govern the self-organization of complex biological systems, including living cells and tissues. The novel feature of our approach is that that we realized that the question of how biological structures form is not a question of chemical composition or even pattern, rather it is a problem in architecture. We therefore took into account the mechanics of the system as well as the spatial relationships between the different interacting components in our analysis of self-organizing systems. In other words, we shifted the focus in the field of biological complexity from pattern to structure.

Through this approach, we have discovered that living cells and tissues use a specific form of architecture, known as "tensegrity" [5], to organize themselves and to control their mechanical behavior [8] [9] [2] [12]. Most importantly, self-stabilization, emergence, and hierarchical organization are all properties inherent to the tensegrity building system. Furthermore, the same guiding principles govern pattern formation at all size scales, in both the organic and the inorganic world [12]. This work also has led to the formulation of a mathematical basis for tensegrity-based self-organization and dynamic behavior in cellular systems [19] [4] which may facilitate development of new computational approaches for solving general problems relating to complexity in a range of different disciplines.

2. Complexity in living systems

Life is the ultimate complex adaptive system. It builds hierarchically from the bottom up by connecting large populations of interacting component subsystems that independently exhibit their own dynamic adaptive behavior. Furthermore, at each system jump, new synergetic properties emerge. For example, by covalently linking small chemicals, larger molecules are created that exhibit new functions, such as polymerization behavior or the ability to catalyze a biochemical reaction. These new properties are clearly emergent: there is no way to predict or understand how an enzyme functions by analyzing the properties of its individual amino acids. Molecules then join together to form larger complexes and again novel behaviors are observed. The formation of flexible membrane bilayers from phospholipids and large enzyme complexes with coupled biochemical activities (e.g., G proteins) are two examples. At a larger size scale, connecting together different types of molecular complexes produces organelles with specialized functions, ranging from tiny transport vesicles and energy-generating mitochondria to the entire nucleus which contains the genome and functions as the "brain center" of the cell. These organelles are then joined together by a filamentous molecular framework or "cytoskeleton" and surrounded by a differentially permeable membrane to create the cell, the smallest independently functioning unit of life. In higher organisms, cells also assemble together to form tissues, tissues to form organs, and organs to form the whole living creature. However, self-organization of the cell remains the fundamental riddle.

3. Cellular tensegrity

Most people still view the cell as a viscous cytoplasm surrounded by an elastic membrane, almost like a waterballoon filled with molasses. The reality is that living cells are literally "hard-wired" by a series of molecular struts and cables, known as the cytoskeleton, which stretches from specific receptors on the cell surface to discrete connection points on the nucleus in the cell center. We have shown experimentally that pulling on certain cell surface molecules results in immediate action at a distance within the very center of the cell's nucleus [15] and that all of the individual chromosomes that comprise the human genome are mechanically coupled [14]. We also have shown that complex cell behaviors, such as growth, differentiation, and death can be controlled through cell shape modulation and associated mechanical distortion of this filamentous supporting framework [17] [3]. Most importantly, we have found that cells use a particular form of architecture to self-organize this supporting framework and to create hierarchical assemblies, such as tissues and organs [8] [9] [2] [12]. This building system is based on tensional integrity and is known as "tensegrity" [5].

The term, tensegrity, was first coined by Buckminster Fuller to describe a sculptural building system first developed by the artist, Kenneth Snelson. Fuller recognized that the great strength and mechanical efficiency of his geodesic domes were based on the use of this form of structural stabilization. The in-

272

credible structural efficiency of the dome does not depend on whether it is built using wood or aluminum. Rather, it is based on its design: how the structure distributes and balances mechanical stresses in three dimensional space. Tensegrity structures gain their mechanical stability through transmission of continuous tension which is balanced by local compression and thus, the term tensional integrity. A simple example is a spider web plus the tree branches which act like struts to hold it in an extended form. This form of mechanical stabilization is in direct contract to compression-dependent structures that dominate man-made architecture.

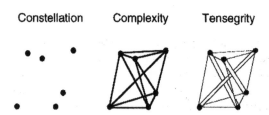

Figure 3.1: Schematic diagrams comparing complexity and tensegrity-based approaches for analysis of multisubunit interacting systems. Constellation, an array of interacting elements which exhibits cooperative behavior (a pattern). Complexity, patterns are often defined in terms of nodes and their interconnecting paths. Tensegrity, patterns represent three dimensional structures which self-stabilize through establishment of an internal balance of continous attractive (tensile) forces, here depicted by thin tension lines, and discontinuous repulsive (compressive) forces, which are depicted by isolated white struts. Only tensegrity incorporates structure and mechanics as well as pattern.

It is difficult to visualize the tensegrity mechanism in fully triangulated structures. However, it becomes fully visible in Snelson-type tensegrity sculptures in which multiple isolated rigid struts self-organize to form a stable three dimensional form as a result of interconnecting the ends of isolated compression struts with a continuous series of tension elements that map out geodesic lines (Fig. 3.1 and Fig. 3.2). Forces are similarly distributed and balanced in fully geodesic structures such as domes, however, the same structural member can bear either tension or compression depending on the loading conditions [2]. The mechanics of these structures is dominated by the buckling modulus of the non-compressible struts and the level of internal tension or "prestress" within the system; the role of bending modulus and torque become relatively insignificant.

Although these structures seem strange and "futuristic", the reality is that this is the way our bodies are constructed. The 206 stiff bones that comprise our skeleton are pulled up against the force of gravity and stabilized in a vertical form by the pull of a continuous series of tensile muscles, tendons and ligaments. Furthermore, the stability of this complex system depends directly on how opposing forces are balanced in the network (arrangement of the elements) and the level of muscular "tone" (prestress) in the system as well as the material properties of the different connecting elements (bone can bear compression whereas muscles only resist tension).

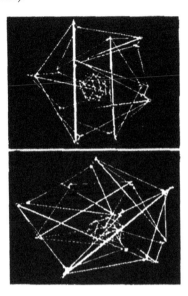

Figure 3.2: Nucleated tensegrity cell models constructed from sticks and elastic string. Top) an unattached model exhibits a round form due to the presence of an internal tension (prestress) which pulls in equally from all sides. Bottom) The same model exhibits coordinated extension and flattening of the cell and nucleus when it is attached to a rigid foundation that can resist its internal tractional forces. This mechanical coupling is due to the presence of black tension lines that connect that cell surface to the inner nucleus (not visible due to black background) and thus, which provide tensional continuity.

Importantly, over the past fifteen years, we have been able to demonstrate that "stick and string" tensegrity models can predict many of the complex behaviors of living cells and tissues, including how cells change shape and polarize when attached to rigid or flexible substrates and how the cell, cytoskeleton, and nucleus are organized hierarchically so as to permit coordinated changes in form [8] [9] (Fig. 3.2).

The smaller and larger spherical structures (here representing the nucleus and the whole cell, respectively) each exhibit their own independent self-stabilizing properties, however, when connected by continous tension (tensional integrity), smooth structural integration between part and whole results. In living cells, contractile actomyosin-based "microfilaments" (the same ones that generate tension in muscle) form a continous porous lattice that transmits tension throughout the cell. This tensile force is resisted and balanced by a series of hollow molecular tubes called "microtubules" which act like compression struts inside the cell. Cross-linked bundles of molecular "actin microfilaments" can similarly resist inward-directed tension in other regions of the cell, particularly at the edge of the cell that leads its forward movement. Another biopolymer system ("intermediate filaments") connects microtubules and contractile microfilaments to each other as well as to the surface membrane and central nucleus and hence, transmits tension to all elements of the cell.

Shape stability in the cell also depends on establishment of a tensegrity force balance at the next larger hierarchical level. Tractional forces generated by the whole cell are resisted by local compression-resistant regions of the underlying molecular attachment substrate, known as the "extracellular matrix", which links together cells in all multicellular tissues. Decrease the rigidity of the matrix and cells spontaneously retract and round up. Thus, once again, shape stability requires maintenance of local compression resistance and global tension. Furthermore, this form of geodesic architecture is also used at the molecular size scale (the next lower hierarchical level) in living cells. For example, we showed that tensegrity structures can predict how a continous three dimensional network of contractile molecular filaments can mechanically rearrange to form linear bundles at the base of the cell and fully triangulated geodesic domes ("actin geodomes") at the top of the same cell, structures that been observed in living cells [9] [13]. The existence of geodesic domes in the cytoskeleton at the molecular level, by definition, confirms that living cells use tensegrity architecture.

Importantly, tensegrity is not just useful for depicting how patterns form in the cytoskeleton. It also can explain the mechanical behavior of living cells. A few years ago, we developed a method using magnetic microbeads bound to cell surface receptors whereby we could apply a controlled mechanical twist (torque) to living cells and simultaneously measure the mechanical response (rotational strain) of the cell. Using this approach, we found that living cells exhibit a mechanical behavior that is also shared by most living tissues, known as a monotonic stiffening response [20]. In simplest terms, the mechanical stiffness of the cell (and tissue) increases in direct proportion as the level of applied stress is raised over a wide range of applied stresses. Although this has been known to be a fundamental property of living tissues for many years, there was no known mechanism for this behavior and it could not be explained starting from first principles. We found that tensegrity could explain this behavior [20]. Working with Dimitrije Stamenovic (Mechanical Engineering Dept., Boston U.), we also were able to develop a mathematical basis for this response starting from first principles [19] [4]. This model reveals that the ability of tensegrity structures to resist shape distortion and self-stabilize depends on two key features: architecture and prestress. We have recently confirmed that the mechanical stiffness of living cells similarly can be modulated by either altering cell spreading (cytoskeletal architecture) [21] or the level of internal tension in the cytoskeleton (actomyosin filament sliding) [7].

4. Mechanochemical control of biochemistry and gene expression

It is important to emphasize that the cytoskeleton is not just a supporting structure. It also plays a key role in the control of cellular biochemistry. For example, work from our lab and many others has revealed that much of the cell's metabolic machinery effectively functions in a "solid-state. Many of the key enzymes and substrates that mediate DNA synthesis, RNA processing, protein synthesis, glycolysis, and signal transduction are physically immobilized on insoluble molecular scaffolds within the cytoplasm and nucleus [10]. This arrangement results in two new emergent properties: 1) the incredible efficiency of biochemical reactions and cross-talk between different chemical pathways that we observe in living cells and can not mimic in a test tube, and 2) the creation of a mechanism by which physical forces and mechanical distortion of cells can alter biochemical reactions. An in depth description of this tensegrity-based mechanism of cellular mechanotransduction has been recently published [11].

We also have demonstrated the importance of mechanics for biochemical regulation directly by developing a system whereby we can artificially engineer cell shape [17] [3]. Using a micropatterning approach to create adhesive islands with defined size and shape on the micron scale, we could literally make cells change shape to take the size and form of their container, including the creation of square cells with 90 degree corners. We then showed that cells can be switched between different gene programs for growth, differentiation, or death in the presence of saturating amounts of soluble growth factors, simply by mechanically distorting (spreading or retracting) the cells. These results bring home a clear message: to understand how the complex adaptive living system that we call the "cell" evolves, we must take into account mechanics and structure as well as form and pattern.

5. The architecture of life

A clue to the puzzle of how life forms comes from the existence of recurring patterns in Nature. For example, the geodesic structure that forms at the molecular level in the cytoskeleton of living cells is an excellent example of a design that nature uses again and again, independent of size scale. Perhaps it is our emphasis on reductionism or our focus on pattern rather than mechanics; however, no one seems to have ever asked why does this particular form appear again and again? Complexity theory can provide a mathematical explanation for why chaotic systems will spontaneously crystallize into order over time and even provide a feel for the time-dependence of this self-organization behavior. However, it will not explain how or why specific three dimensional structures form.

We recently realized that we can gain some insight into the invisible force that guides self-assembly in natural systems by analyzing how natural geodesic structures self-organize [12]. For example, both viruses composed of proteins and Buckyballs composed from carbon atoms exist in the form of fully geodesic spheres. Both structures are often visualized by close packing multiple spherical subunits along the surface of a sphere. However, this is not how they are constructed. In the case of viruses, the tails of the coat proteins often interact to assemble into an internal geodesic protein scaffold. The vertices of this network form as a result of multiple adjacent protein tails overlapping in a consistent manner, with a specific handedness. These overlapping protein tails, in turn, stabilize in space through use of tensegrity: the pull of intermolecular attractive forces (hydrogen bonds) is balanced by the ability of these elongated molecules to repulse one another and to resist buckling (compression) locally. This really is the only way to build at this level since structural elements never really "touch" at this size scale and thus, continuous compression would be impossible. Buckyballs build in the same manner except that rather than overlapping protein strands in the vertices, they overlap stiffened carbon-carbon bonds (the C60 molecules are comprised of 90 similarly oriented C-C bonds). Both of these structures (virus and Buckyball) can be effectively modeled using a series of isolated sticks that overlap with a specific handedness in each vertex and that are then stabilized by interconnecting their ends with tensile filaments (mimicking tensile attractive bonds) [12]. Interestingly, Kenneth Snelson also has published an elegant model of atomic structure based on an internal tensegrity balance between global tension and local compression that offers an exciting alternative to conventional views based on quantum mechanics [18] and Buckminster Fuller offered a geodesic architecture-based explanation for why the natural elemental atoms exhibit eight recurring types of material behavior, as categorized in the Periodic Table [6].

The relation between structure and pattern is clearly evident in fully geodesic forms. However, even irregularly shaped structures, such as proteins, may use tensegrity to self-stabilize [12]. For example, proteins are themselves hiearchical structures that are composed of multiple smaller cassettes (motifs) that exhibit characteristic shape and mechanics. Both the motif elements and the whole protein gain their shape stability through continuous tension and local compression. Multiple portions of the single protein strand are locally stiffened and straightened as a result of secondary structure (e.g., alpha-helix formation). These stiff portions are interconnected by intervening hinge regions of the strand that are relatively flexible. The linear regions rearrange (fold) until they reach a stable configuration in which the hydrogen bonding (tensile) forces between neighboring stiffened regions of the strand are balanced by their ability to resist local compression. Through this process, proteins become prestressed such that local regions may be highly stressed even though the whole structure is in equilibrium. In fact, it is well known that cleaving a single bond in many proteins will cause the entire molecule to splay open, thus indicating the existence of internal tension. Proteins are therefore tensegrity structures. Interestingly, mechanical measurements have been recently carried out on single molecules, such as DNA; these results confirm that molecules also exhibit monotonic stiffening behavior. Taken together, these observations raise the possibility that mathematical approaches based on tensegrity might provide a short-cut for molecular modeling and drug design.

6. The evolution of form

Why tensegrity? Why should this building system guide pattern formation at so many different size scales? Well, because only this type of building system can provide a mechanism for hierarchical integration of many complex adaptive systems and subsystems, such that part and whole function as one. Pulling on one local element in a tensegrity system results in global structural rearrangements (and information transfer) throughout the entire structure. Furthermore, flexibility can be maintained even when rigid building elements (e.g., bones) are incorporated into the system. Use of structural hierarchies and porous networks also maximizes energy dissipation and structural efficiency which means increased stability and minimized metabolic costs [2]. In addition, tensegrity structures function as coupled harmonic oscillators [16] and thus, they provide a means to distribute forces to all interconnected elements and, at the same time, to mechanically couple or "tune" the whole system as one harmonious functioning unit [11].

If tensegrity provides a guiding force for self-assembly, then evolution represents a problem in structural engineering. I propose that life has evolved through self-assembly of hierachical clusters, first of atoms, then molecules, and finally of cells, that exhibit mechanical stability and structural efficiency, yet retain the flexibility necessary to interconvert between different stable forms. Flexibility allows these clusters to explore potential evolutionary space in which adaptation and natural selection can proceed; stiffness is death (rigor mortis). In this context, it is extremely interesting that the presentation by Dr. Hyman Hartman at this meeting raised the possibility that the first life forms emerged from clay whose molecular structure is at once both geodesic and flexible. Perhaps it is through both of these features that the catalytic and self-assembling capabilities of clay first evolved. I refer the reader to Dr. Hartman's work [1] which also explains how clay-based catalysis may have led to the evolution of the first metabolic cycles, CO_2 fixation, and the further acceleration of hierachical clustering.

Thus, whether in the inorganic or organic world, the key elements for self-assembly and structural evolution appear to be very simple: the goal is to self-stabilize in three dimensions while minimizing energy, optimizing structural efficiency, and yet still maintaining flexibility (deformability relevant for that particular size scale). Without flexibility, there can be no effective "joining" and hence, no hierarchical assembly. Once two or more flexible components contact each other in their potential evolutionary landscape, they may either repel or self-organize by similar rules to form a new stable, but flexible structure. If this new structure has a selective or adaptive advantage, its place in the population will be assured. The newly created cluster now changes the local evolutionary landscape and the process of adaptation/natural selection accelerates. From this perspective, it becomes clear that genes are not the driving force behind evolution; they are a product of this process which served to further accelerate the range of possible three dimensional forms that can interact and self-organize.

7. Conclusion: Simplicity in complexity

The major relevant conclusion for this meeting is that there may be simplicity in the field of complexity. For example, the simple tensegrity models shown in Fig. 3.1 and Fig. 3.2 provide an elegant solution to many of the major unanswered questions in the complexity field. Essentially, tensegrity may explain how self-assembly, emergence, and hierarchical integration occur in many different types of complex adaptive systems.

Self-organization results from self-stabilization in three dimensions which, in turn, is determined by a balance of opposing tensile (attractive) and compressive (repulsive) forces. The pattern of interacting dots shown in Fig. 3.1 may be viewed simply as a constellation with this characteristic form or, alternatively, as a three dimensional structure with different connectors linking the different dots or "nodes". Complexity-based approaches often incorporate the latter model without considering the material properties of the connectors or internal force balances. However, it is only when a subset of the connectors function as compression struts and the others as tensile strings that the system will self-stabilize through tensegrity. The flexibility necessary to explore potential evolutionary space is also inherent in prestressed systems such as this, even when highly stiffened struts are utilized. Furthermore, greater flexibility can be obtained without disrupting any of the critical structural relationships, simply by decreasing the internal tone in the system.

Another key feature of self-assembly of this new stable system is the development of emergent properties, such as linear stiffening behavior and shape stability, that could not be predicted by analyzing the material properties of the individual building components in isolation. Tensegrity structures also may be constructed hierarchically, as long as the different subsystems are linked by continuous tension and not compression. In fact, the particular model shown in Fig. 3.2 which demonstrates smooth coupling between large and small tensegrity networks embodies the concept of a "systems jump" within any hierarchical assembly. Thus, tensegrity provides a way to mechanically integrate and tune, part and whole as one. Yet, another advantage is that subsystems may be repaired and replaced without disruption of higher order structure. This is a critical feature in all living systems. The cells and molecules within tissues undergo continual turnover; it is the maintenance of characteristic architectural relationships that we call "life".

One can also envision a tensegrity system in which every strut or cable in the network is itself an independent tensegrity system on a smaller size scale that interconnects with many similarly organized networks to form the larger whole. In essence, this is how the human body is constructed. I suspect that similar rules will guide self-assembly whether it describes interacting societies of atoms, molecules, cells, humans with their characteristic attractions and repulsions, or even the Universe itself with its newly identified porous network of gravitationally-linked galaxies. Only time will tell.

Bibliography

[1] CAIRNS-SMITH, A.G., HARTMAN, H., eds., "Clay Minerals and the Origin of Life", Cambridge University Press, 1986.

[2] CHEN, C., INGBER, D.E., "Tensegrity and mechanoregulation: from skeleton to cytoskeleton", *Osteoarthr. Artic. Cartil.* - in press.

[3] CHEN, C.S., MRKSICH, M., HUANG, S., WHITESIDES, G., INGBER, D.E.," Geometric control of cell life and death", *Science* 276:1425-1428 (1997).

[4] COUGHLIN, M.F., STAMENOVIC, D., "A tensegrity model with buckling compression elements: applications to cell mechanics", *ASME J. Appl. Mech.* 64:480-486 (1997).

[5] FULLER, R.B., "Tensegrity" *Portfolio Artnews Annual* 4:112-127 (1961).

[6] FULLER, R.B., "Conceptuality of fundamental structures", *Structure in Art and in Science*, (KEPES, G., ed.) Braziller:New York, pp. 66-88 (1965).

[7] HUBMAYR, R., SHORE, S.A.,FREDBERG, J.J., PLANUS, E., PANETTIERI, R.A. Jr., MOLLER, W., HEYDER, J., WANG, N., "Pharmacological activation changes stiffness of cultured human airway smooth muscle cells", *Am. J. Physiol.* 271:C1660-C1668 (1996).

[8] INGBER, D.E.,JAMIESON, J.D., "Cells as tensegrity structures: architectural regulation of histodifferentiation by physical forces tranduced over basement membrane", *Gene Expression During Normal and Malignant Differentiation* (ANDERSSON, L.C., GAHMBERG, C.G., EKBLOM, P., eds.) Academic Press (1985) pp. 13-32 .

[9] INGBER, D.E., "Cellular Tensegrity: defining new rules of biological design that govern the cytoskeleton", *J. Cell Sci.* 104:613-627 (1993).

[10] INGBER, D.E., "The riddle of morphogenesis: a question of solution chemistry or molecular cell engineering?", *Cell* 75:1249-1252 (1993).

[11] INGBER, D.E., "Tensegrity: the architectural basis of cellular mechanotransduction", *Ann. Rev. Physiol.* 59: 575-599 (1997).

[12] INGBER, D.E., "The Architecture of Life", *Scientific American* January, 1998- in press.

[13] LAZARIDES, E., "Actin, alpha-actinin, and tropomyosin interactions in the structural organization of actin filaments in nonmuscle cells" *J. Cell Biol.* 68:202-219 (1976).

[14] MANIOTIS, A., BOJANOWSKI, K., INGBER, D.E., "Mechanical continuity and reversible chromosome disassembly within intact genomes microsurgically removed from living cells", *J. Cellul. Biochem.* 65:114-130 (1997).

[15] MANIOTIS, A., CHEN, C., INGBER, D.E., "Demonstration of mechanical connections between integrins, cytoskeletal filaments and nucleoplasm that stabilize nuclear structure", *Proc. Natl. Acad. Sci. U.S.A.* 94:849-854 (1997).

[16] PIENTA, K.J., *Coffey*, D.S., "Cellular harmonic information transfer through a tissue tensegrity-matrix system", *Med. Hypoth.* 34:88-95 (1991).

[17] SINGHVI, R., KUMAR, A., LOPEZ, G., STEPHANOPOULOS, G.N., WANG, D.I.C., WHITESIDES, G.M., INGBER D.E., "Engineering cell shape and function", *Science* 264:696-698 (1994).

[18] SNELSON, K., "The Nature of Structure", *Program from an exhibition by the New York Academy of Sciences* (1989).

[19] STAMENOVIC D., FREDBERG J., WANG N., BUTLER J, INGBER, D.E., "A microstructural approach to cytoskeletal mechanics based on tensegrity", *J. Theor. Biol.* 181: 125-136 (1997).

[20] WANG, N., BUTLER, J.P., INGBER, D.E., "Mechanotransduction across the cell surface and through the cytoskeleton", *Science* 260: 1124-1127 (1993).

[21] WANG, N., INGBER, D.E., "Control of cytoskeletal mechanics by extracellular matrix, cell shape, and mechanical tension", *Biophys. J.* 66:2181-2189 (1994).

Chapter 28

Information transfer between solitary waves in the saturable Schrödinger equation

Mariusz H. Jakubowski

Ken Steiglitz

Princeton University, Computer Science Dept.

Richard Squier

Georgetown University, Computer Science Dept.

In this paper we study the transfer of information between colliding solitary waves. By this we mean the following: The state of a solitary wave is a set of parameters, such as amplitude, width, velocity, or phase, that can change during collisions. We say information is transferred during a collision of solitary waves A and B if the state of B after collision depends on the state of A before the collision. This is not the case in the cubic nonlinear Schrödinger, KdV, and in many other integrable systems. We show by numerical simulation that information can be transferred during collisions in the (nonintegrable) saturable nonlinear Schrödinger equation. A seemingly complementary feature of collisions in this and similar systems is radiation of energy. We give results which show that significant information can be transferred with radiation no greater than a few percent. We also discuss physical realization using recently described spatial solitary light waves in a saturable glass medium.

1. Introduction

A solitary wave can carry information in its envelope amplitude, width, and position; its group and phase velocities; and its carrier phase; and this information can be exchanged in collisions with other solitary waves. This paper is devoted to the question of whether such information transfer can occur in a way that is useful as a basis for computation, while still preserving particle identities. If this is possible, it suggests that general computation can be performed via interacting waves in a uniform medium, such as a nonlinear optical material.

In the usual conception of optical computing, one builds discrete gates based on the propagation of light, and then essentially mimics the construction of a conventional computer. We describe here an alternative approach to building an all-optical computer, using only solitary waves in a homogeneous nonlinear optical medium. In our approach, programs and data are encoded as streams of solitary waves, which are injected into the medium at a boundary, and which compute via the information transfer effected by solitary-wave collisions.

A general Turing-equivalent model for such "gateless" computation, the *particle machine*, was introduced in [22, 23]. By exploiting the fine-grain parallelism of particle systems, this model supports fast and efficient execution of many operations, including arithmetic and convolution. Briefly, particle machines treat solitary waves as particles whose collisions can change particle states, thus performing computation. Such computation requires that if solitary waves A and B collide, then some part of the resulting state of A depend on the state of B; that is, information should be transferred from one wave to the other with "interesting" information-transformation properties.

For our purposes, these "interesting" properties are the opposite of those usually considered interesting in optics: At least some of the collision products must effect a nontrivial transformation of information in the colliding waves. The reason for this is that general computation requires a transform of information in basic logic operations. Unfortunately, many commonly studied systems that support waves do not have this behavior. For instance, because of linear superposition, colliding plane waves in a linear medium do not interact, i.e., do not undergo any state changes and therefore cannot have information interaction among colliding waves.

An example of a system in which collisions do cause change of state but nevertheless cannot transform information in a nontrivial manner is the cubic nonlinear Schrödinger equation (3-NLS). In order to do computation, solitary waves must carry information from one collision to the next; such information must be coded in parameters that are not constant. However, in the 3-NLS system, the state parameters that cause the information transfer are themselves invariant: The only change of state occurs in spatial (or temporal) position and carrier phase, and this change depends only on the amplitudes and velocities of the envelopes of the incoming solitons. We conjecture in [24, 25] that all solitary-wave collisions in integrable systems have this property, and we show that particle machines based on such systems are very limited in computational power. In particular, these particle machines are not Turing-equivalent. We must therefore look to solitary waves in nonintegrable systems for computationally useful collisions.

For solitary waves to carry information, they must also preserve their integrity after a sequence of collisions, and lose negligible energy through radiation. These requirements are apparently antagonistic to the information-transform capability necessary for computation, but our goal is to find systems which meet all these requirements. The results shown here suggest that the saturable nonlinear Schrödinger equation (sat-NLS) describes such a system.

2. Information transfer

To be more precise about the definition of *information transfer*, suppose that a medium supports a set of solitary waves. Then a selected set of properties that can change during a collision define a *state* $S(A)$ of a wave A, whereas a set of constant wave properties that are unaffected by collisions define an *identity* $I(A)$ of A. Note that we may define different types of states S and identities I for the same type of wave. Denote by A' the solitary wave A after a collision with wave B. Then a collision of A with B supports transfer of information if $S(A')$ depends on $S(B)$, for some $S(A)$ and $S(B)$; otherwise, the collision transfers only trivial information (if $S(A')$ depends on only $I(B)$) or no information (if $S(A') = S(A)$).

We illustrate the above definition using the the cubic-NLS and the saturable-NLS systems. The cubic-NLS equation supports solitons whose variable states are phases, and whose constant identities are amplitudes and velocities. Collisions of such solitons transfer only trivial information, since the phase shifts due to soliton collisions are a function of only the amplitudes and velocities, i.e., the identities, of the colliding solitons. On the other hand, the saturable-NLS system gives rise to solitary waves whose variable state includes phases, amplitudes, and velocities. This system supports collisions that transfer nontrivial information, since the state changes due to collisions are a function of the states of the colliding waves.

3. Computational power

To examine how information transfer relates to computational power, we briefly review the notion of Turing equivalence, or computational universality. Informally, a Turing machine is a computational model in which programs and data are stored on an infinite tape of discrete cells. A read-write head processes information by reading cell contents, writing new cell contents, and moving back and forth along the tape, all according to a transition function that considers both the state of the head and the symbol read from underneath the head. The machine can enter a special "halt" state, which signals the end of computation and the presence of the machine's final output on the tape.

It is generally accepted (by virtue of Church's thesis [11]) that given enough time and space, a Turing machine can implement any algorithm; that is, in terms of the results that can be computed, a Turing machine is as powerful as any computer. A computational system is Turing-equivalent, or computation-universal, if it can simulate a Turing machine. While this property is not absolutely necessary for a system to do useful computation, universality nevertheless serves as a good measure of a system's potential to do such computation.

Intuitively, in order for computation to take place in a solitary-wave system, colliding waves should interact and transfer information that is necessary to execute steps of an algorithm. In [25] we show that only at most cubic-time computation can be done on a particle machine that models a system in which collisions transfer at most trivial information. This upper bound on an such a system's computation time proves that this system cannot be Turing-equivalent, since universal computation can take an arbitrarily long time. Moreover, solitary-wave interactions in this system are computationally very limited, and designing algorithms based on these interactions appears tedious and impractical. It is unclear whether or not collisions supporting only trivial information transfer can encode any useful computation at all.

Solitary-wave systems in which collisions transfer nontrivial information are more readily applicable to encoding computation. We have shown in [25] that such a system can be Turing-equivalent, provided that the solitary-wave state changes are sufficiently complex.

4. The NLS equation and its solutions

To study the information-transfer capabilities of NLS solitary waves, we first review the one-dimensional NLS equation and its solutions. We consider the following form of the NLS equation [14, 18]:

$$-i\frac{\partial u}{\partial t} = a\frac{\partial^2 u}{\partial x^2} + N(|u|)u \qquad (4.1)$$

Here x is space, t is time, u is the complex amplitude of waves described by the equation, a is real, and $N(|u|)$ is a nonlinear function of $|u|$. For 3-NLS, $N(|u|) = k|u|^2$, and for sat-NLS, $N(|u|) = m + k|u|^2/(1 + |u|^2)$, where k and m are real constants.

The nonlinearity $N(|u|)$ determines the integrability of eq. 4.1, and the existence of closed-form solitary-wave solutions. To find solitary waves, either analytically or numerically, we assume that each such wave consists of an envelope modulating a sinusoidal carrier wave. Following [17], we make the ansatz

$$u(x,t) = \Phi(x - u_e t)e^{i\theta(x - u_c t)} \qquad (4.2)$$

where $\Phi(x - u_e t)$ is the envelope, $e^{i\theta(x - u_c t)}$ is the carrier, and u_e and u_c are the envelope and carrier velocities, respectively. We find that the carrier function θ

is given by

$$\theta(x - u_c t) = \frac{u_e}{2a}(x - u_c t) + \phi_0 \tag{4.3}$$

where ϕ_0 is an arbitrary constant. The envelope function Φ can be found from

$$x - u_e t = \int_{\Phi(0)}^{\Phi(x - u_e t)} \frac{d\Phi}{\sqrt{\alpha \Phi^2 - 2 \int N(|\Phi|) d\Phi}} \tag{4.4}$$

where $\alpha = (u_e^2 - 2 u_e u_c)/(4a)$.

If the integral in eq. 4.4 can be evaluated analytically and used to solve eq. 4.4 for the envelope function $\Phi(x - u_e t)$, then eq. 4.2 gives an exact expression for a solitary wave, as is the case with the 3-NLS equation. Otherwise the integral and $\Phi(x - u_e t)$ can be computed numerically, using boundary conditions chosen to yield solitary waves. We explain how to do this to obtain the sat-NLS solitary waves used in our numerical simulations.

We consider the following form of the sat-NLS equation:

$$-i\frac{\partial u}{\partial t} = \frac{\partial^2 u}{\partial x^2} + \left(m + \frac{k|u|^2}{1 + |u|^2}\right)u \tag{4.5}$$

Here m and k are real constants. Solitary-wave solutions are of the form of eq. 4.2, where the carrier θ is given by eq. 4.3. The envelope Φ can be found from eq. 4.4, which simplifies to

$$x - u_e t = \int_{\Phi(0)}^{\Phi(x - u_e t)} \frac{d\Phi}{\sqrt{c\Phi^2 + k \log(1 + \Phi^2)}} \tag{4.6}$$

where $c = \alpha - m - k$. We evaluate the above integral numerically using the boundary conditions $\Phi(0) = A$ and $\Phi(\pm\infty) = 0$, where A is the maximum amplitude of the envelope and is determined by u_c and u_e. The integration yields $x - u_e t$ as a function of the envelope $\Phi(x - u_e t)$. We invert the result of the integration to compute the envelope $\Phi(x - u_e t)$ as a function of $x - u_e t$. We multiply this numerical envelope by the exact carrier (eq. 4.3) to plot solitary waves on a discrete one-dimensional grid, and use numerical methods to study the behavior of propagating and colliding waves.

Note that sat-NLS solitary waves are characterized by four parameters: amplitude (A), envelope velocity (u_e), carrier velocity (u_c), and phase (ϕ_0). Using eq. 4.6, it can be shown that any two of A, u_e, and u_c determine the third. We may choose ϕ_0 freely, so that there are three degrees of freedom in choosing the initial state of a sat-NLS solitary wave.

5. Information transfer in collisions of NLS solitary waves

In the integrable 3-NLS system, solitary waves are true solitons whose collisions can change only their envelope position and carrier phase; envelope amplitude

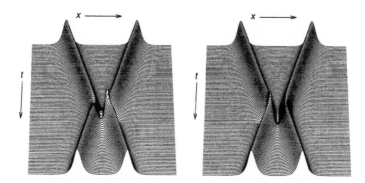

Figure 5.1: Trivial information transfer in collisions of 3-NLS solitons. The initial relative phases of the solitons in the left and right graphs are 0.25π and -0.45π, respectively; velocities are ± 0.2. Phase and spatial shifts, though not apparent from these graphs, are a function of only the constant amplitudes and velocities.

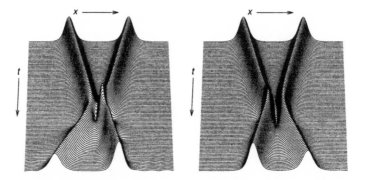

Figure 5.2: Nontrivial information transfer in collisions of sat-NLS solitary waves. The initial relative phases of the waves in the left and right graphs are 0.25π and -0.45π, respectively; velocities are ± 0.2.

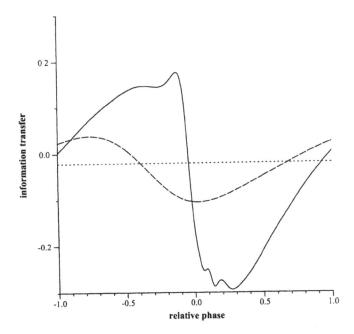

Figure 5.3: Information transfer for collisions of two sat-NLS solitary waves. Here information transfer is defined as the fractional change in the amplitude of one solitary wave; that is, the transfer is equal to $\Delta A_1/A_1$, where A_1 is the initial amplitude of the right-moving wave, and ΔA_1 is the amplitude change due to collisions. The solid, dashed, and dotted curves show information transfer for collisions of two waves with amplitudes 1.0 and velocities ± 0.5, ± 1.5, and ± 10.0, respectively. Relative phase is in multiples of π.

and velocities are conserved. In addition, the spatial and phase shifts of colliding solitons depend only on their constant amplitudes and velocities. Thus, such collisions transmit only trivial information, and are computationally very limited, as we demonstrate in [25]. (See Fig. 5.1.)

The nonintegrable sat-NLS equation gives rise to solitary waves whose collisions support nontrivial information transfer. (See Fig. 5.2.) In particular, phases, amplitudes, and velocities can all change as a function of the parameters of the colliding waves. We have observed that the most computationally useful collisions occur when the solitary waves have a low relative speed (approximately 4.0 and below). The magnitude of information transfer decreases gradually as the relative speed of the waves increases. To estimate this magnitude, we measured the amplitude and velocity changes following collisions of low-velocity waves at various initial phases. In Fig. 5.3, one measure of information transfer is plotted as a function of the relative phase of two colliding solitary waves.

6. Radiation

In general, computation encoded in an NLS system must reuse solitons after they have been involved in multiple collisions. To behave like the particles of a particle machine, these solitary waves should be stable; more specifically, collisions should preserve the identities of solitary waves and generate negligible radiation.

Numerical results reveal that information transfer and radiation often go hand in hand. Soliton collisions in the 3-NLS system are perfectly elastic and generate no radiation, but such collisions support only trivial information transfer, as we have seen. In the sat-NLS system, large amounts of radiation tend to accompany large magnitudes of information transfer. However, much like other known nonintegrable NLS systems [14], the sat-NLS system does support collisions that transfer information and yet generate only small amounts of radiation. More specifically, our numerical studies have revealed the following:

- When at least one of the solitary waves is moving at a high speed (approximately 4.0 and above), their collision generates negligible radiation and supports no measurable information transfer.

- When the relative phase $\phi_0 = \pi$, the collision is the same as in the above case, no matter what the value of the relative speed v_0.

- When both waves have low speeds (below 4.0) and $0 \leq \phi_0 < \pi$, the collision is accompanied by large amounts of radiation and information transfer. However, as ϕ_0 tends towards π, both radiation and the magnitude of information transfer decrease. For $\phi_0 > \pi/2$, very little or no measurable radiation is generated.

The solitary waves that emerge from collisions in the sat-NLS system may or may not be of the form given by eq. 4.2, depending on the initial wave parameters. When wave velocities are very low (< 1.0) and relative phases are approximately in the range 0.0 to 0.3, collisions produce "breathers," or waves whose amplitude pulsates regularly, which cannot arise from eq. 4.2. However, we observed that other collisions result in waves which can be specified by eq. 4.2. To test this, we measured the amplitudes, envelope velocities, and phases at the peaks of waves after collisions; we then used these parameters to plot "fresh" waves and to compare their characteristics with those of the post-collision waves. In particular, we compared the carrier velocities of the "fresh" and post-collision waves, and observed what happens in collisions between two "fresh" waves and between two post-collision waves. The results do suggest that the post-collision waves have the form of eq. 4.2.

We estimated radiation for the collisions of Fig. 5.3 by finding the fixed-size section of the numerical-solution grid with the lowest root-mean-squared (rms) norm of the grid points.[1] Ideally, this rms norm should be very close to zero

---

[1] We use circular boundary conditions in our numerical simulations, so that any radiation generated by collisions remains in the system.

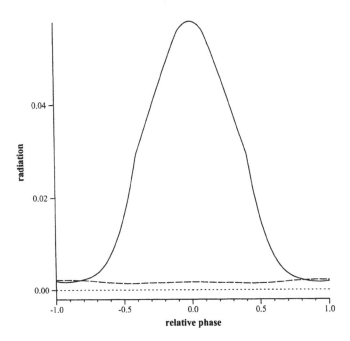

Figure 6.1: Radiation due to collisions in the sat-NLS system. Radiation is computed by finding the section of the numerical-solution grid with the lowest rms norm of grid points, using sections of size $N/10$, where N is the the the size of the entire grid; radiation is taken to be this lowest rms norm. The solid, dashed, and dotted curves show radiation for collisions of two solitary waves with velocities ±0.5, ±1.5, and ±10.0, respectively. Relative phase is in multiples of π.

for solitary waves. Numerical error caused by the discrete nature of time and space in the grid contributes some noise, which we measured for the analytically solvable case of the 3-NLS by comparing numerical results with exact solutions. Based on these investigations, it appears that our simple measure of radiation gives a good general idea of the usefulness of various collisions for computation. In Fig. 6.1, radiation is plotted as a function of the relative phase of two colliding waves.

The sat-NLS solitary waves that appear to hold promise for encoding computation have relative speeds from about 0.2 to 4.0, and relative phases whose absolute values range from about 0.2π to 0.8π.

7. Physical realization

In this section we mention some physical systems that we might expect to be described by the saturable Schrödinger equation. We are continuing research in this direction, with the hope that computationally useful information-transferring collisions will be observed experimentally.

The integrable 3-NLS equation describes soliton propagation in so-called *Kerr materials* — materials in which the operative nonlinearity is due to the Kerr effect. In such materials $\chi^{(2)} = 0$, and the dominant nonlinearity in the dependence of refractive index on electric field intensity is due to $\chi^{(3)} \neq 0$. This is the case for centrosymmetric and isotropic materials [3], and includes optical fibers in which soliton transmission has been demonstrated over long distances [4, 5].

The nonintegrable sat-NLS equation is applicable to simulating various physical phenomena, including the nonlinear effects of laser beam propagation in various media [14]. Sat-NLS also describes the recently discovered $1 + 1$-dimension photorefractive optical spatial solitons in steady state [18, 21], and the optical spatial solitons in atomic media in the proximity of an electronic resonance [28].

Both 3-NLS and sat-NLS describe temporal solitons; with the transformation $t \rightarrow z$, both equations also describe spatial solitons, with x and z being the transverse and longitudinal directions [9, 1]. In practice, spatial solitons appear better suited for computation, because temporal solitons require long distances to propagate. In addition, spatial solitons also exist in dimension $2 + 1$ [21, 28], offering an additional degree of freedom and suggesting the possibility of implementing two-dimensional universal systems such as the billiard-ball model of computation [15].

8. Conclusions

The analytically solvable 3-NLS equation supports only soliton interactions that transfer only trivial information, and is thus unlikely to support a useful computational system such as the particle machine [22, 23]. The nonintegrable sat-NLS equation supports solitary-wave collisions that transfer nontrivial information and generate acceptable radiation, and offers promise for encoding general computation through the particle-machine model. The next step in this line of work will likely involve searching for configurations of collisions that can be used for simple computations, such as ripple-carry addition. Such an algorithm was implemented using the solitons of a filter automaton [26], and we believe that spatial sat-NLS solitary waves support sufficiently general interactions to implement this algorithm in systems of dimension $1 + 1$. Spatial solitary-wave systems of dimension $2 + 1$ also offer possibilities for encoding computation.

Acknowledgments

We are greatly indebted to Dr. Mordechai Segev for invaluable help with questions of physical realization of NLS equations, and for showing us real sat-NLS solitary waves in his laboratory. Vivek Mathew, a summer research student, performed extensive numerical experiments with the NLS equation with quadratic nonlinearity. We also thank Drs. Herschel Rabitz, Jack Gelfand, Keren Bergman, and Stephen Lyon.

Bibliography

[1] AITCHISON, J. S, A.M. WEINER, Y. SILBERBERG, M. K. OLIVER, J. L. JACKEL, D. E. LEAIRD, E. M. VOGEL and P. W. SMITH, "Observation of spatial optical solitons in a nonlinear glass waveguide," *Opt. Lett.* **15** (1990).

[2] HASEGAWA, A., *Optical Solitons in Fibers*, (second edition), Springer-Verlag, Berlin, 1990. See pp. 14*ff*.

[3] NEWELL, A. C., and J. V. MOLONEY, *Nonlinear Optics*, Addison-Wesley, Redwood City, CA, 1992.

[4] MOLLENAUER, L. F., and K. SMITH, "Demonstration of soliton transmission over more than 4,000 km in fiber with loss periodically compensated by Raman gain," *Opt. Lett.*, **13** (1989), pp. 675-677.

[5] MOLLENAUER, L. F., R. H. STOLEN, and J. P. GORDON, "Experimental observation of picosecond pulse narrowing and solitons in optical fibers," *Phys. Rev. Lett.* **45** (1980).

[6] YAGI, T., and A. NOGUCHI, "Experimental studies on modulational instability by using nonlinear transmission lines," *Electronics and Communications in Japan*, **59-A** (11), 1976, pp. 1-6.

[7] BIAŁYNICKI-BIRULA, I. and J. MYCIELSKI, "Nonlinear wave mechanics," *Ann. of Phys.* **100** (1976), pp. 62-93.

[8] BIAŁYNICKI-BIRULA, I., and J. MYCIELSKI, "Gaussons: Solitons of the logarithmic Schrödinger equation," *Phys. Scripta* **20** (1979), pp. 539-544.

[9] CHIAO, R. Y, E. GARMIRE and C. H. TOWNES, "Self-trapping of optical beams," *Phys. Rev. Lett.*, **13** (1964).

[10] DRAZIN, P. G., and R. S. JOHNSON, *Solitons: An introduction*, Cambridge University Press, Cambridge, UK, 1989.

[11] HOPCROFT, J. E., and J. D. ULLMAN, *Introducton to Automata Theory, Languages, and Computation*, Addison-Wesley Publishing Company, Reading, MA, 1979.

[12] ISLAM, N., J.P. SINGH, and K. STEIGLITZ, "Soliton phase shifts in a dissipative lattice," *J. Appl. Phys.* **62** 2 (1987), pp. 689-693.

[13] ISLAM, M. N., *Ultrafast fiber switching devices and systems*, Cambridge University Press, Cambridge, UK, 1992.

[14] MAKHANKOV, V. G., *Soliton phenomenology*, Kluwer Academic Publishers, Norwell, MA, 1990.

[15] MARGOLUS, N., "Physics-like models of computation," *Physica* **10D** (1984) 81–95.

[16] OFICJALSKI, J., and I. BIAŁYNICKI-BIRULA, "Collisions of gaussons," *Acta Phys. Pol.* **B9** (1978), pp. 759-775.

[17] SCOTT, A. C., F. Y. F. CHU, and D. W. MCLAUGHLIN, "The soliton: A new concept in applied science," *Proceedings of the IEEE* **61** 10 (1973), pp. 1443-1483.

[18] SEGEV, M., G. C. VALLEY, B. CROSIGNANI, P. DIPORTO and A. YARIV, "Steady-state spatial screening solitons in photorefractive materials with external applied field," *Phys. Rev. Lett.* **73** (1994).

[19] SHEN, S. S., *A course on nonlinear waves*, Kluwer Academic Publishers, Norwell, MA, 1993.

[20] SHIH, M., and M. SEGEV, "Incoherent collisions between two-dimensional bright steady-state photorefractive spatial screening solitons," *Opt. Lett.* **21** (19) (1996).

[21] SHIH. M., M. SEGEV, G. C. VALLEY, G. SALAMO, B. CROSIGNANI and P. DIPORTO, "Observation of two-dimensional steady-state photorefractive screening solitons," *Elect. Lett.* **31** (10) (1995).

[22] SQUIER, R., and K. STEIGLITZ, "Programmable parallel arithmetic in cellular automata using a particle model," *Complex Systems* **8** (1994), pp. 311-323.

[23] SQUIER, R., K. STEIGLITZ and M. H. JAKUBOWSKI, "Implementation of parallel arithmetic in a cellular automaton," 1995 Int. Conf. on Application Specific Array Processors, Strasbourg, France, July 24-26, 1995.

[24] JAKUBOWSKI, M. H., K. STEIGLITZ, R. K. SQUIER, "Relative computational power of integrable and nonintegrable soliton systems," Fourth Workshop in Physics and Computation: PhysComp96, Boston, Nov. 22-24, 1996.

[25] JAKUBOWSKI, M. H., K. STEIGLITZ, and R. K. SQUIER, "When can solitons compute?" *Complex Systems*, to appear.

[26] STEIGLITZ, K., I. KAMAL, and A. WATSON, "Embedding computation in one-dimensional automata by phase coding solitons," *IEEE Trans. on Computers*, **37** 2, 1988, pp. 138-145.

[27] TAYLOR, J. R., Ed., *Optical solitons: Theory and experiment*, Cambridge University Press, Cambridge, UK, 1992.

[28] TIKHONENKO, V., J. CHRISTOU and B. LUTHER-DAVIES, "Three-dimensional bright spatial soliton collision and fusion in a saturable nonlinear medium," *Phys. Rev. Lett.*, **76** (1996).

[29] ZAKHAROV, V. E., and B. A. MALOMED, in *Physical Encyclopedia*, Great Russian Encyclopedia, Moscow, 1994.

Chapter 29

An integrated theory of nervous system functioning embracing nativism and constructivism

Brian D. Josephson

University of Cambridge, Department of Physics
Cambridge CB3 0HE, UK

Neural constructivists have proclaimed, on the basis of arguments which it is suggested in the following are unconvincing, a view of the nervous system which has little place for innate knowledge relating to specific domains. In this paper an alternative position is developed, which by instantiating innate knowledge in a flexible manner provides a more credible alternative. It invokes dynamical systems whose behaviour initially accords with specific algorithms but is modifiable during development, in combination with an object-oriented programming architecture that provides a natural means of specifying systems that have special knowledge of how to approach specific classes of situation.

1. Introduction

The algorithmic approach to describing the mind that characterised the discipline of artificial intelligence has proved inadequate in contexts such as the recognition of handwriting. In recent years an alternative, called neural constructivism, has received considerable attention [1, 2]. It is based on neural network models that are treated not merely as models of skill acquisition but also as models of the developmental process overall.

A number of constructivists (whom I shall refer to as evangelical constructivists) have proclaimed in their writings the falsity of nativism, which is the hypothesis that there exist innate neural systems dedicated to specific domains of activity such as language. They hypothesise instead that flexible learning networks, not oriented towards any specific domain, can achieve all that can be achieved by domain-specific networks.

Close examination of the arguments given against nativism indicates that they are based on a restricted view of the possible forms that innateness might take whilst, on the other hand, the arguments in favour of flexible systems rather than domain-specific ones are based on optimistic extrapolations of what has been achieved by the existing constructivist simulations. The comprehensive case that has been built up by workers such as Pinker [3] for there being a language instinct is dismissed on the basis of a small number of possible flaws in particular arguments which, in the author's opinion, affects very little the cogency of the overall case made by these workers. Crucially, the critics fail to address the central question: if, as the arguments of workers such as Pinker suggest, specific devices can considerably facilitate the acquisition of language, why should nature employ general-purpose networks alone?

It must nevertheless be acknowledged that the arguments in support of domain-specific mechanisms for language have been spelt out within the paradigm of algorithms as opposed to that of neural networks, and thus cannot give the full picture, which should be subject neither to the familiar limitations of the algorithmic point of view nor to the limitations of the evangelical constructivist position which denies the presence in any form in the initial system of algorithmic structures related to specific domains. The question poses itself how we may get beyond both categories of limitation, thereby combining the specificity and precision of an algorithm with the flexibility of constructivist neural networks. Models with this feature can be expected to have a validity and potentiality considerably in excess of both the standard algorithmic and the evangelical constructivist points of view. This paper proposes initial steps in such a direction.

2. Fundamentals of an integrated theory

2.1. Algorithmic and constructivist approaches compared

We begin by discussing in somewhat more detail the two basic approaches referred to above, the one based on algorithms and the other based on neural networks. The basic idea behind neural constructivism is that of taking as primary the level of the neuron, and looking for network architectures and weight-changing rules, often associated with an optimisation process, that can give a system a powerful ability to acquire a range of skills. The rules that a trained network instantiates, unlike the kinds of rules normally associated with the term algorithm, are submerged in the details of the network, and to the extent that rules that emerge as a result of training can be codified, they reflect primarily the nature and structure of the environment.

The alternative, consisting of a system that works purely on the basis of algorithms, is very often conceptualised and/or implemented in the context of digital computation, in which case the algorithm concerned is instantiated by a piece of code written in an appropriate language. This kind of model implies a discreteness and a corresponding lack of flexibility that contrasts significantly with neural network models.

2.2. A third option

The problems of discreteness and lack of flexibility can be circumvented by changing the basic concept from algorithms as such to dynamical systems whose behaviour simulates or equivalently is approximated by an algorithm as conventionally understood. The possibility of such systems undergoing transformations over time introduces a kind of flexibility not present in the traditional algorithmic models such as those of artificial intelligence. As with traditional neural network models, such transformations could come about in response to assessments of the observed performance. The mechanics of change might involve either changing weights in the style of traditional neural network models or, more creatively, linking in an appropriate new process as a response to a problem or a challenge, in the manner of traditional artificial intelligence models.

We thus have these three main ways of implementing a rule:

(i) as a piece of code running on a digital computer. This is the most rigid way and corresponds to traditional artificial intelligence;

(ii) as something that emerges through problem-solving activity in an environment, owing its structure almost entirely to the environment and the nature of the solutions to problems. This is the method favoured by the evangelical constructivists;

(iii) as a consequence of the initial dynamics of a modifiable structure. This provides a flexible mechanism for generating adaptive behaviour that can be modified by problem-solving activity, and combines the merits of (i) and (ii). The existence of universal behaviour patterns, apparently not learnt, in the developing child, which gradually transform into activity well-adapted to the solution of problems, provides a strong argument in favour of (iii) being the actuality.

2.3. Dynamical systems perspective

The essential requirement for a neural network to implement an algorithm is for there to be a limited number of degrees of freedom, corresponding to a specific state space that is not too complex. The parameters of the system (which govern the dynamical laws governing the dynamics of this state space) have then to be adjusted so that the resulting behaviour accords with the algorithm concerned.

We are in reality concerned not with just a single algorithm but with a large collection of algorithms. Correspondingly, our concern is with a collection of interacting dynamical systems. These systems are autonomous to the extent that they can be considered in isolation for the purposes of analysis, but are not necessarily physically separate from each other.

2.4. The object-oriented programming architecture and other concepts from computer science

The idea of practically relevant computations being performed by a collection of subsystems sending messages to each other is found both in Minsky's *Society of Mind* model [4] and in the powerful methodology of *object oriented programming* [5]. The *class construct* of the latter seems particularly relevant in regard to the way it treats the creation of new objects or subsystems. Objects belong to classes that are associated with specific types of situation, and when a new instance of that class is encountered a new object of that class is constructed, according to rules associated with the given class. This is, if we like, a form of innateness in the world of computer programs, this parallel being enhanced by the fact that there normally resides on the computer system concerned a large collection of classes with their defining algorithms and protocols.

What would correspond to this in the nervous system would be a modification of the architecture described by Quartz and Sejnowski [2], incorporating a collection of modules that can play the role of objects. These objects may not be initially domain specific, but mechanisms may exist which will permit domain-specific systems to take them over at critical times and use them for memory purposes related to the needs of the domain. For example, when one was listening to language, processes specific to language would take over a number of such systems and modify them so as to represent the information contained in the instances of language involved, for example by creating structures homologous to the deep structures of language. Such processes would work more effectively if the information that they were dealing with had been organised in a way that reflected the specific kinds of linguistic entities. Neural counterparts of the class construct of object-oriented programming can be expected to facilitate such activity.

The above account, with its reference to "objects belonging to classes that are associated with specific types of situation", may appear to imply that our model contains *only* systems primed with expectations specific to specific domains. This is not the case, because of the possibility of there being an *uncommitted class*. Objects in this class are constructed whenever a new situation of interest arises that has no previous parallels.

In the above it has been postulated that the architecture of the nervous system parallels in particular respects the design of conventional computer programs. This would be a very logical state of affairs, since modern computer programs and the nervous system are both systems designed to perform tasks successfully in very complex situations. It might be expected that, subject to the constraints imposed by hardware, both kinds of system would have evolved similar solutions to problems. Creating structures reflecting very general regularities in nature, as done with the classes of object-oriented programming, is one example of a kind of evolution in computer science that was dictated by the important requirement of minimising the unnecessary duplication of computational structures (which is assisted by means of object-oriented programming's *inheritance mechanisms*, something equally relevant to nervous system design). Two other examples from computer science that may similarly be of value are *assignment operations* and *threads*.

In the computing context, assignment of a value to a variable is a routine component of a computation. Its instantiation in the form of a pointer from a register representing the variable to a register containing a value is one that might be readily implementable in the nervous system.

A thread [5] is a computational mechanism representing a process that has continuity over time but has dormant phases, and protocols for starting up again. A neural equivalent would be useful in a number of ways, for example to implement a process which is interrupted until a problem connected with it is resolved, or in the context of delayed reinforcement, where the success or otherwise of a process may not be known until some time after the process is complete, and a thread may be programmed to wake up at the appropriate time. Again, in the context of working memory, a thread could hold the current state of a search process and start up the search process again if the current trial proved to be unsuccessful.

It is hypothesised, then, that inventions from computer science may be of considerable relevance in modelling and understanding nervous system design and function.

2.5. Algorithms, mappings and order

[section added following the talks and discussions at the ICCS conference]
It is useful to think of an algorithm as implementing a function, not only in the usual informal sense (a specialised means of achieving an end) but also in the sense of a *mapping* from an initial state to a goal state. Furthermore, the functionality in a biological context generally takes the form of a restriction of possibilites or local reduction of entropy (which does not contradict the second law of thermodynamics since there is a corresponding source of negentropy elsewhere). Thus the nervous system structure is a special kind of structure having this capacity to reduce entropy locally. The idea can be taken one step further: *classes* and their corresponding constructors are systems specialised to the *generation* of locally entropy-reducing structures.

2.6. Relevance of hyperstructure theory

In §2.2 there was reference to 'transformations coming about in response to assessments of the observed performance'. Baas [6], and private communication, has suggested that processes of this kind may lead to the generation of new systems through what he calls *observational emergence*, essentially the emergence of systems associated with a process or structure of behaviour in some sense selected or induced by a particular observation process, acting together with the constraints imposed by the environment. This construction process may be very specific to the details of the domains concerned and build up in a cumulative, multi-level manner.

2.7. Illustrative analogy

A very basic account of the kinds of processes that might be involved is contained in the following simple analogy, which makes no attempt to fit accurately the demands of actual situations. In it, note carefully that the engineer, manual and instructions are simply devices to aid the imagination and do not in general correspond to actual component systems. The system as a whole behaves *as if there were a person following instructions*, in the same way as a thermostatic system behaves as if there were a person carrying out instructions to turn on the heating if the room got too cold, but in neither case is there a person, or explicit instructions that a person might follow (cf. [7]).

In the analogy, an engineer is in charge of a complicated electronic system which he is constantly upgrading, following the rules given in a manual. The system is composed of a large number of modules which come successively into use over time. Many of the modules (in accord with the experimental evidence relating to the cerebral cortex) are initially general purpose, but as they become functional the engineer marks them with tags where appropriate to indicate their function. Lights on the modules light up to indicate which ones are currently in use or have recently been in use.

A buzzer indicates the occurrence of a new event, but generally an existing system is able to handle such an event and the buzzer is then switched off. If this is not the case, a hitherto unused module responds to the buzzer and puts on a special light to indicate its availability. The engineer then connects it with other illuminated modules in accord with the rules in the manual. When this has been done the new module can start to play its functional role. It remains illuminated in a special manner for some time afterwards to indicate its status as a newly operational module deserving special attention, and the behaviour of the system at the times when it is performing its function is evaluated and appropriate adjustments are then made by the engineer. This activity is supported by systems which can remember the situation in which the new system was functioning, which allows for the corresponding activity to be repeated as often as is necessary in order that the processes occurring during learning can take place over a shorter period of time. How the above proposals are likely to work in the case of language is discussed in §3.

2.8. Evolutionary aspect

The existence of systems that work harmoniously together to yield a high level of proficiency in a given domain can be understood on evolutionary grounds.

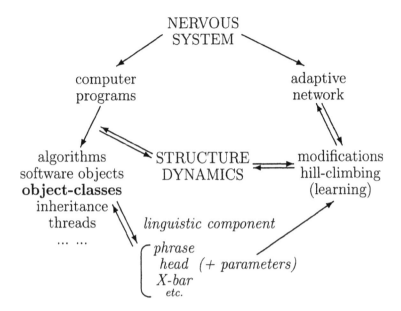

Figure 2.1: Diagram showing relationships between the various concepts discussed in this paper (see text for detailed explanations). Upper case is used for physical entities, and italics for constructs relating to language.

Mutations give rise to simple capabilities in a new domain which offer some increase in fitness. The way is then opened for a succession of further mutations which extend the existing domain or improve functioning in that domain. At every step the new features have to be coordinated with the existing ones (cf. the robot models of Brooks [7]), so at every step one has a properly coordinated system.

3. The case of language

How the above proposals may work in the case of language is indicated by an adaptation of a proposal of Elman et al. [1]:

> "If children develop a robust drive to solve [the problem of mapping non-linear thoughts on to a highly constrained linear channel], and are born with processing tools to solve it, then the rest may simply follow because it is the natural solution to that particular mapping process."

If the possibility of innate systems dedicated to language is not ruled out in accord with the dictates of the evangelical constructivists, then we can change the above account to allow for a collection of *specific drives relevant to aspects of language acquisition*, and for *specific tools that take into account universals of linguistic structure*, and the corresponding classifications.

The existence of drives equates to specific mechanisms being liable to be activated. When this happens the outcome is monitored and weight changes made that amount to learning. When one kind of skill has been developed, another level of process can spring into action.

Some proposals regarding how this might be implemented in practice for language have been made by Josephson and Blair [8], whose proposals for the acquisition of linguistic skills are consistent with the specific facts about language described for example by Pinker, and with the general picture of systems dedicated to handling specific *classes of situation* and *entities* developed here.

4. Diagrammatic representation of relationships discussed

Fig. 2.1 shows digrammatically the various relationships discussed in the paper. First of all, the nervous system has been considered in its two complementary aspects, as an adaptive network and as a system functionally equivalent in some respects to a computer program. The concept of a computer program can be opened out into the various aspects listed (algorithms, software objects, etc.). These aspects are related to the structure and dynamics of the nervous system (for example, the software objects are dependent upon the structure and its interconnectivities, and the functioning of an algorithm to the dynamics of the structure). On the other hand, the adaptive network aspect of the nervous system reveals itself as modifications of the structure which may be related to hill-climbing.

One section of the diagram is concerned specifically with language. It lists a number of constructs of language, such as the phrase, which may have specific correlates on the computing side, such as specific object-classes. According to this schema, the linguistic capacities of the nervous system arise in a process that involves two steps: implicit in descriptions of language such as those of universal grammar there are certain *classes* with corresponding algorithms, and these computational constructs are then implemented in the neural hardware. The parameters of universal grammar may enter via a different mechanism, for example as the outcome of a particular tendency of a learning mechanism's activities to continue in a way corresponding to the initial learning.

5. Summary

It has been shown that a dynamical systems approach can integrate the algorithmic and neural network approaches to development, permitting the respective advantages of both schemes to be both retained and integrated. This appears to correspond more closely to the reality than do the anti-nativist ideas popular among neural constructivists. Methodologies deriving from computer science that are likely to have correlates in the design of the nervous system, such as object-oriented programming with its way of handling specific categories of situation and threads, were discussed. It is to be hoped that the generic proposals of the present paper can be confirmed in the future by more detailed specification and model building.

Acknowledgements

I am grateful to Profs. Nils A. Baas and Michael Conrad for illuminating discussions on the concept of hyperstructure and questions of the flexibility of algorithms respectively.

Bibliography

[1] ELMAN, Jeffrey L., Elizabeth A. BATES, Mark H. JOHNSON, Annette KARMILOFF-SMITH, Domenico PARESI, and Kim PLUNKETT, *Rethinking innateness: A connectionist perspective on development*, MIT (1996).

[2] QUARTZ, Steven R. and Terrence J. SEJNOWSKI, "The neural basis of cognitive development: A constructivist manifesto", *Behavioral and Brain Sciences*, to be published.

[3] PINKER, Steven, *The language instinct*, Penguin (1994).

[4] MINSKY, Marvin, *The society of mind*, Heinemann (1987).

[5] CAMPIONE, Mary and Kathy WALRATH, *The Java tutorial: object-oriented programming for the Internet*, Addison-Wesley (1996), or on the World Wide Web at http://java.sun.com/docs/books/tutorial/index.html

[6] BAAS, Nils A., "Emergence, hierarchies and hyperstructures", *Artificial Life III, SFI Studies in the Sciences of Complexity, vol. XVII*, (Christopher G. LANGTON ed.), Addison-Wesley, (1994), 515–537.

[7] BROOKS, Rodney A., "A robust layered control system for a mobile robot", *IEEE Journal of Robotics and Automation* 2(1) (1986), 14–23.

[8] JOSEPHSON, B.D. and D.G. BLAIR, *A holistic approach to language* (1982), World Wide Web, at URL http://www.tcm.phy.cam.ac.uk/~bdj10/language/lang1.html

Chapter 30

Toward the physics of "death"

David M. Keirsey

R&D Intelligent Systems

The role of the "process of death" in evolution of the universe and the creation of complexity is discussed. Where concept of "death" includes the meaning of a thermodynamic-like death of an organized entity. This process of death is viewed in terms of major levels of complexity within the context of massive dissipative structures. We hypothesize that within a level of major complexity there are three regimes : chaotic, ordered, and the edge-of-chaos. The role of "death" is a form of information feedback from order to chaos via an edge-of-chaos between levels of major complexity. Death can release stored information that is key to the further evolution of complexity of a surrounding dissipative structure. It is further hypothesized that in the increasing complexity of our existence, there are successive levels of selection processes, and in understanding these processes can help predict the structure of our future.

1. Introduction

What is the role of the "process of death" in evolution of the universe and the creation of complexity? This is the question this paper seeks to address. The concept of death we are addressing here is a more generic notion than just the death of living things, rather we are including in the concept of death, a more thermodynamic-like death of an organized entity, such as the death of an atom, a death of a star, and a death of a empire.

There is the age old question of how does new phenomena, such as life, arise. This question has been investigated and discussed for centuries. The area of the science of complexity is starting to address this question by understanding the general processes underlying the production of novelty. Perspectives on natural processes exemplified by the work of Feigenbaum [5], Smolin [18], Prigogine [15], Crutchfield [4], and Langton [11] are starting to lead to new explanations on how novelty is created. More specific to the creation of life, the work of Kauffman [8, 9], Fontana and Buss [7] have begun to shed some light on some the underlying processes that must have taken place.

Kauffman's beginning of a theory of increasing complexity for life will be primarily assumed in this paper. This theory has concentrated on emergence of complexity of life through the notion of autocatalytic sets, but demonstrates that that underlying phenomena is generic many other physical processes characterized by boolean networks. Kauffman primarily assumed in the notion of autocatalytic sets that there are a set of reactants in sufficient quantities to start the autocatalytic process off. Kauffman only briefly discusses the creation of pre-autocatalytic organic molecules. The reactants themselves are simpler than the emerging process of life, but they are complex in their own right. Smith and Szathmary [17], taking on a broader range of the evolutionary steps of life, have outlined in detail the probable steps in the evolution of life.

In the evolution from pre-life to life, the detailed steps are complex. However, the question of how the building of prebiotic complexity occurred, focuses the problem of finding mechanisms for creation of diverse and complex molecules in sufficient quantities via a non-reproductive environment. Kauffman's autocatalytic sets provide an essential process in the formation of complex molecules, but does not provide an *explicit* link to essential notions of the preservation and accumulation of novelty to increase complexity in simpler environments. We hypothesize that one process essential this increase of complexity is the "process of death".

2. Death

The main hypothesis of this paper is the following. The role of "death" as a type of process is essential in the creation of complexity. The "process of death" is a generic process that is inherent in massive dissipative structures. Thermodynamic death of complex structures will occur in dissipative structures. However, "death" is not as complete as it implies. Even in "death", there is something remaining of the original entity. Some of the parts of the entity still remain and those released parts will interact with the surrounding environment at a lower level of complexity. The potential diversity of the lower level of complexity is increased with the death. Thus, death is a form of information feedback between levels of complexity.

With the "parts" left over from death, they can be used in new functional ways that could not occurred in the original creating context. Without reproduction in the living sense, novel molecules created by novel chemical processes must survive many different environments after the chemical environment that created it had long ago died. Novelty must be stored in stable dynamic structures that can survive death. Eventually enough novel molecules will form the basis for the creation of a new level of complexity.

This process is generic to other levels of complexity. For example, we, as human organisms, are made of atoms formed as nuclei in a red-giant star that has long been dead. We could not exist without the death of those types of stars. The red-giants and supernovae serve as far-from-equilibrium dissipative structures creating stable micro-entities within a galaxy (itself a dissipative structure) that are necessary for further evolution of complexity [[18], chapter 9].

Massive death can provide a great deal of dynamic material for the further evolution. Another example of this process is the accumulation of petroleum in the earth crust. Human civilization depends on oil, which is the result of the massive accumulation complex organic molecules of the Carboniferous period, when plants dominated the land. Other examples of this process may include the formation of galaxies by the death of antimatter, accumulation of organic molecules in soil for development of land plants, and the biogenesis origin of key metal deposits for the use by man.

The challenge is to understand exactly how these example evolutions occurred and how other evolutions are similar in the underlying process of increasing complexity. The process and structure of the universe has evolved via multiple levels of dissipative structures and our very existence is built upon multiple levels of complexity arising from this situation [10]. The prediction is that massive "deaths" in the evolution of complexity are necessary.

3. Levels of major complexity

In our use of the concept of complexity we are using a naive definition similar to Richard Dawkins' use in the *Blind Watchmaker*: something is complex if it is made of parts. However, we include in our notion of complexity a surrounding context, namely a dissipative structure. We define loosely a *major level of complexity* [10] as demarcated by a massive, self-organized dissipative structure, hereby called a *macrosystem*[1]. Our universe, galaxies, the solar system, earth, and Gaia are some of the most notable macrosystems. The macrosystem consists of self-organized "parts" that are hereby called *microsystems*. The complexity of a macrosystem is based one the complexity of the majority of microsystems that compose it. Also, a macrosystem is surrounded by another macrosystem of a lower level of complexity. Lastly, if a macrosystem is massive, stable enough, and has enough diversity in its microsystems, then other macrosystems can occur within it, such as the earth within the solar system.

Microsystems are the natural occurring building blocks that compose things. They are self-organized, dissipative structures at the microscale relative to its surrounding macrosystem. Leptons, baryons, atoms, molecules, prokaryotic cells, multi-cellular organisms, human families, and corporations are some of

---

[1] While discussing macrosystems and microsystems in the context of major complexity, there is no denying that there are levels of complexity between major levels. Unfortunately, the definition and the understanding of levels of complexity within a major level several orders of magnitude more difficult. Therefore, the many of issues of regarding complexity within a major level of complexity are not directly addressed here.

the primary examples of microsystems at different levels of complexity.

Since a microsystem is within a macrosystem, the death of the microsystem usually releases its components back to a lower level of complexity, so the information contained in the components will be reincorporated in the lower level. But in addition those components have a chance to be reincorporated into a different microsystem, possibly at the higher level again. What the structure of this process of reincorporation is the key question. We hypothesize this process of reincorporation is intimately related to the notions of chaos and order.

Much of the progress in understanding the interplay between chaos and order regimes has been in simplified situations at one level of complexity using mathematical modeling. It is true that deterministic (i.e., mathematical) chaos and deterministic order regimes do reflect a great deal of the properties of the natural world. However, it is fairly evident that the natural world has multiple levels of chaos and order. So are the laws the same for non-deterministic chaos and non-deterministic order? Are there any missing properties or laws that are essential in understanding the evolution of universe that involve multiple levels of complexity?

Three areas of research: chaos theory, non-equilibrium chemical thermodynamics, and artificial life have given some clues into the process of increasing complexity. The route from order to chaos has been explored by chaos theory. Non-equilibrium chemical thermodynamics [15] have shown the beginning of a route from chaos to order. In the area of artificial life, Langton [11] has hypothesized that between chaos and order there is an "edge of chaos." This edge-of-chaos is where physical systems can exhibit a high degree of physical computation. Langton has asserted that at the edge-of-chaos is where complexity arises and is akin to a phase transition. Crutchfield [3, 14] further argues that all physical systems can compute to some degree, and the degree of computation at an edge of chaos forms an information-rich peak, higher than the ordered or chaotic regimes. In other words, there is no distinct boundary, in the conventional sense, encompassing or demarcating an edge-of-chaos. (Note, however, that neither Langton nor Crutchfield have studied natural, massive dissipative structures using their analysis methods.)

If physical systems are performing computation by their very own dynamics, then what happens when several of these systems couple or decouple their dynamics? The dynamics of coupling most likely are related to development or growth. The dynamics of decoupling should be related to decay or death. The death of a "complex" of physical systems must also play a part in transfer of information between the levels of complexity.

When observing vast range of the macrosystems and microsystems of our natural world, we have noticed the particular patterns of chaos, order, and the edge-of-chaos. Therefore, we hypothesize that there are three basic conceptual regimes of matter at a major level of complexity from a physical computation perspective: chaotic, ordered, and an "edge of chaos" as depicted in Fig. 3.1. The actual size of the edge-of-chaos region is very small compared to chaotic and ordered regimes. The chaotic region has most of the mass and energy, but

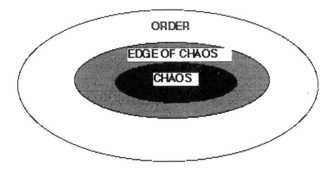

Figure 3.1: A Macrosystem: A Level of Major Complexity

the ordered region occupies most of the space. The chaotic region supplies the evolutionary energy for the growth of the edge-of-chaos. The order region is the "material" area for the storage of dynamic structures (i.e., microsystems).

The chaotic regime is where nothing can be predicted about the position of the microsystems after a duration of time. The chaotic region is centered at the highest physical temperature. On the other hand, the order region is the result of the dissipation of energy and material of the chaotic region and the debris of previous existing dissipative structures. The order regime is where the positional state of entities repeats after a duration of time. The edge-of-chaos regime is the phase transition between the ordered region and the chaotic region. Our solar system is one specific example of a dissipative system as in Fig. 3.2.

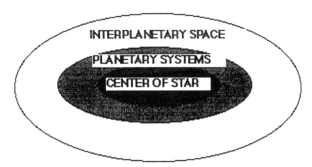

Figure 3.2: A Level of Major Complexity 2: A Star System

Another view of a major level complexity is the characterization of the phases of dynamic organization of the *dominant* microsystems within a macrosystem. Fig. 3.3 represents a phase space of a level of complexity.

Fig. 3.4 represents the phase space of complexity level 2. The particles the sun: protons and neutrons form the chaotic center. The nuclei and ions in sun and planets form the edge-of-chaos. The atoms of the planets, asteriods, comets, and Ort cloud compose the order region. The "morphogenesis" of simpler microsystems combining together into a dynamic structure represent the transition

310

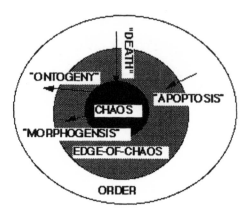

Figure 3.3: The Phase Space of Major Complexity

from chaos towards order via an edge-of-chaos.

The word morphogenesis is generalized to include phase transitions at major levels of complexity including: quarks to baryon, baryons to nuclei, ions to molecular ion, molecular autocatalyic sets to cell (e.g., procaryote cell division), gametes to organism, parents to family, families to "societies" (e.g., corporations). The word "apoptosis" is generalized to include the phase transitions from order to chaos via an edge-of-chaos which include nuclei to baryons(e.g. radioactive decay), atom to ion, molecule to molecular cations/ions (e.g. chemical decomposition), cell to molecular sets, organism to cells (e.g., molting), family to organisms (e.g., divorce), "societies" to families (e.g. corporate restructuring).

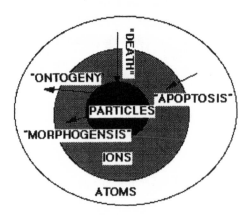

Figure 3.4: The Phase Space of Major Complexity Level 2

At the next level of major complexity above atoms is the transition from atoms to molecules. Planetary systems are the typical massive dissipative structure in creating more complex molecules (microsystems). Fig. 3.5 illustrates the major complexity level three. For example, the earth's center is a center for atomic chaos. The nuclei in the center of earth are relatively stable, but too hot to form molecules in the conventional sense. The planet's crust is an edge-of-chaos, where atomic ions can combine or decompose into. The planet's atmosphere and planetary space are the order region, where relative numbers and types of molecules such as oxygen and hydrocarbons are relatively stable compared to the surface.

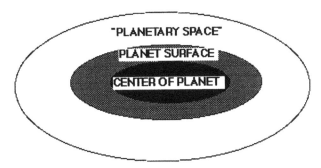

Figure 3.5: A Level of Major Complexity 3: A Planet System

The planet's biosphere (Gaia) is the next level of major complexity (as in Fig. 3.6), where molecules combine to make cells. The planet surface is the chaotic center. The individual molecules are chaotic in the long run. The edge-of-chaos region is composed of the autocatalytic sets of molecules that are involved in life at the moment. The order region are all the life-based organic molecules in the biosphere.

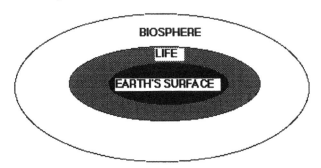

Figure 3.6: A Level of Major Complexity 4: A Biological System

We hypothesize that all massive dissipative structures have an edge-of-chaos as in Fig. 3.1. (Although the actual size of the edge-of-chaos is extremely small compared to the chaotic and order regions.) If a further set of conditions exist, an edge-of-chaos within a dissipative structure can be substantial in absolute terms (but not relative) and contain another major level of complexity. Fig. 3.7 illustrates the embedding of one major complexity level within another.

312

Within an edge-of-chaos region there can be multiple levels of major complexity. If an edge-of-chaos region (level N) is large, the underlying stable microsystems (level N) are diverse in type, and the chaos region (level N) of the surrounding dissipative structure is stable, then the next level of major complexity (level N+1) can form. In the next level of major complexity (level N+1), there is a chaotic region (level N+1), an order region (level N+1), and an edge of chaos (level N+1).

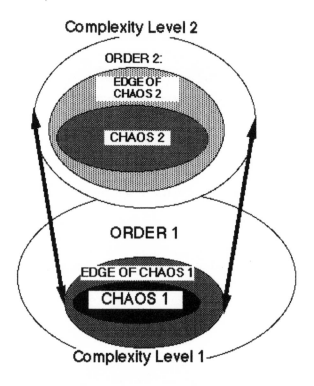

Figure 3.7: Two major levels of complexity

In the natural world, galaxies, star systems with planets, and the earth are the most readily observable examples of massive dissipative structures with multiple levels of complexity. There are also higher-level dissipative structures that have been hypothesized, such as Gaia [12], Hypersea ("organismal entity" that extended life onto land) [13], and Metaman (global mankind with electronic communication as "organismal entity") [19] that will be useful in understanding the evolution of complexity.

In more concrete terms as in Fig. 3.7 for example, the star systems in the middle periphery of Milky Way galaxy is a large region, the underlying charged particles are diverse enough to form nuclei. The center of galaxy has a black hole which serves as a chaotic center (level 1). The interaction of living star systems, the dynamic debris of the dead star systems, and the formation of our solar system create a part of an edge-of-chaos (level 1) in our galaxy. Our solar system is on the edge-of-chaos (level 1) and comprises of a stable chaotic region (level 2), our sun, and the order region (level 2), namely interplanetary space and the Ort cloud. The edge-of-chaos (level 2) are the geologically active planets and moons of the solar system.

Although in a natural chaotic regime, the position of microsystems cannot be predicted after a certain amount of time, the microsystems themselves are mostly stable and hence predictable. There is "order within chaos". For example, our sun is chaotic relative to atoms (atomic nuclei with a complete set of electrons), but is relatively stable for particles. Within the sun, the number of protons, neutrons, and electrons is statistically stable. Also, relatively stable within the sun are the nuclei that were included in the matter coalesced by the sun from the debris of previous deaths of stars. (Whereas, nuclei are not preserved in within a black hole, which is a dissipative structure at a lower level of major complexity.) On the other hand, any matter, such as an asteroid in the form of atoms, molecules, or anything more complex that is incorporated into the sun is reduced to the complexity of atomic ions.

In the order regime, the position of the microsystem can be predicted to a degree. But, within an order regime there can be some chaos, depending on the degree of symmetry breaking and bifurcation of microsystems. For example, in the solar system, most of the matter in interplanetary space and the Ort cloud is in the form of atoms and is frozen and relatively inactive. However, the microsystems in the order regime themselves although mostly stable, there is no predictability in the sudden death of any particular microsystem. (For example, a neutrino colliding with a proton can start the destruction). There is "chaos within order". Because of this ultimate instability of it's component microsystems, no order regime is completely ordered. In addition, an order regime is physically next to a chaotic regime, which subjects the order region to the energy fluctuations of the chaotic region.

Between chaos and order, within an edge-of-chaos there can be multiple levels of complexity. Growth is the generic word for indicating the process of going from chaos to order. Whereas, death is the generic word for indicating the process of going from order to chaos. The feedback of the chaos and the order regions into the edge-of-chaos drive an edge-of-chaos to higher complexity.

Fig. 3.8 illustrates the possible current path of macrosystems in the evolution of complexity in our human existence: hence called *involution*.

When a macrosystem (e.g., the earth) contains higher-level complexity, not only will it exhibit the processes of lower complexity, most notably, thermodynamic dissipation, but the macrosystem also will exhibit higher forms of process order, such as non-equilibrium chemical thermodynamics.

Is there any structure and process to this evolution in the higher levels of complexity? Unfortunately, the standard use the concept of "evolution" has been overused and includes two significantly different processes. The first process is adaptation. Algae are said to evolve or "adapt" as time progresses, even though they do not change in significant complexity. When one talks of the

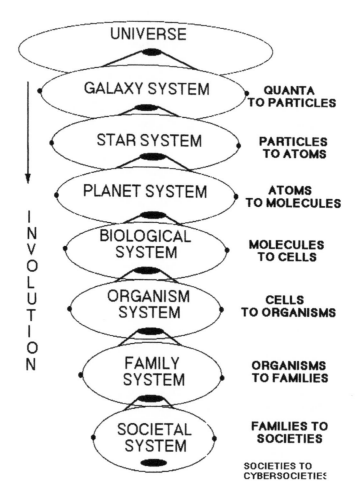

Figure 3.8: The involution of our existence: the major levels of complexity

evolution of algae from their creation to the current age, there is very little implication of increasing complexity. But on other hand, the other significant process of evolution is increasing of complexity. It is generally accepted that some common ancestors of prokaryotes had evolved into multi-celled eukaryotes by the processes involving parasites, symbioses, and epigenetic programs of the genomes. This type of evolution is an increase in complexity. Stephen Jay Gould argues that the evolving to more complex organisms is only one strategy in adapting to the environment, which is true; however, there is more to the story of the evolution of complexity besides multi-cellular organisms.

4. Involution and levels of selection

The word "*involution*" will be used to signify the "evolution of complexity," and exclude the concept of evolution as adaptation within one level of major complexity. We assert the basic difference between evolution and involution is that involution requires multiple levels of major selection processes (see Buss [2]) whereas evolution as adaptation does not.

If there are levels of complexity in the involution of our universe, then what are the selections laws that govern the involution? Ilya Prigogine has a more general notion of selection than neo-Darwinian natural selection. He views the second law of thermodynamics as a thermodynamic selection law ([15] pp 285). In the realm of physics, Lee Smolin in his recent book [18] has opened the debate to whether there is a selection process at the universe level.

Nevertheless, the most evident and understood process of selection is "natural selection" based on the neo-Darwinian biological theory. However, the Darwinian theory has centered on the study of eukaryotic multi-cellular evolution. On the other hand, the evolution of prokaryotes and the epigenetic inter-cellular mechanisms in the development of animals has been mostly shrouded in mystery. The debate is beginning in the biological community how much the conventional neo-Darwinian evolutionary principles can explain the biological phenomena at multi-cellular level[6]. Only recently has there been any discussion in the biological community about possible incompleteness of the Darwinian principles (see Buss [2], Smith and Szathmary [17]). Besides the evolution of bacteria and the transition to multi-cellular organisms, there is also the cultural evolution of the human race, which may involve a more complex selection process, such as the Baldwin effect [1].

If one can accept a more general notion of selection in involution, then the next question is what is the structure of the selection processes. Given that each level of complexity builds upon the previous levels, it seem plausible that the simple selection laws at the lowest level put constraints on the higher levels. As a higher-level macrosystem involutes, higher-level selection laws begin to operate. A plausible sequence of selection processes would be as in Fig. 4.1.

The involution of carbon-based molecules(proteins, lipids, nucleic acids, plastics), "single-cells" (viriods, viruses, prokaryotes), "organisms" (multi-celled eukaryotes), "families" (e.g., ant colonies and human families), and human "societies" (tribes, corporations, nations) surely have similar selection processes. The key question is what are the differences in selection processes at each level of complexity.

The non-equilibrium thermodynamic process uses time and space as it main separation criteria as a selection difference. It exhibits the three regimes of chaos, order, and edge-of-chaos. All it requires is that the macrosystem be a dissipative structure. If all of the macrosystems are thermodynamic entities, then even the most complex proposed macrosystem, MetaMan, should exhibit properties of this process. Indeed, it has been noted that mankind has dissipated the earth's energy to a larger extent than simpler organisms. Moreover, modern

316

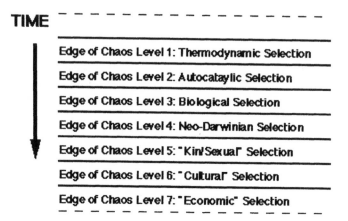

Figure 4.1: Involution of selection processes in a macrosystem

civilized man has shown even more predilection for consuming energy than less technological humans. Given that MetaMan, Hypersea, Gaia, and Earth are first thermodynamic systems as they involute, then it should follow that the higher-level microsystems, such as organisms, with initially behave as simple dissipative elements in a thermodynamic process. Indeed, when studying the history of both physical systems and living systems the recurring themes of chaos, order, and the edge-of-chaos appear in broad patterns.

Just as the thermodynamic macrosystems, such as galaxies, first exhibit thermodynamic selection, the more complex macrosystems will evolve from first from a simple thermodynamic system to having a more complex selection process of "chemical" selection. Hence, these more complex macrosystems will exhibit secondly the non-equilibrium chemical thermodynamics characterized by the notion of autocatalytic sets. It stands to reason that the more complex macrosystems will continue to involute to exhibiting in succession: simple biological selection (one level reproduction), Darwinian selection (two level reproduction), and possibly cultural selection.

If this notion of succession of selection processes is correct, then this gives us a powerful tool in predicting and understanding the future. As involution of Metaman proceeds via the development of the Internet, the structuring of corporations, and the advance of technology, it should exhibit the stages of selection processes. Hence, the detection and analysis of analogous lower-level processes will help predict the overall structural processes and transformation of the Earth, life, and mankind in the future.

Despite the inherent difficulties in making conceptual analogies between seemly disparate phenomena [16] such as atoms, molecules, cells, and organisms, there is much to be gained by attempting to precisely understand the underlying regularities. The primary reasons for this are to better understand both the general laws at each level of complexity and the historical quirks that happened in our particular path of involution.

Bibliography

[1] BALDWIN, J.M. "A new factor in evolution", *American Naturalist* **30** (1896) 441-451.

[2] BUSS, Leo W., *The Evolution of Individually*, Princeton University Press, (1987).

[3] CRUTCHFIELD, James P. and Karl YOUNG, "Computation at the Onset of Chaos", In *Complexity, Entropy, and the Physics of Information, SFI Studies in the Sciences of Complexity*, Vol VIII, Ed. W.H. ZUREK, Addison-Wesley, (1990).

[4] CRUTCHFIELD, James, "The Calculi of Emergence: Computation, Dynamics, and Induction", *Physica D* **75**, (1994) 11-54.

[5] FEIGENBAUM, Mitchell, "Quantatitive Universality for a Class of Nonlinear Transformations", *Journal of Statistical Physics*, **19**, (1978), 25-52.

[6] GOODWIN, Brian, *How the Leopard Changed its Spots*, Touchstone Books, (1994).

[7] FONTANA, Walter and Leo W. BUSS. "What would be conserved if 'the tape were played twice'?" *Proc. Natl. Acad. Sci.* **91** USA, (1994), 757-761.

[8] KAUFFMAN, Stuart, *The Origins of Order*, Oxford University Press, (1993).

[9] KAUFFMAN, Stuart, *At Home in the Universe*, Oxford University Press, (1995).

[10] KEIRSEY, David, "Involution: On the Structure and Process of Existence", Einstein meets Magritte, International Conference, Brussels, (1995).

[11] LANGTON, Christopher, "Computation at the Edge of Chaos, Phase Transitions and Emergent Computation", *Physica D*, **4**,(1990) 12-37.

[12] LOVELOCK, James, *The Ages of Gaia: A Biography of Our Living Earth*, Norton and Company, (1986).

[13] MCMENAMIN, Dianna and Mark MCMENAMIN, *Hypersea, Life on Land*, Columbia University Press, (1996).

[14] MITCHELL, Melanie, James CRUTCHFIELD, and David HRABER, "Dynamics, Computation, and the Edge of Chaos: A Re-Examination", G. COWAN, D. PINES, and D. MELZNER (editors), Integrative Themes, Santa Fe Institute Studies in the Sciences of Complexity, Proceedings Volume 19. Reading, MA (1994).

[15] PRIGOGINE, Ilya and Irene STENGERS, *Order out of Chaos*, Bantam Books, (1984).

[16] SERENO, Martin I., "Four Analogies Between Biological and Cultural/Linguistic Evolution", *Journal of Theoretical Biology*, **151**, (1991) 467–507.

[17] SMITH, John Maynard and Eors SZATHMARY, *Major Transitions in Evolution*, W.H. Freeman, (1995).

[18] SMOLIN, Lee, *The Life of the Cosmos*, Oxford University Press, (1997).

[19] STOCK, Gregory, *Metaman*, Simon and Schuster, (1993).

Chapter 31

Ragnar Frisch at the edge of chaos

Francisco Louçã
UECE-ISEG, Lisbon University

The paper investigates a simple nonlinear dynamic model set by Ragnar Frisch in 1935 and never published until now. The model describes a simple economy with two agents, each producing one commodity and consuming the other's one, and is designed to discuss whether capitalism is doomed to collapse or not. Although Frisch did neither solve nor simulate the behaviour of the system, he eventually understood that at least some very complicated dynamics emerged from it.

Although this is not the only instance of Frisch's concern with the wild side of the street, all along his life he carefully avoided publishing any nonlinear model and argued that linear specifications were satisfactory. Yet, evidence shows that he looked around for something else. The current model is the proof that he found complexity, although he could not deepen the study of the problem.

Section §1 presents the author and the problem, section §2 summarizes some of the findings about the numerical simulation, and finally section §3 indicates some conclusions.

1. Will capitalism collapse or equilibrate?

Ragnar Frisch (1895-1973) was one of the more brilliant mathematical economists of this century. He was deeply involved in statistical theory, in political advice and in planning, and he made major contributions to business cycle research, to economics of development as well as to other topics. In recognition to this life work, Frisch was awarded the first Nobel Prize in economics, in 1969, ex-aequo with his close friend Jan Tinbergen.

The paper which gave him the Nobel Prize is a landmark in the history of economics: it was written and published in 1933, presenting an ingenuous three dimensional mixed system of difference and differential equations to account for several modes of oscillation[1]. Slutsky random shocks were also added to the equations, and this model introduced the probabilistic approach into economics and defined for a long time the dominance of linear systems in the early econometric programs[4].

The model investigated in this paper is rather different. It was included in a three-page typewritten document, dated the 1st October 1935, under the suggestive title "The Non-Curative Power of the Capitalistic Economy—A Non-Linear Equation System Describing how Buying Activity Depends on Previous Deliveries"[3]. As Frisch tells the story, when he met Tinbergen an Koopmans at the Namur meeting of the Econometric Society, the three engaged in a discussion about the equilibrium properties of the market economies. Frisch argued that contraction was unavoidable, whereas Koopmans answered that flexibility of prices could eventually constitute an adaptive system preventing the collapse of the system. On the contrary, Frisch thought that they could even aggravate the problems. In order to settle the question, the three scientists adopted quite a singular procedure: Koopmans would formulate the assumptions and Frisch would represent the mathematical form of the model and discuss its solutions. Indeed, Koopmans did his part of the job, since he indicated the economic relations to be embodied in the model. Frisch defined it, although he did not explore the behaviour of the system: "This I plan to do on a later occasion", which never occurred.

From all points of view, this is an exceptional document. Here we have three founders of modern mathematical economics and econometrics—all to be rewarded at some point with the Nobel Prize—discussing the structure of capitalism and exploring new mathematical insights. Furthermore, in order to define a more realistic model of a simple economy of production and exchange, they constructed a nonlinear model, a quite uncommon feature by that time and definitively unusual in their previous and future work. In this framework, Frisch and eventually Koopmans understood they were forced to recur to numerical simulations in order to uncover its dynamics. Although there is no evidence that Frisch took the issue again with his challenger after the formulation of the model, the paper confirms that at least the author suspected the emerging properties of the model.

Frisch had previously illustrated his argument with a simple model of an economy with one shoemaker and one farmer, each one producing to the other's consumption, and assuming that decision on production was taken on the basis of the sales in the previous period[2]. Accordingly, each one's sales determined his level of expense. What ruled this very simple interrelationship was what Frisch called the "coefficients of optimism": if the agents were in an expanding mood, trade would increase; otherwise, if they were in a saving mood, "the whole system would gradually dwindle down to nothing"[2, p.263]. Therefore, cycles and the eventual collapse of the economic system were related to its mode of trading, to the concrete form of organization of the "liberalistic" society[2, p.272].

And this is how the 1935 model came about. The model defines two agents ("primus" and "secundus"), producing and exchanging much in the same way as in the 1934 model. The very simple nonlinearity is introduced with the definition of sale for each one of them: price times quantity (there are no stocks). This defines the two first equations, where the superscripts identify *primus* or *secundus*:

$$s_t^p = p_t^p q_t^p \tag{1.1}$$
$$s_t^s = p_t^s q_t^s \tag{1.2}$$

Afterwards, Frisch assumed that supply of *primus*'s good was a (negative) function of the price of his own good and a (positive) function of the previous growth of sales of *secundus*. A minimum quantity is always traded, $\hat{a}$ and $\hat{e}$. Therefore, quantity is fixed by the market conditions:

$$q_t^p = \hat{a} - a p_t^p + g(s_{t-1}^s - s_{t-2}^s) \tag{1.3}$$
$$q_t^s = \hat{e} - b p_t^s + g(s_{t-1}^p - s_{t-2}^p) \tag{1.4}$$

Finally, Frisch hypothesized that growth of prices, fixed by the seller, was a proportion of the previous growth in sold quantities:

$$p_t^p - p_{t-1}^p = y(q_{t-1}^p - q_{t-2}^p) \tag{1.5}$$
$$p_t^s - p_{t-1}^s = x(q_{t-1}^s - q_{t-2}^s) \tag{1.6}$$

This quite elementary nonlinear six-dimensional system of difference equations encapsulated Koopmans and Frisch's argument about the nature of the evolution of a liberalistic economy.

2. A shared judgement

In order to check the model, a numerical simulation is run for the following values of the parameters and initial conditions (P, Q and S being price, quantities and sales for 1 and 2, *primus* and *secundus*):

Notice that the system is not bounded and therefore the variables may eventually have negative values: the interpretation is that, under some circumstances, the agent does not sell and is forced to buy necessary inputs or other products in order to go on trading in the future. Assuming first that p_1 remains the same in $t - 2$ and $t - 1$ (0.09), the model generates large cycles at first and both *primus* and *secundus* eventually dominate the market for a short time, but then it stabilizes with *primus*'s dominance. The oscillations in *secundus*' sales are larger than in *primus*'s.

The graphical evidence of the attractor in *primus*'s sales (Fig. 2.1B) confirms the result: equilibrium is quickly reached.

Table 2.1: Initial conditions and values of the parameters.

Initial conditions	$t-1$	$t-2$	Parameters	Values
$S1$	0.21784	0.09	$\hat{e}$	1.1
$S2$	0.10926	0.09	$\hat{a}$	2
$Q1$	2.1784	1	a	0.2
$Q2$	1.214	1	b	0.4
$P1$	0.1	0.09	g	1.1
$P2$	0.09	0.09	l	3
			x	0.3
			y	0.2

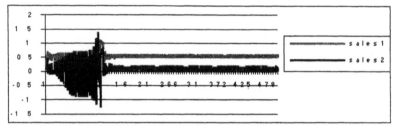

A

N: all graphs for $t = 0, \ldots, 500$, except if noticed otherwise.

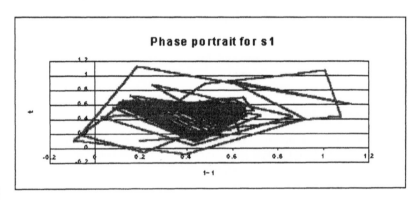

B

Figure 2.1: The behaviour of the model. A: The evolution of sales. Phase portrait of the sales of *primus*.

Now, if the initial conditions are modified as stated in Table 2.1, with *primus* taking the initiative of increasing his price 11% from $t-2$ to $t-1$, both agents dominate the emergent cyclical behaviour of the economy, with alternating positions. But the cycles are irregular, and indeed we have a chaotic regime for these values of the parameters.

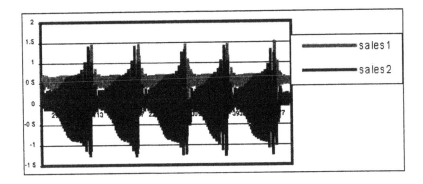

Figure 2.2: An aggressive intervention by *primus* at $t - 1$.

Let's suppose now that *primus* still sharpens his strategy and adopts a moderately inflationary policy, slightly increasing the parameter y (measuring the impact of previous growth of sales in the following growth of price). The cyclical regime is deeply altered.

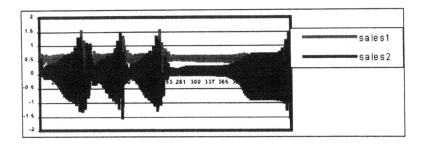

Figure 2.3: A further increase in prices of *primus*. An irregular cyclical regime ($y = 0.202102994$).

But if the same strategy is followed through, it will result in the destruction of the structured market relation, since after irregular but shorter cycles the system collapses (after $y > 0.204455702$).

Suppose instead that *secundus* reacts to the original change in initial conditions, taking a parallel inflationary measure, so that x is increased with $y = 0.2$. Notice that increases in y and in x are the most accessible interventions by the agents in order to change their relative position, since they settle the prices whereas quantities are defined by the market conditions. *Secundus* gets a larger part of the market sometimes, but *primus* still dominates for most of the time. And, as y is increased, both parts are harmed, since large cycles and eventually a collapse of the market (after $x = 0.310531146$) are generated. Aggressive competitive strategies based on inflationary action lead to the destruction of trade.

The simulation shows that profit-maximizing behaviour may easily lead to the collapse of trade in this model.

3. Conclusions

It is quite obvious that evidence provides a shared judgement about Frisch-Koopmans argument. In the framework of the model—and that does not allow for any meaningful claim about reality itself—, it is true that for some values of the parameters we may have an almost stationary situation, representing an equilibrium in the market, therefore vindicating Koopmans' opinion. But it is also true that, for other values of the parameters, cycles dominate; moreover, if the agents are profit maximizing, their action may eventually lead to the collapse of trade, vindicating Frisch's opinion. Still, collapse is brought about not by the lack of sales—by contraction, as Frisch expected—but by the too severe oscillations that imply at some point the bankruptcy of one or of both agents. As in the case of Frisch's original research, no inquiry was here pursued about the likelihood of the different sets of parameters.

For these and other reasons, this 1935 paper is very important to the history of economics. It provided an early framework for an investigation about emerging behaviour in a very simple model, and namely of the conditions for equilibrium, for cycles, for chaos and for a catastrophe. And, last but not least, it also proves that Frisch, the apostle of linear, computable and parsimonious models, also dared to travel at the edge of chaos.

Bibliography

[1] Frisch, Ragnar (1933), Propagation Problems and Impulse Problems in Dynamic Economics, in Koch, K. (ed.), Economic Essays in Honour of Gustav Cassel, London: Unwin, 171-205

[2] — (1934), Circulation Planning—Proposal for a National Organization of a Commodity and Service Exchange, Econometrica, 2, July, 258-336 (with a mathematical appendix in Econometrica, 2, October, 422-35)

[3] — (1935), The Non-Curative Power of the Liberalistic Economy—A Non-Linear Equation System Describing how Buying Activity Depends on Previous Deliveries, manuscript at Oslo University, Frisch's Archive

[4] Louçã, Francisco (1997), Turbulence in Economics—An Evolutionary Appraisal of Cycles and Complexity in Historical Processes, Lyme, US, Cheltenham, UK: Edward Elgar

Chapter 32

Programming complex systems

Philip Maymin

Long-Term Capital Management

Classical programming languages cannot model essential elements of complex systems such as true random number generation. This paper develops a formal programming language called the lambda-q calculus that addresses the fundamental properties of complex systems. This formal language allows the expression of quantumized algorithms, which are extensions of randomized algorithms in that probabilities can be negative, and events can cancel out. An illustration of the power of quantumized algorithms is the ability to efficiently solve the satisfiability problem, something that many believe is beyond the capability of classical computers. This paper proves that the lambda-q calculus is not only capable of solving satisfiability but can also simulate such complex systems as quantum computers. Since satisfiability is believed to be beyond the capabilities of quantum computers, the lambda-q calculus may be strictly stronger.

1. Introduction

The purpose of this paper is to introduce a formalism for expressing models of complex systems. The end result is that modelling any complex system such as human society, evolution, or particle interactions, may be reduced to a programming problem.

In addition to the modelling functionalities it provides, a programmable complex system also allows us to see, in its specification, what the distilled and essential elements of a complex system are. In particular, as we will see, interactions like those in a cellular automaton need not be explicit in the formalism, as they may be simulated.

Classical programming languages are not strong enough to model complex systems. They do not allow for randomized events and are completely predictable and deterministic, features rarely found in complex systems. Some problems that may be quickly solved on quantum computers, which is a complex system, have no known quick solutions on classical computers or with classical programming languages.

In this paper we extend the λ-calculus, the logical foundation of classical programming languages. The first extension, the λ^p-calculus, is a new calculus introduced here for expressing randomized functions. Randomized functions, instead of having a unique output for each input, return a distribution of results from which we sample once. The λ^p-calculus then provides a formal method for computing distributions. More useful, however, would be the ability to compute conditional distributions. The second extension, the λ^q-calculus, is a new calculus introduced here for expressing quantumized functions. Quantumized functions also return a distribution of results, called a *superposition*, from which we sample once, but λ^q-terms have signs, and identical terms with opposite signs are removed before sampling from the result. Quantumized functions can then compute conditional distributions. The effect is that of applying some filter to a superposition to adjust each of the probabilities according to its fitness. One example is the quick solution of satisfiability: by merely filtering out the logical mappings of variables that do not satisfy the given formula, we are left only with satisfying mappings, if any. The λ^q-calculus is the most general of the three calculi.

One of the results of this paper is that the λ^q-calculus is at least as powerful as quantum computers. Although much research has been done on the hardware of quantum computation (c.f. [5], [6], [10]), none has focused on formalizing the software. Quantum Turing machines [5] have been introduced but there has been no quantum analogue to Church's λ-calculus. The λ-calculus has served as the basis for many programming languages since it was introduced by Alonzo Church [4] in 1936. It and other classical calculi make the implicit assumption that a term may be innocuously observed at any point. Such an assumption is hard to separate from a system of rewriting rules because to rewrite a term, you must have read it. One of the goals of these calculi is to make observation explicit.

The λ^p- and the λ^q-calculi allow the expression of algorithms that exist and operate in the Heisenberg world of *potentia* [7] but whose results are observed. To this end, collections (distributions and superpositions) should be thought of with the following intuition. A collection is a bunch of terms that co-exist in the same place but are not aware of each other. Thus, a collection of three terms takes up no more space than a collection of two terms. A physical analogy is the ability of a particle to be in a superposition of states. When the collection is observed, at most one term in each collection will be the result of the observation. The key point is that in neither calculus can one write a term that can determine if it is part of a collection, how big the collection is, or even if its argument is part of a collection. Despite this inability, the λ^q-calculus is powerful enough to efficiently solve problems such as satisfiability that are typically believed to be beyond the scope of classical computers.

2. The lambda calculus

This section is a review of the λ-calculus and a reference for later calculi. For more details see e.g. [1].

The λ-calculus is a calculus of functions. Any computable single-argument function can be expressed in the λ-calculus. Any computable multiple-argument function can be expressed in terms of computable single-argument functions. The λ-calculus is useful for encoding functions of arbitrary arity that return at most one output for each input. In particular, the λ-calculus can be used to express any (computable) *algorithm*. The definition of algorithm is usually taken to be Turing-computable.

2.1. Syntax

The following grammar specifies the syntax of the λ-calculus.

$$
\begin{array}{lll}
x & \in \textit{Variable} & \text{Variables} \\
M & \in \textit{LambdaTerm} & \text{Terms} \\
w & \in \textit{Wff} & \text{Well-formed formulas} \\
\\
M & ::= \quad x & \text{variable} \\
& \mid \quad M_1 M_2 & \text{application} \\
& \mid \quad \lambda x.M & \text{abstraction} \\
\\
w & ::= \quad M_1 = M_2 & \text{well-formed formula}
\end{array}
\tag{2.1}
$$

To be strict, the subscripts above should be removed (e.g., the rule for well-formed formulas should read $w ::= M = M$) because M_1 and M_2 are not defined. However, we will maintain this incorrect notation to emphasize that the terms need not be identical.

With this abuse of notation, we can easily read the preceding definition as: a λ-term is a variable, or an application of two terms, or the abstraction of a term by a variable. A well-formed formula of the λ-calculus is a λ-term followed by the equality sign followed by a second λ-term.

We also adopt some syntactic conventions. Most importantly, parentheses group subexpressions. Application is taken to be left associative so that the term MNP is correctly parenthesized as $(MN)P$ and not as $M(NP)$. The scope of an abstraction extends as far to the right as possible, for example up to a closing parenthesis, so that the term $\lambda x.xx$ is correctly parenthesized as $(\lambda x.xx)$ and not as $(\lambda x.x)x$.

2.2. Substitution

We will want to substitute arbitrary λ-terms for variables. We define the substitution operator, notated $M[N/x]$ and read "M with all free occurences of x replaced by N." The definition of the free and bound variables of a term are standard. The set of free variables of a term M is written $FV(M)$. There are six rules of substitution, which we write for reference.

1. $x[N/x] \equiv N$
2. $y[N/x] \equiv y$ for variables $y \not\equiv x$
3. $(PQ)[N/x] \equiv (P[N/x])(Q[N/x])$
4. $(\lambda x.P)[N/x] \equiv \lambda x.P$
5. $(\lambda y.P)[N/x] \equiv \lambda y.(P[N/x])$ if $\begin{array}{l} y \not\equiv x \text{ and} \\ y \notin FV(N) \end{array}$ (2.2)
6. $(\lambda y.P)[N/x] \equiv \lambda z.(P[z/y][N/x])$
$\qquad y \not\equiv x,$
$\qquad$ if $y \in FV(N),$ and
$\qquad z \notin FV(P) \bigcup FV(N)$

This definition will be extended in both subsequent calculi.

2.3. Reduction

The concept of *reduction* seeks to formalize rewriting rules. Given a relation R between terms, we may define the one-step reduction relation, notated $\rightarrow_R$, that is the contextual closure of R. We may also define the reflexive, transitive closure of the one-step reduction relation, which we call R-reduction and notate $\twoheadrightarrow_R$, and the symmetric closure of R-reduction, called R-interconvertibility and notated $=_R$.

The essential notion of reduction for the λ-calculus is called β-reduction. It is based on the β-relation, which is the formalization of function invocation.

$$\beta \triangleq \left\{ \begin{array}{c} ((\lambda x.M)N, M[N/x]) \\ \text{s.t. } M, N \in LambdaTerm, x \in Variable \end{array} \right\}$$ (2.3)

There is also the α-relation that holds of terms that are identical up to a consistent renaming of variables.

$$\alpha \triangleq \left\{ \begin{array}{c} (\lambda x.M, \lambda y.M[y/x]) \\ \text{s.t. } M \in LambdaTerm, y \notin FV(M) \end{array} \right\}$$ (2.4)

We will use this only sparingly.

2.4. Evaluation semantics

By imposing an evaluation order on the reduction system, we are providing meaning to the λ-terms. The evaluation order of a reduction system is sometimes called an operational semantics or an evaluation semantics for the calculus. The evaluation relation is typically denoted $\rightsquigarrow$.

We use call-by-value evaluation semantics. A *value* is the result produced by the evaluation semantics. Call-by-value semantics means that the body of an abstraction is not reduced but arguments are evaluated before being passed into abstractions.

There are two rules for the call-by-value evaluation semantics of the λ-calculus.

$$\frac{}{v \rightsquigarrow v}(\text{Refl}) \qquad (\text{for } v \text{ a value})$$

$$\frac{M \rightsquigarrow \lambda x.P \quad N \rightsquigarrow N' \quad P[N'/x] \rightsquigarrow v}{MN \rightsquigarrow v}(\text{Eval})$$

2.5. Reference terms

The following λ-terms are standard and are provided as reference for later examples.

Numbers are represented as Church numerals.

$$\underline{0} \equiv \lambda x.\lambda y.y \tag{2.5}$$

$$\underline{n} \equiv \lambda x.\lambda y.x^n y \tag{2.6}$$

where the notation $x^n y$ means n right-associative applications of x onto y. It is abbreviatory for the term $\underbrace{x(x(\cdots(x\,y)))}_{n \text{ times}}$. When necessary, we can extend Church numerals to represent both positive and negative numbers. For the remainder of the terms, we will not provide definitions. The predecessor of Church numerals is written $\underline{P}$. The successor is written $\underline{S}$.

The conditional is written $\underline{IF}$. If its first argument is truth, written $\underline{T}$, then it returns its second argument. If its first argument is falsity, written $\underline{F}$, then it returns its third argument. A typical predicate is $\underline{0?}$ which returns $\underline{T}$ if its argument is the Church numeral $\underline{0}$ and $\underline{F}$ if it is some other Church numeral.

The fixed-point combinator is written $\underline{Y}$. The primitive recursive function-building term is written $\underline{\text{PRIM-REC}}$ and it works as follows. If the value of a function f at input n can be expressed in terms of $n-1$ and $f(n-1)$, then that function f is primitive recursive, and it can be generated by providing $\underline{\text{PRIM-REC}}$ with the function that takes the inputs $n-1$ and $f(n-1)$ to produce $f(n)$ and with the value of f at input 0. For example, the predecessor function for Church numerals can be represented as $\underline{P} \equiv \underline{\text{PRIM-REC}}\,(\lambda x.\lambda y.x)\,\underline{0}$.

3. The lambda-p calculus

The λ^p-calculus is an extension of the λ-calculus that permits the expression of *randomized* algorithms. In contrast with a computable algorithm which returns at most one output for each input, a randomized algorithm returns a *distribution* of answers from which we sample. There are several advantages to randomized algorithms.

1. Randomized algorithms can provide truly random number generators instead of relying on pseudo-random number generators that work only because the underlying pattern is difficult to determine.

2. Because they can appear to generate random numbers arbitrarily, randomized algorithms can model random processes.

3. Given a problem of finding a suitable solution from a set of possibilities, a randomized algorithm can exhibit the effect of choosing random elements and testing them. Such algorithms can sometimes have an *expected* running time which is considerably shorter than the running time of the computable algorithm that tries every possibility until it finds a solution.

3.1. Syntax

The following grammar describes the λ^p-calculus.

$$
\begin{array}{lll}
x & \in Variable & \text{Variables} \\
M & \in LambdaPTerm & \text{Terms} \\
w & \in WffP & \text{Well-formed formulas} \\
\\
M & ::= \quad x & \text{variable} \\
& \quad | \quad M_1 M_2 & \text{application} \\
& \quad | \quad \lambda x.M & \text{abstraction} \\
& \quad | \quad M_1, M_2 & \text{collection} \\
\\
w & ::= \quad M_1 = M_2 & \text{well-formed formula}
\end{array}
\tag{3.1}
$$

Since this grammar differs from the λ-calculus only in the addition of the fourth rule for terms, all λ-terms can be viewed as λ^p-terms. A λ^p-term may be a collection of a term and another collection, so that a λ^p-term may actually have many nested collections.

We adhere to the same parenthesization and precedence rules as the λ-calculus with the following addition: collection is of lowest precedence and the comma is right associative. This means that the expression $\lambda x.x, z, y$ is correctly parenthesized as $(\lambda x.x), (z, y)$.

We introduce abbreviatory notation for collections. Let us write $\left[M_i^{i \in S} \right]$ for the collection of terms M_i for all i in the finite, ordered set S of natural numbers. We will write $a..b$ for the ordered set $(a, a+1, \dots, b)$. In particular, $\left[M_i^{i \in 1..n} \right]$ represents $M_1, M_2, \dots, M_n$ and $\left[M_i^{i \in n..1} \right]$ represents $M_n, M_{n-1}, \dots, M_1$. More generally, let us allow multiple iterators in arbitrary contexts. Then, for instance,

$$
\left[\lambda x.M_i^{i \in 1..n} \right] \equiv \lambda x.M_1, \lambda x.M_2, \dots, \lambda x.M_n
$$

and

$$\left[M_i^{i\in 1..m} N_j^{j\in 1..n}\right] \equiv \begin{array}{c} M_1 N_1, M_1 N_2, \ldots, M_1 N_n, \\ M_2 N_1, M_2 N_2, \ldots, M_2 N_n, \\ \vdots \\ M_m N_1, M_m N_2, \ldots, M_m N_n \end{array}.$$

Note that $\left[\lambda x.M_i^{i\in 1..n}\right]$ and $\lambda x.\left[M_i^{i\in 1..n}\right]$ are not the same term. The former is a collection of abstractions while the latter is an abstraction with a collection in its body. Finally, we allow this notation to hold of non-collection terms as well by identifying $\left[M_i^{i\in 1..1}\right]$ with M_1 even if M_1 is not a collection. To avoid confusion, it is important to understand that although this "collection" notation can be used for non-collections, we do not extend the definition of the word *collection*. A *collection* is still the syntactic structure defined in grammar (3.1).

With these additions, every term can be written in this bracket form. In particular, we can write a collection as $\left[\left[M_i^{i\in S_i}\right]_j^{j\in S}\right]$, or a collection of collections. Unfortunately, collections can be written in a variety of ways with this notation. The term M, N, P can be written as $\left[M_i^{i\in 1..3}\right]$ if $M_1 \equiv M$ and $M_2 \equiv N$ and $M_3 \equiv P$; as $\left[M_i^{i\in 1..2}\right]$ if $M_1 \equiv M$ and $M_2 \equiv N, P$; or as $\left[M_i^{i\in 1..1}\right]$ if $M_1 \equiv M, N, P$. However, it cannot be written as $\left[M_i^{i\in 1..4}\right]$ for any identification of the M_i. This observation inspires the following definition.

Definition 3 *The cardinality of a term M, notated $|M|$, is that number k for which $\left[M_i^{i\in 1..k}\right] \equiv M$ for some identification of the M_i but $\left[M_i^{i\in 1..(k+1)}\right] \not\equiv M$ for any identification of the M_i.*

Note that the cardinality of a term is always strictly positive.

3.2. Syntactic identities

We define substitution of terms in the λ^P-calculus as an extension of substitution of terms in the λ-calculus. In addition to the substitution rules of the λ-calculus, we introduce one for collections.

$$(P, Q)\,[N/x] \equiv (P\,[N/x], Q\,[N/x]) \tag{3.2}$$

We identify terms that are collections but with a possibly different ordering. We also identify nested collections with the top-level collection. The motivation for this is the conception that a collection is an unordered set of terms. Therefore we will not draw a distinction between a set of terms and a set of a set of terms.

We adopt the following axiomatic judgement rules.

$$\frac{}{M, N \equiv N, M}(\text{ClnOrd})$$

$$\frac{}{(M, N), P \equiv M, (N, P)}(\text{ClnNest})$$

With these axioms, ordering and nesting become innocuous. As an example here is the proof that $A, (B, C), D \equiv A, C, B, D$. For clarity, we parenthesize fully and underline the affected term in each step.

$$
\begin{aligned}
\underline{A, ((B, C), D)} &\equiv (\underline{(B, C)}, D), A && \text{(ClnOrd)} \\
&\equiv (\underline{(C, B)}, D), A && \text{(ClnOrd)} \\
&\equiv \underline{(C, (B, D))}, A && \text{(ClnNest)} \\
&\equiv \underline{A, (C, (B, D))} && \text{(ClnOrd)}
\end{aligned}
$$

It can be shown that ordering and parenthesization are irrelevant in general. Aside, it no longer matters that we took the comma to be right associative since any arbitrary parenthesization of a collection does not change its syntactic structure.

Because of this theorem, we can alter the abbreviatory notation and allow arbitrary unordered sets in the exponent. This allows us to write, for instance, $\left[M_i^{i \in 1..n - \{j\}} \right] \equiv M_1, M_2, \ldots, M_{j-1}, M_{j+1}, \ldots, M_n$ where $a..b$ is henceforth taken to be the unordered set $\{a, a + 1, \ldots, b\}$ and the subtraction in the exponent represents set difference.

This also subtly alters the definition of *cardinality* (3). Whereas before the cardinality of a term like $(x, y), z$ was 2, because of this theorem, it is now 3.

We may now also introduce a further abbreviation. We let $[(M_i : n_i)]$ be a rewriting of the term $\left[N_i^{i \in I} \right]$ such each of the M_i are distinct and the integer n_i represents the count of each M_i in $\left[N_i^{i \in I} \right]$.

3.3. Reductions

The relation of collection application is called the γ-relation. It holds of a term that is an application at least one of whose operator or operand is a collection, and the term that is the collection of all possible pairs of applications.

$$
\gamma^p \triangleq \left\{
\begin{array}{l}
\left(\left[M_i^{i \in 1..m} \right] \left[N_j^{j \in 1..n} \right], \left[M_i^{i \in 1..m} N_j^{j \in 1..n} \right] \right) \\
\text{s.t. } M_i, N_j \in LambdaPTerm, \, m > 1 \text{ or } n > 1
\end{array}
\right\} \tag{3.3}
$$

We will omit the superscript except to disambiguate from the γ-relation of the λ^q-calculus.

It can be shown that the γ-relation is Church-Rosser and that all terms have γ-normal forms. Therefore, we may write $\gamma(M)$ for the γ-normal form of M.

We extend the β-relation to apply to collections.

$$
\beta^p \triangleq \left\{
\begin{array}{l}
\left((\lambda x.M) \left[N_i^{i \in S} \right], \left[M \left[N_i^{i \in S} / x \right] \right] \right) \\
\text{s.t. } M, \left[N_i^{i \in S} \right] \in LambdaPTerm, \, x \in Variable
\end{array}
\right\} \tag{3.4}
$$

where $\left[M \left[N_i^{i \in S} / x \right] \right]$ is the collection of terms M with N_i substituted for free occurrences of x in M, for $i \in S$.

3.4. Evaluation semantics

We extend the call-by-value evaluation semantics of the λ-calculus. We modify the definition of a value v to enforce that v has no γ-redexes.

$$\frac{}{v \rightsquigarrow v}(\text{Refl}) \qquad (\text{for } v \text{ a value})$$

$$\frac{\gamma(M) \rightsquigarrow \lambda x.P \quad \gamma(N) \rightsquigarrow N' \quad \gamma(P[N'/x]) \rightsquigarrow v}{MN \rightsquigarrow v}(\text{Eval})$$

$$\frac{\gamma(M) \rightsquigarrow v_1 \quad \gamma(N) \rightsquigarrow v_2}{(M, N) \rightsquigarrow (v_1, v_2)}(\text{Coll})$$

3.5. Observation

We define an observation function Θ from λ^P-terms to λ-terms. We employ the random number generator $RAND$, which samples one number from a given set of numbers.

$$\Theta(x) = x \tag{3.5}$$
$$\Theta(\lambda x.M) = \lambda x.\Theta(M) \tag{3.6}$$
$$\Theta(M_1 M_2) = \Theta(M_1)\Theta(M_2) \tag{3.7}$$
$$\Theta\left(M \equiv \left[M_i^{i \in 1..|M|}\right]\right) = M_{RAND(1..|M|)} \tag{3.8}$$

The function Θ is total because every λ^P-term is mapped to a λ-term. Note that for an arbitrary term T we may write $\Theta(T) = T_{RAND(S)}$ for some possibly singleton set of natural numbers S and some collection of terms $\left[T_i^{i \in S}\right]$.

We can show that observing a λ^P-term is statistically indistinguishable from observing its γ-normal form.

3.6. Observational semantics

We provide another type of semantics for the λ^P-calculus called its *observational semantics*. A formalism's observational semantics expresses the computation as a whole: preparing the input, waiting for the evaluation, and observing the result. The observational semantics relation between λ^P-terms and λ-terms is denoted $\multimap$. It is given by a single rule for the λ^P-calculus.

$$\frac{M \rightsquigarrow v \quad \Theta(v) = N}{M \multimap N}(\text{ObsP}) \tag{3.9}$$

3.7. Examples

A useful term of the λ^P-calculus is a random number generator. We would like to define a term that takes as input a numeral $\underline{n}$ and computes a collection of numerals from $\underline{0}$ to $\underline{n}$. This can be represented by the following primitive recursive λ^p-term.

$$\underline{R} \equiv \underline{PRIM\text{-}REC} \; (\lambda k.\lambda p.\,(k,p)) \; \underline{0} \qquad (3.10)$$

Then for instance $\underline{R}\,\underline{3} = (\underline{3}, \underline{2}, \underline{1}, \underline{0})$.

The following term represents a random walk. Imagine a man that at each moment can either walk forward one step or backwards one step. If he starts at the point 0, after n steps, what is the distribution of his position?

$$\underline{W} \equiv \underline{PRIM\text{-}REC} \; (\lambda k.\lambda p.\,(\underline{P}p, \underline{S}p)) \; \underline{0} \qquad (3.11)$$

We assume we have extended Church numerals to negative numbers as well. This can be easily done by encoding it is a pair. We will show some of the highlights of the evaluation of $\underline{W}\,\underline{3}$. Note that $\underline{W}\,\underline{1} = (\underline{-1}, \underline{1})$.

$$
\begin{aligned}
\underline{W}\,\underline{3} \;=\;& \underline{P}\,(\underline{W}\,\underline{2}), \underline{S}\,(\underline{W}\,\underline{2}) \\
=\;& \underline{P}\,(\underline{P}\,(\underline{W}\,\underline{1}), \underline{S}\,(\underline{W}\,\underline{1})), \underline{S}\,(\underline{P}\,(\underline{W}\,\underline{1}), \underline{S}\,(\underline{W}\,\underline{1})) \\
=\;& \underline{P}\,(\underline{P}\,(\underline{-1}, \underline{1}), \underline{S}\,(\underline{-1}, \underline{1})), \underline{S}\,(\underline{P}\,(\underline{-1}, \underline{1}), \underline{S}\,(\underline{-1}, \underline{1})) \\
=\;& \underline{P}\,((\underline{-2}, \underline{0}), (\underline{0}, \underline{2})), \underline{S}\,((\underline{-2}, \underline{0}), (\underline{0}, \underline{2})) \\
=\;& ((\underline{-3}, \underline{-1}), (\underline{-1}, \underline{1})), ((\underline{-1}, \underline{1}), (\underline{1}, \underline{3})) \\
\equiv\;& (\underline{-3}, \underline{-1}, \underline{-1}, \underline{1}, \underline{-1}, \underline{1}, \underline{1}, \underline{3})
\end{aligned} \qquad (3.12)
$$

Observing $\underline{W}\,\underline{3}$ yields $\underline{-1}$ with probability $\frac{3}{8}$, $\underline{1}$ with probability $\frac{3}{8}$, $\underline{-3}$ with probability $\frac{1}{8}$, and $\underline{3}$ with probability $\frac{1}{8}$.

4. The lambda-q calculus

The λ^q-calculus is an extension of the λ^p-calculus that allows easy expression of *quantumized* algorithms. A quantumized algorithm differs from a randomized algorithm in allowing negative probabilities and in the way we sample from the resulting distribution.

Variables and abstractions in the λ^q-calculus have *phase*. The phase is nothing more than a plus or minus sign, but since the result of a quantumized algorithm is a distribution of terms with phase, we call such a distribution by the special name *superposition*. The major difference between a superposition and a distribution is the observation procedure. Before randomly picking an element, a superposition is transformed into a distribution by the following two-step process. First, all terms in the superposition that are identical except with opposite phase are cancelled. They are both simply removed from the superposition. Second, the phases are stripped to produce a distribution. Then, an element is chosen from the distribution randomly, as in the λ^p-calculus.

The words *phase* and *superposition* come from quantum physics. An electron is in a superposition if it can be in multiple possible states. Although the phases of the quantum states may be any angle from $0°$ to $360°$, we only consider binary phases. Because we use solely binary phases, we will use the words *sign* and *phase* interchangeably in the sequel.

A major disadvantage of the λ^p-calculus is that it is impossible to compress a collection. Every reduction step at best keeps the collection the same size. Quantumized algorithms expressed in the λ^q-calculus, on the other hand, can do this as easily as randomized algorithms can generate random numbers. That is, λ^q-terms can contain subterms with opposite signs which will be removed during the observation process.

4.1. Syntax

The following grammar describes the λ^q-calculus.

S	$\in Sign$	Sign, or phase
x	$\in Variable$	Variables
M	$\in LambdaQTerm$	Terms
w	$\in WffQ$	Well-formed formulas
S	$::= +$	positive
	$\mid -$	negative
M	$::= Sx$	signed variable
	$\mid M_1 M_2$	application
	$\mid S\lambda x.M$	signed abstraction
	$\mid M_1, M_2$	collection
w	$::= M_1 = M_2$	well-formed formula

(4.1)

Terms of the λ^q-calculus differ from terms of the λ^p-calculus only in that variables and abstractions are *signed*, that is, they are preceded by either a plus (+) or a minus (-) sign. Just as λ-terms could be read as λ^p-terms, we would like λ^p-terms to be readable as λ^q-terms. However, λ^p-terms are unsigned and cannot be recognized by this grammar.

Therefore, as is traditionally done with integers, we will omit the positive sign. An unsigned term in the λ^q-calculus is abbreviatory for the same term with a positive sign. With this convention, λ^p-terms can be seen as λ^q-terms all of whose signs are positive. Also, so as not to confuse a negative sign with subtraction, we will write it with a logical negation sign ($\neg$). With these two conventions, the λ^q-term $+\lambda x. + x - x$ is written simply $\lambda x.x \neg x$.

Finally, we adhere to the same parenthesization and precedence rules as the λ^p-calculus. In particular, we continue the use of the abbreviatory notations $[M_i^{i \in S}]$ and $[(M_i : n_i)]$ for collections of terms. In addition, we can also $[(M_i : n_i)]$ as $[(M_i : a_i, b_i, n_i)]$ such that $M_i \not\equiv M_j$ and $M_i \not\equiv \overline{M_j}$ for $i \neq j$, all of the M_i are of positive sign, the integer a_i denotes the count of M_i, the integer b_i denotes the count of $\overline{M_i}$, and $n_i = a_i - b_i$.

4.2. Syntactic identities

We will call two terms *opposites* if they differ only in sign.

We define substitution of terms in the λ^q-calculus as a modification of substitution of terms in the λ^p-calculus. We rewrite the seven rules of the λ^p-calculus to take account of the signs of the terms. First, we introduce the function notated by sign concatenation, defined by the following rule in our abbreviatory conventions.

$$\neg\neg \mapsto \epsilon \qquad (4.2)$$

We also note that the concatenation of a sign S with ϵ is just S again. Now we can use this function in the following substitution rules.

1. $(Sx)\,[N/x] \equiv SN$
2. $(Sy)\,[N/x] \equiv Sy$ for variables $y \not\equiv x$
3. $(PQ)\,[N/x] \equiv (P\,[N/x])\,(Q\,[N/x])$
4. $(S\lambda x.P)\,[N/x] \equiv S\lambda x.P$
5. $(S\lambda y.P)\,[N/x] \equiv S\lambda y.\,(P\,[N/x])$ if $\begin{array}{l} y \not\equiv x,\text{ and} \\ y \notin FV(N) \end{array}$ $\qquad (4.3)$
6. $(S\lambda y.P)\,[N/x] \equiv S\lambda z.\,(P\,[z/y]\,[N/x])$
   $\qquad y \not\equiv x,$
   $\quad \text{if}\ \ y \in FV(N),\text{ and}$
   $\qquad z \notin FV(P)\bigcup FV(N)$
7. $(P,Q)\,[N/x] \equiv (P\,[N/x]\,,Q\,[N/x])$

The use of the sign concatenation function is hidden in rule (1). Consider $(\neg x)\,[\neg\lambda y.y/x] \equiv \neg\neg\lambda y.y$. This is not a λ^q-term by grammar (4.1) but applying the sign concatenation function yields the term $\lambda y.y$.

4.3. Reduction

The γ-relation of the λ^q-calculus is of the same form as that of the λ^p-calculus.

$$\gamma^q \triangleq \left\{ \begin{array}{l} \left(\left[M_i^{i\in 1..m}\right]\left[N_j^{j\in 1..n}\right], \left[M_i^{i\in 1..m}N_j^{j\in 1..n}\right]\right) \\ \text{s.t. } M_i, N_j \in LambdaQTerm,\ m > 1 \text{ or } n > 1 \end{array} \right\} \qquad (4.4)$$

We omit the superscript when it is clear if the terms under consideration are λ^p-terms or λ^q-terms. We still write $\gamma(M)$ for the γ-normal form of M.

We extend the β-relation to deal properly with signs.

$$\beta^q \triangleq \left\{ \begin{array}{l} ((S\lambda x.M)\,N,\, SM\,[N/x]) \\ \text{s.t. } S \in Sign,\text{ and } S\lambda x.M, N \in LambdaQTerm \end{array} \right\} \qquad (4.5)$$

4.4. Evaluation semantics

We modify the call-by-value evaluation semantics of the λ^p-calculus.

$$\frac{}{v \rightsquigarrow v}(\text{Refl}) \qquad (\text{for } v \text{ a value})$$

$$\frac{\gamma(M) \rightsquigarrow S\lambda x.P \quad \gamma(N) \rightsquigarrow N' \quad \gamma(SP[N'/x]) \rightsquigarrow v}{MN \rightsquigarrow v}(\text{Eval})$$

$$\frac{\gamma(M) \rightsquigarrow v_1 \quad \gamma(N) \rightsquigarrow v_2}{(M,N) \rightsquigarrow (v_1, v_2)}(\text{Coll})$$

4.5. Observation

We define an observation function Ξ from λ^q-terms to λ-terms as the composition of a function Δ from λ^q-terms to λ^p-terms with the observation function Θ from λ^p-terms to λ-terms defined in (3.5). Thus, $\Xi = \Theta \circ \Delta$ where we define Δ as follows.

$$\Delta(Sx) = x \qquad (4.6)$$

$$\Delta(S\lambda x.M) = \lambda x.\Delta(M) \qquad (4.7)$$

$$\Delta(M_1 M_2) = \Delta(M_1)\Delta(M_2) \qquad (4.8)$$

$$\Delta([M_i : a_i, b_i, n_i]) = \left[\Delta\left(M_i^{i \in \{i \mid n_i \neq 0\}} : |n_i|\right)\right] \qquad (4.9)$$

Note that unlike the observation function Θ of the λ^p-calculus, the observation function Ξ of the λ^q-calculus is not total. For example, $\Xi(x, \neg x)$ does not yield a λ-term because $\Delta(x, \neg x)$ is the empty collection, which is not a λ^p-term.

Although observing a λ^p-term is statistically indistinguishable from observing its γ-normal form, observing a λ^q-term is, in general, statistically distinguishable from observing its γ-normal form.

4.6. Observational semantics

The observational semantics for the λ^q-calculus is similar to that of the λ^p-calculus (3.9). It is given by a single rule.

$$\frac{M \rightsquigarrow v \quad \Xi(v) = N}{M \multimap N}(\text{ObsQ}) \qquad (4.10)$$

4.7. Examples

We provide one example. We show how satisfiability may be solved in the λ^q-calculus. We assume possible solutions are encoded some way in the λ^q-calculus and there is a term CHECK$_f$ that checks if the fixed Boolean formula f is satisfied by a particular truth assignment, given as the argument. The output from this is a collection of T (truth) and F (falsity) terms. We now present a term that will effectively remove all of the F terms. It is an instance of a more general method.

$$\underline{\text{REMOVE-F}} \equiv \lambda x. \underline{\text{IF}}\, x\, x\, (x, \neg x) \qquad (4.11)$$

We give an example evaluation.

$$
\begin{aligned}
\underline{\text{REMOVE-F}}\ (\underline{F},\underline{T},\underline{F}) \quad &\equiv \quad (\lambda x.\,\underline{\text{IF}}\,x\,x\,(x,\neg x))\,(\underline{F},\underline{T},\underline{F}) \\
&\rightarrow_\gamma \quad \begin{pmatrix} (\lambda x.\,\underline{\text{IF}}\,x\,x\,(x,\neg x))\,\underline{F}, \\ (\lambda x.\,\underline{\text{IF}}\,x\,x\,(x,\neg x))\,\underline{T}, \\ (\lambda x.\,\underline{\text{IF}}\,x\,x\,(x,\neg x))\,\underline{F} \end{pmatrix} \\
&\twoheadrightarrow_\beta \quad ((\underline{F},\neg\underline{F}),\underline{T},(\underline{F},\neg\underline{F})) \\
&\equiv \quad (\underline{F},\neg\underline{F},\underline{T},\underline{F},\neg\underline{F})
\end{aligned}
\tag{4.12}
$$

Observing the final term will always yield $\underline{T}$. Note that the drawback to this method is that if f is unsatisfiable then the term will be unobservable. Therefore, when we insert a distinguished term into the collection to make it observable, we risk observing that term instead of $\underline{T}$. At worst, however, we would have a fifty-fifty chance of error.

Specifically, consider what happens when the argument to $\underline{\text{REMOVE-F}}$ is a collection of $\underline{F}$'s. Then $\underline{\text{REMOVE-F}}\,\underline{F} = (\underline{F},\neg\underline{F})$. We insert $\underline{I} \equiv \lambda x.x$ which, if we observe, we take to mean that either f is unsatisfiable or we have bad luck. Thus, we observe the term $(\underline{I},\underline{F},\neg\underline{F})$. This will always yield $\underline{I}$. However, we cannot conclude that f is unsatisfiable because, in the worst case, the term may have been $(\underline{I},\underline{\text{REMOVE-F}}\,\underline{T}) = (\underline{I},\underline{T})$ and we may have observed $\underline{I}$ even though f was satisfiable. We may recalculate until we are certain to an arbitrary significance that f is not satisfiable.

Therefore, applying $\underline{\text{REMOVE-F}}$ to the results of $\underline{\text{CHECK}_f}$ and then observing the result will yield $\underline{T}$ only if f is satisfiable.

5. Simulation to quantum computers

We show that the λ^q-calculus can efficiently simulate the *one-dimensional partitioned quantum cellular automata* (1d-PQCA) defined in [11]. By the equivalence of 1d-PQCA and quantum Turing machines (QTM) proved in [11], the λ^q-calculus can efficiently simulate QTM.

To show that 1d-PQCA can be efficiently simulated by the λ^q-calculus, we need to exhibit a λ^q-term M for a given 1d-PQCA A such that A after k steps is in the same superposition as M after $P(k)$ steps, with P a polynomial.

We assume for now that the 1d-PQCA has transition amplitudes not over the complex numbers, but over the positive and negative rationals. It has been shown [3] that this is equivalent to the general model in QTM.

To express A in M, we need to do the following things.

1. Translate states of A into λ^q-terms that can be compared (e.g. into Church numerals).

2. Translate the acceptance states and the integer denoting the acceptance cell into λ^q-terms.

3. Create a λ^q-term $\mathbf{P}$ to mimic the operation of the permutation σ.

4. Translate the local transition function into a transition term. For 1d-PQCA this means translating the matrix Λ into a term **L** comparing the initial state with each of the possible states and returning the appropriate superposition.

5. Determine an injective mapping of configurations of A and configurations of M.

Although we will not write down M in full, we note that within M are the mechanisms described above that take a single configuration, apply **P**, and return the superposition as described by **L**.

We recall that the contextual closure of the β^q-relation is such that $M, N \to_\beta M', N'$ where $M \to_\beta M'$ and $N \to_\beta N'$. Thus there is parallel reduction within superpositions. By inspection of the mechanisms above it follows that k steps of A is equivalent to a polynomial of k steps of M.

Steps 1, 2, and 3 are straightforward. Then for step 5, the λ^q-superposition $[(M_i : a_i, b_i, n_i)]$ (let $n = \sum n_i$) will be equivalent to the 1d-PQCA-superposition $\sum \frac{n_i}{n} |c(M_i)\rangle$, where c takes λ^q-terms and translates them into 1d-PQCA configurations. Essentially this means stripping off everything other than the data, that is to say, the structure containing the contents. Note that c is not itself a λ^q-term. It merely performs a fixed syntactic operation, removing extraneous information such as **P** and **L**, and translating the Church numerals that represent states into the 1d-PQCA states. This is injective because the mapping from states of A into numerals is injective. Thus, step 5 is complete.

Step 4 requires translating the Λ matrix into a matrix of whole numbers, and translating an arbitrary 1d–PQCA superposition into a λ^q-superposition. The latter is done merely by multiplying each of the amplitudes by the product of the denominators of all of the amplitudes, to get integers. We call the product of the denominators here d. We perform a similar act on the Λ matrix, multiplying each element by the product of all of the denominators of Λ. We call this constant b. Then we have that $T = b\Lambda$ is a matrix over integers. This matrix can be considered notation for the λ^q-term that checks if a given state is a particular state and returns the appropriate superposition. For instance, if

$$\Lambda = \begin{pmatrix} \frac{2}{3} & \frac{1}{3} \\ 0 & 1 \end{pmatrix}$$

then

$$T = b\Lambda = 9\Lambda = \begin{pmatrix} 6 & 3 \\ 0 & 9 \end{pmatrix}$$

which we can consider as alternate notation for

$$\mathbf{Q} \equiv \lambda s. \; \mathbf{IF} \; (\mathbf{EQUAL} \; s1) \; (1,1,1,1,1,1,2,2,2)$$
$$(\mathbf{IF} \; (\mathbf{EQUAL} \; s2) \; (2,2,2,2,2,2,2,2,2))$$

Then it follows that if c is a superposition of configuration of A, applying Λ k times results in the same superposition as applying T k times to the representation of c in the λ^q-calculus.

6. Conclusion

We have seen two new formalisms. The λ^p-calculus allows expression of randomized algorithms. The λ^q-calculus allows expression of quantumized algorithms. In these calculi, observation is made explicit, and the notion of superposition common to quantum physics is formalized for algorithms.

This work represents a new direction of research. Just as the λ-calculus found many uses in classical programming languages, the λ^p-calculus and the λ^q-calculus may help discussion of randomized and quantum programming languages.

It should not be difficult to see that the λ^p-calculus can simulate a probabilistic Turing machine and we have shown that the λ^q-calculus can simulate a quantum Turing machine (QTM). However, as we have shown, the λ^q-calculus can efficiently solve NP-complete problems such as satisfiability, while there is widespread belief (e.g. [2]) that QTM cannot efficiently solve satisfiability. Thus, the greater the doubt that QTM cannot solve NP-complete problems, the greater the justification in believing that the λ^q-calculus is strictly stronger than QTM.

It should also follow that a probabilistic Turing machine can (inefficiently) simulate the λ^p-calculus. However, it is not obvious that a quantum Turing machine can simulate the λ^q-calculus. An answer to this question will be interesting. If quantum computers can simulate the λ^q-calculus efficiently, then the λ^q-calculus can be used as a programming language directly. As a byproduct, satisfiability will be efficiently and physically solvable. If quantum computers cannot simulate the λ^q-calculus efficiently, knowing what the barrier is may allow the formulation of another type of computer that can simulate it.

Acknowledgements

Thanks to Stuart Shieber for helpful comments.

Bibliography

[1] BARENDREGT, Hendrik Pieter, *The lambda calculus: its syntax and semantics*, North-Holland (1981).

[2] BENNETT, Charles H., Ethan BERNSTEIN, Gilles BRASSARD, and Umesh VAZIRANI, "Strengths and Weaknesses of Quantum Computing," available online as quant-ph/9701001 at http://xxx.lanl.gov/abs/quant-ph/9701001.

[3] BERNSTEIN, E. and U. VAZIRANI, "Quantum complexity theory," *Proceedings of the 25th Annual ACM Symposium on Theory of Computing* (1993), 11-20.

[4] CHURCH, Alonzo, "An unsolvable problem of elementary number theory", *American Journal of Mathematics* **58** (1936), 345-363.

[5] DEUTSCH, David, "Quantum theory, the Church-Turing principle and the universal quantum computer", *Proc. R. Soc. Lond.* **A400** (1985), 97-117.

[6] DEUTSCH, David, "Quantum computational networks", *Proc. R. Soc. Lond.* **A425** (1989), 73-90.

[7] HEISENBERG, Werner, *Physics and philosophy*, Harper & Bros. (1958).

[8] MAYMIN, Philip, "Extending the Lambda Calculus to Express Randomized and Quantumized Algorithms," available online as quant-ph/9612052 at http://xxx.lanl.gov/abs/quant-ph/9612052. Many of the proofs omitted from the current paper because of space considerations can be found here.

[9] MAYMIN, Philip, "The lambda-q calculus can efficiently simulate quantum computers," available online as quant-ph/9702057 at http://xxx.lanl.gov/abs/quant-ph/9702057.

[10] SIMON, Daniel, "On the power of quantum computation", *Proc. 35th Annual Symp. FOCS* (1994).

[11] WATROUS, John, "On One-Dimensional Quantum Cellular Automata," *Proceedings of the 36th IEEE Symposium on Foundations of Computer Science* (1995), 528-537.

Chapter 33

Towards the global: Complexity, topology and chaos in modelling, simulation and computation

David A. Meyer[1]

Center for Social Computation, Institute for Physical Sciences, and
Project in Geometry and Physics, Department of Mathematics
University of California/San Diego, La Jolla, CA 92093-0112

Topological effects produce chaos in multiagent simulation and distributed computation. We explain this result by developing three themes concerning complex systems in the natural and social sciences: (*i*) Pragmatically, a system is complex when it is represented *efficiently* by different models at different scales. (*ii*) Nontrivial topology, identifiable as we scale towards the global, *induces* complexity in this sense. (*iii*) Complex systems with nontrivial topology are typically chaotic.

1. Introduction

Although the concerns of modelling and simulation can be quite different [1], both support a pragmatic definition of complexity: *A system is complex if it is represented efficiently by different models at different scales.* This idea is

---

[1]This paper reports on part of a collaborative project with Thad Brown, Gary Doolen, Brosl Hasslacher, Ronnie Mainieri and Mark Tilden. I also thank Mike Freedman, Melanie Quong, Leslie Smith and Peter Teichner for discussions on various aspects of this work, and Bruce Boghosian for inviting me to speak in the modelling and simulation session at the ICCS. I gratefully acknowledge support from the John Deere Foundation and from DISA.

commonplace, reflected in the way we organize our understanding of the world around us into physics, chemistry, biology, psychology, economics and political science at (roughly) increasing size scales. The goal of this paper is to explain a recently demonstrated difficulty with multiagent simulations of complex systems at the social science end of this spectrum [2] by placing it in the context of (possibly more familiar) models and simulations of complex systems at the natural science end.

We begin by showing that this definition of complexity is more than subjective. In §2 we consider the hierarchical algorithm of Barnes and Hut for simulation of gravitationally interacting particles [3]. Their multiscale algorithm is more efficient, in a very precise sense, than naïve direct simulation. Large scale states are described by total mass and average position of particles, so the model is similar at different scales. We describe this situation as *simplicity*. For more complicated systems, as we increase in scale towards the global, we may identify states with nontrivial topology. The presence of such states can lead to a different and more efficient model at the larger scale; *topology induces complexity*.

The purest example of this phenomenon might be topological geons in quantum gravity [4]: prime components of the spacelike 3-manifold comprise localized nontrivial topology and can be modelled as particles. The intellectual heritage of this model includes Lord Kelvin's theory of atoms as knotted vortex lines [5]. That theory was mistaken, of course, but it was motivated by Helmholtz's mathematical analysis of hydrodynamical vortices [6]. Vortices alone are topologically nontrivial states, in terms of which the equations for fluid mechanics can be recast and which are utilized in very practical vortex simulations of fluid flow [7]. We explain this in §3, remarking on the efficiencies gained by use of a hierarchical algorithm. While demonstrations of improved simulation efficiency are not available for all the systems we consider, these two examples reenforce the belief, based on the conceptual adequacy of different models at different scales, that the systems are indeed complex.

Multiagent systems are discrete, in contrast to PDE models for fluid mechanics. To develop our theme of topology induced complexity for application in the former, we must explain how it can occur in non-continuum systems. In §4 we consider the homogeneous sector of a multispecies reaction-diffusion model in chemistry/population biology. This model, analyzed by Ruijgrok and Ruijgrok [8], illustrates our first two themes: (discrete) topology induces complexity in the sense of a different efficient model at the global scale.

The same nontrivial discrete topology—a cycle—is ubiquitous in formal models of economics and political science: In §5 and §6 we describe two fundamental results—Sonnenschein, Debreu, and Mantel's excess demand theorems [9] and Arrow's voting theorem [10]—which guarantee the existence of cycles in market and voting models, respectively. We also expand our third theme: systems with topologically induced complexity are typically chaotic. In particular, we explain the precise mathematical sense in which aggregation by voting makes multiagent simulations chaotic [2].

But the aggregation processes which scale multiagent simulations towards the global are usually market mechanisms or voting rules; thus such systems are typically complex, and, according to the results described in §6, chaotic. In the final section we discuss the consequences of this phenomenon and mechanisms which might be implemented to control it.

2. Hierarchical efficiency

Simulation of interesting natural or social systems compels careful attention to the efficiency of the algorithms used. Algorithms which are exponential in the size of the system are essentially useless, and even for polynomial algorithms, decreasing the leading exponent reduces runtime by orders of magnitude. The simulations of concern here typically consist of a large number N of fundamental objects interacting according to rules which model the dynamics of the system under investigation. Algorithmic efficiency is thus determined by runtime as a function of N.

Consider the problem of simulating the (Newtonian) gravitational dynamics of N particles. Each particle exerts a force on every other particle and is subject to the sum of the forces exerted on it by all the other particles. The most straightforward algorithm would compute the $N(N-1)/2$ pairwise forces, sum the forces on each particle, and then evolve each particle accordingly, at each timestep. This algorithm involves no approximations beyond the finite precision computer representation of real numbers; it provides an accurate $O(N^2)$ description of the dynamics.

Barnes and Hut showed, however, that by aggregating the particles into a hierarchy of clusters of increasing sizes, the average run time can be reduced to $O(N \log N)$ with bounded error [3]. Their algorithm works by dividing up the volume of space containing the particles into a tree of cubical cells: Starting with a cube large enough to contain all the particles, at each timestep consider the particles in some (arbitrary) sequence. While the particle lies in the same cube as any previous particle, subdivide that cube into eight cubes of half the linear size. Now assign to each non-empty cube a 'cluster-particle' with the total mass of the particles in that cell, and locate the cluster-particle at the center of mass of those particles. On average, constructing the tree of cells and cluster-particles requires $O(N \log N)$ steps.

The force on a given particle is now computed recursively by working down the tree: If a cluster-particle lies in a cube of size l, is distance d from the particle, and $l/d < \theta$ (a constant), compute the force it would exert; if not move down to the next smaller cluster-particles in the tree and repeat. This algorithm approximates the force on a particle, with bounded error (depending on θ), in an average of $O(\log N)$ steps. The average runtime per timestep is thus $O(N \log N)$ for the whole algorithm.

In this description 'average' refers to possible particle configurations with respect to a uniform probability distribution on the cube. When the particles are literally clustered, as in the simulation Barnes and Hut present of two interacting clusters, additional efficiencies obtain. Even in this case, however, the larger scale representation is essentially similar to the smaller scale one; the system is simple.

3. Topology induces complexity

Fluid flow provides our first example of a system with different representations at different scales. Disregarding the fact that real fluids are composed of molecules, which are composed of atoms, ..., and should, by virtue of that multiscale description alone, be complex systems, let us take the system to be defined by the Navier-Stokes equations:

$$\partial_t v + v \cdot \nabla v = -\nabla p + \nu \nabla^2 v$$
$$\nabla \cdot v = 0,$$

together with appropriate boundary conditions. Here v is the velocity field of the (incompressible) fluid, p is the pressure, and ν is the viscosity. Immense effort has, of course, gone into simulating the Navier-Stokes equations [11], whose solutions range from laminar to turbulent flow.

The configuration variable in this system, the velocity field, can be topologically nontrivial: there may be, for example, *vortices*, *i.e.*, closed loops in the flow. Their presence is measured by the *vorticity* $\omega := \nabla \wedge v$ [6], in terms of which the first equation above can be rewritten as

$$\partial_t \omega = \nabla \wedge (v \wedge \omega) + \nu \nabla^2 \omega,$$

eliminating the pressure. The inverse of the Poisson equation defining vorticity is the Biot-Savart law; it gives v as a function of ω and allows the system to be reformulated entirely in terms of the vorticity. This is a different, although entirely equivalent, model for fluid flow.

When the flow is restricted to two dimensions the vorticity has only a single (orthogonal) component so the configuration variable for the model is a (pseudo-)scalar field. As such it can be approximated by a superposition of delta functions, or more physically, by a collection of point vortices with circulations c_i:

$$\omega(r, t) \approx \sum_{i=1}^{N} c_i \delta(r - r_i(t)).$$

Even before the advent of digital computers Rosenhead simulated fluid flow using this approximation [12]: For inviscid flow, the velocity of each vortex is the value of the velocity field at its present location; this is given in terms of the vorticity field by the Biot-Savart law. Notice that a straightforward algorithm for this computation would be $O(N^2)$. Just as in the simulation of gravitating particles considered in §2, the vortex positions can then be integrated forward in time.

There are several problems with this procedure: While it is correct for inviscid flow, the singularities in the vorticity field make the errors in time integration difficult to control when the point vortices pass too close to one another. (We anticipate the results of §6 by remarking that such systems with only a few vortices have been shown to be chaotic [13].) This problem can be resolved to some extent by implementing a 'cloud-in-cell' technique [14] which also improves the runtime for solving the Poisson equation to $O(N) + O(M \log M)$, where M is the size of an approximating spatial mesh. Alternatively, and at the same time modifying the model so as to be able to simulate viscous flow, the point vortices can be replaced with 'vortex blobs' in which the vorticity has a fixed but non-singular distribution [15]. Hald has shown that the error can be controlled in simulations of inviscid flow with such models [16].

These simulations demonstrate that two dimensional hydrodynamics is therefore a complex system according to our operational definition: Not only are there efficient algorithms at different size scales, as presaged by our discussion of gravitating particles, but the system is modelled differently at each scale—by a velocity field, by a collection of vortices, and by a distribution of vorticity on a mesh. Topology will play this role—inducing new models at more global scales—for the rest of the discussion.

4. Finite topology

Vortices in fluids contain cycles—closed continuous flow lines. But continuity of this sort is not necessary for nontrivial topology. Moving away from physics to a model for a system in chemistry or population biology, consider three species (types of particles) with interactions[2]

$$A + B \to 2A, \qquad B + C \to 2B, \qquad C + A \to 2C.$$

We notice immediately that there is a kind of cyclicity inherent in this set of reactions. Ruijgrok and Ruijgrok have analyzed a system of such particles which are simultaneously interacting according to

$$A + 2B \to 3B, \qquad B + 2C \to 3C, \qquad C + 2A \to 3A,$$

with relative rate α [8].

Since each of the reactions conserves particle number we can normalize to population densities $A + B + C = 1$. Since A, B and C are all positive, the configurations of the system can be represented by points in the interior of an equilateral triangle with side $2/\sqrt{3}$: for any point inside this triangle the sum of the altitudes A, B and C to the three sides is 1. The evolution of the system is a flow on the triangle, described by the system of three ODEs obtained from

$$\dot{A} = A(B - C) + \alpha A(AC - B^2)$$

by cyclic permutations of (ABC).

---

[2] Delightfully, Sinervo and Lively have observed populations of lizards competing according to very similar rules [17].

Notice that the center of the triangle $A = B = C = 1/3$ is an equilibrium point. When $\alpha = 0$, the product $P := ABC$ is left invariant by the flow: all the solutions are periodic orbits along the closed curves $P = \text{const}$. In general,

$$\frac{1}{P(t)} \frac{dP(t)}{dt} = -\frac{9}{4} \alpha r^2(t),$$

where $r(t)$ is the distance from the center of the triangle. Thus for $\alpha < 0$, the center $(1/3, 1/3, 1/3)$ is a stable equilibrium, while for $\alpha > 0$ it is unstable; the $\alpha = 0$ model is not structurally stable.

Although we have not introduced the formalism for finite topological spaces, this analysis motivates identification of the cyclicity of the two sets of reactions as topological nontriviality; certainly this cyclicity can induce cycles in the evolution of the system. Moreover, in the sense that the system can be represented globally as a cycling 'ecology', or one at a stable or unstable equilibrium, rather than only in terms of individual, interacting particles, it is complex.

5. Economics and politics

Thinking of the preceding system as a model in population biology leads us towards social science models of multiple interacting agents. By now we should expect cycles to occur in these systems—and they do, in very similar ways. We begin by considering an example, due to Scarf [18], which indicates difficulties with the general equilibrium model in economics [19].

In this model there are multiple agents with initial endowments $w_i \in \mathbb{R}^k_{\geq 0}$, where the components of w_i represent amounts of k goods. The prices of these goods are represented by another vector $p \in \mathbb{R}^k_{\geq 0}$, so the initial wealth of agent i is $p \cdot w_i$. By exchanging goods according to prices p, agent i can afford any commodity bundle $x \in \mathbb{R}^k_{\geq 0}$, provided $p \cdot x \leq p \cdot w_i$.

The agents also have preferences among the different goods, modelled by *utility functions* $u_i : \mathbb{R}^k_{\geq 0} \longrightarrow \mathbb{R}$; $u_i(y) \geq u_i(x)$ means that agent i prefers commodity bundle y to bundle x. We assume that for each agent i, $\{y \in \mathbb{R}^k_{\geq 0} \mid u_i(y) \geq u_i(x)\}$ is a strictly convex set and that $\nabla u_i \in \mathbb{R}^k_{\geq 0}$. At given prices p, agent i maximizes utility with a commodity bundle $x_i(p)$ satisfying $\lambda p = \nabla u_i(x_i(p))$, for some $\lambda > 0$; this bundle is the agent's *demand*. Notice that any rescaling of p can be absorbed into λ; we follow tradition and rescale so that $p \cdot p = k$, although we could equally well rescale p so that $p \cdot (1, \dots, 1) = 1$; in either case we refer to the *price simplex* of possible price vectors. The *excess demand* of agent i at prices p is $x_i(p) - w_i$; the *aggregate excess demand* $v(p) := \sum_i (x_i(p) - w_i)$ is a continuous vector field on the price simplex.

The aggregate excess demand defines a flow on the price simplex according to $\dot{p} = v(p)$. That there is always an equilibrium point $v(p) = 0$ for such a flow is a consequence of Brouwer's fixed point theorem [20]; this is the *topological* content [21] of the Arrow-Debreu approach to the existence of competitive equilibria [22]. Scarf demonstrated, however, that such equilibria can be globally unstable [18]: Consider the utility function in a $k = 3$ good economy defined by $u_1(x) =$

$\min(x_1, x_2)$. Suppose this is the utility function for agent 1 and that there are two more agents with utility functions obtained from u_1 by cyclic permutations of (x_1, x_2, x_3). Let the initial endowments of the three agents be $(1, 0, 0)$ and its cyclic permutations, respectively. Then the first component of the aggregate excess demand is

$$v_1(p) = \frac{-p_2}{p_1 + p_2} + \frac{p_3}{p_3 + p_1};$$

the other two components are obtained by cyclic permutation of (p_1, p_2, p_3).

The center point $p_1 = p_2 = p_3 = 1$ is an equilibrium point and, just as in the $\alpha = 0$ case of the example discussed in the preceding section, the product $p_1 p_2 p_3$ is left invariant by the flow. Thus the system has no stable equilibrium. While the utility functions used in this example are convex, but not strictly convex, and the initial endowments are at an extreme, Scarf showed that there is a family of such models (just as there was in §4) which have strictly convex utility functions and inital endowments nonzero in each good, but which are also globally unstable [18].

In this economic model the discrete cycle of utility functions leads to *continuous* cyclic orbits just as the discrete cycle of reaction equations did in Ruijgrok and Ruijgrok's model. But suppose there are only a finite number of states for the system. This is the situation in political science models where each agent's utility function is replaced with a *preference order* among a finite set of alternatives: a relation, denoted $\geq$, which is *complete* (for all pairs of alternatives $a \geq b$ or $b \geq a$) and *transitive* (if $a \geq b$ and $b \geq c$ then $a \geq c$) [10]. When $a \geq b$ and $b \geq a$, the agent with this preference order is *indifferent* between a and b; when only $a \geq b$, say, the agent *strictly* prefers a and we write $a > b$.

The analogue of the aggregate excess demand is a map f from preference *profiles* (lists of the agents' preference orders) p to directed graphs f_p. A directed edge $a \leftarrow b$ in f_p indicates that for profile p the map f chooses alternative a over alternative b. We call f a *voting rule* if for all profiles p, f_p is *complete* (for all pairs of alternatives $a \leftarrow b$ or $b \leftarrow a$) and Pareto (if $a \geq b$ in each preference order in p then $a \leftarrow b$ in f_p).

More than 200 years ago Condorcet recognized that there are potential problems with voting rules, namely that aggregation might produce cycles rather than a definitive outcome [23]. For example, suppose that there are three alternatives $\{a, b, c\}$ and three agents rank them in the orders $a > b > c$, $b > c > a$, and $c > a > b$. Given a choice between b and a, a 2:1 majority prefers a; if they are offered the opportunity to switch from a to c, again a majority will vote to do so; finally a majority also prefers b to c, completing a cycle. This cycle exists in the directed graph f_p corresponding to the majority voting rule; it is the discrete analogue of the continuous cyclic orbits we have seen in the two preceding systems.

6. Complexity and chaos

These social science models, therefore, have features in common with the natural science models we considered in §2, §3 and §4. At the more global scale defined by the aggregation mechanisms, the set of agents may be replaced by a cyclic (sub)market and a cyclic decision process, respectively. Of course, if the utility functions of the economic agents are such that the aggregate excess demand is representable as the excess demand for a single agent then the market has a unique stable equilibrium. Similarly, given a voting rule, the preferences of the agents may cohere to the extent that there is a definitive outcome; even more simply, the voting rule might be *dictatorial*, *i.e.*, dependent only on the preferences of a single, specified agent. In all of these cases the model at the global scale is similar to the individual agent model; the topology is trivial and the systems are simple.

We emphasize, however, that complexity is inherent in social aggregation: The theorems of Sonnenschein, Mantel, and Debreu show that for two or more goods and at least as many agents, the aggregate excess demand can be *any* continuous vector field on the price simplex [9]. In particular, the flow need not have a stable equilibrium: As we have seen already in two dimensions, even though the Poincaré-Bendixson theorem precludes more complicated behaviour [24], the limit set may include cycles. If there is any time dependence from external effects, and in higher dimensional markets, the Kupka-Smale theorem [25] implies that *chaotic* dynamics is structurally stable; systems with a unique or stable equilibrium point are far from generic.

For completely discrete voting models, Arrow's theorem imples that for more than two agents, under a reasonable condition[3] on voting rules, any which are not dictatorial must contain cycles [10]. We recently showed that in the latter situation, not only is the system complex, but it is also chaotic in the mathematical sense [2]: The *topological entropy* [26], defined to be

$$\lim_{n \to \infty} \frac{1}{n} \log(\text{number of } n\text{-periodic orbits in } f_p)$$

is positive exactly when there is a cycle in f_p; beyond identifying the existence of chaos, it quantifies 'how chaotic' the system is. Averaged over the space of voter preferences, the topological entropy measures the complexity of a voting rule, and averaged over voting rules, it measures the (lack of) coherence among voter preferences [2].

7. Consequences

Aggregation rules define the way models of social systems scale towards the global. The larger scale model may be similar to the smaller scale one, as is the case for a dictatorial voting rule, for example. In equivalently simple physical

---

[3]This is the *independence of irrelevant alternatives*, namely that the relation between a and b in f_p depend only on the relations of a and b in the preference orders in p.

models, rescaling simply renormalizes the state variables and their interactions, but introduces no new ones. Complexity emerges when nontrivial topology exists at the more global scale. This is a cycle in each of our examples, and it induces qualitatively different models in both natural and social systems.

The attendant chaos has profound consequences for simulations of these systems: Fine grained prediction is impossible—only certain statistical properties of the evolution are robust. This fact is appreciated heuristically in some of the economics and sociology literature [27], but seems to receive less emphasis in discussions of specific multiagent simulation platforms [28]. It will, nevertheless, determine to which problems such programs can be applied successfully.

Systems of multiple software agents, interacting to effect some (un)intentional distributed computation, are subject to exactly the same analysis. Whether their interactions are decision theoretic [29] or market oriented [30], designers and users must be aware of the possibilities for complexity and chaos. Even when the artificial agents are physical—autonomous robots [31]—attempts at designing coordinated action [32] should be informed by these same considerations.

Particularly in the context of deploying interacting software or hardware agents for some specific task, but also in the context of modelling or simulating multiagent social systems, success may depend on controlling the degree of chaos in the resulting complex system. As we noted at the end of the last section, the topological entropy, for example, measures how chaotic a system is and allows us to identify the sources of complexity, but this alone is insufficient for control. There are several possibilities: We have primarily considered immutable agents; there are also models with adaptive agents [33] which one might hope to have better generic behaviour. The population biology example of §4, however, models a particular kind of adaptation, so chaos is probably no less generic in complex adaptive systems. Agents with 'higher' rationality have also been modelled [34]; these are agents which make internal models for the behaviour of the other agents with which they interact. This approach, while possibly more realistic for modelling human agents, seems to flirt with self-reference and undecidability [35], and must, at best, constrain the depth of the internal models to achieve acceptable computational efficiency [36]. Most generally, it seems that models for complex social systems should include some form of 'back reaction' from the more global aggregation scale to the local individual agent scale. Such a back reaction might affect the agents' preferences or the way they interact, but must be carefully implemented to model interesting complex systems which balance precariously between simplicity and extreme chaos.

Bibliography

[1] SEGEL, Lee A., "The theoretician grapples with complex systems", SFI working paper 93-05-032 (1993).

[2] MEYER, David A., and Thad A. BROWN, "Statistical mechanics of voting". CSC/IPS/UCSD/UM preprint (1997).

[3] BARNES, Josh, and Piet HUT, "A hierarchical $O(N \log N)$ force-calculation algorithm", *Nature* **324** (1986) 446–449.

[4] SORKIN, Rafael D., "Introduction to topological geons", in Peter G. BERGMANN and Venzo DE SABBATA, eds., *Topological Properties and Global Structure of Space-Time*, proceedings of the NATO Advanced Study Institute, Erice, Italy, 12–22 May 1985 (New York: Plenum 1986) 249–270.

[5] KELVIN, William H. THOMSON, Baron, "On vortex atoms", *Proc. Roy. Soc. Edinburgh* **VI** (1867) 94–105.

[6] HELMHOLTZ, Herman L. F. von, "*Über Integrale der hydrodynamischen Gleichungen, welche den Wirbelbewegung entsprechen*", *J. Reine Angew. Math.* **LV** (1858) 25–55.

[7] LEONARD, Anthony, "Vortex methods for flow simulation", *J. Comput. Phys.* **37** (1980) 289–335.

[8] RUIJGROK, Th., and M. RUIJGROK, "A reaction-diffusion equation for a cyclic system with three components", *J. Stat. Phys.* **87** (1997) 1145–1164.

[9] SONNENSCHEIN, Hugo, "Do Walras' identity and continuity characterize the class of community excess demand functions?", *J. Econom. Theory* **6** (1973) 345–354;
MANTEL, Rolf R., "On the characterization of aggregate excess demand", *J. Econom. Theory* **7** (1974) 348–353;
DEBREU, Gerard, "Excess demand functions", *J. Math. Econom.* **1** (1974) 15–21.

[10] ARROW, Kenneth J., *Social Choice and Individual Values* (New York: Wiley 1951).

[11] BORIS, Jay P., "New directions in computational fluid dynamics", *Ann. Rev. Fluid Mech.* **21** (1989) 345–385;
PEYRET, Roger, ed., *Handbook of Computational Fluid Mechanics* (San Diego: Academic Press 1996).

[12] ROSENHEAD, L., "The formation of vortices from a surface of discontinuity", *Proc. Roy. Soc. Lond.* A **134** (1931) 170–192.

[13] AREF, Hassan, "Integrable, chaotic, and turbulent vortex motion in two-dimensional flows", *Ann. Rev. Fluid Mech.* **15** (1983) 345–389.

[14] CHRISTIANSEN, J. P., "Numerical simulation of hydrodynamics by the method of point vortices", *J. Comput. Phys.* **13** (1973) 363–379.

[15] CHORIN, Alexandre J., "Numerical study of slightly viscous flow", *J. Fluid Mech.* **57** (1973) 785–796.

[16] HALD, Ole H., "Convergence of vortex methods for Euler's equations. II.", *SIAM J. Numer. Anal.* **16** (1979) 726–755.

[17] SINERVO, Barry, and Curt M. Lively, "The rock-paper-scissors game and the evolution of alternative male strategies", *Nature* **380** (1996) 240–243.

[18] SCARF, Herbert, "Some examples of global instability of the competitive equilibrium", *Int. Econom. Rev.* **1** (1960) 157–172.

[19] WALRAS, Leon, *Elèments d'économie politique pure, ou, Théorie de la richesse sociale* (Lausanne: Corbaz 1874–7).

[20] BROUWER, Luitzen E. J., "*Über Abbildung von Mannigfaltigkeiten*", *Math. Ann.* **71** (1912) 97–115.

[21] UZAWA, Hirofumi, "Walras' existence theorem and Brouwer's fixed point theorem", *Econom. Studies Quart.* **12** (1962) 59–62.

[22] ARROW, Kenneth J., and Gerard DEBREU, "Existence of an equilibrium for a competitive economy", *Econometrica* **22** (1954) 265–290.

[23] CONDORCET, Marie-Jean-Antoine-Nicolas DE CARITAT, Marquis de, *Essai sur l'application de l'analyse à la probabilité des décisions rendues à la pluralité des voix* (Paris: *l'Imprimèrie Royale* 1785).

[24] POINCARÉ, Jules Henri, "*Mèmoire sur les courbes définies par les équations différentielles, I.*", *J. Math. Pures Appl.*, 3. série **7** (1881) 375–422; *II.* **8** (1882) 251–286; *III.* 4. série **1** (1885) 167–244; *IV.* **2** (1886) 151–217; BENDIXSON, Ivar, "*Sur les courbes définies par les équations différentielles*", *Acta Math.* **24** (1901) 1–88.

[25] KUPKA, Ivan, "*Contributions à la théorie des champs générique*", *Contrib. Diff. Eq.* **2** (1963) 457–484; SMALE, Steve, "Stable manifolds for differential equations and diffeomorphisms", *Ann. Scuola Norm. Sup. Pisa, ser. III* **17** (1963) 97–116.

[26] SHANNON, Claude E., "A mathematical theory of communication", *Bell System Tech. J.* **27** (1948) 379–423; 623–656; PARRY, William, "Intrinsic Markov chains", *Trans. Amer. Math. Soc.* **112** (1964) 55–66; ADLER, Roy L., Alan G. KONHEIM and M. H. MCANDREW, "Topological entropy", *Trans. Amer. Math. Soc.* **114** (1965) 309–319.

[27] SÉROR, Ann C., "Simulation of complex organizational processes: a review of methods and their epistemological foundations", in Nigel GILBERT and Jim DORAN, eds., *Simulating Societies: The Computer Simulation of Social*

Phenomena (London: UCL Press 1994) 19–40;
BYRNE, David, "Simulation—a way forward?", *Sociological Research Online* **2** (1997)
http://www.socresonline.org.uk/socresonline/2/2/4.html.

[28] MINAR, Nelson, Roger BURKHART, Chris LANGTON and Manor ASKENAZI,
"The Swarm simulation system: a toolkit for building multi-agent simulations", SFI working paper 96-06-042 (1996);
EPSTEIN, Joshua M., and Robert AXTELL, *Artificial Societies: Social Science from the Bottom Up* (Cambridge, MA: MIT Press 1996).

[29] ZLOTKIN, Gilad, and Jeffrey S. ROSENSCHEIN, "Mechanism design for automated negotiation, and its application to task oriented domains", *Artif. Intell.* **86** (1996) 195–244;
EPHRATI, Eithan, and Jeffrey S. ROSENSCHEIN, "Deriving consensus in multiagent systems", *Artif. Intell.* **87** (1996) 21–74.

[30] WELLMAN, Michael P., "The economic approach to artificial intelligence", *ACM Comput. Surveys* **27** (1995) 360–262;
CHENG, John Q., and Michael P. WELLMAN, "The WALRAS algorithm: a convergent distributed implementation of general equilibrium outcomes", to appear in *Comput. Econom.*

[31] HASSLACHER, Brosl, and Mark W. TILDEN, "Living machines", *Robotics and Autonomous Systems* **15** (1995) 143–169.

[32] BENI, Gerardo, and Jing WANG, "Theoretical problems for the realization of distributed robotic systems", in *Proceedings of the 1991 IEEE International Conference on Robotics and Automation*, Sacramento, CA 9–11 April 1991 (Los Alamitos, CA: IEEE Computer Society Press 1991) 1914–1920;
MATARIC, Maja J., "Distributed approaches to behavior control", in Paul S. SCHENKER, ed., *Sensor Fusion V*, proceedings of the SPIE conference, **1828**, Boston, MA, 15–17 November 1992 (Bellingham, WA: SPIE 1992) 373–382.

[33] HRABER, Peter T., and Bruce T. MILNE, "Community assembly in a model ecosystem", SFI working paper 96-12-094 (1996).

[34] GMYTRASIEWICZ, Piotr J., and Edmund H. DURFEE, "A rigorous, operational formalization of recursive modelling", in *ICMAS-95*, Proceedings of the First International Conference on Multi-Agent Systems, San Francisco, 12–14 June 1995 (Menlo Park, CA: AAAI Press 1995) 125–132;
TESFATSION, Leigh, "A trade network game with endogenous partner selection", in Hans M. AMMAN, Berc RUSTEM and Andrew B. WHINSTON, eds., *Computational Approaches to Economic Problems* (Boston: Kluwer Academic 1997) 249–269.

[35] GÖDEL, Kurt, "*Über formal unentscheidbare Sätze der Principia Mathematica und verwandter System I.*", *Monat. Math. Phys.* **38** (1931) 173–198.

[36] VIDAL, José M., and Edmund H. DURFEE, "Using recursive models effectively", in Michael J. WOOLDRIDGE, Jörg P. MÜLLER and Milind TAMBE, eds., *Intelligent Agents II: Agent Theories, Architectures, and Languages*, proceedings of the IJACI'95 Workshop (ATAL), Montréal, 19–20 August 1995, LNCS **1037** (New York: Springer-Verlag 1996) 171–186.

Chapter 34

An effect of scale in a non-additive genetic model

Jason G. Mezey
Günter P. Wagner
Yale University, EE Biology Dept.

A model of non-additive gene effects was analyzed with respect to evolutionary dynamics. A major result was the evolutionary dynamics predicted by the model scales with character value. Such a scale effect is absent in typical population genetic models which consider only additive gene effects. We think that scale effects are a generic property of population genetic models which include gene interaction effects.

1. Introduction

Aspects of a phenotype, the properties which constitute the structural and functional character of an organism [13], can often be attributed to possession of a particular gene or set of genes. Such genes, termed alleles, may therefore be thought of as having an effect on a phenotype. As an example, a male *Drosophila melanogaster* will have unpigmented eyes if it possesses one of the white alleles at the white-eye genetic position or "locus". The white allele can therefore be thought of as having an effect on the fruitfly phenotype.

Phenotypic evolution is a process which may be analyzed in terms of allelic effects. This process is constrained by changes in frequency of alleles in a population and by creation of new alleles [7]. The latter occurrence is called mutation. Several population genetic models have been studied to analyze phenotypic evolution in terms of frequency changes and mutations (see [15] for review). One of the major assumptions of many of these models is that the allelic contributions are additive in nature (e.g. [2]). In such models, the value assessed to a phenotypic character of an individual is dependent on the sum of the allelic contributions. The effect of an allele substitution is not changed by the presence or absence of other alleles, since allelic effects combine in a linear manner. The justification for this approach is that the response to selection of a character is determined by the variance of the additive effects, i.e. the linear component of a regression of genotypes ("genetic effects") to phenotypes.

Though additive models have shown to be powerful approximations [11], empirical evidence shows that the assumption of additivity is generally not met. Epistasis, the effect of the substitution of an allele at one locus on the additive effect of a substitution at another locus [6], has been shown to be widespread phenomenon [18]. Even if the immediate response to natural selection is not influenced by epistatic effects, epistasis has been shown to be important in a variety of evolutionary scenarios. Among them are the genetic consequences of population bottlenecks [1; 4; 9; 10], speciation [3; 5; 8; 12; 14; 16; 19] and the evolution of the genotype-phenotype map [17].

To investigate some of the consequences of non-linear effects on evolutionary dynamics, a simple, non-linear population genetic model is considered. One major result of considering non-linearities is a scale effect, a phenomenon not present in purely additive models. These preliminary results suggest that the mathematical analysis of non-linear models requires different modeling techniques than the theory of additive models.

2. Methods

2.1. The model:

The effects of an allele present at a genetic position or locus j are represented by the variable y_j and i variables of the type ρ_{ij}, where i is the index of phenotypic character number. The variable y_j reflects the effect of an allele which has some physiological property relevant to the phenotype. The variable ρ_{ij} reflects an effect of this same allele on the effect of an allele present at a different locus with respect to a character i.

To determine which alleles have an effect ρ_{ij} on which other alleles, a matrix $\mathbf{R}$ is utilized to determine the "interaction scheme". The matrix $\mathbf{R}$ is a Boolean matrix with a "1" at entry r_{jk} reflecting an "interaction" and a "0" reflecting "non-interaction". The structure of the $\mathbf{R}$ matrix is a parameter of the model.

The effects of alleles at n loci determine the genotypic value of a phenotypic character. In terms of the above notation, the genotypic value of a specific character i is determined by the equation:

$$x_i = \sum_{j=1}^{n} y_j \sum_{k=1}^{n} r_{jk} \, \rho_{ik} \qquad (2.1)$$

The vector of genotypic effects $\mathbf{G} = (x_1, ..., x_n)$ has i components determined by equations of the form (1). The genotypic effect vector $\mathbf{G}$ and the environmental effect vector **E** determine the vector of phenotypic effects **P**:

$$\mathbf{P} = \mathbf{G} + \mathbf{E} \tag{2.2}$$

In this model, an individual is represented by a phenotypic vector **P** in multidimensional phenotypic character space.

The environmental vector **E** accounts for random deviations of **P** from the value of the genotypic vector **G**. It is a standard assumption of quantitative genetic models that such deviations are normally distributed. This assumption is consistent with empirical observations (see [7] for examples). Each of the entries of vector **E** are therefore drawn from a normal distribution with expected value zero and variance V_E. It is a further assumption of this model that the vector **E** combines with the vector **G** in a linear manner.

To consider evolutionary dynamics, a population of individuals with phenotypes determined by the above equations must be considered. Simulations of such a population accounts for mating, genetic recombination and probability of survival with respect to a particular fitness function, the domain of which is the phenotype vector of an individual. Mutation occurs at a rate μ and the effect of a mutation changes the y_j, and the ρ_{ij} values by values chosen from constant distributions with expected values zero and variances V_y and V_{Pij} respectively.

2.2. The experiment:

The objective of the experiment was to determine the effect of scale on the evolution of pleiotropy. Pleiotropy is defined as the state where a locus affects two or more characters, such that if alleles of the locus are segregating, it causes simultaneous variation in these characters [7]. In terms of the model, the pleiotropic effects of a locus j are determined by the angle of the vector:

$$\left[y_j \sum_{k=1}^{n} r_{jk}\, \rho_{1k}, y_j \sum_{k=1}^{n} r_{jk}\, \rho_{2k}, ..., y_j \sum_{k=1}^{n} r_{jk}\, \rho_{nk} \right] \tag{2.3}$$

For example, in a two character system, the closer the angle of the vector described by equation (3) to 45°, the greater the pleiotropic effect.

The pleiotropic effect of a particular locus is not only dependent on the allele present at the locus, but on alleles present at other loci. This is due to the effects of the ρ_{ij} values. The pleiotropic effects of a particular locus are therefore dependent on which alleles are present at other loci.

A set of experiments were performed with a two character system. In the initial population, each locus had no pleiotropic effects. The angle of half the loci was 0°, the other half was 90°.

Populations of these individuals were studied under a corridor fitness function. The result of applying this function was that individuals with a large phenotypic value for the first character and a phenotypic value of the second character near the starting value, had a better chance of survival.

To keep track of how pleiotropic effects evolved, the changes in the angle of locus vectors were tracked with a statistic which determined the fraction of loci with angles in a particular range. As an example, one range included any loci with angles 75° to 105°. Any loci in this range, called the "orthogonal" range, pleiotropically affected character "2" to a much greater degree than character "1". At the start of the simulations 50% of the loci were in the orthogonal range.

Each set of simulations started with the same number of loci in a particular angular range. To test the effects of scale, the initial values of the two characters were increased for each set of runs. This was accomplished by increasing the mean value of the y values of the alleles. Though the value of the characters increased, the initial pleiotropic effects of each locus remained the same. The populations were allowed to evolve for thirty generations and then the change in pleiotropic effects in the orthogonal range was determined. The results of twenty simulation runs were averaged for each character value.

3. Results

Evolution of pleiotropic effects occurs in this model if the values of ρ_{ij} change. A mathematical analysis of the model was performed to determine the fitness effects of mutations which changed the values of $\rho_{ij} \rightarrow \rho'_{ij} = \rho_{ij} + \alpha_{ij}$:

$$\alpha_{1l} > \frac{k\,V_\delta \left(\sum_{k=1}^{n} r_{lk}\,\bar{p}_{2k} \right)^2}{s\,n\,\bar{y}} + \frac{k\,V_\alpha\,n\,\bar{y}}{s} \qquad (3.1)$$

The important terms to note in this equation are α_{1l}, $\bar{y}$ and "s".

The term α_{1l} is the effect of a mutation on a value ρ_{ij} of some allele. If the value of mutation α_{1l} is greater than the value of the right hand side of the equation, this mutation will be propagated and the result will be a change in pleiotropy of a locus. For pleiotropic evolution to occur, the right side of the equation must be small relative to the average mutational effect α. Given the dynamics of the model, the value of the right side of the equation is most dependent on the values of $\bar{y}$ which is the mean y value of the alleles across loci. The closer the value of $\bar{y}$ to zero, the larger the first term and the smaller the second. For terms near zero therefore, the first term is the more "dominant". As an example, if the value of $\bar{y}$ starts near zero and then increases, the amount of pleiotropic evolution increases. However, as the value of $\bar{y}$ increases the second term becomes larger and beyond a certain point becomes dominant. Beyond this point, the value of $\bar{y}$ continues to increase, the amount of pleiotropic evolution begins to *decrease*.

The term "s" reflects the strength of directional selection. The larger this term, the greater an individual's chance of survival is dependent on the value of its first character. For the dynamics of the system, the values of "s" only have an important effect on the second term. As the strength of directional selection decreases (as the term "s" decreases), the second term becomes larger. For very small values of "s", it would be expected that the second term becomes more dominant with smaller $\bar{y}$ values. Having a small "s", i.e. weak directional

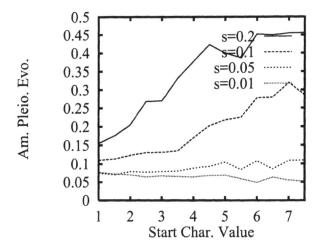

Figure 3.1: Plots depicting the dependency of pleiotropic evolution on starting character value. The amount of pleiotropic evolution is calculated by considering the fraction of allelic vector effects which exit the orthogonal angular range after thirty generations (see text). Initially, half of the total number of allelic vectors are in the orthogonal range. The units on the character value axis are standard deviations of environmental effects.

selection, therefore has the same effect as if the values of $\bar{y}$ were very large. In the cases of very small "s", increasing the values of $\bar{y}$ from near zero could actually result in a decrease in pleiotropic evolution as the second term would initially be more dominant. These expected results are borne out in the simulations, the results of which are summarized in Fig. 3.1. The graphs of Fig. 3.1 compare increasing character values, which correspond to increasing values of $\bar{y}$ with the amount of pleiotropic evolution. Each line reflects different strengths of directional selection, the greatest value of "s" reflects the strongest directional selection. As expected, the amount of pleiotropic evolution increases with increasing character values. For strong directional selection (s = 0.2) increasing character size results in increased pleiotropic evolution (slope of the least squares line: $\beta = 0.0498$). The effect becomes less pronounced as "s" decreases ($\beta = 0.0341$ for s = 0.1 and $\beta = 0.0056$ for s = 0.05) Under the weakest directional selection (s = 0.01), the amount of pleiotropic selection decreases ($\beta = -0.0031$). The weakest directional selection produces an effect similar to the case of strong directional selection where character values become very large.

4. Discussion

In the present model, the amount of pleiotropic evolution scales with character value. This is the result summarized in Fig. 3.1. The reason for this result is that the effect of a particular allele is dependent on the alleles present at other

362

loci. Since the same alleles determine the genotypic values of the characters, the effect of mutations and the state (size or scale) of the character are no longer independent variables. A non-additive genetic system therefore allows the same mutation to have different effects depending on the state of the characters. This means that changing character values, which is the direct result of changing allelic effects, results in different mutational effects.

This result contrasts with the properties of additive genetic models. In an additive model, the same mutation changes allelic effects by the same amount, irrespective of the state of other alleles. Such a model does not produce a scale effect, since the effects would be constant for a particular mutation.

The results of this study has some implications for biological investigations if non-linear allelic effects are important with respect to evolutionary dynamics. First, since the mutational effect scales with character value, studies attempting to determine the average mutational effect on a character will be specific to a particular state of the character.

Second, even with a deep understanding of the genetic system, parameters of the fitness function and by extension, the structure of the fitness function, must be known to predict the evolution of genetic architecture. As summarized in Fig. 3.1, different values of the directional selection parameter "s" result in different scaling properties, even scaling effects with opposite signs.

Third, if with this simple non-additive model such an obvious scaling effect occurs, it implies many other effects may be uncovered by studying such models. The challenge will be to develop nonlinear genetic models which are good representations of biological systems. The present study simply highlights one of the implications if non-linear genetic interactions are taken into account.

Bibliography

[1] BRYANT, E., S. MCCOMMAS, and L. COMBS, "The effect of an experimental bottleneck upon genetic variation in the house fly", *Genetics* **114** (1986), 1191–1211.

[2] BULMER, M., "The effect of selection on genetic variability", *The Am. Natur.* **105** (1984), 201–211.

[3] CARSON, H. and Alan TEMPLETON, "Genetic revolutions in relation to speciation phenomena: the founding of new populations", *Ann. Rev. Ecol. Syst.* **15** (1986), 97–131.

[4] CHEVERUD, James and Eric ROUTMAN, "Epistasis and its contribution to genetic variance components", *Genetics* **130** (1995), 1455–1461.

[5] COYNE, Jerry, "Genetics and Speciation", *Nature* **355** (1992), 511–515.

[6] CROW, James and Motoo KIMURA *An introduction to population genetics*, Harper and Row (1970).

[7] FALCONER, Douglas and Trudy MACKAY *Introduction to population genetics*, Longman (1996).

[8] GAVRILETS, S. and J. GRAVNER, "Percolation on the fitness hypercube and the evolution of reproductive isolation", *J. Ther. Biol.* **184** (1997), 51–64.

[9] GOODNIGHT, Charles, "On the effect of founder events on epistatic genetic variance", *Evolution* **41** (1987), 80–91.

[10] GOODNIGHT, Charles, "Epistasis and the effect of founder events on the additive genetic variance", *Evolution* **42** (1988), 441–454.

[11] LANDE, Russell, "The maintenance of genetic variability by mutation in a polygenic character with linked loci", *Genetic Res.* **26** (1975), 221–235.

[12] ORR, H. and Jerry COYNE, "The genetics of postzygotic isolation in the *Drosophila virilis* group", *Genetics* **121** (1989), 527–537.

[13] STRICKBERGER, Monroe, *Evolution*, Jones and Barlett (1990).

[14] TEMPLETON, Alan, "Genetic architectures of speciation", *Mechanisms of Speciation* (C. BARAGOZZI, Alan LISS ed.), (1982), 105–121.

[15] TURELLI, Michael, "Phenotypic evolution, constant covariances, and the maintenance of additive variance", *Evolution* **2** (1988), 1342–1347.

[16] WAGNER, Andreas, Günter WAGNER, and Philippe SIMILION, "Epistasis can facilitate the evolution of reproductive isolation by peak shifts: a two-locus, two-allele model", *Genetics* **138** (1994), 533–545.

[17] WAGNER, Günter and Lee ALTENBERG, "Complex adaptations and the evolution of evolvability", *Evolution* **50** (1996), 967–976.

[18] WHITLOCK, M., F., PHILIPS, F., MOORE and S. TONSOR, "Multiple fitness peaks and epistasis", *Ann. Rev. Ecol. System* **26** (1995), 601–629.

[19] WU, Chung and Michael PALOPOLI, "Genetic of postmating reproductive isolation in animals", *Annu. Rev. Genet.* **27** (1994), 238–308.

Chapter 35

Parallel computational complexity in statistical physics

Kenneth J. Moriarty and Jonathan L. Machta[1]
Department of Physics and Astronomy, University of Massachusetts
Amherst, Massachusetts 01003
Raymond Greenlaw
Department of Computer Science, University of New Hampshire
Durham, New Hampshire 03824

A model of non-additive gene effects was analyzed with respect to evolutionary dynamics. A major result was the evolutionary dynamics predicted by the model scales with character value. Such a scale effect is absent in typical population genetic models which consider only additive gene effects. We think that scale effects are a generic property of population genetic models which include gene interaction effects.

1. Introduction

Parallel computational complexity theory is the branch of theoretical computer science in which problems are classified according to the time and processor requirements of their parallel solutions. From a practical standpoint, parallel complexity analysis will become a valuable tool in simulation physics as parallel computers become more widely available. On a more conceptual level, parallel computational complexity seems to capture important features of the intuitive concept of physical complexity.

[1] Supported in part by NSF grants Nos. DMR-9311580 and DMR-9632898.

Bennett[2, 3] suggests that an object should be regarded as complex if it contains structures that are unlikely to have arisen quickly. In this view, the presence of unavoidable history dependence is the signature of physical complexity. We believe that the intrinsic history dependence of a physical process may be isolated by considering a massively parallel computer simulation of the process. The central idea is that superficial history dependence can be eliminated through parallelism. Specifically, we suggest that the physical complexity of a system may be quantified by the time complexity of the fastest feasible parallel algorithm for simulating its states.

Parallel complexity theory has been applied to a variety of equilibrium and non-equilibrium models in statistical physics. For example, Eden growth, invasion percolation, ballistic deposition, and solid-on-solid growth have all been shown[8] to have highly parallel algorithms; that is, these systems may be simulated in a time that scales as some power of the logarithm of the system size (*polylog time*) using a number of processors polynomial in the system size. Mandelbrot percolation is somewhat less complex, requiring only constant parallel time for a polynomial number of processors[9]. On the other hand, a model of fluid invasion in porous media as well as a restriced version that mimics diffusion-limited aggregation (DLA) have been shown[7] to belong to the class of **P**-complete problems. Problems in this class are suspected of being inherently sequential; therefore, it is unlikely that fluid-invasion patterns or DLA clusters can be generated in polylog time when restricted to a polynomial number of processors. **P**-completeness has also been established[9] for the original random-walk dynamics for DLA and several schemes for equilibrating the Ising model. Both the FHP and HPP III lattice gases have been shown[11] to be **P**-complete, whereas the Lorentz lattice gas can be simulated in highly parallel fashion[10].

The property of **P**-completeness and other central concepts of parallel computational complexity theoryare discussed in more detail in §2. As an example of an application in statistical physics, in §3 we examine DLA from a somewhat different angle than [7] and [9]. Although the **P**-completeness results indicate that a highly parallel DLA algorithm probably does not exist, we demonstrate that, on average, DLA clusters can still be produced in a time sub-linear in the cluster mass by a polynomial-processor algorithm. We calculate the algorithm's dynamic exponent z, which gives the scaling of the average running time with cluster radius, and argue that z is a measure of the intrinsic history dependence of DLA. §4 is a summary of our results.

2. Parallel complexity theory

This section provides some background on parallel computational complexity theory. The reader is referred to [5, 4, 14] for further details. The objective of computational complexity theory is to classify problems according to how the computational resources needed to solve them scale with the size of the problem. For parallel complexity the primary resources are hardware (consisting of memory and processors, or their equivalents) and time. One of the strengths of complexity theory is that resource requirements are comparable within a diverse group of computational models including parallel random-access machines, Boolean circuits, and systems of formal logic. Time requirements for a wide class of computational models differ by only a logarithmic factor when the models are required to use polynomially related amounts of hardware. Complexity results thus have a rather fundamental status independent of the computational model adopted. This fact supports the belief that a complexity analysis of simulating a physical system reveals intrinsic properties of the system.

In the example in §3, we employ the parallel random-access machine (PRAM) model of parallel computation. A PRAM consists of a number of processors, distinguished by positive-integer labels, which execute the same program and may access any cell of a global random-access memory in a single time step. Due to the finiteness of signal speeds and hardware density, PRAM performance cannot be achieved in a scalable parallel computer. Nonetheless, the PRAM model is useful from both practical and theoretical standpoints. On the practical side, it serves as a guide to the implementation of algorithms on real parallel machines. On a conceptual level, PRAM time provides a measure of a fundamental feature of a computation that may be called *logical depth*[2]. Logical depth is the minimum number of logical operations that must be carried out in sequence in order to complete a computational process. Thus, the greater a problem's logical depth, the less significant the speed-up that can be achieved through parallelism.

According to parallel complexity theory, problems that can be solved in $n^{O(1)}$ time using $n^{O(1)}$ processors for an instance of size n are considered *feasible* and constitute the complexity class **P**. Problems that can be solved in $(\log n)^{O(1)}$ time using $n^{O(1)}$ processors are in the class **NC** of problems with *highly parallel* solutions. An important unproven conjecture in computer science is that there exist *inherently sequential* problems in **P** that have no highly parallel solutions. **P**-*complete* problems are the best candidates. If indeed **NC** $\neq$ **P**, then **P**-complete problems are inherently sequential.

Strictly speaking, the complexity classes mentioned here refer to *decision problems*, i.e., problems with "yes" or "no" answers. Computational statistical physics, on the other hand, typically deals with *sampling problems*, for which the goal is to generate a representative member of a statistical ensemble, e.g., a configuration of Ising spins at a given temperature or a DLA cluster. Sampling problems require a supply of random numbers. Rather than confront the subtle issues related to generating random or pseudorandom numbers, we assume here that our PRAM is augmented with a device that generates ideal random bits. An alternative, more rigorous approach involves first associating a given sampling problem with a "natural" decision problem (see [8, 9]) and then analyzing the complexity of the latter.

3. Example: Parallel algorithm and dynamic exponent for DLA

Diffusion-limited aggregation (DLA)[17] is a growth model that gives rise to highly branched clusters of particles. These clusters bear a strong resemblance to structures observed in experiments on electrodeposition, viscous fingering, crystallization, and the growth of bacteria colonies[15]. A DLA cluster begins as a single stationary seed particle and grows by the addition of diffusing particles that stick to the cluster upon contact. A diffusing particle (random walker) is released a large distance from the growing cluster and either joins the cluster by sticking to it or is discarded if it journeys very far away. In either case, a new walker is released as soon as the fate of the preceding one has been determined. It is important to note that only one diffusing particle is present in the system at any given time. Allowing multiple diffusing particles alters the resulting cluster distribution[6, 16]; therefore, it is not obvious how to take advantage of parallel computation in DLA simulations.

In accord with the apparent structural intricacy of DLA clusters and the observed importance of the sequential nature of the growth process in determining the cluster distribution, decision problems associated with DLA have been shown to be **P**-complete[7, 9]. Therefore it is unlikely that DLA clusters of mass M can be generated in $(\log M)^{O(1)}$ time using $M^{O(1)}$ processors. Nevertheless, we have shown[12] that a more modest speedup relative to the $O(M)$ time required by the best sequential algorithms can still be achieved using $M^{O(1)}$ processors. In describing our method, we find it useful to define an *interference* within a group of random walkers to be the attachment of one of the walkers to another member of its group that has already joined the cluster. Our algorithm is based on identifying and processing in parallel successive groups of non-interfering walkers. By analyzing the complexity of the individual steps of the algorithm and obtaining an expression for the interference probability, we determine the algorithm's dynamic exponent z, defined by $T \sim R^z$, where T is the average
running time and R is the cluster radius.

3.1. Parallel DLA algorithm

Here we present a PRAM algorithm for generating off-lattice DLA clusters. The algorithm's main loop consists of the following steps:

1. Generate W random walks, each of K equal-length steps. Number the walks $1, \dots, W$ to indicate the order in which they would be performed in a sequential algorithm.

2. Determine the fate of each walker, temporarily ignoring interferences with the others in the group.

3. Identify the first interference that would occur if the walks were performed sequentially in their specified order.

4. Attach any walkers that stick to the cluster up to and including the second member of the interfering pair identified in Step 3. Disregard any remaining walks. Update the cluster mass M and radius R accordingly.

A detailed analysis (see [12]) of the individual steps of the loop reveals that only Steps 1 and 4 require more than constant parallel time when restricted to $M^{O(1)}$ processors. In Step 1, we choose $W \sim M^{1+\epsilon}$ for some small ϵ to keep the interference probability near unity as M increases. We argue in [12] that by choosing $K \sim R^2$ the ideal DLA distribution can be approximated to any desired degree of accuracy. Since $R \sim M^{1/D}$, where D is the fractal dimension of DLA, Steps 1 and 4, and hence an entire iteration of the loop, can be carried out by means of standard parallel techniques in $O(\log M)$ time using $M^{O(1)}$ processors.

3.2. Determination of dynamic exponent

The expected change in cluster mass ΔM during one iteration of the parallel algorithm's main loop depends on the interference probability. We define $P(M, n)$ to be the probability for an interference to occur amongst the next n walkers that stick to a cluster of mass M. Simple arguments suggest the scaling form $P(M, n) = F(nM^{-\beta/2})$, where β is defined via $P(M, 2) \sim M^{-\beta}$ and is expected to have the value $\beta = D_2/D$, where D_2 is the second generalized dimension[1] of DLA. Results shown in Fig. 3.1 of simulations performed using the sequential algorithm of [13] yield an experimental value $\beta_{\text{expt}} \approx .53$.

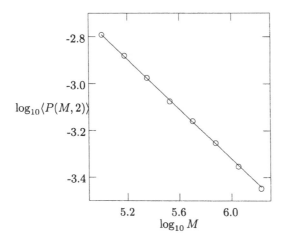

Figure 3.1: $\log_{10}\langle P(M, 2)\rangle$ vs. $\log_{10} M$, where $\langle P(M, 2)\rangle$ is the mean probability, from a sample of 20 DLA clusters of mass M, for an interference to occur between the next two walkers that stick. The solid line is a linear fit to the data and has slope $-.53$.

Simulation results supporting the assumed scaling form of $P(M, n)$ are shown in Fig. 3.2. This scaling hypothesis implies $\Delta M \sim M^{\beta/2}$. Since the expected time ΔT to complete an iteration of the algorithm's main loop scales according to $\Delta T \sim \log M$ it is easy to show that the average running time T to grow a cluster of radius R scales as $T \sim R^z$, with dynamic exponent $z = D - D_2/2$. Using our value $\beta_{\text{expt}} \approx .53$ yields $z \approx 1.3$ as compared with $z = D \approx 1.7$ for the best possible sequential algorithm.

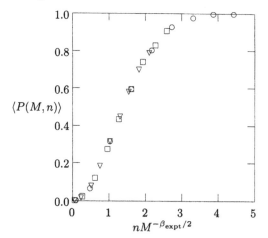

Figure 3.2: $\langle P(M,n)\rangle$ vs. $nM^{-\beta_{\text{expt}}/2}$ plotted for $M = 1 \times 10^5$ ($\circ$), 5×10^5 ($\square$), and 1.7×10^6 ($\triangledown$). $\langle P(M,n)\rangle$ is the mean probability, from a sample of 20 DLA clusters of mass M, for an interference to occur amongst the next n walkers that stick, and $\beta_{\text{expt}} = .53$ is minus the slope of the line shown in Fig. 3.1.

Interferences between random walkers seem to provide the fundamental limitation to parallelizing DLA. Since our algorithm works by processing in parallel, at each stage, the initial maximal group of non-interfering walkers, it seems plausible that the algorithm attains the minimum possible value of z. Because of the equivalence, up to logarithmic factors in the time, of differing models of parallel computation, the minimum value of z is a well-defined quantity that characterizes the DLA distribution. Assuming that we have actually found the fastest parallel DLA algorithm, then we have measured the intrinsic history dependence (logical depth) of DLA.

4. Summary

We have analyzed a number of models in statistical physics from the perspective of parallel complexity theory. While this type of analysis can serve as a guide in designing practical parallel algorithms, we are primarily interested in exploring the merits of parallel computational complexity as a gauge of physical complexity. In the case of DLA, **P**-completeness results for the known dynamics suggest that the DLA growth process is inherently sequential. This is in contrast to several other growth models that can be simulated by polynomial-processor algorithms in polylog time. By developing a parallel DLA algorithm and calculating its dynamic exponent we have placed an upper bound on the speedup attainable by parallelization. In addition, we have related the algorithm's dynamic exponent to static scaling exponents of DLA.

Parallel complexity analysis of the type presented here for DLA allows us to characterize more precisely the intrinsic history dependence of a given model system. Since non-trivial structures seem to require a long time to emerge, establishing a quantitative measure of history dependence could be an important step in developing a useful definition of physical complexity.

Bibliography

[1] AMITRANO, C., and A. CONIGLIO, "Growth probability distribution in kinetic aggregation processes", *Phys. Rev. Lett.* **57** (1986), 1016.

[2] BENNETT, C. H., "How to define complexity in physics, and why", *Complexity, Entropy and the Physics of Information* (W. H. ZUREK ed.), SFI Studies in the Sciences of Complexity, Vol. 7, Addison-Wesley (1990), 137.

[3] BENNETT, C. H., "Universal computation and physical dynamics", *Physica D* **86** (1995), 268.

[4] GIBBONS, A., and W. RYTTER, *Efficient Parallel Algorithms*, Cambridge University Press (1988).

[5] GREENLAW, R., H. J. HOOVER, and W. L. RUZZO, *Limits to Parallel Computation: P-completeness Theory*, Oxford University Press (1995).

[6] KAUFMAN, H., A. VESPIGNANI, B. B. MANDELBROT, and L. WOOG, "Parallel diffusion-limited aggregation", *Phys. Rev. E* **52** (1995), 5602.

[7] MACHTA, J., "The computational complexity of pattern formation", *J. Stat. Phys.* **70** (1993), 949.

[8] MACHTA, J., and R. GREENLAW, "The parallel complexity of growth models", *J. Stat. Phys.* **77** (1994), 755.

[9] MACHTA, J., and R. GREENLAW, "The computational complexity of generating random fractals", *J. Stat. Phys.* **82** (1996), 1299.

[10] MACHTA, J., and K. MORIARTY, "The computational complexity of the Lorentz lattice gas", *J. Stat. Phys.* **87** (1997), 1245.

[11] MOORE, C., and M. G. NORDAHL, "Lattice gas prediction is P-complete", Available as comp-gas/9609001 (1997).

[12] MORIARTY, K., J. MACHTA, and R. GREENLAW, "Parallel algorithm and dynamic exponent for diffusion-limited aggregation", *Phys. Rev. E* **55** (1997), 6211.

[13] OSSADNIK, P., "Multiscaling analysis of large-scale off-lattice DLA", *Physica A* **176** (1991), 454.

[14] PAPADIMITRIOU, C. H., *Computational Complexity*, Addison Wesley (1994).

[15] VICSEK, T., *Fractal Growth Phenomena*, World Scientific Singapore (1992).

[16] VOSS, R. F., "Multiparticle diffusive fractal aggregation", *Phys. Rev. B* **30** (1984), 334.

[17] WITTEN, T. A., and L. M. SANDER, "Diffusion-limited aggregation, a kinetic critical phenomenon", *Phys. Rev. Lett.* **47** (1981), 1400.

Chapter 36

Statistical models of mass extinction

Mark Newman
Santa Fe Institute, 1399 Hyde Park Road, Santa Fe, NM 87501.
U.S.A.

Recently, a number of authors have pointed out that several distributions of quantities extracted from fossil extinction data appear to follow power laws, including the distributions of the sizes of mass extinction events, the power spectrum of mass extinction events, and the distribution of the lifetimes of taxa. In this paper we review briefly these results and then discuss a number of models which have been put forward as possible explanations.

1. Introduction

Species extinction has played an important role in the development of life on the Earth [1, 2, 3]. The continual extinction and replacement of species allows for a much higher rate of biotic turnover than the comparatively slow process of phyletic transformation and pseudoextinction in which species are replaced by their own fitter offspring. The typical lifetime to extinction of a species is about four million years, which means that the entire biopshere has turned over many times in the 3.5 billion year history of life on the planet, and this has, without doubt, been an important influence on the current diversity and level of adaptation of terrestrial organisms.

Traditionally the study of extinction has focussed on the extinctions of particular species or on the causes of mass extinction events, of which there have been a number in the history of the planet. More recently however, a number of researchers have begun examining the statistical properties of species extinction focussing primarily on the the fossil record from the Phanerozoic, which is

374

approximately the last 600 million years. There are two reasons for this. First, there have become available in the last decade good compilations of data on species extinction drawn from a wide variety of paleontological investigations. The databases compiled by Sepkoski [4, 5] and Benton [3, 6] have proved particularly useful. Second, a number of intriguing theories about the possible causes of mass extinction have been put forward, based on statistical models of evolution and on computer simulations of coevolving ecosystems. These models only make predictions about average trends in the extinction profile, and not about particular extinction events, and testing them therefore requires statistical analysis of the fossil data.

In this paper, we first review briefly a number of trends visible in the fossil extinction record (§2) and then discuss a number of models which have been put forward in explanation of these trends (§3). In §4 we give our conclusions.

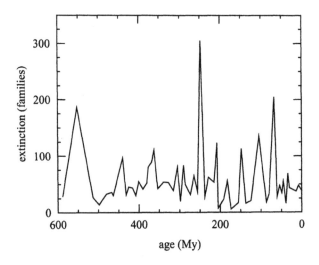

Figure 1.1: The number of families of marine invertebrates becoming extinct in each stratigraphic stage of the Phanerozoic.

2. The fossil data

The total number of species which have ever lived on the Earth is estimated to be on the order of a few billion. Of these we have fossil remains of only about a quarter of a million. Although this is a small sample of the total, we nonetheless have sufficient data to perceive some clear trends in the extinction profile over the course of time. Fig. 1.1 shows the extinction intensity of known marine invertebrate families over the course of the Phanerozoic. The figure shows that the extinction rate is strongly heterogeneous, with clear peaks corresponding to mass extinction events which cannot be accounted for by mere Poissonian fluctuations in the sampling. Tentative explanations have been put forward to account for some of the largest of these peaks (particularly the late-Permian and end-Cretaceous peaks at 245 and 65 Ma respectively), but for the majority of species extinction we have no idea of the cause.

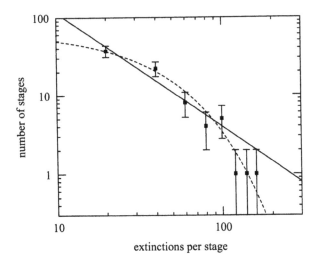

Figure 2.1: Histogram of the number of family extinctions per stage in the Phanerozoic. The scales are logarithmic and the error bars are simple Poissonian $\sqrt{N}$ estimates, which assume that extinction in successive stages is statistically independent. The solid line is the best power-law fit to the data and the dotted one is the best exponential fit.

In Fig. 2.1 we show a histogram of the number of families becoming extinct in each of the stratigraphic stages of the Phanerozoic. (The length of the stages varies, but they are typically about four or five million years long.) The plot is on logarithmic scales and it has been suggested that the data fall approximately on a straight line (the solid line in the figure) indicating that the data are roughly power-law in their distribution. In fact, the data are equally well (perhaps better) fit by an exponential distribution (the dotted line in the figure), and despite the interest that this distribution has attracted recently, we must, if we are honest, admit that the data are not complete enough to draw any firm conclusions from the distribution.

A much more convincing result has been pointed out by Solé *et al.* [7], who calculated the power spectrum of the extinction profile shown in Fig. 1.1. Their results are reproduced in Fig. 2.2 and show a very clear power-law distribution. The power-law is close to $1/f$ in form. Although the $1/f$ functional form appears in data generated by a wide variety of dynamical process, Solé *et al.* suggested the possibility that the power-law power spectrum arises in this case as the result of a so-called self-organized critical dynamics. We discuss this possibility further in §3.2.

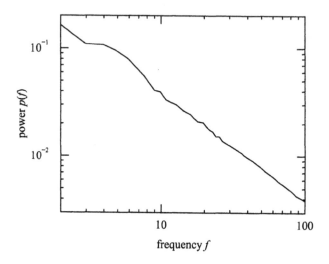

Figure 2.2: Power spectrum of familial extinction during the Phanerozoic.

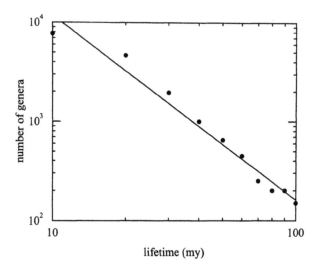

Figure 2.3: Histogram of the lifetimes of genera in the fossil record plotted on logarithmic scales. The line is the best power-law fit to the data. After Sneppen *et al.* [8].

Another striking statistical trend which has been observed in the fossil record is shown in Fig. 2.3. This figure shows a histogram of the lifetimes of fossil genera, plotted once more on logarithmic scales. The distribution is again clearly power-law in form.

The power law is a very distinctive mathematical form, which arises only from certain types of processes, and many people have taken its appearance here as evidence of some special type of dynamics taking place on long timescales in the evolving ecosystem. In the next section we discuss what this dynamics might be.

3. Models of extinction

Explanations for the causes of extinction fall into two basic classes. The traditional view, still held by most paleontologists, is that extinction is caused by environmental stresses, such as climate change, or changes in sea level. There is good evidence to support this view; a number of the larger mass extinction events in the Earth's history coincide with stresses of one kind or another which could well have been the immediate cause of the extinction. The most famous example is the KT boundary event 65 million years ago, which may have been caused by the impact of a meteor or comet in eastern Mexico [9, 10]. However, most of the smaller extinction events, as well as all "background" extinction (which in fact comprises the majority of extinction events), have no known cause, and it has been proposed that these events may have a different origin. Starting with the work of Kauffman [11, 12], a number of authors have suggested that these extinctions maybe the result of the evolutionary dynamics of the ecosystem, and nothing to do with any external factors. When it first appeared, this idea was quite revolutionary, implying as it does, that mass extinction may take place even in the absence of any external pressures on the ecosystem, that extinction may be a natural part of the dynamics of the system regardless of anything else that happens. Indeed, it is still not a widely accepted view, resting as it does on the comparison of the fossil data with a number of simple models of species coevolution. In this section we examine some of these models and ask whether they do indeed form persuasive evidence of an intrinsic extinction mechanism.

3.1. Kauffman's NK models

The idea that extinction might be caused by the internal evolutionary dynamics of an ecosystem was first proposed by Kauffman, who proposed a specific class of models for studying coevolutionary dynamics. These models are commonly known as NK models. The basic idea, as described by Kauffman and Johnsen [11], is as follows.

Each species in the model is represented by its genotype, which consists of a fixed number N of genes, each with the same number A of alleles. This of course is a simplification of the true situation. However, as with all of the models discussed here, it is hoped that by simplifying the system we will be able to understand what governs its behavior, without throwing out the interesting features of the real system. With each genotype is associated a fitness. Presumably this arises through some kind of genotype/phenotype map, but in the NK models this is simplified to a direct mapping from genotype to fitness. Each gene makes an additive contribution to the fitness of the whole. The fitness contribution of the gene is dependent on the allele of the gene and, crucially, on the alleles of K other genes, chosen at random. This represents epistatic inter-action between the functions of the different genes. The fitness contributions for each of the configurations of a gene and its K epistatic "neighbors" are chosen at random in the interval between zero and one. The fitness of the organism is then the average of the fitness contributions of the individual genes. This creates a fitness "landscape" over the space of genotypes, which has a certain degree of ruggedness, dictated by the value of the parameter K. For $K = 0$, the landscape is "smooth", possessing only a single fitness peak. For $K = N - 1$, the maximum possible value, each genotype has a fitness which is an independent random number and the landscape has the maximum possible number of local fitness maxima, which is, on average, $A^N/(N + 1)$. For values of K in between, the number of local fitness maxima lies somewhere between these two extremes.

Kauffman and Johnsen represented species in their model by single geno-types, rather than by populations of individuals, an approximation which is valid in the regime in which selection pressure acts on much shorter timescales than the timescale for mutation.

In general, one expects species to perform adaptive walks to fitness peaks in the landscape, and Kauffman and Johnsen studied a number of different types of dynamics which achieve this, including random mutation and selection, in which a new genotype is generated at random from those close to the present one and the species adopts the new genotype if it has a higher fitness than the present one, and greedy dynamics in which the species moves on each step to the neighboring genotype which offers the greatest improvement in fitness over the current genotype, if such a genotype exists. In general, they found very much what one might expect; the species in the system evolve to local fitness peaks via adaptive walks whose length decreases with increasing K, regardless of the type of dynamics employed.

Then however, Kauffman and Johnsen introduced a further level of sophisti-cation into the model, by simulating many species simultaneously. The species affect one another through coupling between the genes of different species; in ef-fect, there are epistatic interactions between the genes of different species. Under these circumstance a new behavior is seen: when K is high and the landscapes are rugged, the system behaves in a manner similar to a set of non-interacting species, each one finding a local fitness maximum near its starting genotype and then staying there. However, if K is low, the number of genes which a species must change in order to reach such a maximum is large, and this means that the adaptation of one species to a local fitness maximum has a greater effect on the other species with which that one interacts. The practical consequences of this are that the evolution of one species causes the peaks to move in the fitness land-scapes of other species, which causes them to evolve, even if they were previously sitting at a peak. The evolution of these other species can then affect the land-scapes of still other species, causing them to evolve, and so on, in a domino effect which has been dubbed a "coevolutionary avalanche". Kauffman and Johnsen found that, depending on the value of K, their model showed two regimes—chaotic and ordered—in which the coevolutionary avalanches either came to a halt after a while, or continued indefinitely. The dividing line between these two regimes appears to be an ordinary second-order phase transition, characterized by fluctuations (avalanches) with a power-law size distribution. It is these avalanches which Kauffman has associated with mass extinction events.

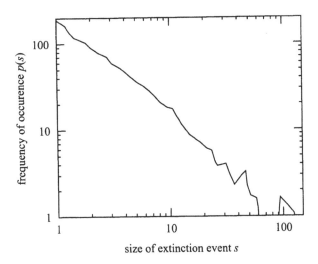

Figure 3.1: Histogram of extinctions generated during a simulation of the extinction model of Kauffman and Neumann [13]. The distribution is approximately power-law in form with an exponent close to −1. After Kauffman [14].

The relationship between coevolutionary avalanches and extinction was not made clear in the work of Kauffman and Johnsen. However, later work by Kauffman and Neumann [13, 14] has extended the NK model to include extinction. In this work, the authors included a competitive replacement mechanism whereby organisms could compete for niches in the ecosystem, the fittest winning control of the niche and the least fit becoming extinct. When tuned to the phase transition between the ordered and chaotic regime, the model then shows a distribution of extinction events whose size follows a power law. In Fig. 3.1 we show the distribution of extinctions generated by the model.

3.2. The model of Bak and Sneppen

The disadvantage of the models proposed by Kauffman and co-workers is that in order to show power-law distributions they must be tuned to a special point— the phase transition between the ordered and chaotic regimes. Although they offered some suggestions about how nature might perform this tuning, they were unable to incorporate these ideas into the model. In 1993 however, Bak and Sneppen [15] proposed a model, based on the ideas of Kauffman along with extensions of their own, which "self-organizes" to the phase transition without the need of tuning, perhaps providing an explanation for the way in which nature generates power laws in this system.

The model proposed by Bak and Sneppen incorporates a much simplified representation of the fitness landscapes of the NK model. In this model, species are assumed to spend most of their time at fitness peaks, and only to evolve on rare occasions by "excitation" over fitness barriers to new peaks, driven presumably by mutation. The only property of a species which is recorded is the height of the barrier it has to cross in order to reach a new peak, which is initially chosen at random, and it is assumed that the species with the lowest barrier is the first to mutate. When it does so, it is given a new barrier height, also chosen at random. In addition, epistatic interactions between species, which are not explicitly modeled, change the landscape of a number K of neighboring species, altering the height of the barriers which they have to cross, the new values being chosen again at random.

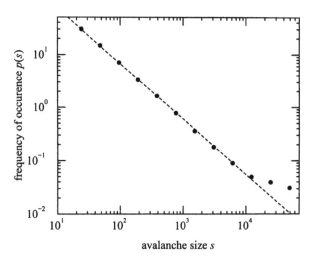

Figure 3.2: Histogram of the sizes of coevolutionary avalanche generated during simulation of the model of Bak and Sneppen [15].

When the model is set running, there is an initial equilibration period in which the lowest barriers in the system are removed one by one, until all remaining species are relatively stable, in the sense of possessing high barriers to mutation. In this situation it is relatively likely that the new barrier height chosen for the neighbor of a species which has just evolved will be one of the lowest in the system, since there are very few species with low barrier height. Thus there is an increased chance of a species evolving if it is the neighbor of another species which has just evolved. Again therefore, the model shows avalanches of coevolution. Furthermore, there is in this model, no parameter to tune in order to vary the distribution of the sizes of these avalanches, but Bak and Sneppen found that the model produces a distribution with follows a power law closely, even without such tuning. The results are duplicated in Fig. 3.2.

However, as with the model of Kauffman and Johnsen [11], Bak and Sneppen did not make an explicit connection between the sizes of the avalanches in their model and the sizes of extinction events, making it difficult to compare the results directly with the fossil data. However, a variation on the model was proposed by Newman and Roberts [16, 17] which does make this connection.

3.3. The model of Newman and Roberts

Newman and Roberts proposed a model which attempts to reconcile the fact of extinctions caused by environmental stress with the possibility of a power-law distribution of extinction sizes, which suggests that the causes of extinctions are related to the internal dynamics of the ecosystem. Their model is an extension of the Bak-Sneppen model which includes extinctions caused by external stresses.

The basic idea behind the model is that coevolutionary avalanches take place in the same way as in the Bak-Sneppen model, and that the species which take part in such an avalanche tend to have lower fitness on average after the avalanche than before, which makes them more susceptible to extinction by environmental stress. Thus the coincidence of a large avalanche and a large stress could cause a large extinction event, even though either one on its own would not. In a little more detail, the model is as follows.

The species in this model are characterized by two numbers, a barrier height for mutation as with the model of Bak and Sneppen, and a new number representing their fitness at the current fitness peak. The dynamics of the model is the same as for the Bak-Sneppen model, except for two things. First, when a species evolves, new values for *both* the barrier and the fitness are chosen at random. Second, at each time step a stress is applied to the system, represented by a single scalar random number. It is assumed that the less fit species will become extinct when this stress is applied, and this process is modeled by making all species whose fitness is numerically less than the stress level extinct. Extinct species are replaced with an equal number of new ones with randomly chosen barriers and fitnesses, to keep the number of species in the model constant.

In Fig. 3.3 we show the distribution of the sizes of extinction events which is generated by this model. As we can see, the distribution is again power-law in form, in good agreement with the fossil data. Furthermore, the exponent of this power law is close to -2, which agrees with the value of 2.0 ± 0.2 extracted from the fossil data [18]. The model proposed by Kauffman and Neumann by contrast gives a value of -1 for this exponent, which does not agree well with the data.

This certainly seems very satisfying—both the mechanism and the results appear very plausible. However, it turns out that the model is misleading in one important respect: the coevolutionary avalanches in fact have nothing to do with the appearance of a power-law extinction distribution in this model. This result can be seen clearly if we consider another later model proposed by Newman [18, 19].

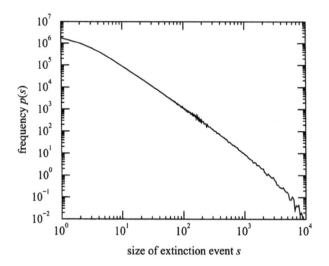

Figure 3.3: Histogram of the sizes of extinction events generated during a simulation of the model of Newman and Roberts [16].

3.4. The extinction model of Newman

Newman [18, 19] suggested a model of extinction which disposes altogether of the coevolutionary avalanches introduced in the models of Kauffman *et al.* and of Bak and Sneppen, including only the process of extinction by environmental stress introduced in the previous section. In this model species are characterized only by a fitness measure, which represents their tolerance for stress. At each time-step a random stress is applied to the ecosystem as before, and those species with fitness lower than the stress level become extinct and are replace by new species with randomly chosen fitnesses. In addition, it turns out to be necessary to include a "evolution" process under which the fitness of species drifts around over time, rather than staying still; this process is necessary to prevent the model from stagnating. The net result is that the model produces a power-law distribution of extinction sizes, even though it has no avalanches and the extinctions are entirely produced by the external stresses on the system and are nothing to do with any internal dynamics. It was found that the exponent of this power law is again close to −2, regardless of the distribution of the applied stresses. The mathematical developments required to demonstrate these results are quite lengthy. The interested reader is referred to the appendix of Ref. [19] for a detailed discussion.

The conclusion is that it is not after all necessary that mass extinction be related to some internal dynamics of the ecosystem in order to produce the observed power law in the extinction size distribution. It is possible to get results in good agreement with the fossil data without any such internal dynamics at all. Furthermore, this model also produces a power-law extinction power spectrum and power-law distributions of species lifetimes, in agreement with the data shown in Fig. 2.2 and Fig. 2.3.

4. Conclusions

In this paper, we have discussed a number of data distributions drawn from the fossil record which appear to show power-law functional forms. We have then described a number of models which have been put forward as possible explanations of the origins of these power laws. Most of these models hinge around the idea that power-law distributions are generated by critical dynamics within the ecosystem. However, it turns out also to be possible to generate distributions in good agreement with the fossil data using a model in which all extinction is caused by external environmental stress. Thus, it is not possible to conclude from models such as these whether the observed distributions are the result of internal dynamics of the evolving ecosystem or simply the result of external stress on the ecosystem. In order to tell these two possibilities apart, we need to look for new indicators in the fossil data. Various such indicators have been suggested, such as the structure of taxonomic trees, or the time series of speciation events. At present however, these are still issues which have not been tackled with any rigor.

Bibliography

[1] RAUP, D. M., *Extinction: Bad Genes or Bad Luck?* Norton, New York (1991).

[2] RAUP, D. M., "Biological extinction in Earth History", *Science* **231** (1986), 1528–1533.

[3] BENTON, M. J., "Diversification and extinction in the history of life", *Science* **268** (1995), 52–58.

[4] SEPKOSKI, J. J., "Ten years in the library: New data confirm paleontological patterns", *Paleobiology* **19** (1993), 43–51.

[5] SEPKOSKI, J. J. and C. F. KOCK, "Evaluating paleontologic data relating to bio-events", *Global Events and Event Stratigraphy,* (O. H. WALLISTER, ed.), Springer, Berlin (1996).

[6] BENTON, M. J. (ed.), *The Fossil Record 2*, Chapman and Hall, London (1993).

[7] SOLÉ, R. V., S. C. MANRUBIA, M. BENTON and P. BAK, "Self-similarity of extinction statistics in the fossil record", *Nature* **388** (1997), 764–767.

[8] SNEPPEN, K., P. BAK, H. FLYVBJERG and M. H. JENSEN, "Evolution as a self-organized critical phenomenon", *Proc. Nat. Acad. Sci.* **92** (1995) 5209–5213.

[9] ALVAREZ, L. W., W. ALVAREZ, F. ASARA and H. V. MICHEL, "Extraterrestrial cause for the Cretaceous-Tertiary extinction", *Science* **208**, (1980) 1095–1108.

[10] GLEN, W. "What the impact/volcanism/mass-extinction debates are about", *The Mass Extinction Debates*, (W. GLEN, ed.), Stanford University Press, Stanford (1994).

[11] KAUFFMAN, S. A. and S. JOHNSEN, "Coevolution to the edge of chaos: Coupled fitness landscapes, poised states, and coevolutionary avalanches", *J. Theor. Biol.* **149** (1991), 467–505.

[12] KAUFFMAN, S. A., *The Origins of Order*, Oxford University Press, Oxford (1992).

[13] KAUFFMAN, S. A. and K. NEUMANN, unpublished results (1994).

[14] KAUFFMAN, S. A., *At Home in the Universe*, Oxford University Press, Oxford (1994).

[15] BAK, P. and K. SNEPPEN, "Punctuated equilibrium and criticality in a simple model of evolution", *Phys. Rev. Lett.* **71** (1993), 4083–4086.

[16] NEWMAN, M. E. J. and B. W. ROBERTS, "Mass extinction: evolution and the effects of external influences on unfit species", *Proc. R. Soc. Lond.* b **260** (1995), 31–37.

[17] ROBERTS, B. W. and M. E. J. NEWMAN, "A model for evolution and extinction", *J. Theor. Biol.* **180** (1996), 39–54.

[18] NEWMAN, M. E. J. "Self-organized criticality, evolution and the fossil extinction record", *Proc. R. Soc. Lond.* B **263** (1996), 1605–1610.

[19] NEWMAN, M. E. J. "A model of mass extinction", *J. Theor. Biol.*, **189** (1998) 235–252.

A dual processing theory of brain and mind: Where is the limited processing capacity coming from?

Danko Nikolic[1]

The University of Oklahoma, Department of Psychology

Dynamical neural activity, which involves oscillatory and activity synchronization in neural firing, seems to have significant implications for understanding how the brain produces cognitive behavior. Computations based on dynamical activity require much more feedback information than is employed in traditional neural network approaches. The traditional neural network approach typically involves computation based on feedforward mathematical mapping from input to output. Several findings, including the limited correlational dimensionality of EEG activity, suggest that if dynamical neural activity contributes to information processing in the brain, than its processing capacity is very limited. I review the evidence suggesting that dynamical neural activity might be responsible for the limited processing capacity observed in voluntary/conscious processes in human cognition and propose the following hypothesis: the brain changes its information processing strategy from heavily relying on dynamical activity toward processing that relies on unidirectional mapping as learning and the development of skills increase (i.e. processing becomes more automatic). This transfer results in a decrease in the need for limited dynamically-based resources. Computations based on such a transfer might provide powerful information processing properties that optimize the complexity of the computational tasks in the brain.

---

[1] I thank my advisor Scott Gronlund for his help on previous versions of this manuscript.

1. Introduction

Our understanding of how the cognitive behavior of the brain emerges from its computational elements (neurons) has profited from theoretical work using neural network simulations. Research on neural networks has had considerable success in recent decades. Neural nets have provided us with explanations of several brain/mind problems as well as a commercial technology that provides information processing based on the computational power of the brain. Perhaps, the most important mechanism employed in neural networks is the ability of several layers of units to provide mathematical mapping functions between input and output. A proper mapping between input and output is achieved by the synaptic weights between the units.

However, the neural networks did not succeed in providing answers to all the questions about the brain and mind that we wish to answer. Although almost any phenomena on the behavioral level has at least one corresponding model on the neural level, neural network models still are not sufficiently powerful theoretical tool for neuropsychological research. They still do not govern most of the research in either neurophysiology or psychology. One possible reason is that they do not include all the important neural mechanisms that the brain utilizes.

This paper suggests a possible important role of dynamical neural activity for information processing in the brain. However, there is a need for caution regarding the use of the term "dynamical" when referring to neural activity. There is a long tradition of studying the dynamics of neural networks that is describable by dynamical system theory (e.g., [9] [12] [37]). However, I will use the term "dynamical" to refer to mechanisms that involve both oscillations and synchrony in neural firing. In this way, I broaden the definition of dynamical neural activity. Through this paper, I attempt to contrast dynamical neural activity with mapping in neural networks. I assign to them different processing roles in the brain's adaptive mechanisms.

2. Mapping in neural networks

The proof that neural networks have powerful mapping capabilities follows from Kolmogorov's theorem which shows that a three-layer feed-forward neural network with n input units, 2n+1 hidden units and m units in the output layer can implement any continuos function of a type:

$$f : [0, 1]\mathbb{R}^n \to \mathbb{R}^m$$

Later, it was shown that some neural networks can perform mapping from $\mathbb{R}^n$ to $\mathbb{R}^m$ with a sufficient number of units and/or hidden layers [11]. A particular mapping function is provided by fine tuning of the weights of connections between the units. In this respect, the brain is typically viewed as a highly parallel, multi-layered, mapping machine that maps input to output. In models that attempt to explain brain mechanisms, this mapping power is typically accompanied with many additional mechanisms that provide additional computational power. One example is the adaptive resonance theory [8]. The adaptive resonance theory includes top-down feedback connections, orienting and gain control mechanisms. The dynamical activity of the brain, as defined here was, in most of the cases, neglected or considered an epiphenomena.

For such a highly parallel mapping machine, as the brain was viewed to be, there does not seem to be any obvious reason why the processing would be limited in capacity. In other words, the architecture is such that a bottleneck in information flow does not exist [24]. The only reason for limited capacity in processing could be a need for sharing the same processing resources [18]. This paper, however, suggests that interesting information processing advantages that might emerge from the dynamical neural activity must have a very narrow bottleneck. In other words, unlike mapping, the brain is probably not able to perform information processing based on the dynamical neural activity in a highly parallel fashion.

3. Oscillations and synchrony in neural firing

Oscillatory neural activity has been observed for more than a century but until recently has been considered as an important processing mechanism by only a few scientists [15][6]. The synchrony in neural firing, on the other hand, is a much more recent finding [30][31][32][36], and only recently has been the importance of both oscillatory and synchronous activity seriously recognized [7][13][22][23]. Although it is not clear how much the oscillatory and synchronous activity are related, in this paper it is assumed that there is a correlation between the firing rhythm and the oscillatory activity observed in EEG [30]. The relevance of synchrony in neural firing for information processing has been demonstrated by showing that the neurons that code for properties of the same object have a high level of synchronization of action potentials. It has also been found that just before a motor response on a visual stimulus is executed, cells in the visual, parietal, and motor cortex synchronize [27]. The best explanation of the binding problem in human perception is the synchrony in firing [4][36].

3.1. Computational advantages of dynamical activity

One advantage of synchronous input to a cell is that the input becomes more effective (see Fig. 3.1). If cells A and B are synchronized and cell C fires with the same intensity but off-synchrony, than cell D could win the competition with cell E although cell E has stronger connections [30]. It has been also proposed and shown that synchrony in firing might enhance the synaptic learning process [30][4].

An important computational advantage of the process depicted in Fig. 3.1 is that a specific pattern in synchrony in neural firing could temporarily override the learned synaptic weights and redirect the information flow in a new direction that has not been experienced or trained before. This effect is also a character-

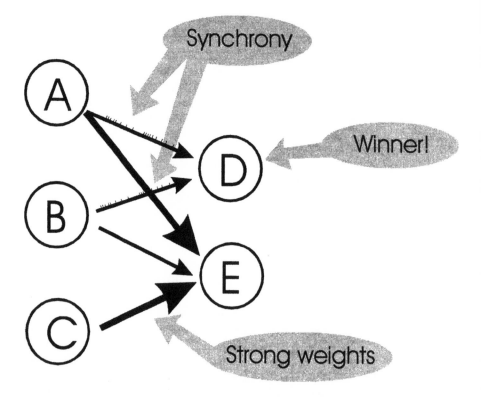

Figure 3.1: An example of a possible computational role of synchrony in neural firing. The neuron that has weaker connections could win the competition if the input arrives in synchrony.

istic of voluntary or conscious processes that often provide new behavior that has never before been experienced.

4. Controlled and automatic processes in the brain

The difference between controlled and automatic processes in psychology has been known for long time [33][34]and is closely related to conscious versus unconscious processes [14]. It could be exemplified by comparing the cognitive processes while driving a car for the first time and driving it after years of experience. When doing it for the first time, most of the actions are consciously controlled and the processing has very limited capacity. After prolonged training, the processing becomes in large part unconscious (i.e. automatic) and the demand on the limited processing capacity significantly decreases. This enables the driver to conduct other activities in parallel while driving (e.g., conversation). In this case, the processing moved from conscious or controlled to unconscious or automatic. The result of making processing unconscious is the freeing-up of processing capacity so that additional activities could be performed.

It is hard to provide a good measure of the size of the capacity space that is available for controlled processes. The best estimate of the capacity size, that psychologists have, is probably the Miller's "magic" number 7 +/- 2 [19]. This number pertains to the number of items that one can hold in working memory, the number stimuli that one can attend to at the same time, and the number of categories that one can distinguish in absolute judgments. It has also been shown that with extensive practice people can overcome this limitation. Significant increases in processing capacity with practice have been shown for memory of numbers [2], restaurant orders [5], and chess pieces [3].

Besides the amount of the occupied processing capacity, controlled and automatic processing differ in other respects. The response time for automated reactions is much shorter than for controlled actions. A skilled driver, for example, reacts much faster to a new, possibly dangerous, event than a novice. Controlled processes seem to take more of capacity, and are slower to execute. The disadvantage of automatic processes is that they are much less adaptive than controlled ones. This means that it takes relatively long time to develop automatic processing and a long time to unlearn them when they need to be changed. Conscious processes, on the other hand, are very adaptive and they allow us to perform behavioral actions that we have no previous experience with.

A simple stimulus-response learning situation can demonstrate the difference between automatic and controlled processes. Both processes can result in similar behavior, say, closing an eyelid after a tone. One way to acquire an automatic response is classical conditioning where the tone could be paired with an air-puff. In this case, several hundreds of parings are necessary to establish the response on the stimulus. On the other hand, one can give an instruction to a person: "close your eyelid whenever you here the tone". Both manipulations will lead to the same result: the eyelid will close after the tone. However, these two processes are fundamentally different. The first difference is in the time it takes to establish each. Whereas the automatic processes employing conditioning need hundreds of repetitions, the controlled process needs only one instruction or decision. Thus, the adaptability of the controlled process is obvious: one can quickly produce new behavior that has never been experienced before. This is not the case with the automatic processes. Controlled processes have the same adaptability advantage if the behavior has to cease. Again, a simple instruction to quit responding is sufficient for controlled processes whereas automatic processes need again a number of extinction trials.

Finally, controlled processes can control automatic processes. Controlled processes can trigger automatic ones, shut them off, or retrieve them into consciousness when the situation requires it (i.e. novel situation). This adaptive feature of controlled processes is probably one of the most powerful computational mechanisms that the brain has.

5. Is dynamical neural activity responsible for controlled processes?

A comparison of dynamical processing in the brain with the mapping shows the same three basic differences in controlled and automatic processes: speed, adaptability and limited capacity.

5.1. Adaptability

The patterns of oscillatory activity and synchrony in neural firing change quickly [31]. A new oscillatory pattern can be produced in several hundred milliseconds. Structural changes, on the other hand, take much longer time. They could take anywhere from several minutes to several days [29]. Therefore, the fast changes in behavior due to controlled processes are more likely to be caused by changes in some dynamic oscillatory patterns rather than by structural synaptic changes. In contrast, the relatively slow changes in behavior, due to the development of an automatic process, will likely result in structural changes in the synapses.

5.2. Speed in processing

The processing that takes place in a dynamical system includes feedback information. Processing with feedback information is slower than one-directional, feed-forward processing that takes place in mapping processing. Therefore, the slow processing in controlled processes could be attributed to the time needed for a dynamical system to reach a new pattern of oscillations or synchrony patterns after a stimulus is presented.

5.3. Limited processing capacity

Mapping processing does not have an obvious limitation on processing capacity. In other words, many separate neural network modules can receive input from the same input and process it in parallel, without disturbing each other. They can also submit their results to the same output module without slowing down each other [24]. On the other hand, there are several sources of evidence that suggest that the dynamically based processing is narrowly limited in its processing capacity. In other words, it seems that there is always a small number of separate neural, concurrently existing, groups which contain synchronized neurons. Attempts to simulate synchronous neural firing result in a small number of separate groups [13]. More direct evidence is provided from EEG recordings. This evidence comes from the phase space reconstruction of EEG time series and the estimation of the embedding dimension. The embedding dimension represents the minimal number of differential equations necessary to reproduce the time series. In our case, a group of neurons that oscillate together could be regarded as one differential equation in the system. The embedding dimension, therefore, gives a rough estimate of the number of separate neural groups oscillating at the same time. Analyses have shown that the embedding dimension varies between 3 and 10 [1][28]. If one group of oscillating neurons provides a temporary change in overall processing, than the number of those changes in the same time is limited to a one-digit number. It is interesting to note that the embedding dimension for EEG is in the same range as the psychological measure for the maximum capacity for controlled processes (i.e., 7 +/- 2). From the comparison of the processing characteristics of controlled and automatic processes on the behavioral scale and mapping and dynamical processing on the cell scale of brain's behavior it follows that: controlled/conscious processes rely more on the dynamical neural activity than do automatic/unconscious processes.

6. Derived hypothesis

There are several hypotheses that follow from the proposed role of dynamical activity for controlled processes.

6.1. Harder tasks have larger dimensionality

The tasks that are more difficult require more voluntary effort, and should require more temporary changes in the neural nets implemented by dynamical processes. The embedding dimension of the underlying activity (e.g., the number of oscillating neural groups) should therefore be larger for more difficult tasks. There is indirect support for this hypothesis that comes from recent research on the embedding dimension of the reconstructed phase space of repetitive hand movements. Mitra, Riley and Tuvey [20]have shown that repetitive hand movements produce chaotic activity with embedding dimension between 3 and 4. Swinging a heavier stick produces a smaller dimension than swinging a light stick. The participants, however, judge swinging the heavier stick as being a much easier task than swinging the light stick [26]. This finding provides only indirect support for the hypothesis because dynamical neural activity was not measured (they only measured the dynamical activity of hand movements). The notion that the brain couples with the environment through its dynamical activity [7][17][35]suggests that the complexity of the dynamical activity of the movements reflects the complexity of the neural activity that produces those movements.

6.2. Transfer from dynamical to mapping processing

Because controlled processes rely more on dynamical processing, and automatic processes rely more on mapping, during practice when controlled processing becomes automatic, the dynamical processing should be replaced with mapping. In other words, with automation of an activity, the ratio of involvement of dynamical and mapping processing should change so dynamical processing becomes involved to a lesser extent. Greater reliance on mapping frees the limited processing capacity of dynamical processing for additional parallel processes. This prediction receives, albeit indirect, support from the hand movement research.

Mitra, Amazeen and Turvey [21]have found that with practice the dimensionality of repetitive hand movements decreases.

6.3. Automatic processes that are under conscious control

Some automatic processes are under voluntary control, meaning that they can become conscious if it is necessary. Eye-blink and breathing are examples of processes that are very automatic but that can be brought under conscious control and therefore be controlled by a conscious decision. Other automatic processes (like patellar reflex, many processes in autonomous nervous system (ANS) and emotional responses) cannot be directly under conscious control. There also probably is a difference in the neural processes that underlay consciously controllable and not controllable automatic processes. The view on the controlled and automatic processes proposed here suggests an explanation: If voluntary processes perform control through dynamical activity than automatic processes that can be consciously controlled should employ, to some extent dynamical activity (i.e. oscillations and synchrony in neural firing). If they employ dynamical processes, than they should be subjected to the limited processing capacity, at least to a small extent. The support for this prediction comes from eye-blink research. It has been found that the eye-blink reflex decreases its activity if a second hard task is given to participants. Stern [25]had participants hold six digit numbers in working memory while eye-blinks were recorded. The average time between two eye-blinks significantly decreased compared to a situation where participants had to memorize only two items. The eye-blink reflex is therefore subjected to limited processing capacity. However, the reflexes that are not under conscious control, should not be affected by a secondary task. There are no experimental data that I am aware of that would suggest and answer to this prediction.

7. Conclusions

The differences in some of the processing characteristics between one-directional mapping in neural networks and the dynamical activity in the brain seem to match the difference in processing characteristics between automatic and voluntary processing. In addition, experimental support exists for several hypotheses derived from an assumption that this match is not coincidental. It appears, therefore, that the brain uses heavily dynamical processing of information in novel situations where conscious processing is necessary. Familiar, previously experienced, processing on the other hand seems to rely much less on dynamical activity but more on the learned one-directional mapping. The transfer from dynamical to mapping processing with experience has implications on the optimization of computational complexity: The brain adapts so that in familiar situations it executes computations by involving a small amount of resources. In novel situations, however, the brain seems to use larger amount of resources by combining the already learned mappings with the flexibility of the dynamical processing.

Bibliography

[1] BASAR,E., C. BASAR-EROGLU, and J. ROSCHKE, "Do coherent patterns of the strange attractor EEG reflect deterministic sensory-cognitive states of the brain?", *From chemical to biological organization*. (M. MARKUS, S. MULLER, and G. NICOLIS eds.), Springer (1988), 297-306.

[2] CHASE, W. G. and , K. A. ERICSSON, "Skilled memory". *Cognitive skills and their acquisition*, (J. R. ANDERSON ed.), Erlbaum (1981), 141-189.

[3] CHASE, W. G., and H. A.SIMON, "The mind's eye in chess", *Visual information processing*, (W. G. CHASE ed.), Academic Press (1973), 215-281.

[4] DAMASIO, Antonio R. "The brain binds entities and events by mutiregional activation from convergence zones", *Neural Computation* **1** (1989), 123-132.

[5] ERICSSON, K. A., and P. G. POLSON, "A cognitive analysis of exceptional memory for restaurant orders", *The nature of expertise*, (M T. H.CHI, R. GLASER, and M. J. FARR eds.), Erlbaum. (1988), 23-70.

[6] FREEMAN, Walter. J., "Neural networks and chaos", *Journal of Theoretical Biology* **171** (1994), 13-18.

[7] FREEMAN, Walter. J. and C. A. SKARDA, "Chaotic dynamics versus representationalism", *Behavioral and Brain Sciences* **13** (1990), 167-168.

[8] GROSSBERG, Steven, "Competitive learning: From interactive activation to adaptive resonance", *Cognitive Science* **11** (1987), 23-63.

[9] GROSSBERG, Steven and Daniel S. LEVINE, "Some developmental and attentional biases in the contrast enhancement and short-term memory of recurrent neural networks", *Journal of Theoretical Biology* **53** (1975), 341-380.

[10] HEATHCOTE, A., and MEWHORT, D. J. K. (submitted). Measuring the Law of Practice.

[11] HECHT-NIELSEN, R., *Neurocomputing*, Addison-Wesley, (1990).

[12] HOPFIELD, J. J., "Neural network and physical systems with emergent collective computational abilities", *Proceedings of the National Academy of Science* **79** (1982), 2554-2558.

[13] HUMMEL, J. E. and K. J. HOLYOAK, "Distributed representations of structure: A theory of analogical access and mapping", *Psychological Review* **104** (1997), 427-466.

394

[14] KIHLSTROM, J. F., "The cognitive unconscious", *Science* **237** (1987), 1445-1451.

[15] sc Lebedev, A. N. "A mathematical model for human visual information perception and storage", (R.F. NEURAL MECHANISMS OF GOAL DIRECTED BEHAVIOR AND LEARNING THOMPSON et al. eds.) Academic Press (1980).

[16] MARKRAM, Henry, Joackim LUBKE, Michael FROTSCHER, and Bert SAKMANN, "Regulation of synaptic efficacy by coincidence of postsynaptic APs and EPSPs", SCIENCE **275** (1997), 213 - 215.

[17] MATURANA, H. R. and F. J. VARELA, *Autopoiesis and Cognition, The Realization of the Living*, D. Reidel Publishing Company (1980).

[18] MEYER, D. E. and D. E. KIERAS, "A computational theory of executive cognitive process and multiple-task performance: Part 1. Basic mechanisms", *Psychological Review* **104** (1997), 3-65.

[19] MILLER, G. A., "The magical number seven, plus or minus two: Some limits on our capacity for processing information", *Psychological Review* **63** (1953), 81-97.

[20] MITRA, S., M. A. RILEY, and M. T. TURVEY, "Chaos in human rhythmic movement", *Journal of Motor Behavior* **29** (1997), 195-198.

[21] MITRA, S., P. G. AMAZEEN, and M. T. TURVEY, "Intermediate motor learning as decreasing active (dynamical) degrees of freedom" *Human Movement Science* (In press)

[22] NEWMAN, J. "Putting the puzzle together, Part I: Towards a general theory of the neural correlates of consciousness", *Journal of Consciousness Studies* **4** (1997), 47-66.

[23] NEWMAN, J. "Putting the puzzle together, Part II: Towards a general theory of the neural correlates of consciousness", *Journal of Consciousness Studies* **4** (1997), 100-121.

[24] NEUMANN, O. "Beyond capacity: A functional view of attention", *Perspectives on perception and action*, (H. HEUER and A. F. SANDERS eds., Erlbaum (1987), 361-394.

[25] STERN, J. A. "The eye blink: Affective and cognitive influences", *Anxiety: Recent developments in cognitive, psychological, and health research* (Donald FORGAYS, G., Tytus SOSNOWSKY, and Kazimierz WRZESNIEWSKI, eds.), Hemisphere Publishing Corp. (1992), 109-128.

[26] RILEY, M. A., personal communication.

[27] ROELFSEMA, P. R., A. K. ENGEL, P. KIG, and Wolf SINGER, "Visuomotor integration is associated with zero time-lag synchronization among cortical areas", *Nature* **385** (1997), 157-161.

[28] ROSCHKE, J., and E. BASAR, "Correlation dimensions in various parts of cat and human brain in different states", *Brain Dynamics: Progress and Perspectives* (E. Basar and T. H. Bullock eds.), Springer (1989), 131-148.

[29] ROSE, Steven, "How chicks make memories: the cellular cascade from c-fos to dendritic remodeling", TINS **14** (1991), 390-397.

[30] SINGER, Wolf, "Synchronization of cortical activity and its putative role in information processing and learning", *Annual Review in Physiology* **55** (1993), 349-74.

[31] SINGER, Wolf, "The organization of sensory motor representations in the neocortex: A hypothesis based on temporal coding", *Attention and performance 15: Conscious and nonconscious information processing, Attention and performance series* (Carlo UMILTA, Morris MOSCOVITCH, eds.), MIT Press (1994) .), 77-107.

[32] SINGER, Wolf, "Development and plasticity of cortical processing architectures", *Science* **270** (1995), 758-763.

[33] SCHNEIDER, W and R. M. SHIFFRIN, "Controlled and automatic human information processing: I. detection, search, and attention", *Psychological Review* **84** (1977), 1-66.

[34] SHIFFRIN, R. M. and W. SCHNEIDER, "Controlled and automatic human information processing: II. perceptual learning, automatic attending, and a general theory", *Psychological Review* **84** (1977), 127-190.

[35] VON STEIN, Astrid, "Does the brain represent the world? evidence against the mapping assumption", *Proceedings of the International Conference New Trends in Cognitive Science*, Vienna, Austria (1997).

[36] sc Stryker, M. P., "Is grandmother an oscillation?" *Nature* **338** (1989), 297-298.

[37] WILSON, H. R and J. D. COWAN, "Excitatory and inhibitory interactions in localized populations of model neurons", BIOPHYSICAL JOURNAL **12** (1973), 1-24.

Chapter 38

Evolutionary strategies of optimization and the complexity of fitness landscapes

Helge Rosé

Humboldt Universität zu Berlin, Institut für Physik
Invalidenstr. 110, D-10115 Berlin, Germany

Evolutionary Algorithms have proved to be a powerful tool for solving complex optimization problems. The underlying physical and biological strategies can be described by the Master Equation formalism. Combination of both strategies creates a new basic class of Evolutionary Algorithms where robustness as well as performance of the optimization process improve substantially. To characterize the complexity of an optimization problem one may introduce a measure which remains invariant with regard to different schemes of representation: the density of states. It is the probability that an arbitrary chosen state has a certain fitness value. The knowledge of this probability makes it possible to estimate the optimal fitness value and the computational effort to find a better solution of the problem. A general method is presented which allows to approximate the density of states during the optimization process. This is demonstrated for frustrated sequences, road networks and especially for the secondary structures of RNA.

1. Introduction

The process of optimization is a fundamental principle which is reflected in the dynamical laws of physics, the progress of biological evolution and the technological development. The question about *optimal solutions* becomes more and more important in many spheres of our life. The fruitful confluence and interaction of various fields of science led to the idea of solving complex optimization problems by means of thermodynamical and biological principles. The last decade was marked by a great development of stochastic optimization algorithms which may be collected by the concept of *Evolutionary Algorithms* consisting of: Simulated Annealing [15], Evolutionsstrategien [19], Genetic Algorithms [14] and Evolutionary Programming [10]. The search for optimal solutions may be regarded by a motion $x(t)$ in a search space visiting global optima of a function $F(x)$. The motion $x(t)$ is not deterministic – there is a stochastic *generation* of *alternatives* for further motion. The realization of one of them takes place by an *assessment* related to the *fitness* $F(x)$ which denotes the quality of the state x of the optimization object. The class of all algorithms realizing the same probabilistic dynamics is called an *Evolutionary Strategy* [2].

Any information about the complexity of an optimization problem is very fruitful to design a good representation and algorithm for practical optimization. The general issue is the classification of the problems to understand the dynamics of optimization leading to optimal search algorithms.

An *optimization problem* provides the set X of all admitted states of the objects which has to be optimized and a fitness function F which quantifies the aim of the problem. But it tells us nothing about it how we should move from one state to an other. As part of the used strategy, we have to choose a mutation operator generating new states from the current one. At this point, the pure set of all possible states becomes a search space (X, N_Ξ) with a *neighborhood structure* N_Ξ consisting of all pairs of states connected by one mutation. The *fitness landscape* is the surface of F on (X, N_Ξ) and the dynamics of the optimization strategy reflects the motion of the searchers on this landscape. It is important to realize that the landscape depends on the problem by X and F as well as the used mutation operator generating N_Ξ.

To classify the optimization problem for any mutation operator and strategy we need measures depending on X and F but remains *invariant* related to the neighborhood structure. The *density of states* is the probability to find any state on a certain fitness level. Obviously this number is invariant with regard to N_Ξ and depends on X and F only. By means of the density it is possible to approximate the optimum of the problem and to assess the probability to find it by pure random search.

Using a Boltzmann strategy one may approximate the density of any optimization problem [1, 21]. In this article a improved method shall be presented which may approximate the density of states en passant during the optimization process.

First we consider the frustrated game proposed by Engel [6] and secondly we determine the density of graphs for the problem of road network optimization. Finally the method shall be applied to the folding problem of secondary RNA structures.

2. Evolutionary strategies

The stochastic motion of searchers governed by the algorithm can be described by the dynamics of the probability $P(x, t)$ to find a searcher with state x at time t. Different algorithms may realize one and the same probabilistic dynamics. The class of all algorithms realizing the same dynamics is called an *Evolutionary Strategy* [2]. Evolutionary Algorithms may be divided according to their physical or biological origin into the classes of the Boltzmann strategy (BS) or Darwin strategy (DS). They can be described by the master equation of $P_x = P(x, t)$ [4, 1]

$$\frac{dP_x}{dt} = \gamma \left(\langle F \rangle - F \right) P_x + \sum_{y \in X} \left(W_{xy} P_y - W_{yx} P_x \right) . \tag{2.1}$$

The first term represents the selection process and the second one the mutations of an Evolutionary strategy *minimizing* $F(x)$. In the case of a Simulated Annealing algorithm [17, 15, 16] we have to set $\gamma = 0$ and the transition probability to

$$W_{xy} = \begin{cases} S_{xy} & : \triangle F \leq 0 \\ S_{xy}\, e^{-\beta \triangle F} & : \triangle F > 0 \end{cases} \tag{2.2}$$

with $\triangle F = F(x) - F(y)$ and the probability S_{xy} of a mutation of y to x. The dynamics (2.1) describes the Boltzmann strategy and its stationary distribution is the Boltzmann distribution

$$\lim_{t \to \infty} P(x, t) \sim e^{-\beta F(x)}.$$

Evolutionary Algorithms using a selection scheme adopted from the natural selection of biological systems may realize a Darwin strategy with $\gamma = 1$ and $W_{xy} = S_{xy}$. The first term of (2.1) was introduced by FISHER [9] and EIGEN [8] to explain a simple model of Darwinian selection with a reproduction rate related to the mean fitness of a population $\langle F \rangle = \sum F(x) P_x$. The combination of the strategies $\gamma > 0$ (2.2) generates a new class of Evolutionary Algorithms – the *Mixed Strategy* – which shows a improvement in robustness and velocity of optimization. For details of the simulation of the algorithms and the analytical investigation of the strategies see [4, 2, 22].

3. The density of states

The concept of the density of states originates from statistical physics. It is widely used in solid state physics and also to describe thermodynamic optimization processes [18]. For the explanation, let us use a *minimization problem* defined by a finite set X of N possible states x of the optimization object and

400

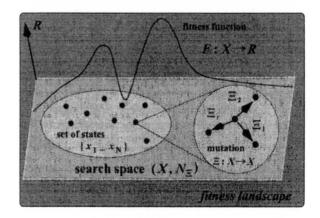

Figure 3.1: The concept of fitness landscapes encloses three entities: the set of states, a group of mutations and the fitness function.

a fitness function $F : X \to \mathbb{R}$. The solutions of such a problem are the optimal states minimizing F.

If we do not know the explicit form of F all we can do is moving from state to state and looking for the optimal one. To do that we have to choose a group of mutation operators $\Xi_r : X \to X$ transforming one state in a neighboring one. The set of all pairs of neighbors builds the neighborhood structure N_Ξ of the search space (X, N_Ξ). The *fitness landscape* is a surface $F = F(x)$ in $X \times \mathbb{R}$ (see Fig. 3.1). An optimization problem can be represented by different fitness landscapes which depends on the choice of the mutation operator generating different neighborhood structures.

A general way to classify the landscape for any neighborhood structure is to build equivalence classes of the states which are invariant due to a change of N_Ξ. We have to choose an equivalence function yielding the same value for all states of a class. The only function we have given by the problem without further knowledge is the fitness function.

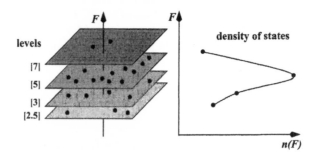

Figure 3.2: The numbers of states at the fitness levels of the problem form the density of states.

Using this fact we are able to define (see Fig. 3.2): The *equivalence classes* of the landscape are the fitness levels $[\epsilon] = \{x : x \in X, F(x) = \epsilon\}$. The number of states on the level $[\epsilon]$ is called *weight* W_ϵ and the probability

$$n(\epsilon) = \frac{W_\epsilon}{\sum_\epsilon W_\epsilon} \qquad (3.1)$$

to find any state on the level $[\epsilon]$ is the *density of states* (DOS).

The DOS depends by definition on X and F but it does not depend on N_Ξ. In this sense we may say: The DOS is a classifying measure for the optimization problem. It tells us how sparsely states of a given quality are distributed. Most of the realistic problems possess only few optimal states which corresponds to a very small probability of the tail of the DOS. Thus, in the limit[1] $n(F) \to \frac{1}{N}$, the optimal fitness value can be estimated [21].

Another important fact is the possibility to derive informations about the computational effort of optimization. For problems characterized by a very steep decay in $n(F)$, one can almost be sure that an extensive enhancement of the computational effort does not necessarily result in better optimization values. Most interesting, however, are problems which display slow decay of the DOS in the region of optimal fitness values. Here, the number of better solutions remains sufficiently large, which means that more effort in the optimization has a good chance to yield better results. In case of pure random search, the effort needed to improve a solution can be derived from the DOS. This gives us an upper bound for all more effective optimization strategies. The time to find a solution by random search can be estimated by $\tau(F) \sim \frac{1}{n(F)}$, thus the effort to improve the solution by $\triangle F$ can be characterized by time ratio

$$\eta_r = \frac{\tau(F + \triangle F)}{\tau(F)} = \frac{n(F)}{n(F + \triangle)}.$$

If the time ratio η of the used optimization strategy remains smaller then η_r in the range of good fitness values, the enhancement of the computational effort has to be advised. Otherwise one should spend time to design a more suitable strategy for the problem.

3.1. Methods of determination of the density of states

It is a nice fact that the DOS can be approximated en passant during the optimization process. One method is to determine the DOS by simulating a ensemble of searchers according to the Boltzmann strategy [1, 21]. The probability distribution $P(F, t)$ of the ensemble , which is approximated by the frequency distribution of the searcher, tends towards the equilibrium density $P(F) \sim n(F) \exp(-\beta F)$. After a sufficiently long simulation time we obtain the DOS from the frequency distribution by multiplication with $\exp(\beta F)$. This

---

[1]In case of one optimum F^* we have $n(F^*) = \frac{1}{N}$.

method can be applied to any optimization problem but is restricted according the choice of the strategy.

To find a method without this restriction let us answer the question: How does the neighborhood structure influence the transition from one fitness level to another one? The neighborhood structure can be represented by its adjacency matrix which is $A_{xy} = 1$ for all pairs (x, y) connected by one mutation and otherwise $A_{xy} = 0$. With the assumption that all mutations have the same chance, the *transition probability of the states* with respect to one mutation is simply $S(x, y) = A_{xy} / \sum_y A_{xy}$, i.e. the transitions of the states are completely determined by the neighborhood structure.

To obtain the probability $T(\epsilon, \gamma)$ of a transition from a fitness level γ to the level ϵ by one mutation we introduce the *level structure* of the landscape by

$$\Phi_{\epsilon x} = \begin{cases} 1 & : F_\epsilon = F(x) \\ 0 & : F_\epsilon \neq F(x) \end{cases}, \quad \Phi^\top = (\Phi_{x\epsilon}). \tag{3.2}$$

This matrix formalizes the affiliation of states to the fitness levels. Inserting the definition it is easy to prove the relations

$$\left(\Phi\Phi^\top\right)_{\epsilon\epsilon} = W_\epsilon, \ \text{tr}\Phi\Phi^\top = \sum_\epsilon W_\epsilon, \tag{3.3}$$

$$\sum_x S(x, y) = 1 \implies \sum_\epsilon \left(\Phi S\Phi^\top\right)_{\epsilon\gamma} = W_\gamma, \tag{3.4}$$

leading to the *transition probability of the fitness levels*

$$T(\epsilon, \gamma) = \frac{\Phi S\Phi^\top}{\left(\Phi\Phi^\top\right)_{\gamma\gamma}}, \ \sum_\epsilon T(\epsilon, \gamma) = 1. \tag{3.5}$$

Multiplying (3.5) with W and using (3.4) it becomes clear that the density of states is an eigenvector with eigenvalue one of the transition probability of the fitness levels.

$$TW = W, \ Tn = n, \ n(\epsilon) = \frac{\left(\Phi\Phi^\top\right)_{\epsilon\epsilon}}{\text{tr}\Phi\Phi^\top} \tag{3.6}$$

We can see that the DOS depends only on the level structure Φ and is independent of S, e.g. it is invariant with respect to N_Ξ.

The eigenvector property (3.6) makes it possible to approximate $n(F)$ from the transition probability of the levels. We just have to count every transition caused by a mutation in a frequency matrix during the optimization and to calculate its eigenvectors numerically . This was already been mentioned in [18] in the context of Simulated Annealing. But in fact, the derivation of (3.6) shows that there is no reason to restrict the method to a particular strategy. Every strategy which is able to reach all points of the search space from any starting point, e.g. every ergodic strategy, may be used.

4. Examples

The first example is to compare the two methods described above. It was introduced by Engel [6] as a simple model of a frustrated optimization problems. A detailed description can be found in [1, 21].

The search space X consists of all sequences x of the length L with $\lambda = 4$ letters. For any sequence there are $(\lambda - 1)L$ point mutations generating the group of mutation operators. The number $\alpha(x)$ of letters with a alphabetic successor and the number $\pi(x)$ of letters which recur with a period $p = 5$ has to be maximized. These two requirements standing opposed to one another cannot be solved at the same time. Therefore this kind of problem is called frustrated. In order to formulate the problem as a minimization task the fitness function is chosen by $F = -\alpha(x) - b\pi(x)$, $b > 0$. It can be shown [20] that for $L = np$, $n = 1, 2, ...$ a bifurcation takes place if the parameter b becomes the critical value

$$
b_c = \left\{ \begin{array}{ll} \frac{1}{p} & : \frac{L}{\lambda} \, integer \\ \frac{1}{p} - \frac{1}{L} & : \frac{L}{\lambda} \, rational \end{array} \right. . \tag{4.1}
$$

For $b < b_c$ only the complete alphabetic sequence is the optimum. For $b \geq b_c$ optimal sequences are mostly periodic with a few breaks in the alphabetic order, e.g $L = 20$: **BBCDABBCDABBCDABBCDA**.

Fig. 4.1 shows the DOS of sequences with $L = 20$ which were obtained with help of both methods. The DOS for the Engel problem has a Gaussian like shape. This is not surprising since the fitness function is composed of a number of additive contributions.

The second method was used both with the Boltzmann strategy (BS) and the Mixed strategy (BDS). Whereas the first method is able to approximate the DOS up to a fitness value of $F = -17$ the second one reached the the optimum $F = -20$.

The obtained results are very similar one to another in all of the three cases. One may conclude that the second method does not depend on the used optimization strategy. Thus the determination of the density of states is not confined by using the Boltzmann strategy anymore.

The second application is the optimization of road networks [7, 22]. As configuration 39 points of the street network of Martina Franca Fig. 4.2 were chosen.

The search space is the set of all connected graphs x of the point configuration. Deletion and insertion of a link are used as elementary mutations. There are two obvious factors to be minimized:

1. The distance from each point to each other point shall be as small as possible.

404

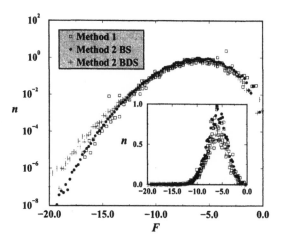

Figure 4.1: Density of states of the Engel problem of frustrated sequences with $L = 20$, $\lambda = 4$, $p = 5$, $b = 0.2$ obtained by the first method using the Boltzmann strategy and by the second method applying the Boltzmann strategy (BS) as well as the Mixed strategy (BDS).

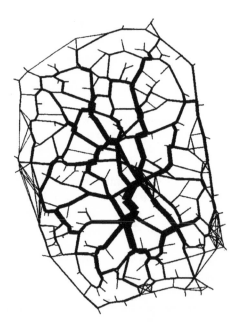

Figure 4.2: Road network of Martina Franca (Italy).

2. The network shall be as short as possible to minimize the cost of construction.

Thus the fitness function is given by

$$F(x) = (1 - \lambda)d(x) + \lambda l(x), \ 0 \leq \lambda \leq 1.$$

The first part d is the detour, i.e. the difference between the shortest and direct path from one point to an other averaged over all pairs. l is the length of the network. The two contradicting goals are weighted by the price factor λ, which allows us to choose the importance of shortness or construction cost, respectively.

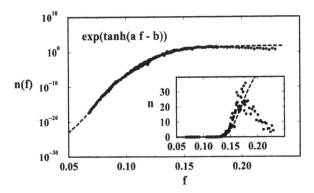

Figure 4.3: Density of states of the network optimization.

The obtained DOS for the network optimization (see Fig. 4.3) is an example for a non Gaussian DOS (for details see [21, 22]) .

5. Secondary RNA structures

A central problem in biopolymer design and research is the folding of sequences into structures. In this field of research, the investigation of RNA sequences is very important because of their catalytic capacities and perfectly tailored properties [23, 3]. The first step of folding can be modelled by folding into secondary structures [11]. The DOS of the secondary RNA structures is important to understand the versatility of RNA molecules [12].

The number of structures increases exponentially with the sequence length L [11], therefore an exhaustive search becomes quickly infeasible for $L > 40$. The DOS can be determined by a specialized dynamic programming method [5] for longer sequences ($L = 76$), but this technique does not provide the explicit secondary structure.

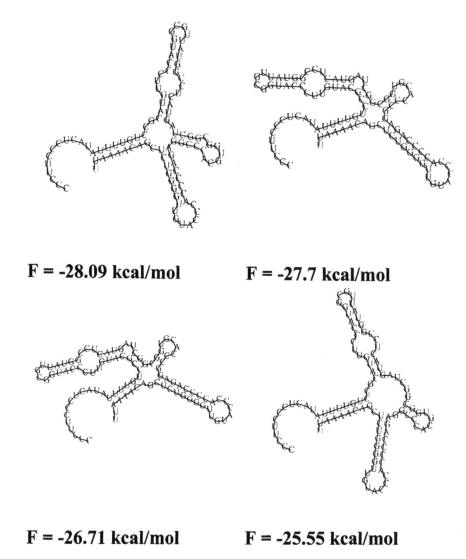

F = -28.09 kcal/mol   **F = -27.7 kcal/mol**

F = -26.71 kcal/mol   **F = -25.55 kcal/mol**

Figure 5.1: The secondary structures of a part ($L = 100$) of the polio virus type 1 Mahoney (AC V01148) optimized by the Boltzmann strategy. The strategy was able to find the minimal free energy structure as well as suboptimal structures.

The application of an Evolutionary Strategy overcomes this limitation. Using this optimization strategy the suboptimal structures, which are relevant for the biological interpretation, as well as an approximation of the DOS are provided.

The search space of the folding problem can be represented by the set of sequences consisting of balanced parentheses and dots. The pairs of parentheses has to be compatible with the possible Watson-Crick pairs (AU, GC) and GU

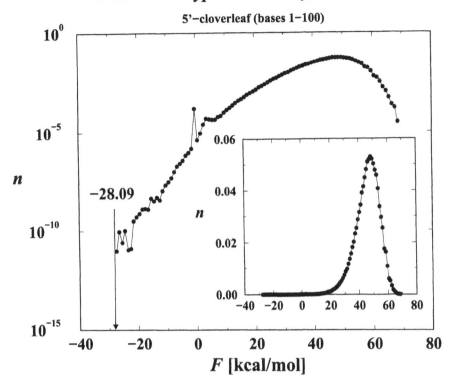

Figure 5.2: The density of states of the secondary structures of the polio virus.

of the given RNA sequence and the dots denotes unpaired nucleotides. For the optimization procedure the following mutations operators were chosen:

bind	.... → .(.....)...
dissolve	.(.....)... → ....
move	.(.....)... → ..(....)...
pull tight	...(...)... → .(((...))).
pull up	.(((...))). → ...(...)...

The fitness function is the free energy of the structure which was calculated by a subroutine of the VIENNA RNA package 1.1 beta [13] .

The minimization was carried out for the first part (5'-cloverleaf, $L = 100$) of the polio virus type 1 Mahoney (AC V01148) by the Boltzmann strategy. The Evolutionary strategy was able to obtain both the minimal free energy structure with $F = -28.09\,\text{kcal/mol}$ and the suboptimal structures of the RNA Fig. 5.1. The DOS was determined by the eigenvector method Fig. 5.2.

The obtained density for the Polio virus is overall Gaussian like the DOS for yeast obtained by the dynamic programming method in [5]. The slow decay of the DOS suggests that also the optimization of longer sequences may to be feasible by application of Evolutionary Strategies.

6. Conclusions

The density of states and in particular its tail at optimal fitness values is a classifying measure of the complexity of an optimization problem. It is invariant to any choice of the neighborhood structure of the fitness landscape and characterizes only the problem. The knowledge of this density allows the approximation of the optimal fitness value of the given problem and to assess the benefit of the further optimization process. This important fact makes it feasible to define a termination criterion which is relevant for practical optimizations. For wide a class of optimization problems the density of states can be approximated by the eigenvector method during the optimization process.

Acknowledgment

This work has been supported by the BMBF, *Verbundprojekt* EVOALG: FKZ 01 IB 403 B3. I thank Prof. W. Ebeling, T. Asselmeyer, U. Erdmann, C. Forst and J. Palinkas for ideas, help and discussions.

Bibliography

[1] ASSELMEYER, T., W. EBELING, and H. ROSÉ, "Smoothing representation of fitness landscapes - the genotype-phenotype map of evolution", *BioSystems* **39** (1996), 63–76.

[2] ASSELMEYER, T., W. EBELING, and H. ROSÉ, "Evolutionary strategies of optimization", *Phys. Rev. E* **56** (1997), 1171–1180.

[3] BEAUDRY, A., and G. JOYCE, "Directed evolution of RNA enzyme", *Science* **257** (1992), 635–641.

[4] BOSENIUK, T., W. EBELING, and A. ENGEL, "Boltzmann and Darwin Strategies in Complex Optimization", *Phys. Lett.* **125** (1987), 307–310.

[5] CUPAL, J., C. FLAMM, A. RENNER, and P. STADLER, "Density of states, metastable states, and saddle points. exploring the energy landscape of RNA molecule", *in preparation* (1997).

[6] EBELING, W., A. ENGEL, and R. FEISTEL, *Physik der Evolutionsprozesse*, Akademie-Verlag Berlin (1990).

[7] EBELING, W., H. ROSÉ, and J. SCHUCHHARDT, "Evolutionary strategies for solving frustrated problems", *Proceedings of the First World Congress of Computational Intelligence* (Orlando,), IEEE (June 27-29 1994), 79–81.

[8] EIGEN, M., "The selforganization of matter and the evolution of biological macromolecules", *Naturwiss.* **58** (1971), 465.

[9] FISHER, R.A., *The Genetical Theory of Natural Selection*, Oxford University Press Oxford (1930).

[10] FOGEL, D.B., *Evolutionary computation – toward a new philosophy of machine intelligence*, IEEE Press Piscataway NJ (1995).

[11] FONTANA, W., P.F. STADLER, E.G. BORNBERG-BAUER, T. GRIESMACHER, I. HOFACHER, M. TACKER, P. TARAZONA, E.D. WEINBERGER, and P. SCHUSTER, "Rna folding and combinatory landscapes", *Phys. Rev. E* **47**, 3 (1993), 2083–2099.

[12] HIGGS, P.G., "Thermodynamic properties of transfer RNA: A computational study", *J.Chem.Soc.Faraday Trans.* **91** (1995), 2531–2540.

[13] HOFACKER, I., W. FONTANA, P. STADLER, L. BONHOEFFER, M. TACKER, and P. SCHUSTER, "Fast folding and comparison of rna secondary structures", *Monatshefte f. Chemie* **125** (1994), 167–188.

[14] HOLLAND, J.H., *Adaptation in Natural and Artificial Sytstems*, University of Michigan Press Ann Arbor (1975).

[15] KIRKPATRICK, S., C.D. GELATT JR., and M.P. VECCHI, "Optimization by simulated annealing", *Science* **220** (1983), 671–680.

[16] LAARHOVEN, P.J.M., and E.H.C. AARTS, *Simulated Annealing: Theory and Applications*, Reidel Dordrecht (1987).

[17] METROPOLIS, N., A. ROSENBLUTH, M. ROSENBLUTH, A. TELLER, and E. TELLER, "Equation of state calculations by fast computing machines", *J. Chem. Phys.* **21** (1953), 1087–1092.

[18] NULTON, J.D., and P. SALAMON, "Statistical mechanics of combinatorial optimization", *Phys. Rev. A* **37**, 4 (1988), 1351–1356.

[19] RECHENBERG, I., *Evolutionsstrategien - Optimierung technischer Systeme nach Prinzipien der biologischen Information*, Friedrich Frommann Verlag (Günther Holzboog K.G.) Stuttgart - Bad Cannstatt (1973).

[20] ROSÉ, H., *Evolutionäre Strategien und Multitome Optimierung*, PhD thesis Humboldt Universität zu Berlin (in preparation 1997).

[21] ROSÉ, H., W. EBELING, and T. ASSELMEYER, "The density of states - a measure of the difficulty of optimisation problems", *Parallel Problem Solving from Nature IV. Proceedings of the International Conference on Evolutionary Computation* (Heidelberg,) (H.-M. VOIGT, W. EBELING, I. RECHENBERG, AND H.-P. SCHWEFEL eds.), Springer (1996).

[22] SCHWEITZER, F., W. EBELING, H. ROSÉ, and O. WEISS, "Optimization of road networks using evolutionary strategies", *subm. to Evolutionary Computation* (1997).

[23] SYMONS, R., "Small catalytic RNAs", *Ann.Rev.Biochem.* **61** (1992), 641–671.

Chapter 39

Conformational switching as assembly instructions in self-assembling mechanical systems

Kazuhiro Saitou[1]
University of Michigan, MEAM Dept.

A study of one-dimensional self-assembly of a type of mechanical conformational switches, *minus devices*, is presented where assembly occurs via the sequential mating of a random pair of parts selected from a part bin, referred to as *sequential random bin-picking*. Parametric design optimization of the minus devices maximizing the yield of a desired assembly, and rate equation analyses of the resulting designs, reveal that the minus devices facilitate the *robust* yield of a desired assembly against the variation in the initial fraction of the part types, by specifying a fixed assembly sequence during the self-assembling process. It is also found that while the minus devices can "encode" some assembly sequences, encoding other assembly sequences requires the use of another type of conformational switches, *plus devices*. To investigate the "encoding power" of these conformational switches, a formal model of self-assembling systems, one-dimensional *self-assembling automaton*, is introduced where assembly instructions are written as local rules that specifies conformational changes realized by the conformational switches. It is proven that the local rules corresponding to the minus and plus devices, and *three* conformations per each component, can encode *any* assembly sequences of a one-dimensional assembly of distinct components with *arbitrary* length.

---

[1]This work was carried out in part as the author's PhD study at MIT supported by NSF under DDM-9058415 funded to Prof. Mark Jakiela.

412

1. Introduction

Many complex biological structures arise from a process of self-assembly – assembly via random interactions among components. Biologists believe that assembly instructions for such self-assembly are built in each component molecules in the form of *conformational switches*, a mechanism that changes component shape as a result of local interactions among other components. In a protein molecule with several bond sites, for instance, a conformational switch causes the formation of a bond at one site to change the conformation of another bond site. As a result, a conformational change occurred at an assembly step provides the essential substrate for assembly at the next step, realizing a fixed assembly sequence during self-assembly [9].

Designing self-assembling mechanical systems with such built-in assembly instructions is of great interest from an engineering point of view. If assembly of a mechanical system can occur via a bulk random agitation, *e.g.* shaking, it would be ideal for assembly of very small parts where direct part grasping is extremely difficult. Also, since assembly instructions are distributed among each parts, the assembly is more robust against unknown disturbances to the systems. This would make the systems suitable to automated assembly in unstructured/unmodeled environment. As in their biological counterparts, conformational switches would also play an important role in such self-assembling mechanical systems.

This paper summarizes our first attempt toward the fundamental understanding of the role of conformational switches in self-assembling mechanical systems. Section §3 discusses a case study of one-dimensional self-assembly of a type of mechanical conformational switches, *minus devices*, via the sequential mating of a random pair of parts selected from a part bin, referred to as *sequential random bin-picking*. Section §4 introduces a formal model of self-assembling systems, one-dimensional *self-assembling automaton*, where assembly instructions are written as local rules that abstracts conformational changes realized by the conformational switches. Before proceeding to these results, Section §2 briefly discusses some related work on on the use of conformational switches in self-assembling mechanical systems.

2. Related work

Although not termed as mechanical conformational switches, a few work incorporated them in attempts to develop and analyze self-assembling mechanical systems. Penrose [4], suggested several designs of mechanical conformational switches that are used in devices that "self-reproduce". These conformational switches cause a bond at one location to break a bond existing at another location or prevent a bond from occurring at another location. When the correct number and arrangement of sub-devices are linked, the conformational switches cause the entire chain to cleave into two copies of the original self-reproducing device in a process akin to cell division.

Another example is found in Hosokawa *et al.* [2]. They developed triangular parts employing switches realized with movable magnets that allow parts to bond together to form hexagons. The switches allow a part to be either in an active or inactive state. An activated part can bond to an inactivated part, turning the part to an activated state. These parts are assembled in a rotating box randomizer. The amounts of each intermediate subassembly achieved agreed reasonably well with the predicted values obtained by techniques analogous to chemical kinetics. These work, however, neither addresses the optimization of conformational switch design to maximize yield, nor the assembly sequences encoded by conformational switches.

3. A case study

This case study discusses the role of a type of mechanical conformational switches, *minus devices* [6], in one-dimensional self-assembly via a sequential mating of a random pair of parts selected from a part bin called *sequential random bin-picking*. A part can form a "bond" with another part *only* at the left or right *bond sites* (hence forming a one-dimensional assembly) when the two mating bond sites have complimentary shapes. By changing the shape of one of the bond sites, the minus device allows the formation of a bond *only* after the conformational change.

3.1. Minus devices

Fig. 3.1 (a) shows a part with a minus device [6]. It consists of right and left bars that can slide horizontally, and one inner sliding block that can slide vertically. The left and right pictures in Fig. 3.1 (a) show the minus device *before* and *after* conformational change, respectively. Before conformational change, the left bar is free to be pushed in whereas the right bar *cannot* be pushed in due to the position of the inner block. Conformational change of the device can be induced by pushing in the left sliding bar, which causes the the inner block to slide down, which in turn creates a space for the right bar to slide in. As a result, the right bar can pushed in after conformational change. A simplified notation of the device is shown in Fig. 3.1 (b), where an arrow indicates the direction in which one of the two horizontal bars can be pushed in to induce the conformational change. Note that in the simplified notation, the right sliding bar after conformational change is drawn as the "pushed-in" state, representing that the bar is "free" to be pushed in.

Fig. 3.1 (c) illustrates how conformational change of the minus device can enforce an order of bond formation. Since only the left bar can be pushed in before the conformational change, a hatched rectangular part must come from the left in order to form a bond with the part with a minus device (top figure of Fig. 3.1 (c)). Then it is only after the conformational change when another hatched rectangular part can form a bond from the right (bottom figure of Fig. 3.1 (c)). Since the right bar cannot be pushed in before the conformational

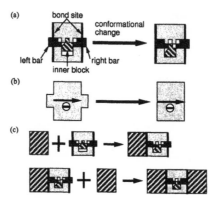

Figure 3.1: A part with a minus device. (a) mechanism, (b) simplified notation, (c) interaction with another part.

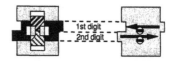

Figure 3.2: A part with "two-digit" bonding sites.

change, first part *must* come from the left, enforcing the left-right order of bond formation. Since the minus device is not "spring-loaded", it is impossible to reverse the conformational change. This implies that once a part changes its conformation and a bond is formed as a result, it will *never* be destroyed.

The part design presented below has "two-digit" bonding sites as shown in Fig. 3.2. The two-digit bonding sites are introduced in order to increase the number of possible shapes they can take, necessitated by the need of increasing the number of distinguishable part types. The shape of bond sites are represented by a pair of integers (a_1, a_2) referred to as *bond configuration*. Each component of a bond configuration takes 1 if the corresponding "digit" of the bond site has convex shape, -1 if the the digit is concave, and 0 if it is flat. When bond sites of two parts meet, they form either 1) a stable bond, 2) an unstable bond, or 3) no bond. The occurrences of these cases depend on the shape of the two mating bond sites, or equivalently, the bond configurations of the sites. Let (a_1, a_2) and (b_1, b_2) be the bond configurations of two bond sites contacting each other. These sites form a stable bond if they are complementary to each other, i.e $a_1 + b_1 \leq 0$ and $a_2 + b_2 \leq 0$, and form an unstable bond if they are fairly complementary, i.e. $a_1 + b_1 = 1$ and $a_2 + b_2 \leq 0$, or $a_1 + b_1 \leq 0$ and $a_2 + b_2 = 1$. If none of the above applies, the two bond sites are not considered as complementary and therefore no bond is formed. An unstable bond

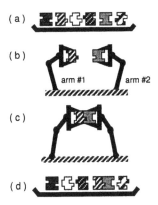

(a)

(b)

arm #1   arm #2

(c)

(d)

Figure 3.3: Sequential random bin-picking.

induces conformational change of the involved bond sites. After the conformational changes, a stable bond is formed if the condition for the stable bond is satisfied. Otherwise, no bond is formed.

3.2. Sequential random bin-picking

In this case study, a simple model of one-dimensional self-assembly called *sequential random bin-picking* is employed. It is a process of sequential mating of a random pair of parts selected from a part bin, which initially contains a random assortment of parts (Fig. 3.3(a)). Mating of a pair of parts is accomplished according to the following three steps, which are repeated until pre-specified conditions are satisfied:

Step 1: Arm #1 *randomly* picks up a part (possibly an assembly) from the bin. Then, arm #2 randomly picks up another part (possibly an assembly) from the bin (Fig. 3.3 (b)).

Step 2: The two parts are pushed against each other, possibly forming of an assembly (Fig. 3.3 (c)).

Step 3: The parts are returned to the bin (Fig. 3.3 (d)), possibly as an assembly.

It is assumed that a defect could occur with a certain probability at step 2 when two chosen parts form an assembly[2]. If the defective assembly is chosen at step 1 in subsequent iterations, no assembly can form at step 2 *regardless of* the mating part, *i. e.* the parts in the defective assemblies are completely wasted.

---

[2]If they cannot form an assembly, no defect occurs and they are simply returned to the bin.

3.3. Design optimization of minus devices

In order to investigate the effect of the minus devices to the yield of a desired assembly during the process of sequential random bin-picking, parametric design optimization of the minus devices is performed. More precisely, the problem is stated as follows:

Given: the number of each type of parts in the bin, (*i.e.* the initial state of the bin), and defect probabilities.

Find: an optimal design of minus devices (and their initial bond configurations) that maximizes the yield of a desired assembly in the process of the sequential random bin-picking.

The above problem is formulated as search by parameterizing the design of the minus devices. Genetic Algorithm [1], in conjunction with computer simulation of sequential random bin-picking, is used to search the parameter space of possible part designs.

The following four parameters uniquely specify a part two minus devices and two-digit bond sites: left_config, right_config, upper_link and lower_link. Left_config and right_config are the initial bond configurations of the left bond site and the right bond site, respectively. Upper_link and lower_link are variables that specify the existence of the minus devices in a part, each of which takes one of the values LEFT, RIGHT or NONE, depending on the direction of the arrow in the simplified notation of the corresponding minus devices. If upper_link is LEFT, for instance, there is a minus device at the "first digits" of the bond sites, with the arrow pointing to the left bond site. If upper_link is NONE, there is no minus device that connects the first digits of the bond sites, so they cannot undergo any conformational changes. Similarly, lower_link specifies the existence of the minus device between the second digits of the bond sites.

To apply Genetic Algorithm, the values of these parameters for *all types of* parts are represented as one binary string. As illustrated in Fig. 3.4, each part type has two bits for each component of left_config and right_config, and two bits for each of upper_link and lower_link. Since each part type requires 12 bits, n types of part requires a bit string with the length $12n$.

3.4. Self-assembly with three part types

First, the design optimization of the minus device as described above is done in the case of $n = 3$. The initial bin contains a random assortment of *three* types of parts, part A, part B and part C, and the desired assembly whose yield is to be maximized is assembly ABC. The yield of ABC is computed as an average of the number of ABC's in the bin after 700 iterations of the sequential robot bin-picking, over 50 of such runs. The Genetic Algorithm in this example uses a crowding population [1] with 10% replacement per generation, fitness proportionate (roulette wheel) selection, and linear fitness scaling with scaling

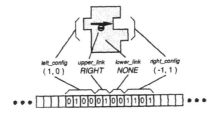

Figure 3.4: A bit string representation of a part design.

coefficient $= 2.0$. Also, the crossover probability is 0.9, the mutation probability is 0.03, the population size is 300, and the number of generations is 200.

In the following results, $n_0 = (n_A(0), n_B(0), n_C(0))$ is the vector of the initial numbers of parts A, B and C in the bin, and $q = (q_{AB}, q_{BC})$ is the vector of defect probabilities of the bonds between AB and BC, respectively[3].

Fig. 3.5 shows the best part designs obtained in the case of $n_0 = (10, 20, 10)$, with three different defect probabilities: (a) $q = (0.0, 0.0)$, (b) $q = (0.2, 0.0)$, and (c) $q = (0.0, 0.2)$. For $q = (0.0, 0.0)$ and $q = (0.0, 0.2)$, a part A can bind to a part B only after part B changes its conformation, which is triggered by the binding of part C. The formation of an assembly ABC, therefore, takes place through the two-step "reactions," $B + C \rightarrow B'C$ and $A + B'C \rightarrow AB'C$, where B' represents a part B after conformational change. Since no other reactions are possible, an ABC assembles only in the fixed assembly sequence $(A(BC))$. On the other hand, $((AB)C)$ is "encoded" in the best design with $q = (0.2, 0.0)$. In this case, a part C can bind to a part B only after the part B changes its conformation, which is triggered by binding of a part A as in $A + B \rightarrow AB'$ and $AB' + C \rightarrow AB'C$. Note that only one conformational link is actually used during the assembly of ABC in all three cases.

The total of nine optimization runs are done for three different cases of n_0 and q. The summary of these nine runs are shown in Table 3.1, where $\{ABC\}$ represents no fixed assembly is specified by the minus devices, *i. e.* parts can be assembled in either $((AB)C)$ or $(A(BC))$. These results indicate the best part design specifies a fixed assembly sequence or no fixed assembly sequences, depending on the values of n_0 and q. It should be noted that Genetic Algorithm searches the space of possible part designs, not the space of assembly sequences. There is no explicit representation of assembly sequences in the above formulation. Rather, assembly sequences in Table 3.1 *emerged* due to a particular design of parts (with minus devices) that maximizes the yield of ABC.

---

[3]It is assumed that the defect probability of a bond depends only on the parts associated to the bond, in particular, $q_{AB} = q_{A(BC)}$ and $q_{BC} = q_{(AB)C}$.

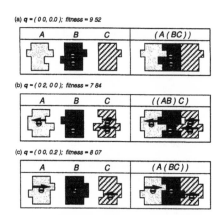

Figure 3.5: Best part designs with $n_0 = (10, 20, 10)$.

Table 3.1: Summary of the results: self-assembly with three part types A, B, and C.

q \ n0	(10,10,10)	(10,20,10)	(20,20,10)
(0.0,0.0)	{ ABC }	(A (BC))	((AB) C)
(0.2,0,0)	{ ABC }	((AB) C)	((AB) C)
(0.0,0.2)	{ ABC }	(A (BC))	(A (BC))

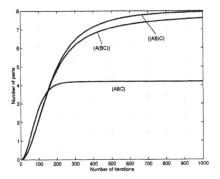

Figure 3.6: Yield of ABC with $\mathbf{n_0} = (10, 20, 10)$ and $\mathbf{q} = (0.2, 0.0)$.

3.5. Rate equation analyses

Although the above results suggest the importance of assembly sequences for maximizing the yield of a desired assembly, it does not provide direct means of comparing the dynamic change of the part counts for different assembly sequences. This can be done by a discrete-time version of rate equation used to describe the kinetics of chemical systems [8, 2]:

$$\mathbf{n}(t+1) = \mathbf{n}(t) + \mathbf{A}\mathbf{p}(t) \tag{3.1}$$

where $\mathbf{n}(t)$ is a vector of the *expected* numbers of each possible assembly (including defected ones) at iteration t, $\mathbf{p}(t)$ is a vector of probabilities for each possible reaction at iteration t, and $\mathbf{A}$ is a matrix of stoichiometric coefficients of all possible reactions.

For each assembly sequences $((AB)C)$, $(A(BC))$, and $\{ABC\}$, Equation (3.1) is numerically solved for 1000 iterations, which corresponds to 1000 iterations in the sequential robot bin-picking, with the nine conditions in Table 3.1. The expected numbers of ABC when assembled in the sequence $((AB)C)$, $(A(BC))$, and $\{ABC\}$, are then plotted in a same plane to allow comparison among the three assembly sequences.

Fig. 3.6 is the resulting plots in the case of $\mathbf{n_0} = (10, 20, 10)$, and $\mathbf{q} = (0.2, 0.0)$. The most noticeable fact in this figure, also observed in other cases with $\mathbf{n_0} = (10, 20, 10)$, is the *very* low yield of $\{ABC\}$ compared to $((AB)C)$ and $(A(BC))$. This is due to the large number of the middle part B, which produces a large number of AB's and BC's in the early stage of iterations if no fixed assembly sequence is specified. This excess production of AB's and BC's then causes the shortage of individual C's and A's later on, which are necessary to complete the final assembly ABC from the assemblies AB and BC, resulting in the low yield. By enforcing a fixed assembly sequence, $((AB)C)$ or $(A(BC))$, this excess production of AB and BC can be avoided.

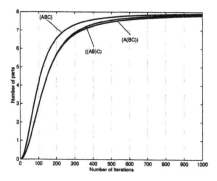

Figure 3.7: Yield of ABC with $\mathbf{n_0} = (10, 10, 10)$ and $\mathbf{q} = (0.2, 0.0)$.

Fig. 3.7 is the resulting plots in the case of $\mathbf{n_0} = (10, 10, 10)$, and $\mathbf{q} = (0.2, 0.0)$. In this case, and also in the other cases with $\mathbf{n_0} = (10, 10, 10)$, $\{ABC\}$ has the best yield, although the yield of other two assembly sequences are almost as good. Since the fraction of the part types in the initial bin is equal to the fraction in the desired assembly, there is no chance of "wasting" subassembly as discussed above, and therefore having no fixed assembly sequence realized fastest production of the final assembly ABC. With the change in the initial fraction of part types, $\{ABC\}$ is outperformed by one of two fixed assembly sequences, as illustrated in Fig. 3.6 as well as in Table 3.1.

The above results suggests an important role of minus devices: facilitating the *robust* yield of a desired assembly against the variation in the initial fraction of the part types, by specifying a fixed assembly sequence during the self-assembling process.

3.6. Self-assembly with four part types

The design optimization of the minus device is also done in the case of $n = 4$. The initial bin contains a random assortment of four types of parts, part A, part B, part C, and part D, and the desired assembly whose yield is to be maximized is assembly $ABCD$. The yield of $ABCD$ is computed as an average of the number of $ABCD$'s in the bin after 1400 iterations of the sequential robot bin-picking, over 50 of such runs. The GA runs described in this section have population size of 600 and the number of generations is 900. In the following results, $\mathbf{n_0} = (n_A(0), n_B(0), n_C(0), n_D(0))$ is the vector of the initial numbers of parts A, B, C and D in the bin, and $\mathbf{q} = (q_{AB}, q_{BC}, q_{CD})$ is the vector of defect probabilities of the bonds between AB, BC and CD, respectively. Note with $n = 4$, five fixed assembly sequences are possible: $(((AB)C)D)$, $((AB)(CD))$, $((A(BC)D))$, $(A((BC)D))$ and $(A(B(CD)))$.

The total of twelve optimization runs are done for three different cases of $\mathbf{n_0}$ and four different cases of $\mathbf{q}$. The summary of these twelve runs are shown in

Table 3.2: Summary of the results: self-assembly with four part types A, B, C, and D.

q \ n0	(10,10,10,10)	(10,20,10,10)	(10,20,20,10)
(0.0,0.0,0.0)	{ ABCD }	{ (AB) CD }	((AB) (CD))
(0.2,0,0,0.0)	{ ABCD }	{ (AB) CD }	((AB) (CD))
(0.0,0.2,0.0)	{ ABCD }	$\overline{\{ (AB) CD \}}$	(A(B(CD)))
(0.2,0,0,0.2)	((AB) (CD))	((AB) (CD))	((AB) (CD))

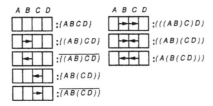

Figure 3.8: Eight possible assembly sequences.

Table 3.2, where $\{(AB)CD\}$ represents either $(((AB)C)D)$ or $((AB)(CD))$, and $\overline{\{(AB)CD\}}$ represents $((A(BC))D)$, $(A((BC)D))$ or $(A(B(CD)))$. Rate equations (3.1) of are formulated in a similar way to the case of $n = 3$. The yield of the final assembly $ABCD$ is then compared for all assembly sequences which can be encoded by two minus devices, which are listed in Fig. 3.8. Among the eight assembly sequences listed, there are five ambiguous (partially fixed) assembly sequences $\{ABCD\}$, $\{(AB)CD\}$, $\overline{\{(AB)CD\}}$, $\{AB(CD)\}$, and $\overline{\{AB(CD)\}}$, and three fixed assembly sequences $(((AB)C)D)$, $((AB)(CD))$, and $(A(B(CD)))$. Note that the minus device cannot encode two of the fixed assembly sequences $((A(BC))D)$ and $(A((BC)D))$. As in the case of $n = 3$, the results of design optimization in Table 3.2 shows the emergence of various optimal assembly sequences depending on n_0 and q, and the associated rate equation analyses[4] confirms the role of minus devices observed in the case of $n = 3$.

3.7. Encoding power of minus devices

An important difference between the cases of $n = 3$ and $n = 4$, as noted earlier, is the existence of "un-encodable" assembly sequences. In particular, one can quickly notice that the two assembly sequences $((A(BC))D)$ and $(A((BC)D))$ are un-encodable with no matter how many minus devices being employed. In other words, these assembly sequences are beyond the encoding power of the minus devices. This suggests there could be an un-encodable assembly sequence which yield better than any encodable assembly sequences. As shown in Fig. 3.9,

---

[4]not shown due to space limit

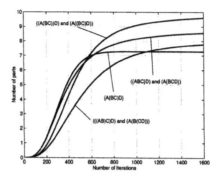

Figure 3.9: Yield of $ABCD$ with $n_0 = (10, 20, 20, 10)$ and $q = (0.0, 0.2, 0.0)$: comparison with un-encodable assembly sequences

in fact,the two un-encodable assembly sequences $((A(BC))D)$ and $(A((BC)D))$ yield better than $(A(B(CD)))$, the best sequence obtained by the GA with $n_0 = (10, 20, 20, 10)$ and $q = (0.0, 0.2, 0.0)$. It should be emphasized, however, that the sequence found by the GA is the best among the assembly sequences encodable to the minus devices. For comparison, Fig. 3.9 also shows the plot for three other un-encodable assembly sequences, $(\{ABC\}D)$, $(A\{BCD\})$ and $\{A(BC)D\}$. The sequences $(\{ABC\}D)$ and $(A\{BCD\})$ also outperform the best encodable sequences $(((AB)C)D)$ and $(A(B(CD)))$.

In order to encode the sequences $((A(BC))D)$, it is necessary to introduce another type of conformational switch, a *plus* device, a simple sliding bar which propagates conformational change *through* multiple parts, and an additional "digit." Fig. 3.10 illustrates such parts with conformational switches that encode $((A(BC))D)$. Note that the part B has both a plus device and a minus device and the part B has *three* conformations B, B', and B'', such that the *second* conformational change from B' to B'' upon the attachment of A is propagated to C through B. Other un-encodable assembly sequences shown in Fig. 3.9 can also be encoded with the help of this plus device.

4. A formal model

In the previous section, it was found that minus devices can encode fixed or partially-fixed assembly sequences, which facilitates the robust yield of a desired assembly against the variation in the initial fraction of part types. It was also found that some assembly sequences are *un-encodable* to minus devices no matter how many devices being employed, and such un-encodable assembly sequences can be encoded by using *both* minus device *and* plus device, another type of conformational switches which propagates conformational changes through multiple components.

These findings raise the following two questions: 1) Is it possible to tell

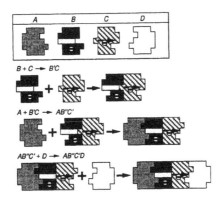

Figure 3.10: Parts with conformational switches that encode $((A(BC))D)$.

whether a given assembly sequence is encodable to minus devices, or to minus *and* plus devices? 2) If so, how many conformations (or switch states) are necessary to encode the given assembly sequence? The relationship between assembly sequences and conformational switches is analogous to the one between languages and "machines" (models of computation) in the theory of computation [3], with an assembly sequence being an instance of a language, and a set of conformational switches that encodes the assembly sequence being a machine that accepts the instance of the language.

This analogy motivated us to develop a formal model of self-assembling systems which abstracts the function of the minus and plus devices, and to identify classes of self-assembling systems based on the classes of assembly sequences in which the components of the systems self-assemble. The model, which we will refer to as an one-dimensional *self-assembling automaton*, is defined as a sequential rule-based machine that processes one-dimensional strings of symbols. Several theorems regarding the classes of self-assembling automaton are provided, although proofs are omitted due to page restrictions. The complete proofs to all theorems are found in [5, 7]. Due to the abstract nature of the model, the presentation of this section is more formal than the previous ones.

4.1. Definition of self-assembling automata

We abstract a component (part) to be an element of a finite set Σ, and an assembly to be a string in Σ^+. Additionally, a component $a \in \Sigma$ can take a finite number of conformations represented by $a, a', a'' \ldots$, and the transition among conformations is specified by a set of assembly rules, which abstracts the function of conformational switches. Each component, therefore, can be viewed as a finite automaton that self-assemble. Note that in the definition below, attaching rules are an abstraction of minus devices, and propagation rules are an abstraction of the plus devices.

Definition 4 *A one-dimensional self-assembling automaton (SA) is a pair $M = (\Sigma, R)$, where Σ is a finite set of* components, *and R is a finite set of* assembly rules *of the form either $a^\alpha + b^\beta \to a^\gamma b^\delta$ (attaching rule) or $a^\alpha b^\beta \to a^\gamma b^\delta$ (propagation rule), where $a, b \in \Sigma$ and $\alpha, \beta, \gamma, \delta \in \{'\}^*$.[5] The* conformation set *of $a \in \Sigma$ is a set $Q_a = \{a^\alpha \mid \alpha \in \{'\}^*, a^\alpha$ appears in $R.\}$. The conformation set of M is the union of all conformation sets of $a \in \Sigma$.*

As in the case of the sequential robot bin-picking, we view an SA as having an associated *component bin*, with an infinite number of "slots" each of which can store an assembly or the null string Λ. Initially, a finite number of the slots contain assemblies and the rest of the slots are filled with Λ. Self-assembly of components proceeds by *selecting* a random pair of assemblies or an assembly in the component bin, and *applying* the rules in R to the selected assemblies. As a result of the rule application, assemblies are *deleted* from and *added* to the component bin, just like assertions are deleted from and added to working memory in rule-based inference systems. The rule application is done according to the following procedure, where $a, b \in \Sigma$ and $\alpha, \beta, \gamma, \delta \in \{'\}^*$:

1. If a pair of assemblies $(x, y) = (za^\alpha, b^\beta u)$ for some $z, u \in \Sigma^*$ is selected, and R contains the rule $r = a^\alpha + b^\beta \to a^\gamma b^\delta$ (r fires), delete x and y and add $za^\gamma b^\delta u$.

2. If an assembly $x = za^\alpha b^\beta u$ for some $z, u \in \Sigma^*$, and R contains the rule $r = a^\alpha b^\beta \to a^\gamma b^\delta$ (r fires), delete x and add $za^\gamma b^\delta u$.

If neither of the above applies, the selected assemblies are simply returned to the component bin, leaving the bin unchanged. Note that the random pick and the rule application model Steps 1–3 of the sequential robot bin-picking. Note also that at any point of self-assembly, the component bin contains a finite number of non-null strings with finite length, since the total number of components in the initial bin is finite and no new components are created when applying the rules to the bin.

We define SEQ(A) as the language generated by the context-free grammar $\forall a \in A, S \to (SS) \mid a$. Note that $A \subset$ SEQ(A). A string x in SEQ(A) is a full parenthesization of a string $u =$ RM-PAREN(x) in A^+, where RM-PAREN is a function that removes parentheses from its argument string. We interpret the parse tree of x as a (binary) assembly tree, *i.e.* a representation of a pairwise assembly sequence of u.

Definition 5 *Let Σ be a component set of an SA. An* assembly sequence *is a string in SEQ(Σ). An assembly sequence x is* basic *if x contains at most one copy of elements in Σ, i.e $\forall a \in \Sigma, N_a(x) \le 1$.*

Definition 6 *Let $M = (\Sigma, R)$ be an SA. A* configuration *of M is a bag $\langle x \mid x \in$ SEQ(Q)$\rangle$, where Q is the conformation set of M. Let $x \in$ SEQ(Σ) be an assembly sequence. A configuration Γ* covers *x if $\Gamma = \langle a \mid a \in \Sigma \rangle$ and $\forall a \in \Sigma, N_a(x) \le N_a(\Gamma)$.*

---

[5]It is assumed $a^\Lambda = a$, where Λ is the null string.

The sequence of self-assembly can be traced by examining the configuration each time the component bin changes as a result of applying the rules in R to the component bin. To keep track of the order of assembly, the non-null strings newly added to the component bin are parenthesized in the new configuration if they were added by an attaching rule.

For two configurations Γ and Φ, we write $\Gamma \vdash_M \Phi$ if the configuration of M changes from Γ to Φ as a result of applying a rule in R to the component bin *exactly once*, reading "Φ is *derived* from Γ at one step." Similarly, $\Gamma \vdash_M^* \Phi$ if the configuration of M changes from Γ to Φ as a result of applying the rules in R to the component bin *zero or more times*, reading "Φ is *derived* from Γ." If there is no ambiguity, $\vdash_M$ and $\vdash_M^*$ are often shortened to $\vdash$ and $\vdash^*$, respectively.

Example 1 *Consider a SA* $M_1 = (\Sigma_1, R_1)$ *where* $\Sigma_1 = \{a, b, c\}$ *and* $R_1 = \{a + b \to ab', b' + c \to b'c\}$. *Let* $\Gamma = \langle a, b, c, c \rangle$ *and* $\Phi = \langle a, b, c \rangle$. *The configurations* Γ *and* Φ *cover the assembly sequence* $((ab)c)$. *Self-assembly of* $((ab)c)$ *from* Γ *proceeds as* $\langle a, b, c, c \rangle \vdash_{M_1} \langle (ab'), c, c \rangle \vdash_{M_1} \langle ((ab')c), c \rangle$.

Note that the SA in Example 1 is an abstraction of the three part self-assembly discussed in §3.4.

Given an SA as defined above, the process of self-assembly eventually terminates when no rule firing is possible, or runs forever due to an infinite cycle of rule firing. It is natural to say an SA self-assembles a given string in a given sequence if the process of self-assembly terminates, and all terminating configurations contain the string that is assembled in the sequence. This is a conservative definition, requiring *stable* and *reliable* production of the string assembled in the sequence. Formally, this can be stated as follows:

Definition 7 *Let* $M = (\Sigma, R)$ *be an SA*, Γ *be a configuration of* M *and* $x \in$ SEQ(Σ) *be a assembly sequence.* Γ *is stable if there is no rule firing from* Γ, *i.e.* $C_M(\Gamma) = \{\Gamma\}$, *where* $C_M(\Gamma) = \{\Phi | \Gamma \vdash_M^* \Phi\}$. M *self-assembles* x *from* Γ *if the both of the followings hold:*
1. *All configurations derived from* Γ *can derive a stable configuration, i.e.* $\forall \Phi \in C_M(\Gamma), \exists \Phi_1 \in C_M(\Phi), C_M(\Phi_1) = \{\Phi_1\}$.
2. $\forall \Phi \in C_M^*(\Gamma), \exists y \in \Phi$ *such that* $x =$ RM-PRIME(y)*, where* $C_M^*(\Gamma)$ *is a set of stable configurations derived from* Γ, *and* RM-PRIME *is a function that removes the prime symbols (') from its argument.*

Example 2 M_1 *in Example 1 self-assembles* $((ab)c)$ *from* $\langle a, b, c, c \rangle$.

4.2. Construction of rule set

Given a basic assembly sequence $x \in$ SEQ(Σ), one can write a procedure to construct a set of assembly rules R such that $M = (\Sigma, R)$ self-assembles x from any configuration Γ that covers x. This corresponds to designing parts with the minus and plus devices that can be self-assembled in a given assembly sequence. The following procedure MAKE-RULE-SET takes as input a basic assembly sequence $x \in$ SEQ(Σ), a flag $\eta \in \{left, right, none\}$, and a rule set R. The flag

η indicates from which side the next assembly would occur, with *none* indicating there is no next assembly. $\texttt{MAKE-RULE-SET}(x, none, \emptyset)$ returns a pair (u,R), where u is the final assembly such that $\texttt{RM-PRIME}(u) = \texttt{RM-PAREN}(x)$ and R is the rule set containing the assembly rules to assemble x from Γ. In the following pseudo-code, $x, y, z \in \texttt{SEQ}(\Sigma)$ are basic assembly sequences, $a, b \in \Sigma$, $\alpha, \beta \in \{'\}^*$, and $u, v \in Q^*$ where $Q = \{a^\alpha \mid a \in \Sigma,\ \alpha \in \{'\}^*\}$, and LEFT and RIGHT are functions that return the symbol at the left and right ends of the argument string, respectively.

$\texttt{MAKE-RULE-SET}(x, \eta, R)$

```
1    if x = a
2       then return (a, R)
3    if x = (yz)
4       then (u, R) ← MAKE-RULE-SET(y, right, R)
5            (v, R) ← MAKE-RULE-SET(z, left, R)
6            aᵅ ← RIGHT(u)
7            bᵝ ← LEFT(v)
8            if η = none
9               then R ← R ∪ {aᵅ + bᵝ → aᵅbᵝ}
10                   return (uv, R)
11           if η = left
12              then R ← R ∪ {aᵅ + bᵝ → a^INC(α)bᵝ}
13                   (u, R) ← PROPAGATE-LEFT(u, R)
14                   return (uv, R)
15           if η = right
16              then R ← R ∪ {aᵅ + bᵝ → aᵅb^INC(β)}
17                   (v, R) ← PROPAGATE-RIGHT(v, R)
18                   return (uv, R)
```

MAKE-RULE-SET (henceforth abbreviated MRS) recursively traverses the left and right subtrees (y and z in line 3), and adds an attaching rule (a minus device) to R that assembles (yz). If a component is assembled from the left at the next assembly step ($\eta = left$ in line 11), propagation rules (plus devices) are added (by the procedure PROPAGATE-LEFT in line 13) that propagate conformational changes through the assembly corresponding to the left subtree. If a component is assembled from the right at the next assembly step ($\eta = right$ in line 15), propagation rules (plus devices) are added (by the procedure PROPAGATE-RIGHT in line 17) that propagate conformational changes through the assembly corresponding to the right subtree. If there is no next assembly step, *i. e.* yz is the final assembly ($\eta = none$ in line 8), no propagation rules are added. The subroutines PROPAGATE-LEFT and PROPAGATE-RIGHT are defined as follows:

$\texttt{PROPAGATE-LEFT}(u, R)$

```
19   if u = aᵅ
20      then return (a^INC(α), R)
21   if u = vaᵅbᵝ
22      then R ← R ∪ {aᵅb^INC(β) → a^INC(α)b^INC(β)}
```

```
23            (u, R) ← PROPAGATE-LEFT(vaᵅ, R)
24            return (ub^{INC(β)}, R)
```

```
PROPAGATE-RIGHT(u, R)
25   if u = aᵅ
26      then return (a^{INC(α)}, R)
27   if u = aᵅbᵝv
28      then R ← R ∪ {a^{INC(α)}bᵝ → a^{INC(α)}b^{INC(β)}}
29         (u, R) ← PROPAGATE-RIGHT(bᵝv, R)
30         return (a^{INC(α)}u, R)
```

where INC is the "conformation incrementor" function which appends the prime symbol (') to its argument string such that for $\alpha \in \{'\}^*$, $\text{INC}(\alpha) = \alpha'$. For example, $\text{INC}(\Lambda) = '$ and $\text{INC}(') = ''$. The correctness of MRS can be proven by induction on $|\text{RM-PAREN}(x)|$. Here we simply state the fact.

Theorem 1 MRS *is correct.*

Example 3 *Consider* $\Sigma = \{a, b, c, d\}$ *and* $x = ((a(bc))d)$. *A call of* MRS$(x, none, \emptyset)$ *returns with* $(ab''c'd, R)$ *where* R *contains the following rules:* $b + c \to b'c$, $a + b' \to ab''$, $b''c \to b''c'$ *and* $c' + d \to c'd$. *It is clear that an SA* $M = (\Sigma, R)$ *self-assembles* x *from the configurations that cover* x, *e.g.* $\langle a, b, c, d \rangle$ *and* $\langle a, a, b, b, c, c, d, d \rangle$.

4.3. Classes of self-assembling automata

The running time of MRS depends on the shape of the parse tree of the input (basic) assembly sequence. The worst case behavior of MRS occurs when, at *every* step of recursion, either PROPAGATE-LEFT or PROPAGATE-RIGHT is called. The best case, on the other hand, is when there is no call of PROPAGATE-LEFT and PROPAGATE-RIGHT. This is the case when the rule set R returned by MRS contains no propagation rules, whereas R contains at least one propagation rule in other cases. Accordingly, two classes of SA are defined based on the presence of propagation rules in the rule set. Note that in the definition below, a class I SA corresponds to a set of parts with *only* minus devices, and a class II SA corresponds a set of parts with *both* plus and minus devices.

Definition 8 *Let* $M = (\Sigma, R)$ *be an SA. M is class I if R contains only attaching rules. M is class II if R contains both attaching rules and propagation rules.*

Definition 9 *An assembly template is a string $t \in \text{SEQ}(\{p\})$. An instance of t on a finite set Σ is a assembly sequence $x \in \text{SEQ}(\Sigma)$ obtained by replacing p in t by $a \in \Sigma$. If x is an instance of t, t is an assembly template of x.*

Example 4 *Two strings $t_1 = ((pp)(pp))$ and $t_2 = ((p(pp))p)$ are assembly templates. Let $\Sigma = \{a, b, c, d\}$. The basic assembly sequences $x_1 = ((ab)(cd))$ and $x_2 = ((b(ad))c)$ are instances of t_1 and t_2 on Σ, respectively.*

428

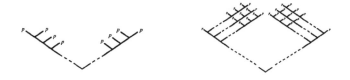

Figure 4.1: Parse tree of an assembly template generated by G_I (left), and by G_{II} (right).

Definition 10 *An* assembly grammar *is a context-free grammar with a language that is a subset of* $\text{SEQ}(\{p\})$. *The class I assembly grammar G_I is defined by the following production rules:*

$$
\begin{aligned}
S &\rightarrow (LR) \\
L &\rightarrow (Lp) \mid p \\
R &\rightarrow (pR) \mid p
\end{aligned}
$$

The parse tree of the assembly templates in $L(G_I)$ is shown in the left of Fig. 4.1. Each of the left and right subtrees is a linear assembly tree, which specifies self-assembly proceeding in one direction. The parse trees of the assembly templates in $\text{SEQ}(\{p\})$ are general binary tree with no special structures.

We can interpret $L(G_I)$ and $\text{SEQ}(\{p\})$ as sets of assembly templates with different numbers of changes in the direction of self-assembly. Let t be an assembly template and x be an instance of t. If $t \in L(G_I)$, the direction of self-assembly does *not* alter during the self-assembly of x. If $t \in \text{SEQ}(\{p\}) \setminus L(G_I)$, the direction of self-assembly alters *at least once* during the self-assembly of x. Based on these observations, the following theorems can be proven. Here we again state only the facts.

Theorem 2 *For any basic assembly sequence x that is an instance of an assembly template $t \in L(G_I)$, there exists a class I SA which self-assembles x from a configuration that covers x.*

Theorem 3 *For any basic assembly sequence x that is an instance of an assembly template $t \in \text{SEQ}(\{p\}) \setminus L(G_I)$, there exists a class II SA which self-assembles x from a configuration that covers x. Further, there exist no class I SA which self-assembles x from a configuration that covers x.*

Note that the assembly sequence $(((AB)C)D)$, $((AB)(CD))$, and $(A(B(CD)))$ listed in Fig. 3.8 are instances of an assembly template $t \in L(G_I)$. Theorem 2 assures that these assembly sequences are indeed encodable with only minus devices. On the other hand, the assembly sequence $(A((BC)D))$ in Fig. 3.10 encodable with both plus and minus devices is an instance of an assembly template $t \in \text{SEQ}(\{p\}) \setminus L(G_I)$. Theorem 3 assures that this assembly sequence is indeed encodable with both plus and minus devices, but un-encodable with minus devices only.

4.4. Minimum conformation SA

In this section, we provide the minimum number of conformations necessary to encode a given assembly sequence based on the classes of basic assembly sequences introduced earlier. Since the number of conformations may vary for each component, the following definition is necessary.

Definition 11 Let M be an SA and Q is the conformation set of M. M is an SA with n conformations if $n = \max\limits_{a^\alpha \in Q} |\alpha|$.

Definition 12 The class II assembly grammar G_{II} is defined by the following production rules:

$$
\begin{aligned}
S &\rightarrow (L_0 R_0) \\
L_0 &\rightarrow (L_0 R_1) \mid R_1 \\
R_0 &\rightarrow (L_1 R_0) \mid L_1 \\
L_1 &\rightarrow (L_1 p) \mid p \\
R_1 &\rightarrow (p R_1) \mid p
\end{aligned}
$$

Note that $L(G_I) \subset L(G_{II}) \subset \text{SEQ}(\{p\})$. The parse tree of the assembly templates in $L(G_{II})$ is shown in the right of Fig. 4.1. The right parse tree in Fig. 4.1 can be obtained from the left parse tree in Fig. 4.1, by replacing leaves at the right branches of the left subtree by a linear assembly tree, and vice versa. Let x be an assembly sequence and t is an assembly template of x. If $t \in L(G_{II}) \setminus L(G_I)$, the direction of self-assembly alters *exactly once*, and if $t \in \text{SEQ}(\{p\}) \setminus L(G_{II})$, the direction of self-assembly alters *more than once* during the self-assembly of x.

Example 5 The assembly template t_2 in Example 4 cannot be generated by G_I but can be generated by G_{II}, for example, through the derivation $S \Rightarrow (L_0 R_0) \Rightarrow ((L_0 R_1) R_0) \Rightarrow ((p R_1) R_0) \Rightarrow ((p(p R_1)) R_0) \Rightarrow ((p(pp)) R_0) \Rightarrow ((p(pp))p)$ and hence $t_2 \in L(G_{II}) \setminus L(G_I)$. An assembly template $t_3 = (p((p(pp))p))$ cannot be generated by G_{II} and hence $t_3 \in \text{SEQ}(\{p\}) \setminus L(G_{II})$.

The minimum number of conformations of SA that are necessary to self-assemble a given basic assembly sequence x depends on whether x is an instance of an assembly template in $L(G_I)$, $L(G_{II}) \setminus L(G_I)$, or $\text{SEQ}(\{p\}) \setminus L(G_{II})$. Since any attaching rules produced by MRS requires at most two conformations for each component, the minimum number is two if x is an instance of an assembly template in $L(G_I)$.

Theorem 4 For any basic assembly sequence x that is an instance of an assembly template $t \in L(G_I)$, there exists class I SA M with two conformations which self-assembles x from a configuration Γ that covers x. For $\mathrm{L}(x) \geq 3$, M is an SA with the minimum number of conformations which self-assembles x from Γ.

The "conformation incrementor" INC used in MRS simply appends the prime symbol (') to its argument string each time it is called. The number of conformations of a component, therefore, could be very large depending on how many times INC is called for the component before MRS returns. Alternatively, we can use a "modulo n" conformation incrementor INC_n such that for $\alpha \in \{'\}^*, |\alpha| \le n$

$$INC_n(\alpha) = \begin{cases} \alpha' & \text{if } |\alpha| < n \\ \Lambda & \text{if } |\alpha| = n \end{cases}$$

For example, $INC_2(\Lambda) ='$ and $INC_2(') = \Lambda$. Using this notation, we can write INC as INC_∞. Running MRS with INC_n, instead of INC_∞ produces assembly rules with *at most* n conformations for a component. Such rules, however, are no longer guaranteed to self-assemble the components in a given assembly sequence. In particular, there could be more than one conflicting propagation rule that specifies different conformational changes for the same two adjacent components. In order to show that MRS run with INC_n instead of INC_∞ is correct, therefore, it suffices to show that no such conflicts among propagation rules are possible. If the rule set R contains at most one propagation rule for each two adjacent components, no conflicts are possible. Therefore, the above statement is true for the smallest possible n, i. e. $n = 2$. This is the case when the assembly sequence x is an instance of an assembly template $t \in L(G_{II}) \setminus L(G_I)$, when the direction of self-assembly alters *exactly once*.

Theorem 5 *For any basic assembly sequence x that is an instance of an assembly template $t \in L(G_{II}) \setminus L(G_I)$, there exists a class II SA M with two conformations which self-assembles x from a configuration Γ that covers x. And M is an SA with the minimum number of conformations which self-assembles x from Γ.*

Example 6 *Consider $\Sigma = \{a, b, c, d\}$ and $x = ((a(bc))d)$. The assembly sequence x is an instance of $t_2 = ((p(pp))p)$ in Example 4. From Example 5, $t_2 \in L(G_{II}) \setminus L(G_I)$. A call of MRS$(x, none, \emptyset)$ run with INC_2 returns with $(abc'd, R)$ where R contains the following rules: $b + c \to b'c$, $a + b' \to ab$, $bc \to bc'$ and $c' + d \to c'd$. It is clear that $M = (\Sigma, R)$ is an SA with two conformations which self-assembles x from the configurations that cover x, e.g. $\langle a, b, c, d \rangle$ and $\langle a, a, b, b, c, c, d, d \rangle$.*

Note R in Example 6 corresponds to the part design in Fig. 3.10, which has three conformations for part B. Three conformations, instead of two as stated in Theorem 5, were needed since a plus device and a minus device cannot be implemented in one "digit."

If R contains more than one propagation rules of the same two adjacent components, n must be large enough to cause no conflicts among the propagation rules. This corresponds to the case where x is an instance of $t \in SEQ(\{p\}) \setminus L(G_{II})$, when the direction of self-assembly alters *more than once*. Here, we claim that only *three* conformations are necessary to encode an arbitrary x. This might sound counter-intuitive since we are claiming that only three conformations can encode basic assembly sequences with arbitrary (possibly very large) sizes. The proof of this claim is done by showing that MRS run with INC_n causes no conflicts among propagation rules of the same adjacent components in the case of $n = 3$. It is based on the observation that there are only two kinds of propagation rules; the rules that propagate conformational changes to the left, and the rules that propagate conformational changes to the right, and that for any given two adjacent components, these two kinds of propagation rules always fires in alternate order. Complete proof of the claim, as stated below, is found in [5, 7].

Theorem 6 *For any basic assembly sequence x that is an instance of an assembly template $t \in \text{SEQ}(\{p\}) \setminus L(G_{II})$, there exists a class II SA M with three conformations which self-assembles x from a configuration Γ that covers x. And M is an SA with the minimum number of conformations which self-assembles x from Γ.*

Example 7 *Consider $\Sigma = \{a, b, c, d, e\}$ and $x = (a((b(cd))e))$. The assembly sequence x is an instance of $t_3 = (p((p(pp))p))$ in Example 5, and $t_3 \in \text{SEQ}(\{p\}) \setminus L(G_{II})$. A call of $\text{MRS}(x, none, \emptyset)$ run with INC_3 returns with $(ab'c''d''e, R)$ where R contains the following rules: $c+d \to c'd$, $b+c' \to bc''$, $c''d \to c''d'$ $d'+e \to d''e$, $c''d'' \to cd''$, $bc \to b'c$, and $a + b' \to ab'$. It is clear that $M = (\Sigma, R)$ is an SA with three conformations which self-assembles x from the configurations that cover x, e.g. $\langle a, b, c, d, e \rangle$ and $\langle a, b, b, c, d, d, e, e \rangle$.*

5. Summary

This paper discussed a case study of one-dimensional self-assembly of a type of mechanical conformational switches, *minus devices*, where assembly occurs via the sequential mating of a random pair of parts selected from a part bin, referred to as *sequential random bin-picking*. Parametric design optimization of the minus devices maximizing the yield of a desired assembly, and rate equation analyses of the resulting designs, revealed that the minus devices facilitated the *robust* yield of a desired assembly against the variation in the initial fraction of the part types, by specifying a fixed assembly sequence during the self-assembling process. It was also found that while the minus devices could "encode" some assembly sequences, encoding other assembly sequences required the use of another type of conformational switches, *plus devices*.

To investigate the "encoding power" of these conformational switches, a formal model of self-assembling systems, one-dimensional *self-assembling automaton*, was introduced where assembly instructions were written as local rules that specified conformational changes of components realized by the minus and plus devices. Classes of self-assembling automata were defined based on three classes of assembly sequences described by assembly grammars. It was proven that the local rules corresponding to the minus and plus devices, and *three* conformations per each component, could encode any assembly sequences of a one-dimensional assembly of distinct components with arbitrary length.

432

Theorem 6 provides the theoretical lower limit to the number of conformations needed to encode arbitrary assembly sequences. A particular physical implementation of the conformational switches, however, may not be able to encode arbitrary assembly sequences with three conformations per component. For instance, R in Example 7 cannot be realized by using the plus and minus devices due to the physical limitation of implementing two types of switches in one digit. Design of mechanical conformational switches which overcome this problem is left for future development.

Bibliography

[1] GOLDBERG, D., *Genetic algorithms in search, optimization and machine learning*, Addison-Wesley (1989).

[2] HOSOKAWA, K., et. al., "Dynamics of self-assembling systems: analogy with chemical kinetics," *Artificial Life* **1(4)**, 413–427.

[3] MARTIN, J., *Introduction to language and the theory of computation*, McGraw-Hill (1991).

[4] PENROSE, L., "Self-reproducing machines," *Scientific American* **200** (1959), 105–114.

[5] SAITOU, K., "Conformational switching in self-assembling mechanical systems: theory and application," Ph. D. thesis, Massachusetts Institute of Technology (1996).

[6] SAITOU, K., et. al., "Subassembly generation via mechanical conformational switches," *Artificial Life* **2(4)** (1995), 377–416.

[7] SAITOU, K., et. al., "On classes of self-assembling automata," *Complex Systems* **10(6)** (1996), 391–416.

[8] STEINFELD, J., et. al., *Chemical kinetics and dynamics*, Prentice Hall (1989).

[9] WATSON, J., et. al., "Molecular biology of the gene," Benjamin/Cummings (1987).

Chapter 40

Aggregation and the emergence of social behavior in rat pups modeled by simple rules of individual behavior

Jeff Schank[1]
Indiana University, Center for the Integrative Study of Animal
Behavior
Jeff Alberts
Indiana University, Department of Psychology

From infancy to adult life, huddling is a major component in the behavioral repertoire of Norway rats. During infancy, rats are severely limited in their sensorimotor capabilities, yet they are capable of aggregating and displaying forms of group regulatory behavior. We show that huddling as aggregative behavior can emerge as a self-organizing process from autonomous individuals following simple sensorimotor rules. In our model, two sets of sensorimotor parameters characterize the topotaxic responses and the dynamics of contact among infant rats. The first set of parameters are conditional probabilities of activity and inactivity given prior activity or inactivity and the second set are attractions for objects in the infant rat's environment. Using computer simulation, we have found that the behavior of the model and of actual rat pups compare very favorably. In particular, synchronized bouts of inactivity in aggregations

---
[1]This work was supported by NIH 1 F32 HD088188-01 to Schank and NIH MH 28355 to Alberts. We also want to thank Catherine Ragsdale for her summer intern work on thermotaxis and Sherry Lifer help with experiments and reading this paper.

of 10-day old rats emerge by spreading deactivation. We discuss the model and the underlying approach, developmental emergence of social interactions, and how group thermotaxis and individual geotaxis may facilitate aggregation and cohesion.

1. Introduction

Norway rats (*Rattus norvegicus*) are social mammals in which sociality is primarily manifested by physical contact [4]. Indeed, physical contact is so prominent a feature of rat behavior that *Rattus norvegicus* has been referred to as a *contact species* [3]. From infancy throughout adult life, no behavior better illustrates the importance of physical contact than does *huddling*.

Huddles of infant rats (pups) are more than mere piles of bodies. Infant rats—despite being blind and deaf at birth—are capable of aggregating in the nest to form huddles that perform group regulatory functions such as thermoregulation and energy conservation [1] and [2]. These functions of huddling are generally understood as group adaptations, which are beyond the capabilities of the individual infant rat. However, there is nothing mysterious about these group functions. They likely emerge, under appropriate conditions, from self-organizing processes in which individuals follow simple rules of sensorimotor behavior [7].

In the nest, pups change positions by rooting and crawling over, under, and around one another, usually forming a dynamic pile. Even without maternal interactions, huddling *in situ* is too complex a starting point for developing individual-based models for investigating aggregative behavior. A simple experimental paradigm—one that is easily mapped into individual-based models—is required. In the next section (§2), our basic experimental and modeling approaches are outlined.

In previous work [7] (also see §3), we have found that at seven days of age, infant rats exhibit autonomy in their sensorimotor activity. Here it is shown that by the age of 10 days—still blind and deaf—their sensorimotor behavior becomes probablistically coupled to the number and activity state of other pups surrounding them. This coupling gives rise to bouts of synchronized inactivity in aggregates of pups via spreading deactivation, indicating a developmental transition from autonomous behavior to social behavior.

The paper concludes by discussing an issue that has emerged from the study of aggregative behavior in infant rats (see §5). On a flat and level surface, infant rats are unimpressive in their ability to aggregate and cohere together. We have recently found other more subtle global cues that the group or the individual can take advantage of to better aggregate. The first is the group's ability to detect and move up temperature gradients that may be difficult or impossible for the individual to detect. This is group thermotaxis. The second is the previously unknown ability of infant rats to detect very small slopes (1 degree and perhaps less), and thereby exhibit positive geotaxis, a mechanism that may help them in locating the nest.

2. Basic strategy

Our basic research goal is a better understanding of the dynamics of huddling by discovering the rules of individual behavior from which aggregative behavior emerges. Our basic strategy is to use individual-based models as theoretical tools for explaining how local interactions among individual pups give rise to the dynamic patterns of aggregation we observe. Specifically, we design experiments and models in parallel. We hypothesize and parameterize rules of individual behavior and then, by computer simulation, attempt to find sets of parameter values that explain the data we observe. If the match between experiments and models is reasonably good, we have achieved a deeper understanding of the underlying rules of individual behavior from which group behavior emerges. In §2.1 and §2.2, the basic experimental and modeling approaches are outlined. Experiments and models described later are variations on these basic paradigms.

Figure 2.1: Arena and 8 pups in the starting rack.

2.1. Experimental methods

Our experimental approach is based on the observation that infant rats (less than 10 days old) are extremely limited in their sensorimotor capabilities. On a flat and level surface, they can easily crawl around in a "swimming" like motion, but they rarely crawl on top of one another. This creates an ideal situation in which to observe their patterns of aggregation and sensorimotor activity over time.

Animals

Several considerations have gone into the selection of the age range of animals. To insure limited and local sensory capacities, pups must be 12-days of age or younger, since after this time their eyes begin to open. Pups also must be able move around on a flat surface, but unable to readily climb on top of each other. We have observed that pups 5- to 10-days of age satisfy these criteria. Thus, all animals came from litters of 8 pups (Sprague-Dawley), 7 or 10-days of age, bred and born at Indiana University. At three days of age, litters were culled to 4 females and 4 males in order to avoid possible sex-contact biases [8]. The methods used were observational and entirely noninvasive.

Arena

We constructed an arena for controlling a variety of complexity inducing variables potentially affecting aggregation. An aluminum arena ($12 \times 8 \times 2$ in) was sealed and immersed in a 3 gallon water reservoir maintained at a constant temperature by a temperature controlled circulator (see Fig. 2.1). The arena surface forms a heat sink, which a motionless pup cannot warm. The leveled surface prevents geotaxic responses from biasing positions (see §5.2). By embedding the arena in a second chamber equipped with small radiators, precise and uniform ambient temperatures are maintained (see §5.1).

Data collection

Pups are either active or inactive at any given time. *Sensorimotor* activity is not merely movement, but coordinated movement required to initiate contact. There are several types of behavior such as grooming, yawns, and even violent twitches that occur when pups sleep, that could be viewed as "active," but since they are not essential elements in the sensorimotor coordination required for initiating contact, we do not classify these behaviors as active. Operationally, we define sensorimotor activity in an infant rat as either (1) moving on the surface of the arena by moving its legs or attempting to move on the surface by moving its legs in a "swimming" like motion or (2) moving its head from side-to-side in a scanning motion [7].

When placed on the level surface of the arena, infant rats form various patterns of physical contact that we call *aggregons* [7]. Since pups need not be in contact with all other pups, several aggregons may form at any given time. A *pattern of aggregons* is a list of the sizes of each aggregon formed in the arena. For example, if none of the pups are in physical contact, then there is a pattern of eight aggregons (1,1,1,1,1,1,1,1), where we define the smallest aggregon as consisting of 1 individual. If all eight are in contact, then there is a single aggregon forming the pattern (8). For groups of 8 infant rats, there are 22 possible aggregon patterns as depicted in Fig. 2.2.[2]

A pup detects objects in its environment with the side-to-side scanning motions of its snout [7]. Thus, contact between snout and an object should be an indicator of an individual's attraction to a type of object. The patterns and dynamics of aggregons over time should be one indicator of the attraction of pups to other pups, and contact with walls should be another. The attractions and movements (positive or negative) of infant rats toward objects in their environment are *topotaxes*.

---

[2]There is no unique way of ordering these aggregon patterns. However, the index we have constructed is biologically plausible and computationally useful [6].

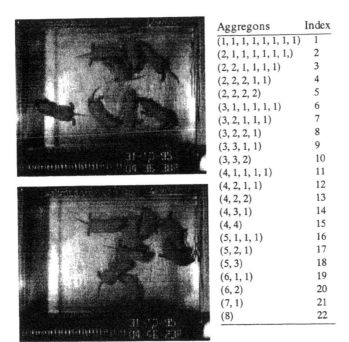

Aggregons	Index
(1, 1, 1, 1, 1, 1, 1, 1)	1
(2, 1, 1, 1, 1, 1, 1,)	2
(2, 2, 1, 1, 1, 1)	3
(2, 2, 2, 1, 1)	4
(2, 2, 2, 2)	5
(3, 1, 1, 1, 1, 1)	6
(3, 2, 1, 1, 1)	7
(3, 2, 2, 1)	8
(3, 3, 1, 1)	9
(3, 3, 2)	10
(4, 1, 1, 1, 1)	11
(4, 2, 1, 1)	12
(4, 2, 2)	13
(4, 3, 1)	14
(4, 4)	15
(5, 1, 1, 1)	16
(5, 2, 1)	17
(5, 3)	18
(6, 1, 1)	19
(6, 2)	20
(7, 1)	21
(8)	22

Figure 2.2: An index of 22 possible aggregon patterns for groups of eight (right) and two examples (left). Top left is a (4, 3, 1) aggregon pattern with index 14, and bottom left is an (8) aggregon pattern with index 22.

Data on these behavioral measures is collected by video recording the pups from overhead (see Fig. 2.1) and by using a version of NIH Image we have modified to record and save data in spread-sheet form. We have written macros for scoring and recording contacts with other pups and walls, and for assessing activity [8].

Procedure

In an experimental session, pups are uniformly spaced on the arena in a rack (as depicted in Fig. 2.1), then simultaneously released to move around the arena. The ambient air temperature, surface and walls of the arena are maintained at 34°C throughout the session. Between each session, the arena is cleaned with a small amount of ethyl alcohol to minimize effects of odors left by previous groups of pups.

2.2. Basic model

In this section, we outline the basic model, simulation procedure, and method of analysis. Model pups occupy a rectangular array of cells (Fig. 2.3). If active, a pup can move at each time step to any unoccupied adjacent cell. The motion

438

of pups entirely depends on (i) their activity state (active or inactive) at each
time step, (ii) their attraction to adjacent cells, and (iii) whether adjacent cells
are empty or occupied. The number of cells is determined by two criteria: the
number of cells should (1) allow the initial spacing of pups to be topologically
similar to the spacing of pups in the experimental arena (see Fig. 2.1), and (2)
be comparable to the area occupied by real pups. A good approximation is 40
cells for 7-day and 30 cells for 10-day old pups.

The basic model assumes discrete time and space whereas this is not true
for real pups (e.g., model pups are confined to cells in a grid, real pups are not
confined to cells in a grid). How then can we compare data from real and model
pups? We do this by first connecting activity in real and model pups in terms
of conditional probabilities, and second, operationally defining contact in model
pups as being in adjacent cells.

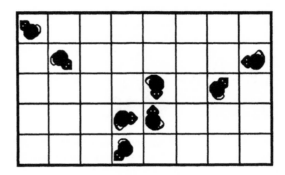

Figure 2.3: Model system with 8 pups. A cell can be occupied by only one pup at a
time and a pup cannot occupy more than one cell.

Activity

The probabilities of activity states (activity, A, or inactivity, I) of pups are modeled with conditional probabilities. We assume that the probability of activity or
inactivity of a pup depends only on its prior activity or inactivity. For 7-day old
pups, the experimental results presented in §3 offer experimental justification for
this assumption, since the activity of pups conformed to a binomial probability
model.

If a pup is active or inactive at time $t - \Delta t$, there are two conditional probabilities that determine whether or not a model pup is active at time t, $p(A_t \mid A_{t-\Delta t})$
and $p(A_t \mid I_{t-\Delta t})$. Similarly, there are two conditional probabilities that determine whether or not the pup is inactive at time t, $p(I_t \mid I_{t-\Delta t})$ and $p(I_t \mid A_{t-\Delta t})$.
Since, $p(A_t \mid A_{t-\Delta t}) + p(I_t \mid A_{t-\Delta t}) = 1$, $p(I_t \mid I_{t-\Delta t}) + p(A_t \mid I_{t-\Delta t}) = 1$, the
conditional probabilities of activity and inactivity are completely determined by:

- $p(A_t \mid I_{t-\Delta t})$ = the probability of activity at time t given inactivity at
  time $t - \Delta t$, and

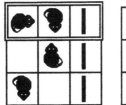 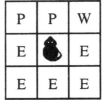

Figure 2.4: Left, focal pup is in the center and its view is outlined. Right, focal pup's attraction for pups (P), walls (W), empty cells (E).

- $p(I_t \mid A_{t-\Delta t})$ = the probability of inactivity at time t given activity at time $t - \Delta t$.

These conditional probabilities make no assumption about the size of the time step Δt. Thus, by considering the limiting case where $\Delta t \to 0$, the rate of change in the frequency of active pups is defined by

$$\frac{df(A)}{dt} = p(A \mid I)(1 - f(A)) - p(I \mid A)f(A) \tag{2.1}$$

By setting the right side of (2.1) to 0, and reintroducing Δt we get the "equilibrium" frequency of activity at time t,

$$\widehat{f}_{A_t} = \frac{p(A_t \mid I_{t-\Delta t})}{p(A_t \mid I_{t-\Delta t}) + p(I_t \mid A_{t-\Delta t})}$$

and by solving (2.1) and reintroducing Δt, we have an equation for the frequency of activity with discrete time changes:

$$f(A_t) = \widehat{f}_{A_t} + (f(A_0) - \widehat{f}_{A_t})e^{-t(p(A_t \mid I_{t-\Delta t}) + p(I_t \mid A_{t-\Delta t}))} \tag{2.2}$$

Attraction

The movement of a model pup is determined by its activity state and its attraction to neighboring cells. A model pup's *view*—based on the side-to-side scanning of the head and snout of actual pups—consists of the three cells in front of it (see Fig. 2.4).

The rules for assigning attraction to cells are

- Cells not in a pup's view are empty: E.

- Cells in a pup's view and occupied by a wall: W.

- Cells in a pup's view occupied by another pup: P.

- Empty cells in a pup's view that are next to another cell in the pup's view which contains another pup: P.

440

- Otherwise: E.

The probability that a pup will attempt to move to any of the eight adjacent cells (given it is active) is calculated for each cell by summing up the attractions in all eight cells and then dividing each cell by the sum.

Simulation procedure

Each simulation of an experimental session consists of a number of time steps matching the data collection procedure. For example, for a 10 min session in which data is recorded every 5 sec, simulations trials would consist of 120 time steps. At the start of a simulation, pups are arranged in a topologically similar fashion to that seen in Fig. 2.1. Each model pup is assumed to be active at the start of a simulation experiment. At each time step, various types of data are recorded: aggregon patterns, the number of pups oriented towards and in contact with a wall (i.e. wall topotaxis), and the number and distribution of active pups.

Pups are assumed to move simultaneously at each time step, but two or more pups may attempt to move into the same empty cell. The program handles this problem by first checking to see which cells each pup has attempted to move into; if more than one pup has attempted to move into a cell, then a pseudo-random number is used to determine which pup will occupy the cell. If a pup attempts to move into a cell that was occupied but later is unoccupied during the same time step, the program checks to see whether this has happened and then moves the pup to the vacated cell.

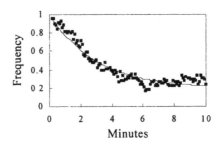

Figure 2.5: Non-linear curve fit of (2.2) to the frequency of pup activity for all 9 litters over the 10 minute session ($p(A_t \mid I_{t-\Delta t}) = 0.005$ and $p(I_t \mid A_{t-\Delta t}) = 0.022$).

Fitness functions

For each type of data collected, a fitness function is constructed between the empirical and models data. For example, aggregon patterns can be represented by frequency distributions, and by taking the absolute difference between empirical and model distributions and dividing by 2, the degree of match between the two distributions falls in the range [0,1]. All such functions are constructed on this principle [7].

3. Aggregation in autonomous individuals

Our first application of this approach was to 9 litters of 7-day old pups [7]. We found that the frequency of activity decreased for the first 5 minutes of trial and then approached an apparent plateau, and (2.2) was fit to this data using non-linear curve fitting (Fig. 2.5). We also found that the activity state of a pup is independent of the activity state of its littermates, since the distribution of active pups over the final 5 minutes did not significantly differ from the corresponding binomial distribution calculated for the expected frequency of pups activity (assuming $p = 0.27$—the mean level of activity over the final 5-minutes, $r =$ the number of active pups, and $N = 8$ the total number of pups, see Fig. 3.1).

Since the conditional probabilities were determined by non-linear curve fitting and we assumed the attraction for other pups is 1.0, this left only two parameters to systematically vary: attraction to wall and empty cells. By systematically varying these parameters by increments of 0.01 and running 10,000 simulations for each set of parameter values, we found that the best fit between simulation and data occurred when the attraction for walls (0.38) was greater than the attraction for empty cells (0.27).

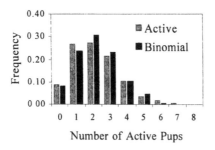

Figure 3.1: Frequency of the number of pups active in a litter of 8 for the last 5 min of a 10-min trial compared with the expected binomial distribution of activity.

The correlation between the model and data aggregon distributions is $r = 0.83$ (Fig. 3.2), but the most striking feature of Fig. 3.2 is how the two distributions follow each other as indicated by the lines connecting the points representing the frequency of each aggregon pattern. This suggests that the movement of pups has a significant random component resulting in the apparent combinatorial character of aggregon patterns.

To separate the contribution of random movement from the contribution of attraction to cells, we compared data and best-fit simulation distributions with a "null" distribution in which the attraction for pups, walls, and empty cells was set to 1 in the model. The correlation between the data and the null model is

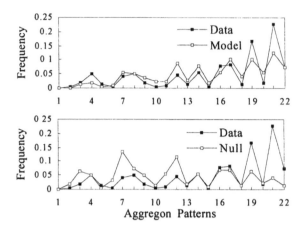

Figure 3.2: Comparisons of aggregon pattern frequencies in the data with the best fit model (top) and the null model (bottom). The correlations between these models and the data are $r = 0.83$ and $r = 0.28$ respectively.

much lower ($r = 0.28$) and the change in the frequency of aggregon patterns for both distributions is typically in the same direction, indicating that the multi-modal character of these distributions is likely due to the random movements of pups (Fig. 3.2). The main effect of the best fit attraction values is to slightly shift the distribution to higher degrees of aggregation (Fig. 3.2).

The fit of the simulated wall contacts to the data is impressive, especially considering that wall contact was not used to select for the best fit. The null frequency distribution over time of wall contacts is similar in form to the best fit preferences except that it is shifted up about 9 percent. Taken together, these results indicate a good match between simulation and data, indicating that our basic approach may be very useful in detecting and explaining other group phenomena.

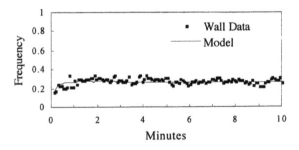

Figure 3.3: Comparison between wall topotaxis frequency distributions of the data and the best fit model.

4. The emergence of synchronized social behavior

At 10 days of age, rat pups are still blind and deaf, but the apparent autonomy in their sensorimotor activity states begins to disappear. We have observed synchronized bouts of inactivity associated with the degree of aggregation (see Fig. 4.2 below). Using 19 litters of 10-day old rat pups, we used the same experimental paradigm, but this time we modeled the conditional probabilities of activity as conditional on the number and activity state of the pups a pup is in contact with. In addition, we assumed that these conditional probabilities may change with time and that for short periods of time, they can be modeled as exponentially decaying. Specifically, we assumed that the probability of activity given prior activity or inactivity should decrease over time

$$p(I_t \mid A_{t-\Delta t}, N_A, N_I, t) = 1 - e^{-r_A t} p(A_t \mid A_{t-\Delta t}, N_A, N_I, 0)$$

$$p(A_t \mid I_{t-\Delta t}, N_A, N_I, t) = e^{-r_I t} p(A_t \mid I_{t-\Delta t}, N_A, N_I, 0)$$

where N_A is the number of active pups, N_I is the number of inactive pups, $-r_A$ is rate of decrease in $p(A_t \mid A_{t-\Delta t}, N_A, N_I, t)$, $-r_I$ is the rate of decrease in $p(A_t \mid I_{t-\Delta t}, N_A, N_I, t)$, and $p(A_t \mid A_{t-\Delta t}, N_A, N_I, 0)$ and $p(A_t \mid I_{t-\Delta t}, N_A, N_I, 0)$ are initial conditional probabilities.

In the new model, there is an explosion in the number of parameters. For (2.2) applied to each of these contact cases, there are 4 parameters: $-r_A$, $-r_I$, $p(A_t \mid A_{t-\Delta t}, N_A, N_I, 0)$, and $p(A_t \mid I_{t-\Delta t}, N_A, N_I, 0)$. For 8 pups, there are 36 possible contact combinations of a given pup with its 7 other active or inactive littermates. Thus, there are 144 parameters governing the conditional probabilities of their activity states.

This increase in the number of parameters, however, does not imply that the model should be easier to fit to the data than the previous model. This is because the contact-dependent conditional probabilities in the new model are required to fit the corresponding contact-dependent frequencies of activity and inactivity in the data.

4.1. Evolutionary optimization

To find values for these parameters that fit and explain the data, we used a form of evolutionary optimization based on the idea of simulated annealing. There are many possible fits of the model to the data, which is an ideal situation for using a simulated annealing approach. However, the individual based-modeling approach we are using is embedded in *Monte Carlo* simulation framework. This means that for any set of parameter values there will be a distribution of goodness of fit between the model parameters and the empirical data. The spread of this distribution is determined by the number of simulated sessions. As the number of simulated sessions for a set of parameters is increased, the spread of the distribution decreases, but the time required for simulation greatly increases. The problem these distributions create is that inevitably highly improbable good fits occur for a given set of parameters. These statistically improbable minima can trap the evolutionary optimization process, preventing it from achieving a good fit. The simulated annealing paradigm provides a mechanism for jumping out of these statistical traps.

Simulated annealing requires calculating the probability, $p(E)$, of jumping out of a minimum energy state (which is based on the Boltzmann probability distribution, $e^{(-E/kT)}$). In order to apply it to non-thermodynamic systems, one has to specify (i) how the model system can be changed (i.e. the parameters of the model system), (ii) how random changes are introduced, (iii) specify an objective function E (the analog of energy), which corresponds to our notion of fitness functions, and (iv) incorporate a temperature analog T.

In our evolutionary optimization system, all parameters except pup attraction (which is held constant at 1.0) are allowed to vary. Mutations are introduced into the parameters at rate r with a Poisson distribution (truncating the 0 case mutations) under constraints for the range of parameter values and step size of mutations. The core objective fitness function is composed of four fitness functions between model and data (each with range [1,0]): the absolute difference in (1) aggregon distributions, (2) aggregon pattern change, (3) frequency of wall topotaxis for each time step, and (4) frequency distributions of active pups. In addition, we also added in the absolute differences between the 9 most frequent conditional probabilities of activity and inactivity.[3] The range that E can take is [0,31], and the optimization process takes E below 1.[4]

4.2. Results

The most striking change in the behavior of 10-day versus 7-day old rats in the arena (§2.1) is the occurrence of synchronized bouts of inactivity in aggregons that form late in a session. In Fig. 4.2 we see an example (middle figure) of the activity of a group of 10-day old rats over fifteen minutes. Between 13 and 14 minutes all pups form a single aggregon and become simultaneously motionless over several observation intervals. Other downward jags in Fig. 4.2 typically indicate that smaller aggregons have become motionless. Our goal in this study was to explain the mechanism of these synchronized bouts.

Fig. 4.1 compares the frequency with which a given pup is surround by pups that are active and inactive, illustrating that the data and model are in close agreement.

Ten-day old rats exhibit a higher degree of aggregation than do 7-day old rats (compare Fig. 4.3 with Fig. 3.2). Model pups are attracted to walls over

---

[3]These fitness functions are only defined for 9 of the conditional probabilities, because as can be seen in Fig. 4.1, most of the possible cases of contact rarely occur and thus are only "higher order" effects in this model.

[4]The parameter T is problematic for optimizing a *Monte Carlo* process. One cannot simply start with a high value and decrement it according to a simple schedule since this inevitably leads to the optimization process becoming stuck at some local optima. Instead, we have found that it is best to explore various T values relative to the number, N, of simulated sessions run for each set of parameter values. The art is to find values for T that allow the process to jump out of statistical minima, while still strong enough to promote the optimization process.

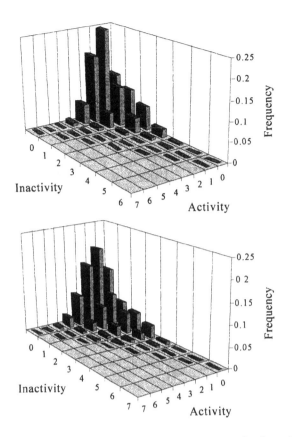

Figure 4.1: Frequency distribution of contacts among pups in the activity and inactivity states (top: data, bottom: best fit simulation).

empty cells by a ratio of 4:1, and they are attracted to other model pups over walls by a ratio of at least 12:1. But, the higher degree of aggregation in 10-day olds is largely a consequence of two other factors revealed by examining the different null distributions in Fig. 4.3 and Fig. 3.2. First, the probability models of activity are different for 7- and 10-day old rats. For 7-day olds, the conditional probabilities of activity are independent of contact with other pups, whereas for 10-day olds these probabilities are both contact and time dependent. Second, 10-day old rats are bigger than 7-day old rats, and so the actual space taken up by 10-day olds in the arena is greater than that taken up by 7-day olds, and so random contact among pups should increase with the increasing size of pups. Thus, the higher degree of aggregation of 10-day old rats largely reflects developmental changes in size and the nature of activity over a 3-day time span.[5]

---

[5]Not shown in Fig. 4.3 is a null model for 10-day olds assuming the same contact-independent activity as in 7-day old rats. The correlation with the data was lower, $r = 0.58$, than the null model shown in Fig. 4.3, but higher than for 7-day olds, Fig. 3.2—as expected if

446

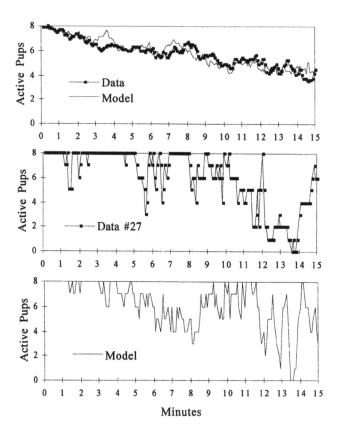

Figure 4.2: Mean activity of pups over time (top). Two examples of activity: data (middle) and simulation (bottom).

Finally, even though attraction among pups and walls plays a smaller role in the distribution of the patterns formed in 10- vs. 7-day olds, it still plays a large role in the frequency with which these patterns change over time in 10-day olds (in the data aggregon patterns change 47.3 percent of the time while in the null model they change 76.5 percent of the time; the best fit simulation agreed with the data pattern change within 2 percent).

Wall topotaxis is exhibited to a higher degree in 10-day old than in 7-day old rats (Fig. 4.4). However, since the null model also agreed well with the empirical data, attraction to walls is not the best explanation for this increase (i.e. we are not observing the development of thigmotaxis—the tendency to orient and move along walls, which is displayed by adult rats). Instead, walls and especially corners, likely form boundaries that "trap" pups allowing them to aggregate—perhaps like the nest a mother builds for her pups. With time there is a tendency to settle into inactivity, manifested by spreading deactivation in

increased size of pups in a fixed arena increases the likelihood of contact.

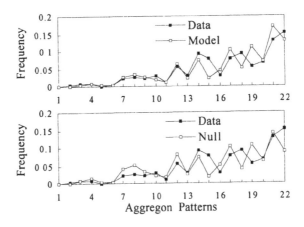

Figure 4.3: Comparisons of aggregon pattern frequencies in the data with the best fit model (top) and the null model (bottom). The correlations between these models and the data are $r = 0.91$ and $r = 0.76$ respectively. A null model assuming independent activities as in Fig. 3.2, exhibited a lower correlation $r = 0.58$.

aggregations of pups.

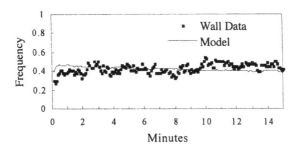

Figure 4.4: Comparisons of wall topotaxis frequencies in the data with the best fit model.

Fig. 4.5 illustrates four examples of the change in activity over time. These graphs illustrate that the activity of pups is conditional on the number and state of surrounding pups. In general, the more inactive pups a pup is surrounded by, the more likely it will become inactive. In addition, the activity of a pup declines with time and declines even more when it is in contact with inactive pups. The consequence is that as time goes by it becomes increasingly likely that aggregons of pups will experience (probabilistic) spreading deactivation, accounting for the observed bouts of synchronized inactivity in aggregations of pups.

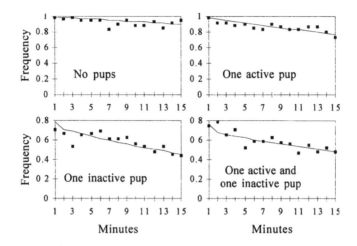

Figure 4.5: Four examples of the change in activity over time.

5. Mechanisms of aggregation

Our analysis of the conditional probabilities of sensorimotor activity in 7 and 10 day-old rats indicates a rapid developmental transition from autonomous behavior to probablistically coupled social behavior. The fact that these models can largely explain the dynamics of aggregation in 7 and 10 day-old rats demonstrates that conditional probabilities of activity and attraction to objects explain the dynamics of aggregation we observed. However, for both 7 and 10 day-old rats, their ability to aggregate and maintain group cohesion is not particularly impressive (see Fig. 3.2 and Fig. 4.3). In the next two sections we briefly discuss two mechanisms we have recently discovered that greatly enhance the aggregative ability of groups of pups.

5.1. Group thermotaxis

Thermotaxis is orientation and movement towards heat and is the behavioral basis for an infant rat's ability to thermoregulate in a huddle. We also believe that group thermotaxis may be a very powerful mechanism for promoting group aggregation and cohesion.

The small literature on individual thermotaxis in infant mammals has yielded mixed results. If infant rats are placed on steep thermogradients (e.g., ranging from 18°C to 45°C) for more than an hour, they clearly exhibit positive thermotaxis in normal room temperatures [5]. However, no one to our knowledge, has ever looked at the behavior of infant mammals on shallow thermoclines. We suspected that on shallow thermoclines there would be differences in the thermotactic abilities of individuals versus groups. If the gradient is shallow enough, then the individual may not be able to detect changes in the gradient as well as the group. We hypothesized, therefore, that if the group is a better detector of shallow temperature gradients (perhaps more characteristic of naturalistic situations), then groups should exhibit greater and more rapid thermotactic responses.

We developed a thermocline that is a variation on our basic experimental arena. The thermocline was a temperature gradient formed across a copper bar (60 × 10 × 0.25 in), enclosed in a Plexiglas chamber (48 × 18.25 × 24 in), which controlled ambient temperature (25°C). The copper bar was heated on both sides by resistor heating bars forming a two-sided thermocline. Each end was heated to 40°C (18 in from the center), with the center at 35°C. In these ambient temperature conditions, 35°C is slightly cooler and 40°C is slightly warmer than the pups prefer. An arena was formed on top of the bar using a rectangular frame (inside measurements: 36 × 8 × 2 in) made of Styrofoam (1 in thick).

Four litters of 8, and sixteen individual 7-day old rats were used. Pups were placed in a rack and released in the middle (temperature valley) of the thermocline (similar to that seen in Fig. 2.1). In sessions with individual pups, a pup was randomly assigned to one of the eight slots, thus all 16 were evenly distributed across slots.

Our results showed that groups exhibit greater and more rapid thermotaxis than did individuals. Individuals tend to stay in the middle with some slowly moving toward warmer temperatures; whereas groups moved up the thermocline rapidly, reaching a significantly higher temperature than individuals within 14 minutes ($F(1, 46) = 5.77, p = 0.02$).

This thermotactic ability of groups was observed to facilitate aggregation and cohesion, since in moving up the thermocline, individuals tended to aggregate more closely together than we have observed in the basic arena with no thermogradients.

5.2. Geotaxis

We have begun investigating the geotaxic abilities of infant rats. We noticed early on that if the surface of the arena was not level, there seemed to be a tendency for pups to aggregate on the down-hill slope of the arena. Shallow slopes may be an indicator of nest location. Thus, we hypothesized that for shallow slopes, infant rats should exhibit positive geotaxis.

To test this hypothesis, we created a 1 degree slope from left to right (and also right to left) in the basic arena (Fig. 2.1). Four groups were tested twice for 15 minutes with the slope in opposite directions. Sixteen individuals were also tested, 8 in each of the two slope conditions. All four groups showed positive geotaxis. Fifteen of the sixteen individuals showed rapid positive geotaxis, with one animal not exhibiting geotaxis (positive or negative) at all because it did not move from the place it was released.

Because individuals display positive geotaxis to shallow grades, we observed that groups tended to exhibit a high degree of aggregation and cohesion with all or nearly all individuals in contact by the end of a fifteen minute session.

Positive geotaxis to shallow grades is likely a global cue for aggregating and maintaining cohesion in the nest when the mother is absent.

6. Conclusions

Although individual infant rats are notoriously inept at basic regulatory functions, by aggregating into huddles, infant rats display "group regulatory behavior" [2]. Our modeling efforts have aimed at explaining aggregation with conditional probabilities of activity and inactivity and attraction to objects. Our models have revealed that local interactions among pups account for many features of aggregation. Social interactions emerge at 10 days of age via the probabilistic coupling of activity of a pup to the activity state of other pups it contacts. Bouts of synchronized inactivity in aggregations of pups are likely due to the probabilistic spreading of deactivation. Our models have also raised the interesting question of how high degrees of aggregation and cohesion can emerge from infant rat interactions. The boundaries of the arena likely play a role, but it is very likely that high degrees of aggregation and cohesion require additional global cues. Group thermotaxis may be one and positive geotaxis on shallow slopes may be another. Future research will aim to better understand these global cues within the theoretical and experimental framework we have described.

Bibliography

[1] ALBERTS, Jeff, "Huddling by rat pups: multisensory control of contact behavior", *J. of Comp. Physiol. Psych.* **92** (1978), 220–230.

[2] ALBERTS, Jeff and David GUBERNICK, "Reciprocity and resource exchange: a symbiotic model of parent-offspring relations", *Symbiosis in Parent-Offspring Interactions* (L. A. ROSENBLUM and H. MOLTZ eds.), Plenum Press (1983), 7–44.

[3] BARNETT, S., *A Study in Behaviour*, Methuen (1963).

[4] CALHOUN, John, *The Ecology and Sociology of the Norway Rat* , U.S. Department of Health, Education, and Welfare (1963).

[5] KLEITMAN, N. and E. SATINOFF ,"Thermoregulatory behavior in rat pups from birth to weaning", *Physiology Behavior* **52** (1982), 537–541.

[6] SCHANK, Jeff,"Problems with dimensionless measurement models of synchrony in biological systems", *Am. J. Primatol.* **41** (1997), 65–85.

[7] SCHANK, Jeff and Jeff ALBERTS,"Self-organized huddles of rat pups modeled by simple rules of individual behavior", *J. of Theoretical Biology* (in press).

[8] SCHANK, Jeff and Jeff ALBERTS, "A connectionist approach to social interactions", (submitted).

Chapter 41

The role of information in simulated evolution

Bernhard Sendhoff
Clemens Pötter
Werner von Seelen
Institut für Neuroinformatik, Ruhr-Universität Bochum, Germany

The definition of macroscopic observables with a microscopic foundation for complex systems is one of the approaches taken to understand such systems and predict their future behaviour. In the biological evolution, the notion of information and entropy has been proposed as a candidate for such a quantity. We will argue that its definition should be hierarchical, that is it will depend on the level of abstraction at which the system is observed. We will propose several definitions of information for different levels of evolution and support our claim with simulations from evolutionary algorithms and structure optimization, where we will concentrate on neural systems.

1. Introduction

The evolutionary process is the designer of the most complex systems – living individuals. At the same time, the process itself can be regarded as complex in terms of the large number of interacting entities and the intrinsic mappings involved. Therefore, the examination of the evolution of structures, especially of neural structures, seems to be an ideal "playground" for researching complex systems. One of the challenges in complex systems research is to define a unifying level of description which characterizes the system macroscopically and at the same time, has a microscopic foundation: a description which can be applied to different levels of abstraction of the system and determines the direction of development. In thermodynamics such a description is given by the entropy of the system. In biological systems, this concept should be extended to include other constraints such as stability, adaptability and feedback, which are not solely statistically ensemble based. Therefore, we are looking for a definition of information which incorporates the above mentioned constraints, which can serve as a applicable level of description of the process and has a foundation in the physics of information and in thermodynamics. The physics of information started with Maxwell's demon, an arbitrary small hypothetical being with a memory which was intended to highlight the statistical character of the microscopic foundation of thermodynamics and in particular the *second law*. The act of the demon was associated with information and as first suggested by Szilard [25] the accumulation of information was considered as a entropy increasing process. Although further work by Landauer [15] and Bennett [1] showed that it is the process of erasure and not of accumulation of information which increases the entropy (and which is necessary for the process to be cyclic) the connection between thermodynamics, entropy and information was established. Since then attempts have always been made to include the life sciences in such an analysis, [2, 3, 4, 18].

In the area of communication theory Shannon [22] proposed a measure of uncertainty which is mathematically equivalent to the entropy of statistical physics although its meaning is very different. The purpose of the Shannon description of information was to provide a measure for the capacity of a channel for communication. The mutual information, equation (4.7), is a measure for the nonlinear correlation among two random variables X and Y.

Both information definitions depend on the statistical nature of the system, they are only applicable to ensembles. The concept of algorithmic complexity introduced by Kolmogorov [14] defines a measure related to information in individual systems, which is based on the minimal description length of the system. The combination of algorithmic complexity or randomness and entropy of a system has been proposed by Zurek [27] to form the physical entropy (in a thermodynamic sense) of the system. The formulation of the entropy as a combination of a measure of what is known about the system and the remaining ignorance is very appealing to our way of envisioning information in the biological system (or simulations of it).

We believe that this combination of approaches is necessary in order to use information as the unifying level of description for evolution. Identifying the conceptual notion of information at an abstract level is important, see [10, 23], however the description of systems with the help of information will depend on the level of abstraction at which the systems are observed. This does not necessarily diminish the fundamentality of information, indeed the energy of a system and its scale depends on the level at which we examine the system.[1] Kampis [12] argued that the "traditional" way of envisaging information as being encoded, stored and transmitted is not appropriate for systems which interact so strongly that there is no common reference frame. We believe that this is related to our argument that there cannot be a global definition of information

---

[1] For example in theoretical elementary particle physics the definition of energy and its scale is different for different space-time levels.

due to the nature of the system. If however the system is split up into different levels, we are able to achieve definitions which are suitable for each level. If we are then able to connect these different definitions we would then have de facto a complete interpretation of the system with the help of information.

2. The information hierarchy

There are several proposals concerning the way information can be organized at different levels, [10, 4]. In our approach, we will follow the basic biological distinction between the genotypes and the phenotypes of the population. We will also make some remarks on *self-organized information acquisition*, which seems to be an important concept when the system actively influences its environment and its own information content. Fig. 2.1 shows the different levels of information which we distinguish in the evolutionary process. For each level we will propose a definition of information. At the lowest level, Fig. 2.1(a), information can be defined with the help of the population distribution, with reference to entropy. The next level, Fig. 2.1(b), is the coding, where the information content has to be defined without reference to a statistical ensemble. Our definition will be based upon Kolmogorov complexity and the variability of the coding. The third level, Fig. 2.1(c), the information of the phenotype, will in our case be defined with respect to neural systems. Information at this level cannot be decoupled from the environment or in artificial systems from the task. At the same time it seems to be sensible to distinguish between information processing and the primary stage of the potential for information processing. Adaptability [6], will play a central role in defining the evaluation of the neural structure. Finally, we will comment on the process in which the system itself generates information by anticipating the outcome of certain actions, which we will term *self-organized information acquisition*, Fig. 2.1(d).

We will clarify and underline our definitions and proposals using simple simulations of the evolution process. For the simulations at the genotype level we use a changing fitness function to highlight the aspect of robustness. The coding and the phenotype level will be analysed with reference to the structure optimization of feed-forward neural networks for prediction of the Lorenz chaotic time series.

3. The population level

The distribution of genotypes at the population level can be analysed by means of statistical physics, which has been successfully carried out by Prügel-Bennet et al. [17]. They were able to predict the probability distribution of states after selection and furthermore in [16] introduced macroscopic observables, the kumulants of the distribution, to describe the process. We can easily define a first measure of information with the help of entropy from statistical physics for our simulations. Let N be the number of states, t the time step, and $p_t(k)$ the probability of state k at the time t. The entropy based information measure

456

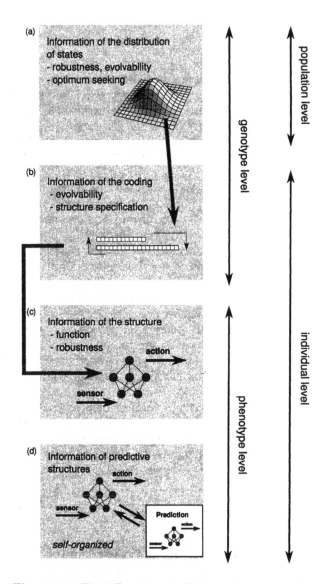

Figure 2.1: The different levels of information in evolution

leads to

$$I_t = S_{max} - S_t = \sum_{k=0}^{N} p_t(k) \log(p_t(k)) + \log(N) \qquad (3.1)$$

Of course the idea of state-space probabilities requires the introduction of a fitness space or landscape. The philosophy of fitness landscapes, advocated e.g. in [13], has been challenged [11] because of the ruggedness of the space due to coevolutionary effects for example. However, we believe that the abstract notion of a landscape can be kept as long as the scale at which the process is analysed is chosen appropriately.

The information gathered during the evolutionary process at the lowest level can be represented by the distribution of occupied states. Conversely, the distribution of the states during the evolutionary process also reveals some insight into how much information the landscape provides for the process. Surely, a constant fitness landscape would provide no information at all and the result would be a drifting population due to mutation whose state probability would fluctuate around a uniform distribution, thus a maximum of entropy and a minimum of information.

The entropy of the population which is well defined through the statistical nature of the process, will vary during evolution and tend to zero due to the selection process (an alternative interpretation is that the selection process stabilizes mutations which decrease the entropy) and to its maximum (logarithm of the population size) due to the average mutation effect. Therefore, after a transition period the entropy will reach a value where the distribution of states corresponds to the *constraints* such as robustness.

We already mentioned the problems which can occur with the purely statistical definition of information such as negentropy. In this case the interpretation of the statistics in order to achieve a meaningful information is achieved by the choice of the state-space which has been used for the definition of statistics. The meaning of information, although purely based on statistical observations, is realized via the definition of the fitness landscape. In this way it is dependent, for example, on the environmental conditions. If these conditions change, the fitness landscape changes and the entropy of the process has to increase and the probability density will be flattened. The definition of large information due to the peaked nature of the probability distribution of the states has lost its meaning, since the *constraints* have changed. Thus, the definition of information depends as much on the mathematical formalism as on the nature of the state-space.

3.1. Information and robustness

As a first proposal for a fitness space dependent definition of information at the population level, we will include a fluctuation term in equation (3.1).

$$
\begin{aligned}
I_t &= S_{max}^2 - (S_t - S_r(\varepsilon))^2 \\
&= \log(N)^2 - \left(-\sum_{k=0}^{N} p_t(k)\log(p_t(k)) - S_r(\varepsilon)\right)^2
\end{aligned}
\tag{3.2}
$$

$S_r(\varepsilon)$ is the entropy of the optimal distribution of the population in the genotype space due to fluctuations or "noise" in the fitness landscape represented by environmental changes. It is tempting to compare the notion of the robustness entropy $S_r(\varepsilon)$ with the concept of quasi-species introduced by Eigen and Schuster, [9, 24]. There the distribution of genotypes for the mutation rate at the error threshold is approximately equal for genotypes with the same Hamming distance from the wild-type; the optimal configuration. This would correspond to an exponential distribution around the best configuration. Therefore, we propose to model $S_r(\varepsilon)$ as the entropy of a Gaussian distribution; the parameter ε now combines the mean and the variance of the distribution $\varepsilon = (\mu, \gamma)$.

$$
S_r(\varepsilon) = \frac{1}{2}\left(1 + \log(2\pi\gamma^2)\right)
\tag{3.3}
$$

In the following simulations we model the Gaussian distribution using a discrete approximation (M is a normalization factor)

$$
S_r(\varepsilon) = \sum_{i=0}^{N} \frac{1}{M} e^{-\frac{1}{2\gamma}(i-\mu)^2} \log\left(\frac{1}{M} e^{-\frac{1}{2\gamma}(i-\mu)^2}\right)
\tag{3.4}
$$

The noisy environment was simulated by a statistically changing fitness function which was averaged over a couple of time-steps. Firstly, we simulated the process of finding the population with the best distribution of configurations with respect to the following fitness function ($\mathcal{P}$=population):

$$
F(\mathcal{P}) = \prod_{t=1}^{T}\sum_{i=1}^{N} p(i) f(i, \eta(t))
\tag{3.5}
$$

The variable i denotes the different states in the population and $p(i)$ their probability, f calculates the fitness of each state depending on a set of parameters $\eta(t)$ which change statistically with time. The values are multiplied to mimic a replication rate. We demand that the best population should maximize the average $\langle F(\mathcal{P})\rangle$ and minimize the variance $\langle\langle F(\mathcal{P})\rangle\rangle$. A smaller variance in the fitness enhances the stability of the population and smoothes the effective fitness function. We started with the simplest possible population consisting of two states with the probabilities p and $(1-p)$ and a state-fitness function, $i \in \{0, 1\}$

$$
f(i, \eta(t)) = \left(\frac{1}{2} + \eta(t)\right) i + 2(1 - \eta(t))(1 - i),
\tag{3.6}
$$

$\eta(t)$ is uniformly distributed in $[0, 1]$ and is drawn every time step (in the following we will denote a uniformly distributed number in $[a, b]$ by $\chi(a, b)$). We derive

$$
\begin{aligned}
F(\mathcal{P}) &= \prod_{t=1}^{T} \left[\left(\frac{1}{2} + \eta(t) \right) p + 2(1 - \eta(t))(1 - p) \right] \\
&= \prod_{t=1}^{T} \left[\eta(t)(3p - 2) - \frac{3}{2}p + 2 \right] \\
\langle F(\mathcal{P}) \rangle &= 1 \\
\langle\langle F(\mathcal{P}) \rangle\rangle &= \left\langle (F(\mathcal{P}) - \langle F(\mathcal{P}) \rangle)^2 \right\rangle = \left(\frac{4}{3} - p + \frac{3}{4}p^2 \right)^T - 1 \\
\frac{\partial \langle\langle F(\mathcal{P}) \rangle\rangle}{\partial p} &= 0 \Rightarrow p = \frac{2}{3}
\end{aligned}
\tag{3.7}
$$

The value for p corresponds to an entropy value $S = 0.636514$, which is the value which is reached in the simulation in Fig. 3.1(a).

Fig. 3.1(b) shows the results for a ten-state population with the fitness function

$$
F(\mathcal{P}) = \prod_{t=1}^{T} \sum_{i=1}^{N} p(i)^2 \delta_{i\eta(t)}
\tag{3.8}
$$

$$
\eta(t) \sim round(\mathcal{N}(\mu = 5.0, \sigma = 2.0))
$$

$$
f(i, \eta(t)) = p(i) \delta_{i\eta(t)}
\tag{3.9}
$$

δ_{ij} denotes the Kronecker symbol, $\mathcal{N}(\mu = 5.0, \sigma = 2.0)$ a Gaussian distribution with mean μ and variance σ and $round()$ rounds to the next integer value. The best distribution is roughly binomial as can be seen from the inlay in 3.1(b).

Thus, a stable distribution of the states within a population is the optimal configuration for a statistically varying fitness function like the one in equation (3.5).

However, the distribution of states in evolving systems is reached by adaptively varying the mutation rate and not the state-probability (we keep the selection constant). In a second simulation, we used only one population and optimized the transition probabilities (similar to the mutation probabilities) between states to achieve the most stable configuration with respect to the same fitness function. We adapted only one probability and used an exponential decay with the distance to estimate the other transition probabilities. In Fig. 3.2(a) we show the entropy and the resulting distribution for the following fitness function before and after selection ($T = 50$).

$$
\prod_{t=0}^{T} f(i, \eta(t)) = \prod_{t=0}^{T} \delta_{i\eta(t)} \chi(1, 2) + (1 - \delta_{i\eta(t)}) \chi(0, 0.01)
$$

$$
\eta(t) \sim round(\mathcal{N}(\mu = 5.0, \sigma = 5.0))
\tag{3.10}
$$

460

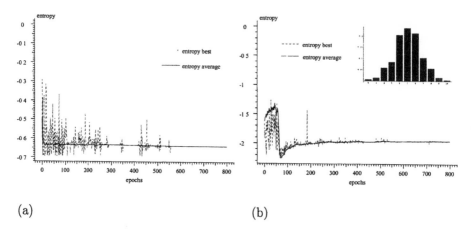

(a)   (b)

Figure 3.1: (a) The best and the average entropy of a population of two state populations with the fitness function given by (3.5) with $T = 100$. The state probability p has been optimized using a $(100, 200)$ standard evolution strategy [19]. (b) Entropies of ten-state populations and the final distribution for the discrete-Gaussian fitness function, the parameters are the same as in (a).

Again the resulting distribution is roughly binomial. However, we note the larger fluctuations of the entropy and the smaller variance of the distribution of states. At the same time the variance of the noise was increased to 5.0. Firstly, the distribution of states is now also dependent on other parameters like the population size or the selection pressure. Furthermore, adaptation of the mutation operator will lead to more fluctuations than direct adaptation of the state probabilities since they introduce another statistical factor.

Fig. 3.2(b) shows that for non-statistical fitness functions the population will indeed all assemble on the fitness peak (besides fluctuations due to mutation), the fitness function assigns value 2.5 to state 5 and values 1.5 to states 4 and 6, all other states have fitness zero. The application of the proposed information measure to our simulations is shown in Fig. 3.3(b). The fitness measure is the same as in equation (3.10), however the values of μ and σ change during the evolution, thus the noise level of the environment changes with time and the mutation rate has to adapt to the environmental changes. Table 3.1 shows the

t	0	104	236	292	394	490
μ	5	5	9	9	1	10
σ	2	3.772	5.559	4.4587	0.946	3.278

Table 3.1: Environmental changes simulated by a change in the mean μ and the variance σ of the rounded Gaussian distribution $\eta(t)$ of the fitness function.

time and the changes of the parameters governing the statistical fitness function. We observe from the entropy diagram the changes in the state probability due to

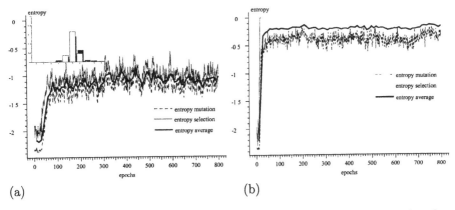

(a)   (b)

Figure 3.2: Entropies before and after selection and averaged over ten generations for the two fitness functions (a) and (b) described in the text. Adaptation of the mutation probabilities with a standard (100,200) evolution strategy. (a) Inlay: Population distribution before and after selection.

the changes in the fitness function. The information measure in Fig. 3.3(b) from equation (3.1) and (3.4) show the transitions in the process. The parameter γ has been set to $\gamma = \frac{1}{5}\sigma$. Additionally, it is possible to read off the impact of the environmental change by the drop in the information I_t, equation (3.2), of the population. For example, for $t = 236$ there is a medium size change and accordingly there is only a slight drop in I_t, whereas for $t = 394$ a large change results in a serious drop in the information content. At the same time the population adapts to the changes very quickly so that the maximal information content is always reached after only a couple of generations.

In this section we introduced an information measure for the genotype level based on the distribution of the states (entropy) combined with a stability term against changing environmental conditions. We analysed the optimal distribution, the entropy and the information measure with the help of simlations.

4. The individual level

4.1. The coding

The second level in our information hierarchy, Fig. 2.1(b) refers to the information of the coding. The coding is the connection between the ensemble based information of the genotype space and the environment dependent information of the phenotype space. It therefore has to fulfill constraints from both levels. It has to guarantee evolvability, see [26], for example its length has to be below the error-threshold [9]. The coding also connects the genotype-level exploration process to the phenotype-level selection process. Therefore, it should satisfy

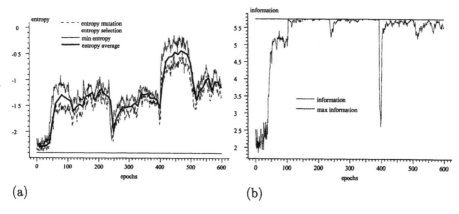

Figure 3.3: The entropies (a) and the information (b) from equation (3.2) for a changing environment

further constraints such as developing a strongly causal[2]

process together with the variation operators, see [21]. At the same time, it codes for the phenotypic structure and therefore has to allow for complex patterns in this structure. Thus, we can summarize several constraints which will play a role in the definition of a measure for the information content in the coding:

- evolvability

- diversity with respect to the structure

- variation-selection constraints

In the following, we will only discuss the implications of the first (confined to a minimal description length of the phenotype) and the second point, although the third is equally worth exploring.

The information content of the coding is related to the complexity of the coding, by this we mean the diversity of different patterns it can realize with the structure. In addition the information has to depend upon the compression rate it achieves compared with a direct encoding of the structure. The compression rate is related to the minimal length of the coding which again is related to the error-threshold and the evolvability. The shorter the code length of the structure, the higher the probability of accurate reproduction and therefore the lower the probability of a loss of information. On the other hand, in a shorter code some parts are used as modules and a mutation in a module will effect every part of the phenotype where the module is used. Thus, in order to preserve information, the

---

[2]Strong causality usually refers to *small variations imply small consequences*. In evolution the variations are due to the mutation/crossover operator and the consequences are the respective fitness values

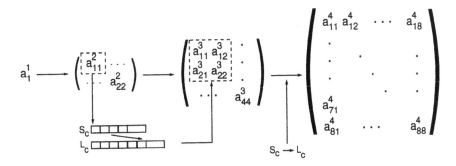

Figure 4.1: Scheme of the recursive development of the connection matrix up to a size of 8 × 8. Each element in each step is replaced by a 2 × 2 matrix via the mapping $S_C \to L_C$

mutation rate is decreased explicitly (molecular repair mechanisms) or mutations are not expressed due to redundancy in the code (64 codons → 20 amino acids).

In order to discuss the coding further, we look at a specific example from the optimization of the structure of neural networks, the recursive encoding method, which we will introduce shortly, see [20] for details.

The coding consists of four chromosomes, where only the first two are important for the building process of the connection matrix. In each iteration step every element of the connection matrix is replaced by a 2 × 2 matrix of new elements. The new elements are specified by a mapping from the small chromosome S_C to the large chromosome L_C. The length of the small chromosome d_{S_C} is variable, the length of the large one is fixed by the condition $d_{L_C} = 4 \cdot d_{S_C}$. At each step i the **first** place $N(y_{nk}^i)$ of each connection matrix element y_{nk}^i in S_C is determined. The element is then replaced by the four elements at the positions

$$\big(4 \cdot (N(y_{nk}^i) - 1) + 1 \quad , \quad 4 \cdot (N(y_{nk}^i) - 1) + 2,$$
$$4 \cdot (N(y_{nk}^i) - 1) + 3 \quad , \quad 4 \cdot (N(y_{nk}^i) - 1) + 4 \,\big) \tag{4.1}$$

in the large chromosome L_C. In case the matrix element y_{nk}^i is not in S_C, it is replaced by four so-called terminal symbols (in the notation of integer strings, the most convenient choice is zero). A terminal symbol is in turn always replaced by another four terminal symbols in an recursion step. Fig. 4.1 shows the evolution of a 8 × 8 connection matrix M_{con} following the introduced rules.

Minimal description length

We will start with the second point, the compression rate of the coding and use the concept of Kolmogorov complexity to define the information content of the coding. See also [8] for a similar approach to estimate the complexity of a protein. The structure of an arbitrarily connected neural network can be fully described by a matrix or a vector of length n^2 if the network has n neurons. In our context, we will allow this vector to have entries from a set of integer values $\{0, N_{sym} - 1\}$. We can binary encode these integer values and end up with a string of length[3] $l = n^2 \lceil \log(N_{sym}) \rceil$. The Kolmogorov complexity K is defined as the length of the shortest program which reproduces this string of length l and stops afterwards, it is therefore the minimal description length of the sequence l. Unfortunately, it is very rarely possible to calculate K for complicated sequences. However, we can do better than simply comparing the encoding length with the direct sequence of length l, since a better upper bound for K for binary sequences with l_1 1-entries is known, see [5] for the derivation

$$
\begin{aligned}
K_u &\leq c + 2\log(l_1) + \log \binom{l_1}{l} \\
&\leq c + 2\log(l_1) + l\, T\left(\frac{l_1}{l}\right) \\
T(p) &= -p\log(p) - (1-p)\log(1-p)
\end{aligned}
\tag{4.2}
$$

Assuming that the network has a connectivity[4] γ and the connections are on average encoded by one half 1-entries, equation (4.2) simplifies to ($l_1 = \gamma\, l/2$)

$$
K_u \leq 2\log\left(\frac{\gamma}{2}n^2 \log(N_{sym})\right) + n^2 \log(N_{sym}) T\left(\frac{\gamma}{2}\right)
\tag{4.3}
$$

We can therefore define the information content in the coding as the relation between the length of the coding sequence and K_u.

$$
\begin{aligned}
I_{code}^1 &= \frac{K_u}{5\, d_{S_C} \log(N_{sym})} \\
&\approx \frac{2\log(\frac{\gamma}{2}n^2 \log(N_{sym})) + n^2 \log(N_{sym}) T(\frac{\gamma}{2})}{5\, d_{S_C} \log(N_{sym})}
\end{aligned}
\tag{4.4}
$$

Diversity of the coding

The compression rate of course does not provide any information about the diversity of the structures which can be specified by the coding. We can use another entropy inspired approach, namely, the logarithm of the number of network structures which can be realized by the coding. Let $\mathcal{Y}$ denote the number of network structures, we define

$$
I_{code}^2 = \log(\mathcal{Y}),
\tag{4.5}
$$

---

[3] $\lceil x \rceil$ is the smallest integer $\geq x$.
[4] The connectivity is the proportion of realized connections to possible connections.

$\mathcal{Y}$ can be calculated as follows. If r is the number of recursions, the size of the matrix is $2^r \times 2^r$ with each entry from the set $\{0, N_{sym} - 1\}$. Therefore, the maximum of generable structures is $N_{sym}^{4^r}$. However, the size of the large chromsome also determines the maximum length, thus $N_{sym}^{4d_{S_C}}$. The third constraint is very specific to this coding. Repeated entries in the small chromsome are ignored, thus the effective chromsome size which maps to the large chromsome is further reduced to $4 N_{sym}$, if $N_{sym} < d_{S_C}$. The number of network structures $\mathcal{Y}$ is then given by the minimal value of the three constraints.

$$\mathcal{Y} = \min\left(N_{sym}^{4^r}, N_{sym}^{4d_{S_C}}, N_{sym}^{4 N_{sym}}\right)$$

$$I_{code}^2 = \min\left(4^r, 4d_{S_C}, 4 N_{sym}\right) \log(N_{sym}) \qquad (4.6)$$

I_{code}^2 serves as the second measure of the information content of the coding and defines the diversity of the possible network structures which can result from the coding. It does not refer to the complexity of one specific code, but to the class determined by the parameters (r, N_{sym}, d_{S_C}).

$I_{code}^{1,2}$ define the information content of the coding with respect to the modularity of the structure (the compression rate) and to the number of possible structures. Fig. 4.2 shows the results of an optimization run. I_{code}^2 does increase during the evolutionary process, thus the coding is chosen so that the diversity of pattern for the phenotype is high. There are two reasons why I_{code}^1 does not show the expected behaviour, the first is that the problem might not favour a very modular solution. More importantly however, the network size of around 10 neurons is not large enough to benefit from a modular construction.

4.2. Information and the evaluation of structures

Information at the phenotype level will be analysed with respect to two different principles. Firstly, the potential of systems to process information and secondly, the actual ability to process information. The definition of quantities which serve as a measure of the information content of the phenotypes is closely connected to the problem of the evaluation of structures in the optimization process. It is clear that the structure provides the framework for information processing and adaptability and does not just refer to the solution of one specific problem.

Potential information processing

For a system to be potentially able to process information, it has to possess some basic properties. For example, the information must propagate through the system at an sufficient, but not too destructive rate. This is what Kauffman [13] examined in boolean networks and phrased the "solid", "liquid" and "gas" phase. While in the "solid" phase, each cell gets less than 3 inputs and errors cannot propagate through the net, in the "gas" phase, each cell gets more than

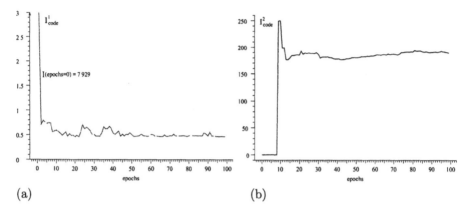

Figure 4.2: The information measures $I_{code}^{1,2}$ for the structure optimization of a feed-forward neural network for the prediction of the Lorenz time series. The population size is $(20, 100)$ with deterministic selection. Each individual network is trained with back-propagation for 50 cycles.

5 inputs and errors can propagate over the whole net. The mutual information

$$I(X;Y) = \sum_{x,y} p(x,y) \log \left(\frac{p(x,y)}{p(x)\,p(y)} \right) \tag{4.7}$$

decreases in the chaotic phase because the neuron activities are decorrelated. Close to the threshold of the chaotic phase the mutual information reaches its maximum, and therefore information is propagated best. We believe for a number of reasons that mutual information is not the best method for investigating the properties for potential information processing. In Fig. 4.3 we show the average mutual information between all neuron units and between target and output values for an optimization run using the coding from the last section. The last quantity is more similar to an error measure than to an information measure. The overall picture is that over several optimization trials the behaviour of the mutual information tends to depend more on other constraints than on the optimization process. For example, the mutual information reaches its maximum when the activation is in the linear range of the transition function and it therefore depends largely on the weight initialization without any information based reason As shown in Fig. 4.3 we were not able to extract a common overall behaviour of the mutual information between neuron units, it is therefore questionable if it represents a sensible measure.

We propose a more basic measure which is directly connected to the argument of Kauffman. The relation between activities of neighbouring neurons a_j should

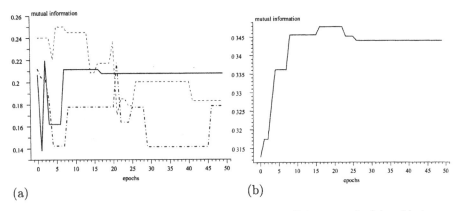

Figure 4.3: The average mutual information between all neuron units (a) and between the target and the output of the network (b). The three curves in (a) are runs for the same optimization problems, but show completely different behaviour. As expected the mutual information in (b) shows a similar behaviour as the squared error (fitness criterion).

be strongly causal with respect to noisy (noise level ε) activities of neuron i, $a_i(\varepsilon)$. The degree of causality is tuned by the parameter n. That is, the probability

$$P\left(\frac{\varepsilon}{n} \le |a_j(a_i) - a_i(\varepsilon)| \le \varepsilon\, n\right) \tag{4.8}$$

should be maximal for the system. Equation (4.8) does not imply any actual information processing in the system but solely defines the overall ability of the system to achieve information processing when a specific task is applied and learned. The nature of such a measure makes it difficult to simulate its behaviour for systems such as the ones which we use, which are of limited complexity with respect to the simple size of the system. We believe measures like the one proposed here can only be used to give a first indication of whether a structure of a large, extended system has such a potential at all. Therefore, we will turn to a more task specific examination of information at the phenotype level.

Information processing in adaptive systems

The ability to process information cannot be decoupled from the environment or in artifical neural networks from the task. The mutual information has been used successfully for describing the ability to process information in neural networks for example by Deco [7]. However, it is not the mutual information within the network which is analysed but the relation between network output and target output. This seems to be too task dependent to be able to serve as a sensible measure of information. The following two hypothesis underline the close linkage between the ability to process information and the environment and at the same time between the adaptability of systems and the definition of information, see also [6].

Hypothesis 1

The information content of an non-adaptive system cannot be defined without explicit reference to the environment.

The sole purpose of non-adaptive systems is to solve one specific task or a well defined class of tasks. If the system is decoupled from this task its structure becomes meaningless.

Hypothesis 2

The information content of adaptive systems, such as neural structures, has to be defined with reference to the probability of reaching systems which perform well in changing environments. If $\Omega(t)$ denotes the expected distribution of optimal systems at time t for changing environments and Λ the structure of the system after the evolutionary process, then the information content can be defined with reference to the minimzation of

$$\frac{1}{T} \sum_{t}^{T} H(\Omega(t)|\Lambda). \tag{4.9}$$

This expresses the uncertainty about the distribution of the optimal systems over time given the structure which is the result of the evolution process. T denotes the length of the adaptation process of the systems, in natural systems T corresponds to the lifetime.

Equation (4.9) serves as a measure of how good the system is prepared by evolution to maximize the expected performance in a stochastically varying environment providing we implicitly assume that the distribution is always around the optimal configuration. Thus minimization of (4.9) alone does not necessarily lead to good performance. However, this is in accordance with the demand to define the information content in structures without explicit reference to a specific task. An upper bound for $\sum_{t} H(\Omega(t)|\omega)$ is $\sum_{t} H(\Omega(t))$, which is the entropy of the distribution of systems with random structures. Thus, we can define the information of the structure as the mutual information between the structure defined by evolution and the distribution of the optimal systems over time.

$$I = \frac{1}{T} \sum_{t=1}^{T} I(\Omega(t); \Lambda) = \frac{1}{T} \sum_{t=1}^{T} H(\Omega(t)) - H(\Omega(t)|\Lambda) \tag{4.10}$$

In order to see how I from equation (4.10) varies over different epochs we simulate the evolution of neural structures with the encoding from §4.1. The stochastic adaptation process does not vary over time ($T = 1$; t is discrete) and can be realized by a stochastic choice of the lerning parameters or by noisy data sets with varying degree of noise influence. $H(\Omega(t)|\Lambda)$ is approximated by

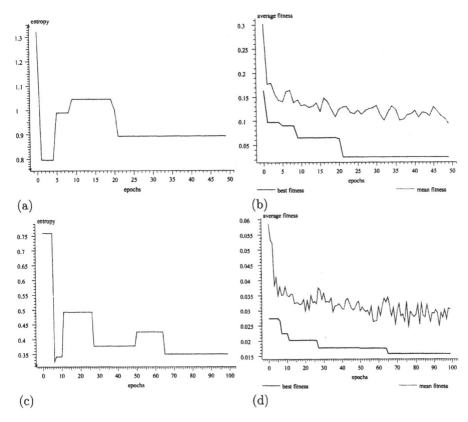

Figure 4.4: The entropy of the fitness values (a), (c) and the average (stochastic adaptation) fitness values (b), (d) of the best individual and the mean (population).

the entropy over the fitness values for several of these adaptation procedures. Fig. 4.4 show the average fitness and the entropy of such a simulation. In each generation, each individual structure was adapted ten times using standard backpropagation (40 cycles). In the first simulation (a, b) a stochastic choice of the learning parameters (learning rate $\sim \chi(0.01, 0.5)$, momentum term $\sim \chi(0, 0.8)$) was applied and in the second simulation (c, d) noise was added to the data with a stochastic signal to noise ratio ($snr \sim \lceil \chi(2, 20) \rceil$). The fitness value equals the average squared prediction error of the normalized Lorenz time series (fitness was minimized). We observe from Fig. 4.4 the minimization of the entropy, although it is not a monotonically decreasing function. The largest drops in the entropy correspond to drops in the average error.

Self-organized information acquisition

In order to complete the picture of definitions of information for the different levels of evolution, we have to include the highest level: modelling and actively influencing the environment. Systems with an internal model of their environment can predict the outcome of certain actions and are therefore able to increase the information which is contained in their structure about the environment. All other definitions of information are passive in nature. The system adapts in one way or the other to environmental constraints such as robustness, evolvability or variability and adaptability. *Self-organized information acquisition* describes a fundamentally different process, in which the system actively creates information by prediction. This concept is connected to systems which use anticipation to guide their behaviour, a field which has recently received much attention. We will not try to define a sensible definition of information for this level, but merely want to highlight that a complete picture of evolution in terms of information content and processing should aim at including this level as well.

5. Discussion

In this paper we set out to propose measures of information for different levels of evolution. The reasons are that on the one hand information seems to be a promising candidate for a macroscopic and unifying description of the evolutionary process and on the other hand the nature of the process seems to demand different definitions for different levels. One of the basic preliminaries is the sensible choice of the hierachy of levels for which a meaningful measure can be defined. Our proposal is close to the obvious biological division into genotypes and phenotypes. We mostly concentrated on the problem of structure optimization of neural networks, firstly since it seems to be an intrinsic enough problem to include the principle difficulties and secondly since we want to advocate the direct examination of information definitions with the help of simulations whenever it seems feasible. We believe that the combination of properties such as robustness, evolvability and adaptability into one concept information can indeed be helpful in further understanding the process.

Bibliography

[1] BENNETT, C.H., "The thermodynamic of computation – a review", *Int. J. Theor. Phys.* **21** (1982), 905–940.

[2] BENNETT, Charles H., "Universal computation and physical dynamics", *Physica D*, 86 (1995), 268–273.

[3] BRILLOUIN, Leon, "Life, thermodynamics and cybernetics", *Am. Sci.* **37** (1949), 554–568.

[4] BROOKS, D.R., and E.O. WILEY, *Evolution as Entropy* 2 ed., The University of Chicago Press (1988).

[5] COVER, Thomas M, and Joy A THOMAS, *Elements of Information Theory*, Wiley Series in Telecommunications, John Wiley (1991).

[6] DE VREE, Johan K., "A note on information, order, stability and adaptability", *BioSystems* **38** (1996), 221–227.

[7] DECO, Gustavo, *Information Theoretic Approach to Neural Computing*, Springer (1996).

[8] DEWEY, T. Gregory, "Algorithmic complexity of a protein", *Phys. Rev. E* **54**, 1 (1996), 39–41.

[9] EIGEN, Manfred, and Peter SCHUSTER, "The hypercycle – a principle of natural self-organization", *Die Naturwissenschaften* **64**, 11 (1977), 541–564.

[10] FLEISSNER, Peter, and Wolfgang HOFKIRCHNER, "Emergent information. towards a unified information theory", *BioSystems* **38** (1996), 243–248.

[11] JONGELING, Tjeerd B., "Self-organization and competition in evolution: A conceptual problem in the use of fitness landscapes", *J. theo. Biol.* **178** (1996), 369–373.

[12] KAMPIS, George, "Self-modifying systems: a model for the constructive origin of information", *BioSystems* **38** (1996), 119–125.

[13] KAUFFMAN, Stuart A., *The Origins of Order*, Oxford University Press (1993).

[14] KOLMOGOROV, A.N., "Three approaches to the quantitative definition of information", *Problems of Information Transmission* **1** (1965), 4–7.

[15] LANDAUER, R., "Irreversibility and heat generation in the computing process", *IBM J. Res. Dev* **5** (1961), 183–191.

[16] PRÜGEL-BENNETT, Adam, "Modelling evolving populations", *J. theo. Biol* **185** (1997), 81–95.

[17] PRÜGEL-BENNETT, Adam, and Jonathan L. SHAPIRO, "Analysis of genetic algorithms using statistical mechanics", *Phys. Rev. Lett.* **72**, 9 (1994), 1305–1309.

[18] SCHRÖDINGER, E., *What is life?*, Cambridge Univ. Press (1967), original german edition 1944.

[19] SCHWEFEL, Hans-Paul, *Evolution and Optimum Seeking*, John Wiley & sons (1995).

[20] SENDHOFF, Bernhard, and Martin KREUTZ, "Evolutionary optimization of the structure of neural networks using a recursive mapping as encoding", *Proc. 3rd Int. Conf. Artificial Neural Networks and Genetic Algorithms* (G. SMITH ed.), Springer Verlag (1997).

[21] SENDHOFF, Bernhard, Martin KREUTZ, and Werner von SEELEN, "A condition for the genotype–phenotype mapping: Causality", *Proc. 7th Int. Conf. Genetic Algorithms* (T. BÄCK ed.), Morgan Kaufman (1997), 73–80.

[22] SHANNON, C.A., and W. WEAVER, *The Mathematical Theory of Communication*, University of Illinois Press (1959).

[23] STONIER, Tom, *Information and the internal structure of the universe*, Springer (1990).

[24] SWETINA, Jörg, and Peter SCHUSTER, "Self-replication with errors – a model for polynucleotide replication", *Biophys. Chem.* **16** (1982), 329–345.

[25] SZILARD, Leo, "Über die Entropieverminderung in einem thermodynamischen System bei Eingriffen intelligenter Wesen", *Zeitschrift f'ur Physik* **53** (1929), 840 – 856.

[26] WAGNER, Günter P., and Lee ALTENBERG, "Complex adaptations and the evolution of evolvability", *Evolution* **50** (1996), 967–976.

[27] ZUREK, W.H., "Algorithmic randomness and physical entropy", *Physical Review A* **40**, 8 (1989), 4731–4751.

Chapter 42

Emergence of complex ecologies in ECHO

Richard Smith
SUNY Binghamton, Systems Science
Portland State University, Systems Science
richards@sysc.pdx.edu
Mark Bedau
Reed College, Philosophy
Portland State University, Systems Science
mab@reed.edu

We present measurements of evolutionary activity, as defined by Bedau and Packard [2, 3], in John Holland's ECHO—a classic artificial model of an evolving ecosystem of interacting agents. We show how evolutionary activity measurements can be used to identify the significant adaptive evolutionary events in complex systems like ECHO. Using this methodology, we provide evidence of the emergence of a complex ecology in ECHO.

1. Motivation and context

One of the most important issues in the study of complex adaptive systems is to determine how complex ecologies spontaneously emerge through the course of evolution [5, 11, 13, 7, 3]. This issue is difficult to pursue, however, in large part because of the lack of objective and feasible ways to measure the relevant aspects of the adaptive dynamics of evolving systems. In the attempt to address this sort of problem, Bedau and Packard have defined statistics that measure the evolutionary activity of evolving systems [2, 1, 3], where these statistics are based on the assumption that, on average, the persistence of a component (e.g., genotype) in an evolving system reflects the component's continual adaptive significance, i.e., its continual adaptive advantage over its competitors.

In the present paper we use these evolutionary activity statistics to study the emergence of ecologies in John Holland's ECHO model [5, 6, 7]. The ECHO model is especially interesting to study in this context because it is one of the models that seems most likely to exhibit the open-ended evolution of complex ecologies. The work we present here shows how evolutionary activity statistics can help us to identify and understand significant evolutionary phenomena in ECHO populations.

2. The statistics

Evolutionary activity (or "activity", as we will sometimes say for simplicity) is computed from data obtained by observing an evolving system. In our view an evolving system consists of a population of components, all of which participate in a cycle of birth, life and death, with each component largely determined by inherited traits. (We use this "component" terminology to maintain generality.) Birth, however, allows for the possibility of innovations being introduced into the population. If the innovation is adaptive, it persists in the population with a beneficial effect on the survival potential of the components that have it. It persists not only in the component which first receives the innovation, but in all subsequent components that inherit the innovation, i.e., in an entire lineage. If the innovation is not adaptive, it either disappears or persists passively.

The idea of evolutionary activity is to identify innovations that make a difference. Generally we consider an innovation to "make a difference" if it persists and continues to be used. Counters are attached to components for bookkeeping purposes, to update each component's current activity as the component persists and is used. If the components are passed along during reproduction, the corresponding counters are inherited with the components, maintaining an increasing count for an entire lineage. Two large issues immediately arise:

1. What should be counted as an innovation? In fact, innovations may be identified on many levels in most evolving systems. We define an innovation as the introduction of a new component into the system. In the case of this study of ECHO, the components we choose to study are entire genotypes. Previous studies have measured innovations on the level of individual alleles [2, 1], genotypes [3], and taxonomic families in the fossil record [3].

2. How should a given innovation contribute to the evolutionary activity of the system? We measure activity contributions by attaching a counter to each component of the system. In the work we present here a component's activity counter is incremented each time step if the component simply exists at that time step. Though there are ways to refine this simple counting method [2, 1], we use this version here because it facilitates direct comparison with many other systems [3].

More formally, let $f_i(t)$ be a characteristic function indicating whether the ith component is present in the record at time t:

$$f_i(t) = \begin{cases} 1 & \text{if component } i \text{ exists at } t \\ 0 & \text{otherwise} \end{cases} . \tag{2.1}$$

Then we define the evolutionary activity $a_i(t)$ of the ith component at time t as its presence integrated over the time period from its origin up to t, provided it exists:

$$a_i(t) = \begin{cases} \int_0^t f_i(t)dt & \text{if component } i \text{ exists at } t \\ 0 & \text{otherwise} \end{cases} . \tag{2.2}$$

Thus, a_i is the ith component's activity counter. Note that a different resolution of the second issue above would result in a different formula for incrementing the activity counters (as in references [2] and [1]). The *total evolutionary activity*, $A(t)$, at time t (which we will often call just "total activity") is simply the sum of the evolutionary activity of all components:

$$A(t) = \sum_i a_i(t). \tag{2.3}$$

The *diversity*, $D(t)$, is simply the number of components present at time t,

$$D(t) = \#\{i : a_i(t) > 0\}, \tag{2.4}$$

where $\#\{\cdot\}$ denotes set cardinality. Then, the *mean total evolutionary activity*, $\bar{A}(t)$, (which we will often call simply "mean activity") is the total evolutionary activity $A(t)$ divided by the diversity $D(t)$:

$$\bar{A}(t) = \frac{A(t)}{D(t)}. \tag{2.5}$$

See references [2] and [3] for further discussion of these statistics.

3. The ECHO model

John Holland created ECHO in response to a request from Murray Gell-Mann to produce a model that would illustrate the creation of complex structures by natural selection [7] (pp. 94-95). Holland based ECHO on what he saw as the essential features common to all complex adaptive systems. Those features consist of four properties (aggregation, nonlinearity, flows, and diversity) and three mechanisms (tags, internal models, and building blocks) [7]. The central explicit focus of the model is to allow natural selection to build and tune the strategies by which the agents in the population engage in various kinds of interactions. That is, ECHO is a model of the evolution of agent interaction strategies.

476

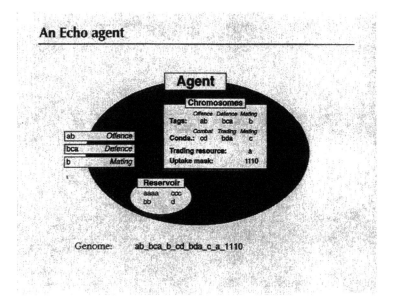

Figure 2.1: Overview of an ECHO agent.

An ECHO world consists of a toroidal lattice of sites each site having a resource fountain and a population of agents. (The ECHO runs we describe here consist of worlds with only one site.) Different letters of the alphabet represents different types of resources that are available in the world. A fixed amount of resources are distributed to each site at each time step. Resources that are not consumed by agents can accumulate at a site up to a fixed ceiling.

An ECHO agent consists of a "chromosome" that is composed of eleven sub-strings of the world's resources (letters of the alphabet) together with a reservoir to store excess resources (see Fig. 2.1). The sub-strings of the chromosome constitute an agents *external tags* and *internal conditions* together with an uptake mask which specifies what resources the agent can take up from the environment. An agent's tags are external in the sense that other agents have access to them. An agent's conditions are not accessible to other agents. The tags and conditions are used to determine the outcome of the three types of interactions that ECHO agents can enter into with one another—combat, trade, and mating. Whether two agents interact and, if so, what type of interaction they have is determined by comparing the agents tags and conditions. A string match of the appropriate tag and condition causes the interaction to take place. The existence of external tags and internal conditions is one of ECHOs critical features as it allows for complex (e.g. non-transitive) relationships to exist between the agents. It is also central to one of the features Holland has emphasized about ECHO—its endogenous fitness function, i.e., a fitness function that is an emergent property of the environment and the other agents [12, 8].

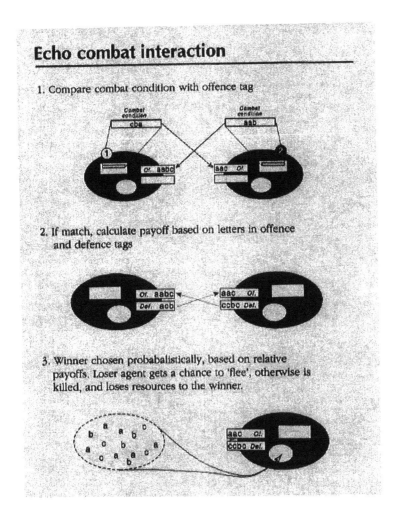

Figure 2.2: Overview of the combat interaction.

The combat interaction is illustrated in Fig. 2.2 and gives a good illustration of how tags and conditions are used. The mating interaction takes place if a bilateral match is found between the mating tags and conditions of two agents chosen to interact. The result of a successful mating interaction is more analogous to the types of genetic exchange seen in bacteria as opposed to sexual reproduction. The two participating agents exchange genetic material via crossover (at a random point in the chromosome) and replace their "parents" in the population. Trading takes place if there is a prefix match between the trading condition of the first agent and the offense tag of the other agent. A trading interaction between two agents results in each agent transferring the excess of its trading resource (the amount of resources in the agents reservoir over and

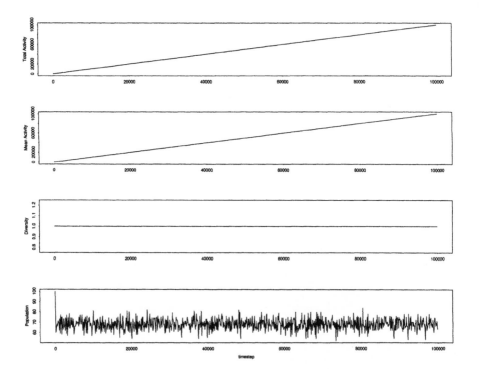

Figure 2.3: A typical ECHO run in which the probability of mutation is zero. The 100 agents in the initial population all have the same genotype. Note that diversity remains exactly 1.0 throughout the run and that total and mean activity are equal to the number of time steps, as expected.

above what it needs for reproduction) to the other agent.

Agents that have acquired enough resources in their reservoir to copy their chromosome are able to reproduce asexually. Asexual reproduction is subject to a probability of mutation which is normally a point mutation but can also be further subject to a probability of mutation by crossover or insertion-deletion within the parent chromosome. As a part of asexual reproduction, parents give a fixed percentage of the resources remaining in their reservoir to their offspring. It is mutation together with the selection pressure due to competition for resources which drive the evolution of the ECHO agents. Agents acquire resources via foraging, fighting, and trading. Agents lose resources via a fixed metabolic tax, trading, and asexual reproduction.

One time step in the ECHO model consists of the following cycle of events: A proportion of the agents are selected to undergo interactions and the interactions take place. Resources at a site are distributed to those agents that can accept them. Agents are taxed probabilistically. Some agents are randomly killed and their resources returned to the environment. Agents that have not collected

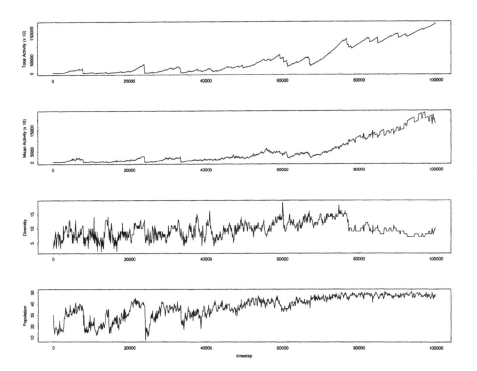

Figure 2.4: An ECHO run in which the mutation rate is positive (0.01). Note the qualitative difference between the statistics before and after time step 50,000. See text for an explanation.

resources migrate to a randomly chosen neighboring site (in multi-site worlds). Finally, agents that have acquired sufficient resources reproduce asexually.

More information about the ECHO model can be found at the ECHO home page at http://www.santafe.edu/projects/echo and in the following references [5, 9, 6, 4, 7, 10].

4. Individual ECHO runs

We have measured evolutionary activity and diversity in many ECHO runs, starting from a variety of different initial conditions. We present time series of these statistics from two runs. Data was collected every 100 generations. The purpose of the first run is solely to develop some feel for the statistics by considering a case in which we know exactly what to expect. The purpose of the second run is to illustrate how the statistics can illuminate the evolution of ecologies.

In the first ECHO run (see Fig. 2.3) mutation is turned off entirely. The initial population consists of 100 agents which all have the same genotype, and all subsequent agents born over the course of the run continue to have this same genotype. Note that the diversity of the population remains pinned at 1.0, as expected. Although the number of agents in the population fluctuates (see population data in Fig. 2.3) all agents share the same genotype. Consequently both total activity and mean activity increase one unit per time-step. Mean activity is the same as total activity here since there is only one genotype. Both of these effects are exactly what is to be expected.

It is worth noting that the calculation of activity in these runs takes into consideration not the concentration of a genotype (i.e. how many agents share it) but only its existence. Thus, the dynamics of the statistics are insensitive to fluctuations in population level, as Fig. 2.3 shows. (Activity could have been calculated taking concentration into consideration, and this sometimes shows qualitatively different results; see [2].) It is also worth noting that the activity curve could be made steeper by seeding a no-mutation world with more distinct genotypes, because the slope of the curve equals the number of fixed genotypes present.

The second run (Fig. 2.4) uses similar parameters as the above run with a couple of exceptions. The paramters for this run are given in the appendix. In this run reproduction is subject to a 0.01 probability of mutation. Also, the initial population is different. The world is seeded with 30 agents representing 10 instances of three distinct genotypes. The three genotypes represent the caterpillar, ant, and fly ecology cited elsewhere in the Echo literature [10]. However, the population is quickly diversified due to mutations, and natural selection produces a succession of distinct ecologies. Finally, the ceiling on resources allowed to accumulate at the site is set at 50% of what it was in the first run (Maxima vector in site file) resulting in a smaller average population.

To analyze this run we group extant genotypes into two classes depending on their ecological significance. Those genotypes that have persisted for 3,000 generations or more are judged to have demonstrated at least a minimal degree of adaptive success, while those that go extinct in fewer than 3,000 generations are regarded as mere "ecological noise"—transitory genotypes that lack sufficient adaptive advantage or sufficient good luck to make a significant stamp on their ecology.

This run has several interesting features. A glance at any of the time series data shows that dramatic evolutionary phenomena are occurring. Here we mainly want to show how these statistics show the evolutionary emergence of two quite different kinds of ecological phenomena. These two kinds of ecologies correspond to the obvious qualitative and quantitative difference between the first two thirds and the last third of the run shown in Fig. 2.4.

For the first roughly 50,000 times steps of this run the activity graphs show a series of periods of moderately increasing activity followed by sharp drops in activity (e.g. generations 8,000 and 24,000). Microanalysis of this portion of the run shows that the bulk of the measured activity is attributable to a small fraction of the population. The first peak in the activity graphs occurs at generation 7,800. At this time the population consists of 11 distinct genotypes.

Of these 11 genotypes 2 are persistent. The most persistent genotype at the time has existed for 5,300 generations. The second persistent genotype has existed for 5,000 generations. The mean persistence of all the genotypes is 1,509 generations. The persistent genotypes account for 18% of the population of genotypes. On the other hand the two persistent genotypes are responsible for 65% of the total activity that has accumulated. Finally, by generation 7,900 (the next point in the data set) *all* of the genotypes that existed at generation 7,800 have gone extinct.

The second sharp drop in activity occurs at generation 24,000. At generation 23,900 the population consists of 13 distinct genotypes, 5 of which are persistent. The oldest genotype has persisted for 6,100 generations and the mean age of the genotypes is 2,523 generations. The mean age of the persistent genotypes is 4,400 generations. The persistent genotypes are responsible for 68% of the total accumulated activity. This population has several notable features over the previously considered population including more persistent genotypes, longer average age, and a larger overall population. But still we see that the persistent genotypes account for less than 40% of the total genotypes while contributing 68% of the total activity. We also see the same phenomena of mass extinction as seen in the above population. All the genotypes present at generation 23,900 have gone extinct by generation 24,000.

To give an idea of what is taking place in the relative valleys of the activity graphs we look at the population at generation 40,000. At the 40,000th generation the population consists of 8 distinct genotypes only one of which is persistent. The mean activity (average age) of the population is 1,150 generations. The single persistent genotype has existed for 4,200 generations and accounts for roughly 46% of the accumulated total activity.

This is a fairly typical pattern in many of the ECHO runs we have observed. A relatively small percentage of the total population of genotypes accounts for the bulk of the activity. The populations are relatively unstable as evidenced by the pattern of mass extinction. It is accelerated or decelerated depending upon how high or low the probability of mutation is. Other features to note about the first 50,000 generations of this run include the variance in the population diversity and the sharp drops in population corresponding to the sharp drops in activity.

The second half of this ECHO run (generations 50,000 to 100,000) is characterized by a more or less uninterrupted increase in both the total and mean activity statistics. By generation 100,000 there are 10 distinct genotypes, 6 of which are persistent. The most persistent genotype emerged at generation 58,600 and so has persisted for over 40,000 generations—an order of magnitude longer than any previous genotype. The remaining five persistent genotypes appear at generations 67,100, 67,200, 69,300, 71,300, and 87,800. The average age of the persistent genotypes is 29,283 generations. The average age of all the genotypes is 18,450 generations. The persistent genotypes represent 60% of the genotype population and are collectively responsible for 97% of the total activity.

There are several striking features to note about this portion of the run. One is simply the striking adaptive success of the genotypes present. The youngest persistent genotype here is twice as old as the oldest persistent genotype we saw in the previous populations, and the average age of the persistent genotypes is nearly five times that of the oldest genotype seen in any of the above populations. The increase in population level shows that this ecology exploits the environment's energetic resources more efficiently than previous ecologies. It is also noteworthy that a minority of the genotypes in this ecology constitute ecological noise—only 40% as compared with 82%, 62%, and 88% respectively for the three previously considered ecologies. A much larger proportion of the genotypes play a significant role in the ecology's persisting structure.

But what is perhaps most striking is the ecology's stability. As the population rises and stabilizes, the diversity falls and stabilizes. In contrast to the previous ecologies, this ecology is robust in the sense that we never see a total recycling of the genotypes (which would be indicated by activity levels returning to 0). Persistent genotypes come and go as indicated by the sawtooth like nature of the climb but the disappearance of one persistent genotype is not accompanied by the demise of all the other genotypes. In fact a longer run of 10^6 generations (see Fig. 4.1) shows that this trend continues through generation 400,000. The stability of the complex ecology is evidence that the persistent genotypes are engaging in significant interactions. This ecology is clearly more complex than the preceding ecologies.

The run of 10^6 generations shows the persistent genotypes going extinct at generations 120,200, 120,300, 136,300, 136,300, 193,500, 205,100, and 413,500. The last genotype to disappear from this ecology persisted for nearly 350,000 generations. Furthermore, this last surviving genotype is joined by two additional genotypes that enter the ecology at generations 137,000 and 205,200 so that a stable three-genotype ecology persists for over 200,000 generations.

This illustrates how the activity statistics illuminate different kinds of adaptive phenomena. First, the qualitative difference in the activity (and diversity and population) statistics calls our attention to the two kinds of ecologies present in this run: simple ecologies in which a few genotypes persist in a setting dominated by a flux of ecological noise, and more stable and complex ecologies dominated by many interacting genotypes. The time series of these statistics clearly shows that a significantly more complex ecology has emerged in the course of evolution.

Second, the quantitative difference between the activity statistics in these two epochs has significance. The higher levels of total and mean activity in the complex ecology can be attributed to two things: there are six genotypes persisting due to being well adapted and a much smaller percent of the extant genotypes are ecological noise. Thus, the higher activity levels are signifying a higher level of continual adaptive success of genotypes, shown through their striking persistence in the face of selection.

A long-standing goal of the ECHO research community has been to demonstrate the emergence of complex ecologies [5, 9, 6, 4, 7, 10]. The statistics in Fig. 2.4 and Fig. 4.1 present the first empirical evidence in the literature of this

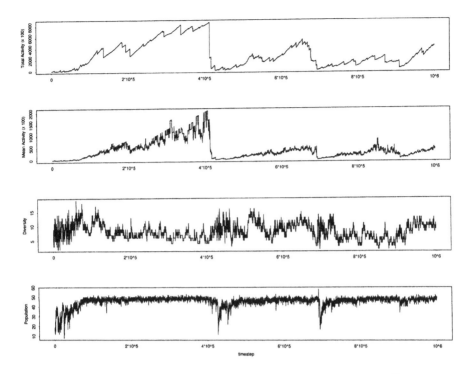

Figure 4.1: The same ECHO run as in Fig. 2.4 but extended out to 10^6 generations

phenomenon. Future work will be directed at articulating how these persistent genotypes emerge and what interactions among them produce their stability. But it is striking that, even in the absence of this future work, the evolutionary activity statistics themselves provide clear evidence for the emergence of this ecological complexity.

5. Conclusion

We believe that the kind of work reported here illustrates a significant new methodology for understanding the behavior of complex adaptive systems such as ECHO. Evolutionary activity statistics quantify the continual adaptive success of genotypes in an evolving population. In so doing they help us to identify when complex ecologies emerge in individual populations. The complex systems community needs such methodologies if it is to mature beyond intuitions and anecdotes.

Acknowledgements

This work is part of a larger collaboration involving Norman Packard, Emile Snyder, and Andreas Rechtensteiner; we are indebted to them for ongoing discussion and other help. We would also like to thank Simon Fraser for his helpful discussion and ongoing contributions to our work, including Fig. 2.1 and Fig. 2.2, and Terry Jones for his help with our initial evolutionary activity measurements in ECHO. Finally, we would like to thank the Santa Fe Institute for the hospitality, financial support, and computation resources which initiated and sustained this work.

Appendix

Parameters in ECHO are specified at three levels—the world, individual sites, and individual agents. They are reproduced here without explanation for those who might wish to reproduce our results.

```
World parameters:
    Name : simple.01mu.wrl
    Trading_Fraction : 0.5
    Interaction_Fraction : 0.5
    Self_Replication_Fraction : 0.5
    Self_Replication_Threshold : 2.0
    Maintenance_Probability : 0.2
    Neighborhood : NONE
    Rows : 1
    Columns : 1
    Number_Of_Resources : 4
    Combat_Matrix : 1 1 1 1 0
    Combat_Matrix : 1 1 1 1 0
    Combat_Matrix : 1 1 1 1 0
    Combat_Matrix : 1 1 1 1 0
    Combat_Matrix : -1 -1 -1 -1 0
    Sites :
        simple.01mu.ste

Site parameters:
    Name : simple.01mu.ste
    Mutation_Probability : 0.01
    Insertion_Deletion_Probability : 0.02
    Crossover_Probability : 0.0
    Random_Death_Probability : 0.0
    Production_Function : 20 20 20 20
    Initial_Resource_Levels : 20 20 20 20
    Maxima : 1000 1000 1000 1000
```

```
   Maintenance : 1 1 1 1
   Agents :
      cat 10
      ant 10
      fly 10

Agent parameters:
Name : cat
Initial_Reserves : 10 10 10 10
Trading_Resource : b
Uptake_Mask : * * * *
Interaction_Tag : *
Offense_String : d
Defense_String : a
Mating_Tag : a
Combat_Condition : c
Trade_Condition : b
Mating_Condition : b

Name : ant
Initial_Reserves : 10 10 10 10
Trading_Resource : a
Uptake_Mask : * * * *
Interaction_Tag : *
Offense_String : b
Defense_String : b
Mating_Tag : a
Combat_Condition : a
Trade_Condition : d
Mating_Condition : b

Name : fly
Initial_Reserves : 10 10 10 10
Trading_Resource : c
Uptake_Mask : * * * *
Interaction_Tag : *
Offense_String : a
Defense_String : b
Mating_Tag : a
Combat_Condition : d
Trade_Condition : c
Mating_Condition : b
```

486

Bibliography

[1] BEDAU, Mark A., "Three Illustrations of Artificial Life's Working Hypothesis," *Evolution and Biocomputation—Computational Models of Evolution*, W. BANZHAF and F. EECKMAN, eds., Springer, (1995).

[2] BEDAU, Mark A. and Norman H. PACKARD, "Measurement of Evolutionary Activity, Teleology, and Life", *Artificial Life II*, Christopher LANGTON, Charles TAYLOR, Doyne FARMER, Steen RASMUSSEN, eds., Addison-Wesley (1992).

[3] BEDAU, Mark A., Emile SNYDER, C. Titus BROWN, Norman PACKARD, "A Comparison of Evolutionary Activity in Artificial Evolving Systems and in the Biosphere", *Proceedings of the Fourth European Conference on Artificial Life*, Phil HUSBANDS and Inman HARVEY, eds., MIT Press/Bradford Books (1997).

[4] FORREST, Stephanie, and Terry JONES, "Modeling Complex Adaptive Systems with Echo," *Complex Systems: Mechanisms of Adaptation*, R. J. STONIER and X. H. YU, eds., IOS Press (1994).

[5] HOLLAND, John H., *Adaptation in Natural and Artificial Systems: An Introductory Analysis with Applications to Biology, Control, and Artificial Intelligence*, 2nd edition, MIT Press/Bradford Books (1992).

[6] HOLLAND, John H., "Echoing Emergence: Objectives, Rough Definitions, and Speculations for Echo-class Models," *Complexity: Metaphors, Models and Reality*, G. A. COWEN, D. PINES, and D. MELTZER, eds., Addison-Wesley (1994).

[7] HOLLAND, John H., *Hidden Order: How Adaptation Builds Complexity*, Helix Books (1995).

[8] HOLLAND, John H., *Talk at ECHO workshop 1997*, New Mexico.

[9] JONES, Terry, and Stephanie FORREST, "An Introduction to SFI Echo," Technical report 93-12-074, Santa Fe Institute, Santa Fe NM.

[10] JONES, Terry, Peter T. HRABER, and Stephanie FORREST, "The Ecology of Echo," *Artificial Life*, in press.

[11] LINDGREN, K., "Evolutionary Phenomena in Simple Dynamics", *Artificial Life II*, Christopher LANGTON, Charles TAYLOR, Doyne FARMER, Steen RASMUSSEN, eds., Addison-Wesley (1992).

[12] PACKARD, Norman, "Intrinsic Adaptation in a Simiple Model for Evolution", *Artificial Life*, Christopher LANGTON, ed. Addison-Wesley (1989).

[13] RAY, T. S., "An Approach to the Synthesis of Life", *Artificial Life II*, Christopher LANGTON, Charles TAYLOR, Doyne FARMER, Steen RASMUSSEN, eds., Addison-Wesley (1992).

Chapter 43

Spatial correlations in the contact process: A step toward better ecological models

Robin Snyder[1]

University of California, Santa Barbara, Physics Dept.

Many ecologists now believe that heterogeneity in the spatial distribution of predator and prey populations may help stabilize population levels over time. This heterogeneity is thought to emerge naturally as a result of the local nature of interactions. As a first step toward developing population models which take spatial correlations into account, I have investigated spatial heterogeneity in the 2-dimensional contact process, a very simple population model. I present data and simple theoretical predictions for several measures of heterogeneity and note which predictions appear to be robust even in the presence of fairly strong correlations.

1. Introduction

The difficulty of explaining the widespread stability of natural populations has plagued ecologists for many years. A predator-prey system which is modeled with two ordinary differential equations typically undergoes large oscillations in population, particularly as the capacity of the environment to support prey in the absence of predation increases (the "paradox of enrichment") [18]. This is in contrast, for example, to the relatively steady population levels seen among

---

[1]Supported in part by NSF grant DEB-9319301

birds, zooplankton, and agricultural pests kept in check by natural predators [10, 9].

Minor model modifications have not been able to resolve this mismatch between theory and observation. For example, positing that the death rate of the water flea *Daphnia* is not a constant but instead depends on the number of *Daphnia* present has not explained the stability of this population [9]. "Age-structured" models represent a more drastic change. These models treat groups of individuals in different life stages separately, noting, for example, that the values of life history parameters such as death rate may differ for juveniles and adults. However, this added realism has not in general explained why stability is so common [13].

Recognizing the spatial distribution of populations, on the other hand, has proven to be a more productive line of development. It has been noted that while populations may go extinct locally, these areas can be recolonized by areas with higher populations, thus stabilizing the average population level. In the laboratory Huffaker's experiments on interacting mite populations remain classic [7]. In the field this effect has been noted among butterflies [21], pikas [20], and mussels [11] in patchily distributed habitat.

A number of spatially-explicit models have arisen to explore this effect. Some have divided space into discrete habitat patches and have modeled the interpatch dynamics ("metapopulation" models) [4]. These models have been useful in studying habitats which are fragmented, such as boggy patches in otherwise dry land or collections of small islands. Others have used cellular automata or coupled map lattices [6, 17, 2, 8]. In these models space is divided into a grid. An individual or small population can exist at each point on the grid, and these interact according to local rules. The local nature of the interactions tends to produce clusters, which stabilize the average population dynamics. These models have been useful in cases where the habitat does not enforce a particular patch size, such as in aquatic environments.

Unfortunately, while computer realizations of these individual-based models provide tantalizing results, there are no analytic descriptions of these models, and so currently, predictions must be made on the basis of numerous simulations. Since clusters of prey and predators are presumed to interact only along their borders (interactions are local), it seems that one could base some sort of approximation on cluster sizes and shapes or on the probability that a site was occupied by prey given that one of the nearest neighbor sites was occupied by a predator, for example. Currently, these configurational quantities have not been much studied. It is for this reason that I have begun looking at clusters and correlations within the contact process, one of the simplest spatial models of population.

2. Introduction to the contact process

The contact process is defined on a lattice, each point of which can be occupied or vacant. Vacant sites become occupied at rate $\lambda\times$ (number of occupied nearest neighbors). Occupied sites become vacant at constant rate d, independent of their local environment. It is traditional to consider d as defining the timescale of the process and to set it equal to one. Thus,

vacant $\Longrightarrow$ occupied at rate $\lambda\times$ (number of occupied nearest neighbors)
occupied $\Longrightarrow$ vacant at rate 1.

One ecological interpretation of this is that occupied sites are prey species which reproduce asexually and establish their offspring nearby. (Consider, for example, strawberry plants spreading by sending out runners.) Their reproduction is limited by local crowding. The predator population on the other hand, is constant. Predators hunt by choosing sites at random. If prey is present, it is consumed. One imagines that all sites are equally accessible and equally likely to be chosen. Thus, prey have a constant probability per unit time of dying. I like to imagine the predators as particularly long-lived and undiscerning birds wheeling over a field of prey.

While the representation of the prey seems reasonable if simplistic, the representation of the predators is clearly ludicrous. Obviously, it will eventually be necessary to let predators occupy sites on the lattice and to allow them to reproduce and die and interact locally with prey. However, the behavior of the contact process is complex enough to be a useful first step.

3. Simulation details

I have used asynchronous update rules in my simulations. A "time of death" or "time of birth" is calculated for each site by noting the number of occupied nearest neighbors and drawing from appropriate exponential distributions. Whichever site's "alarm clock" goes off first is then changed, the system time is updated, and the nearest neighbors of the affected site are given new birth/death times based on their new local environment. All simulations were performed on a 128×128 lattice unless otherwise noted.

4. Measures of heterogeneity

I have considered a number of measures of heterogeneity to describe the clustering I have observed. One is average cluster volume. A cluster is a set of occupied sites which lie adjacent to each other. Each member of a cluster has at least one nearest neighbor which is also a member of the cluster. The volume of a cluster is simply the number of occupied sites it contains. Thus, if the dark squares on a chess board were considered "occupied," then there would be 32 clusters, each of volume 1. Technically, what I mean by "average" is an average over space and over an ensemble of realizations of the contact process. However, I have replaced the ensemble average with a time average, taking data only after the realization has reached a quasi-stationary state. A plot of average cluster volume versus λ can be seen in Fig. 5.2. The dramatic rise in average cluster volume at large λ corresponds to the appearance of a single, large, spanning cluster. This brings to mind site percolation theory, in which each site on the lattice is occupied with constant probability p. A phase transition occurs at p_c, which marks the lowest value of p for which there is a nonzero probability that a cluster will span the system. The system is said to percolate when a spanning cluster is present.

For comparison, I have equated p with the mean occupancy, p_1, and noted the value of λ at which percolation would first occur if this were a percolation system. This is indicated in Fig. 5.2.

I have also measured the conditional probabilities that a site will be occupied given an occupied nearest neighbor and that a site will be occupied given a vacant nearest neighbor. I have labeled these probabilities $q_{1|1}$ and $q_{1|0}$, respectively. It can be shown that $q_{1|0} = (p_1/(1 - p_1))(1 - q_{1|1})$. A graph of conditional probabilities versus λ can be seen in Fig. 5.3, Fig. 5.4.

As a third choice I have measured correlation length, ξ, a measure of heterogeneity heavily used in the physics community. I have defined the correlation length as the spatial separation necessary for the spatial autocorrelation function to drop below e^{-1} (the spatial autocorrelation functions for my data decrease exponentially). Recall that the spatial autocorrelation function is defined as

$$C(R) = \lim_{L \to \infty} \int_0^{2\pi} d\phi \int_0^L r dr f(r) f(r + R) \tag{4.1}$$

for an isotropic function $f(r)$ defined on a plane. The Wiener - Khinchin theorem states that the autocorrelation function is equal to the Fourier inverse of the spectral density. That is,

$$C(R) = \frac{1}{2\pi} \int_{-\infty}^{\infty} dk S_f(k) e^{ikR} \tag{4.2}$$

where

$$S_f(k) = \lim_{L \to \infty} \frac{|\tilde{f}(k)|^2}{L} \tag{4.3}$$

and

$$\tilde{f}(k) = \int_0^{2\pi} d\phi \int_0^L dr f(r) e^{ikr} \tag{4.4}$$

is the Fourier transform of $f(r)$.

I have used the software package Khoros to find the power spectrum of "snapshots" of the contact process. Some azimuthally-averaged power spectra can be seen in Fig. 4.1. The data have been rescaled to have a mean of zero in order

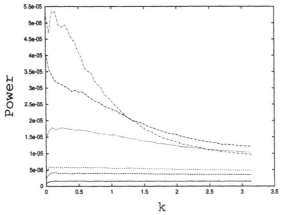

Figure 4.1: Power spectra of "snapshots" of the lattice. Each curve is the average of over 100 spectra. From largest y-intercept to smallest, $1/\lambda = 2.0$, 1.5, 1.0, 0.5, 0.25, 0.1.

to avoid a large spike at $k = 0$. Examples of correlation lengths found from this data can be found in Table 4.1.

The last measures of heterogeneity I have used are join count statistics, measures used most frequently in geography and biology [16, 14]. One imagines placing a point at each site and drawing lines between points that are defined as neighbors of each other. The definition of "neighbor" will depend on what one is investigating. "Joins" are the lines that connect neighbors. Since my system has nearest neighbor interaction rules, I have considered joins between nearest neighbor sites.

I have then asked about the number of joins connecting occupied sites (1-1 joins) and the number of occupied-vacant joins (1-0 joins). Others have computed the expected mean and variance of these join counts in the case of spatial randomness (see, for example, [16, 14]). As a measure of the strength of correlations, I have compared the join counts for my snapshots of the contact processs with the join counts expected under spatial randomness and noted the standard normal deviate, defined as

$$\text{S.N.D.} = \frac{(\text{observed} - \text{expected mean})}{(\text{expected variance})^{1/2}}. \tag{4.5}$$

The standard normal deviate is thus the number of standard deviations aways from the mean that the data would be if the spatial distribution of the occupied sites actually was random. Graphs of the standard normal deviates of join counts can be seen in Fig. 5.5, Fig. 5.6.

$\frac{1}{\lambda}$	ξ
0.1	0.64
0.5	0.67
1.0	0.72
1.5	0.82
2.0	1.10

Table 4.1: Correlation lengths measured from the power spetra shown in Fig. 4.1

5. Theoretical predictions

A simple starting place for theoretical predictions is a mean field approximation (MF), so named because the local environment of any given site is assumed to be equal to the average environment. Thus, if the fraction of occupied sites is denoted by p_1, and each site has z nearest neighbors, then in MF, each site is assumed to have zp_1 occupied nearest neighbors. Thus,

$$\dot{p}_1 = \left\{ \begin{array}{c} \text{probability of} \\ \text{an empty site} \end{array} \right\} \left\{ \begin{array}{c} \text{local birth rate} \\ \text{at that site} \end{array} \right\}$$
$$- \left\{ \begin{array}{c} \text{probability of} \\ \text{an occupied site} \end{array} \right\} \left\{ \text{death rate} \right\}$$

$$= (1 - p_1)(z\lambda p_1) - p_1 \tag{5.1}$$

This predicts a mean occupancy of

$$p_{1MF}^* = 1 - \frac{1}{z\lambda} \tag{5.2}$$

at equilibrium.

It is difficult to use this result to predict mean cluster volume directly, since one must take into account the number of ways a cluster of given size can be configured. However, since MF presumes that there are no spatial correlations, the MF volume prediction should simply be the average volume expected if each site were occupied independently with probability $p_1^*{}_{MF}$. This is just the average volume given by percolation theory, discussed in §4.

I have done a small precolation study on the same size lattice used for the rest of my data, and the predictions for cluster volume are presented in Fig. 5.2. It is interesting to note that both the contact process and the percolation process percolate at roughly the same point. Evidently, spatial correlations are not too significant for the value of p_1 at which the contact process percolates.

While easy, MF often produces rather crude results. As can be seen by examining Fig. 5.1, Fig. 5.2, now is no exception.

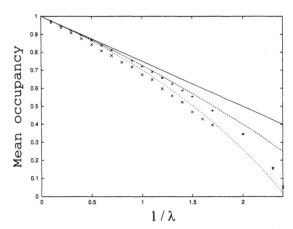

Figure 5.1: Mean occupancy p_1. Solid line: MF, dashed line: PA, shorter dashes: IPA, crosses with error bars: data from 128×128 lattice, $\times$'s: data from 10×10 lattice.

One way of improving upon MF is to make a pair approximation (PA), which considers not just the mean occupancy but also the occupancies of pairs of adjacent sites. Let p_{ab} be the probability that two adjacent sites have values a and b. I treat p_{10} and p_{01} as distinct but equal, so that $p_{11} + 2p_{10} + p_{00} = 1$. Let $q_{a|b}$ be the probability that a site with value b has a neighbor with value a. Finally, let $q_{a|bc}$ be the probability that, given a site with value b and a neighbor with value c, a different neighbor of the b site has value a. The pair approximation consists in assuming the $q_{a|bc} \approx q_{a|b}$.

Now that we are writing down pair probabilities, we can write down the differential equation for p_1 exactly. It is

$$\dot{p}_1 = z\lambda p_{10} - p_1. \tag{5.3}$$

The differential equation for the pair probabilities depend on 3-site probabilities.

$$\dot{p}_{11} = 2\lambda p_{10}(1 + (z-1)q_{1|01}) - 2p_{11}, \tag{5.4}$$

$$\dot{p}_{00} = 2p_{10} - 2\lambda(z-1)q_{1|00}p_{00}, \tag{5.5}$$

$$\dot{p}_{10} = \lambda(z-1)q_{1|00}p_{00} + p_{11} - [1 + \lambda + \lambda(z-1)q_{1|01}]p_{10}. \tag{5.6}$$

However, we truncate the exploding number of differential equations by using the PA to write equations for pair probabilities in terms of other pair probabilities.

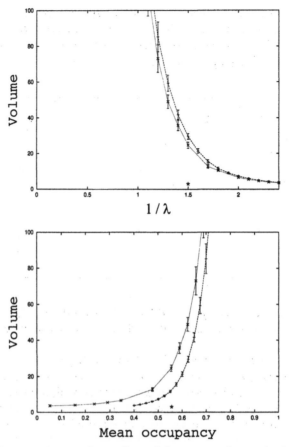

Figure 5.2: Average cluster volume. I noted the average cluster size in various lattice snapshots, then averaged over snapshots. Dotted line with ×'s: data from 128 × 128 lattice, dashed line with crosses: MF prediction based on percolation study on a 128 × 128 lattice. The * indicates the value at which percolation is predicted for the percolation process on an infinite lattice [22].

After some manipulation (see [19] for similar treatment of a three-state model), we arrive at

$$\dot{q}_{1|1} \approx - q_{1|1}[z\lambda(1 - q_{1|1}) - 1] +$$

$$2\lambda(1 - q_{1|1})[1 + (z - 1)(1 - q_{1|1})p_1/(1 - p_1)]. \tag{5.7}$$

This combined with exact Eq. 5.3 gives us a closed system.

Setting both equations equal to zero, we find

$$p^*_{1PA} = \frac{z(\lambda(z - 1) - 1)}{z\lambda(z - 1) - 1} \tag{5.8}$$

and

$$q^*_{1|1} = 1 - \frac{1}{z\lambda} \qquad (exact). \qquad (5.9)$$

These results can then be used to find PA predictions for conditional probabilities, and join counts. These predictions are summarized in Table 5.1 and plotted in Fig. 5.3–Fig. 5.6. They are better than MF but still leave much to be desired.

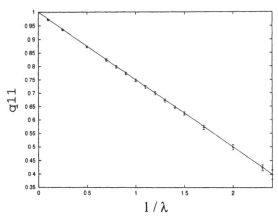

Figure 5.3: Conditional probability of a 1 given a neighboring 1. Solid line: exact prediction, crosses with error bars: data from 128×128 lattice.

One can do a little better without enduring the tedium of including three-state probabilities by using the improved pair approximation (IPA) [19, 5]. The idea here is that because occupied sites tend to cluster, $q_{1|00} < q_{1|01}$. Thus, it is no longer appropriate to approximate them both by $q_{1|0}$. By contrast, other approximation such as $q_{1|11} \approx q_{1|1}$ are still acceptable.

More formally, we assume

$$q_{1|00} \approx \varepsilon q_{1|01}. \qquad (5.10)$$

The value of ε is set by saying that $q_{1|0} = 0$ at the extinction threshold. One finds that

$$\varepsilon = \left(\lambda_c\left(1 - \frac{1}{z}\right)\right)^{-1} < 1, \qquad (5.11)$$

where λ_c is the lowest value of the birthrate at which the contact process has a non-zero probability of surviving indefinitely.

By using the identity $q_{1|01}q_{1|0} + q_{1|00}q_{0|0} = q_{1|01}$ one can derive

$$q_{1|01} \approx 1 - \varepsilon q_{0|0} = 1 - \varepsilon[1 - (1 - q_{1|1})p_1|(1 - p_1)], \qquad (5.12)$$

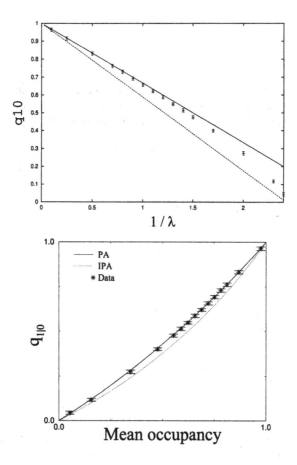

Figure 5.4: Conditional probability of a 1 given a neighboring 0. Solid line: PA, dashed line: IPA, crosses/stars with error bars: data from 128×128 lattice.

which enables one once again to obtain a closed set of equations. A conjecture by Brower, Furman, and Moshe sets $\lambda_c = 0.4119$ [1]. Reworking the calculations made for the PA, one arrives at the improved estimate

$$p^*_{1\,IPA} = \frac{z(\lambda - \lambda_c)}{1 + z(\lambda - \lambda_c)}. \tag{5.13}$$

Predictions for conditional probabilities and join counts are summarized in Table 5.1. Comparisons with data can be found in Fig. 5.3,Fig. 5.4,Fig. 5.5,Fig. 5.6, where I have used Brower et al.'s estimate of λ_c. The IPA predictions often fare somewhat worse than the PA predictions. I do not yet understand this.

A rather different approach to improving mean field theory is to add noise to the mean field equations. One gains no information about spatial correlations, of course. However, by taking into account some of the jitter that results from

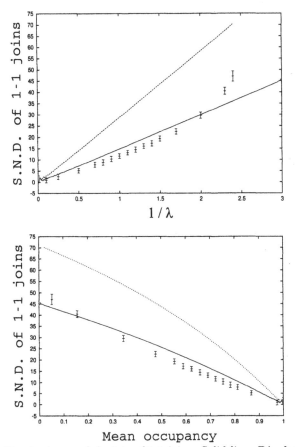

Figure 5.5: Standard normal deviate of 1-1 joins. Solid line: PA, dashed line: IPA, crosses with error bars: data from a 128×128 lattice.

probabilistic birth and death rates, one obtains a better understanding of the occupancy variability. This was the approach taken by Gurney and Nisbet in their 1978 paper on population fluctuations in various simple models, including the contact process [3].

With the addition of noise, Eq. 5.1 becomes

$$\dot{p}_1 = (1 - p_1)(z\lambda p_1) - p_1 + \sqrt{(1 - p_1)(z\lambda p_1)}\, \gamma_1(t) - \gamma_2(t) \qquad (5.14)$$

where $\gamma_1(t)$ and $\gamma_2(t)$ are independent white-noise processes of unit spectral density.

By linearizing about the deterministic equilibrium, Gurney and Nisbet found that the population variance is

$$\sigma^2 = \frac{N}{z\lambda} \qquad (5.15)$$

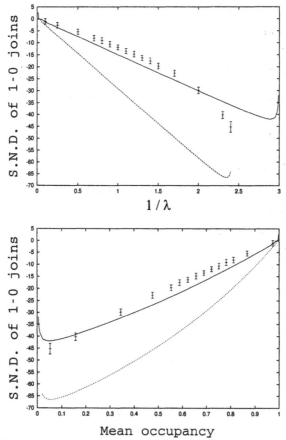

Figure 5.6: Standard normal deviate of 1-0 joins. Solid line: PA, dashed line: IPA, crosses with error bars: data from a 128×128 lattice.

where N is the number of sites. The mean is unchanged from its deterministic value. The whole subject of stochastic additions and approximations that can be made is well treated in [12, 15].

Gurney and Nisbet found that their estimates of population mean and variance did worse as λ decreased, although the ratio of variance to mean as a function of the mean was well predicted. The limited computer power available in 1978 forced them to run all of their simulations on a 10×10 lattice, however, which resulted in significant finite size effects.

Comparing the results from a 10×10 lattice with a 128×128 lattice (Fig. 5.1,Fig. 5.7), one sees that much of the discrepancy with stochastic mean field results is due to finite size effects. It seems probable, however, that even on an infinite lattice, some discrepancy may remain at low λ.

exact:	$q_{1\|1} = 1 - \frac{1}{4\lambda}$
	$q_{1\|0} = \frac{p_1}{1-p_1}(1 - q_{1\|1})$
	number of 1-1 joins $= 2L(L-1)q_{1\|1}p_1$
	where lattice size $= L \times L$
	number of 1-0 joins $= 4L(L-1)q_{1\|0}(1-p_1)$
	number of 0-0 joins $= 2L(L-1)(1-q_{1\|0})(1-p_1)$
MF:	$p_1 = 1 - \frac{1}{4\lambda}$
PA:	$p_1 = \frac{z(\lambda(z-1)-1)}{z\lambda(z-1)-1}$
IPA:	$p_1 = \frac{z(\lambda-\lambda_c)}{1 + z(\lambda - \lambda_c)}$

Table 5.1: Theoretical Predictions

6. Discussion

6.1. Birthrate or density as independent variable?

Ultimately, one would like to use this work to improve the differential equations currently used to model interacting populations. There, population density is the variable of choice. Therefore, I have also included plots of predicted heterogeneity measures versus predicted mean occupancy. In all cases predictions of heterogeneity as a function of occupancy do at least as well as predictions that were functions of birthrate, and in one case, $q_{1\|0}$, the pair approximation as a function of occupancy does noticeably better, predicting the data almost exactly. This is very promising.

6.2. Spatial correlations near extinction

All of the approximations presented here fare worse as λ_c and occupancy approach zero. This suggests that correlations beyond nearest neighbors are becoming significant, which fits in nicely with the observation that ξ increases with decreasing λ (see Table 4.1). One should be cautious in interpreting this. The correlation length, ξ, measures the correlations not only between occupied sites, but also between vacant sites, and it may be that the increasing ξ is actually a measure of the increasingly large vacant areas which occur as λ decreases. Such gaps would occur with decreasing population even if occupied sites were distributed randomly, in accordance with the mean field assumption. That said,

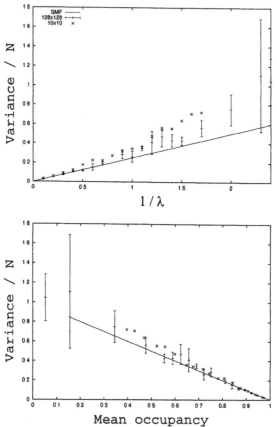

Figure 5.7: Population variance over time normalized by lattice size. Solid line: SMF, crosses with error bars: data from 128×128 lattice, ×'s: data from 10×10 lattice.

I have also looked at some 1-1 join statistics for joins not only between nearest neighbors, but also between second nearest neighbors and so on, all the way out to seventh nearest neighbors. The 1-1 join statistics, which measure only correlations between occupied sites, remain significantly different from the null hypothesis of spatial randomness for increasingly distant neighbors as λ decreases.

More work needs to be done, but I believe that the increasing correlations near extinction are real and I suspect that this will be true not only for the contact process but for a number of other simple birth-death models. My thinking is guided by the continuous phase transitions studied in physics, which exhibit long-range spatial correlations as they pass through critical points. If these lengthening correlations near the extinction threshold prove widespread, it will be interesting to investigate what effect they have on mean times to extinction. In the current model, the variance appears to be greater than one would expect based on a mean field approximation. This suggests that spatial correlations may make it more likely that a small population will experience a fluctuation that brings it down to zero.

Acknowledgements

I express my gratitude to Roger Nisbet for his guidance and to Jun Liu for introducing me to the Khoros software package.

Bibliography

[1] BROWER, R. C., Furman M. A., and M. MOSHE, "Critical exponents for the reggeon quantum spin model", *Physics Letters B* **76B**, 2 (1978), 213–19.

[2] DE ROOS, A M, E MCCAULEY, and W G WILSON, "Mobility versus density-limited predator-prey dynamics on different spatial scales scales", *Proceedings of the Royal Society of London Series B Biological Sciences* **246**, 1316 (1991), 117–122.

[3] GURNEY, W. S. C., and R. M. NISBET, "Single-species population fluctuations in patchy environments", *The American Naturalist* **112**, 988 (1978), 1075–1090.

[4] HANSKI, Ilkka, and Michael E. GILPIN, *Metapopulation biology : ecology, genetics, and evolution*, Academic Press San Diego, CA (1997).

[5] HARADA, Y, H EZOE, Y IWASA, H MATSUDA, and K SATO, "Population persistence and spatially limited social interaction", *Theoretical Population Biology* **48**, 1 (1995), 65–91.

[6] HASSELL, M P, H N COMINS, and R M MAY, "Spatial structure and chaos in insect population dynamics", *Nature (London)* **353**, 6341 (1991), 255–258.

[7] HUFFAKER, C.B., "Experimental studies on predation: dispersion factors and predator-prey oscillations", *Hilgardia* **27**, 14 (Aug. 1958), 343–383.

[8] MCCAULEY, E, W G WILSON, and A M DE ROOS, "Dynamics of age-structured and spatially structured predator-prey interactions individual-based models and population-level formulations", *American Naturalist* **142**, 3 (1993), 412–442.

[9] MURDOCH, W.W., R.M. NISBET, E. McCAULEY, A.M. DEROOS, and W.S.C. GURNEY, "Plankton abundance and dynamics across nutrient levels: tests of hypotheses" (1996).

[10] MURDOCH, W W, "Population regulation in theory and practice", *Ecology (Tempe)* **75**, 2 (1994), 271–287.

502

[11] MURDOCH, W. W., and A. OATEN, "Predation and population stability", *Advances in Ecological Research* **9** (1975), 1–131.

[12] NISBET, R. M., and W. S. C. GURNEY, *Modelling Fluctuating Populations*, John Wiley & Sons New York (1982).

[13] NISBET, R. M., W. S. C. GURNEY, W. MURDOCH, and E. MCCAULEY, *Biological Journal of the Linnaean Society* **37** (1989), 79–99.

[14] ODLAND, John, *Spatial Autocorrelation*, Sage Publications, Newbury Park, CA (1988), ch. Spatial analysis and statistical inference, pp. 7–.

[15] RENSHAW, Eric, *Modelling Biological Populations in Space and Time*, Cambridge University Press New York (1991).

[16] ROBERT R. SOKAL, F.M.L.S., and Neal L. ODEN, "Spatial autocorrelation in biology: 1. methodology", *Biological Journal of the Linnean Society* **10** (1978), 199–228.

[17] ROHANI, P, T J LEWIS, D GRUNBAUM, and G D RUXTON, "Spatial self-organization in ecology: Pretty patterns or robust reality?", *Trends in Ecology & Evolution* **12**, 2 (1997), 70–74.

[18] ROSENZWEIG, M. L., "Paradox of enrichment: destabilization of exploitation ecosystems in ecological time", *Science* **171** (1971), 385–387.

[19] SATO, K, H MATSUDA, and A SASAKI, "Pathogen invasion and host extinction in lattice structured populations", *Journal of Mathematical Biology* **32**, 3 (1994), 251–268.

[20] SMITH, A. T., "The distribution and dispersal of pikas: consequences of insular population structure", *Ecology* **55** (1974), 1112–1119.

[21] THOMAS, Chris D., and Ilkka HANSKI, "Butterfly metapopulations", *Metapopulation Biology: Ecology, Genetics, and Evolution*, (I. HANSKI AND M. GILPIN eds.). Academic Press (1997), pp. 359–386.

[22] VAN DEN BERG, J., and A. ERMAKOV, "A new lower bound for the critical probability of site percolation on the square lattice", *Random Structures and Algorithms* **8**, 3 (May 1996), 199–212.

Chapter 44

Many to one mappings as a basis for life

Roland Somogyi

Molecular Physiology of CNS Development, LNP/NINDS/NIH
36/2C02, Bethesda, MD 20892

While attempting to explain the origins of life, we first need to consider what we mean by life. Since life is a complex phenomenon, a simple, clear-cut definition may not exist. However, there is a list of criteria that help us to determine whether a system is living. This study is centered on one of these criteria, namely that an organism must retain minimal stability and growth in a fluctuating and diverse environment. I refer to this as a many to one mapping. Environmental fluctuations of a wide range of parameter values must map to a narrow range of intracellular parameters compatible with the viability of the organism. For instance, a primitive proto-organism must be able to tolerate the natural range of fluctuations of extracellular building blocks and energy sources. The ability to utilize various species of extracellular chemical sources would allow survival in a more complex environment, also enabling further spread and growth. In addition, complex metabolisms must also be insensitive to internal fluctuations of chemical species. Finally, distributed signaling systems in multicellular, differentiated organisms must exhibit minimal fidelity in inter- and intracellular information transmission. This criterion is a condition for the reliable coordination of the activities of hundreds to billions of cells during development and in the differentiated organism. We will examine several examples of many to one mappings from primordial origins to error correcting codes in distributed information transmission.

1. Criteria for life

A short discussion of some defining criteria for life will be necessary before contemplating origins. A living system

1. must be bounded from the environment; topology—inside versus outside—self vs. non self. Strictly speaking, this boundary must allow for information exchange between inside and outside.

2. must absorb energy and chemical building blocks from environment (dissipative structure; minimal criterion for steady state). Energy source and building blocks were probably united in the first organisms, but are separated through coupled reactions later in evolution. Building blocks loose many degrees of freedom as they become absorbed in organism (many to one mapping).

3. must grow and replicate. Energy source and building blocks not only maintain steady state, but lead to growth and multiplication of the organism.

4. must adapt to changing environments. Either the organism must be able to survive given many energy and building block sources (some bacteria), or must find ubiquitous, reliable sources (e.g. plants).

5. must be able to absorb or identify and avoid external threats (toxins, temperature, mechanical insults, predators).

6. must evolve memory molecule to support greater complexity. Storage of past adaptations in a memory molecule is more reliable than storage in fluctuating concentrations of primitive catalysts. This should significantly quicken the pace of evolution.

7. must be able to absorb fluctuations in its metabolism and signaling networks.

2. The principle of many to one mapping

We assume that life originated in a fluctuating, noisy environment. We certainly know that to be the case for life today. Moreover, complex systems of coevolution generate major long term environmental changes. To a degree, these fluctuations may be carried over into the organism and result in a proportional state change. Some of these fluctuations form the basis of new variability allowing evolution (compare to [7])! However, these internal state changes may never transgress the range of internal parameter values permitting viability (Fig. 2.1). Given the unpredictability of the environment, organisms must learn how to cope with these fluctuations and reduce their impact to tolerable levels of internal parameter values. This is the nature of the many to one mapping.

3. Many to one mappings in the origins of life and evolution of complex networks

Below I will consider each criterion for life with respect to the principle of many to one mapping.

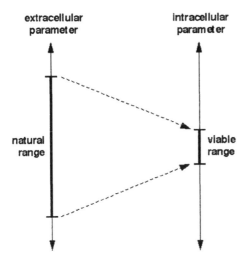

Figure 2.1: Many-to-one mapping of extracellular to intracellular parameter ranges.

1. An organism must be bounded from the environment.

Given a primordial soup scenario, it is conceivable that polar lipids were 'spontaneously' generated. Under these conditions, these building blocks should form vesicles, corresponding to the minimum energy structure in an aqueous environment. The topological requirement for an 'inside vs. outside' or 'self vs. non-self' will thus have been met. On a relatively trivial level, one could view this as a many to one mapping in the sense that many building blocks form one coherent and stable structure. I conjecture that this first boundary allows for limited exchange of extra- and intravesicular components (Fig. 3.1).

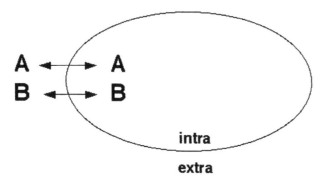

Figure 3.1: A vesicle as the structural basis of a protocell.

2. An organism must absorb energy and chemical building blocks from environment to allow for a primitive, catalytically supported

proto-metabolism.

Again assuming that a primordial soup constitutes a reasonable amount of chemical diversity, one may conjecture that accidental condensations of building blocks could lead to larger molecules with a level of catalytic activity, forming a primitive reaction network or autocatalytic set[1]. If the example of Fig. 3.2, hetero-dimerisation of two monomers is supported by catalytic feedback of the dimer on its own production.

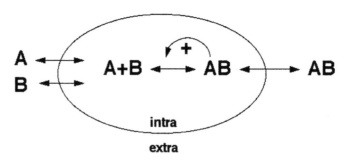

Figure 3.2: A vesicular protocell supporting accumulation and turnover of catalytic products.

I examined the kinetics of this simple catalytic feedback loop taking into account the transport of reactants across the vesicle (Fig. 3.3). The rate equations are simple and will not be considered any further. The point of this analysis is that steady state concentrations of dimer can be sustained by this catalytic feedback process given that the extracellular concentrations of building blocks is sufficiently high. Without catalysis and feedback, steady state concentrations could only be maintained given much higher extracellular monomer concentration. However, the problem with this system is that it has no internal upper boundary. Given very high external monomer concentrations, intracellular dimer concentrations could lead to dangerously high levels. This could threaten the stability of this simple system through osmotic stress or crystallization of dimer. Moreover, this system only allows for a steady state, it cannot grow, and there is no protection from degradation of the vesicular boundary membrane. This system constitutes a many to one mapping, again on a relatively trivial level: many building blocks aggregate to form fewer reaction products. In a sense, there is a reduction of the degrees of freedom of the building blocks as they become integrated into the system.

3. An organism must grow and replicate.

Energy source and building blocks not only maintain steady state, but lead to growth and multiplication of the organism (Fig. 3.4). In the previous example (Fig. 3.2 and Fig. 3.3), the product of the autocatalytic set could only accumulate within an unchanging volume. At worst, this could lead to 'poisoning' of the system if the concentration of extracellular building blocks is too high; and, there is no allowance for growth. Growth requires that the products of the

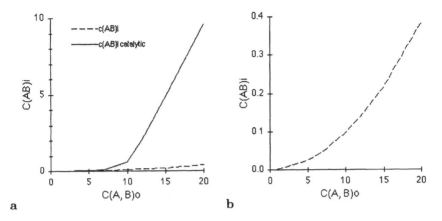

Figure 3.3: Kinetics of autocatalytic set including support of membrane structural components and the resulting spatial considerations. The intracellular concentrations of dimer (AB), C(AB)i, is plotted against the extracellular concentration of monomers (A, B), C(A, B)o. **a)** Comparison of catalytic feedback to basal reaction rates. **b)** Magnified curve for non-catalytic accumulation of c(AB)i (corresponds to lower curve in **a**). The rate constants for synthesis and breakdown of AB were chosen to be 1000 times lower than the membrane transport rates (values symmetric). Catalysis is proportional to 10 times the concentration of AB.

'metabolism' become incorporated in the organism's structure (beginning with the boundary membrane) to increase its size. In the schematic in Fig. 3.4a, part of the XY reaction products are accumulated by the membrane structure. This first leads to a volume increase, and then eventually to budding (conjecture) and growth in the number of organisms.

The kinetics of this simple system are shown in Fig. 3.5. The increase of dimer concentration levels off with increasing extracellular monomer concentration (Fig. 3.5a; catalytic & membrane growth). The risk of 'poisoning' is reduced compared to dimer production without diversion to structural components (Fig. 3.5a; catalytic). However, compared to the non-catalytic case, significant concentrations of dimer can be achieved. Note that there as a significant increase of growth rate with rising extracellular monomer concentration (Fig. 3.5b). First, we again see a simple many to one mapping of large numbers of building blocks being incorporated into one coherent structure. More importantly, we see that the range of intracellular catalytic product concentration narrows with respect to the range of source concentration (Fig. 3.5a; in contrast to the catalytic case without growth). This confers a stability to the internal product concentration with respect to fluctuations in extracellular source availability.

4. An organism must adapt to changing environments.

Either the organism must be able to survive given many energy and building block sources (some bacteria), or must find ubiquitous, reliable sources (e.g.

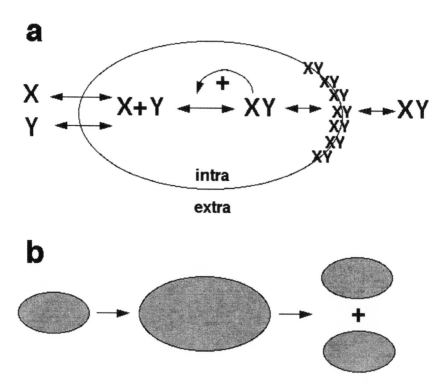

Figure 3.4: Growth and replication protocell harboring a minimal autocatalytic set.

plants). This just extends the many to one mappings from fluctuating concentrations of a small set of chemical sources to a larger set. The point is that some chemical sources are allowed to disappear, causing the organism to generate its structures from alternative species.

5. An organism must be able to absorb or identify and avoid external threats (toxins, temperature, mechanical insults, predators).

Assuming highly non-linear interactions approaching on/off behavior, we can treat the biomolecular parameters in an organism as binary variables. Considering that each binary variable may have multiple inputs and outputs, one can construct a distributed network[1]. Trajectories of such Boolean networks eventually reach attractors. Given biologically plausible network architectures, such attractors can be reached from large numbers of starting and intermediate states. Such extreme many-to-one mappings result in the confluence of hundreds to thousands of different states into small dynamic attractors 10 states or less[2]. These structures are referred to as basins of attraction[5]. If we consider that a subset of the binary variables in this network correspond to environmental parameters, one can envision that selected networks can easily compensate for significant environmental fluctuations.

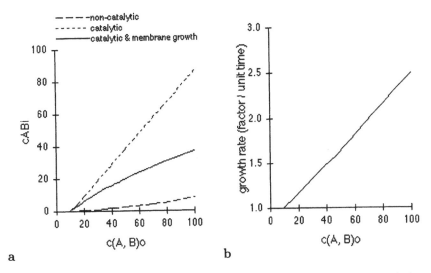

a   b

Figure 3.5: Kinetics of autocatalytic set including growth of membrane. **a)** Accumulation of Abi for non-catalytic, catalytic, and catalytic production of AB including membrane sink. Note that the increase of cABi levels off with increasing extracellular building blocks due to growth of organism. **b)** Growth rate of organisms due to membrane building. Parameters as in Fig. 3.3, except that 1/10 of AB efflux is absorbed by the membrane structure.

6. An organism must evolve a memory molecule to support greater complexity.

Storage of past adaptations in a memory molecule is simpler and more reliable than storage in fluctuating concentrations of primitive catalytic oligomers (Fig. 3.6). A large set of catalytic processes are required to generate a particular polymer in a series of catalytic reactions. Such a system constitutes a distributed memory if the results are reproducible. Many of these processes can be mapped to a single molecular memory structure serving as a template for a wide variety of catalytic polymers. This should significantly quicken the pace of evolution.

7. An organism must absorb fluctuations in its metabolism and signaling network.

There is ample evidence for thresholding behavior in biomolecular signaling systems. I conjecture that thresholding functions not only evolved due to constraints in the generation of biomolecular interactions, but because they provide a many to one mapping that reduce the aberrant fluctuations in the input parameter. This is simply a question of minimal fidelity of information transmission given a noisy source. Networks of thresholding functions generate large many to one mappings (discussed in point 5 above). In terms of a *single* molecular input channel, multiple values of the input parameter map to a reduced set of values of the output parameter (Fig. 3.7).

In terms of *multiple* molecular inputs, the response can be determined by

510

Distributed Memory: autocatalytic sets, primitive metabolism, origins of life

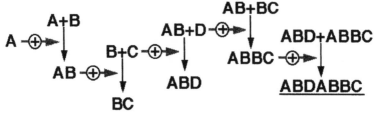

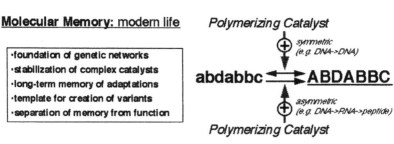

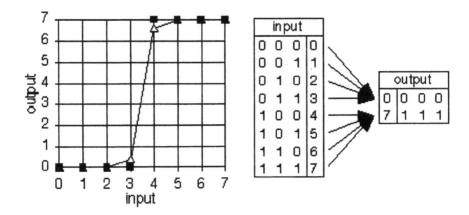

Figure 3.6: Distributed memory vs. molecular memory.

Figure 3.7: Thresholding functions allow a many to one mapping. This allows for decreased sensitivity of fluctuations in the input.

thresholding of the sum of input values (Fig. 3.8). This resembles a simple error-correcting code[4]. This mapping also corresponds to the highest *distributivity* rule for $k = 3$[3]. The *distributivity* measure captures in how far each input element and all combinations of input elements contribute to determining the output, analyzed using the concepts of Shannon entropy and mutual information. The definition of *distributivity* was motivated by the conjecture that it is important (no matter which physical realization of the information transmission process), that if multiple channels contribute to an output, the information of each channel should make an optimal contribution to the output. In other words, the error contribution of each channel to the output should be minimized, and the information contribution democratically maximized.

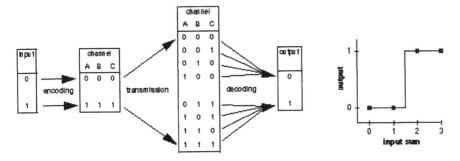

Figure 3.8: Realization of an error correcting code in a 3 element rule (Boolean network idealization).

4. Outlook

I discussed several examples of many to one mappings for several types of parameters. These parameters can be molecules, concentrations of molecules, and, in signaling networks, essentially any physical process that carries information in a biological system. One may wonder which is the appropriate physical scale according to which these parameters may be compared. This is a difficult question. However, at minimum, this scale must take into account the amount of information each parameter carries. But how should this information be quantified. Is it merely relative, or could we find an absolute scale?

Of course, there are frameworks within which information can be quantitatively dealt with, namely information theory and algorithmic information theory. Moreover, these theories are closely related to physical phenomena concerning the thermodynamic concept of entropy and, generally, the complexities of systems. But, there is no proven framework to my knowledge that allows us to consistently translate physical parameters into their information equivalent. Moreover, defining the information content of a physical parameter in terms of its distribution in an ensemble (Shannon entropy) ignores information about how this physical parameter can interact with other physical parameters. In other words, the physical laws that relate one variable to another contain information and are inherently part of any physical parameter. Perhaps these questions may become clearer during the further development of an 'information physics'[6].

The observations discussed above are in accordance with the view that life internalizes the external world. In other words, living organisms absorb information (in many different forms) and make an internal model of the external world. Finally, self awareness allows an organism to make a model of itself. In as much as life has evolved from the 'external world', life could be viewed as a self recognition process from the point of view of the whole system. This is reminiscent point made by Zurek[7], that life absorbs statistical entropy, H, and transforms it into its structural complexity or algorithmic information, K.

Bibliography

[1] Kauffman, S.A. (1993) The Origins of Order, Self-Organization and Selection in Evolution. Oxford University Press.

[2] Somogyi, R. and Sniegoski, C.A. (1996) Modeling the complexity of genetic networks: understanding multigenic and pleiotropic regulation. Complexity 1(6):45-63.

[3] Somogyi, R. and Fuhrman, S. (1997) Distributivity, a general information theoretic network measure, or why the whole is more than the sum of its parts. Proc. International Workshop on Information Processing in Cells and Tissues (IPCAT) 1997. In press.

[4] Thomas J.B. (1969) An Introduction to Statistical Communication Theory. Wiley.

[5] Wuensche, A., Lesser, M.J. (1992) The Global Dynamics of Cellular Automata, SFI Studies in the Sciences of Complexity. Addison Wesley.

[6] Zurek, W.H, ed. (1990) Complexity, Entropy and the Physics of Information, , SFI Studies in the Sciences of Complexity. Addison Wesley.

[7] Zurek, W.H (1990) Algorithmic information content, Church-Turing thesis, physical entropy, and Maxwell's Demon, in Complexity, Entropy and the Physics of Information, Zurek, W.H, ed., SFI Studies in the Sciences of Complexity. Addison Wesley.

Chapter 45

Generic mechanisms for hierarchies

Didier Sornette[1]

LPMC, CNRS and Université de Nice-Sophia Antipolis
06108 Nice Cedex 2, France
and IGPP and ESS, Box 951567
Los Angeles, CA 90095-1567, USA

Common to essentially all complex systems is the existence of a hierarchical organization, with transfers across the different levels from the smallest to the largest ones and vice-versa. Here, we propose that hierarchies are generically formed dynamically due to a spontaneous breakdown of continuous scale invariance into discrete scale invariance. We present a short introduction to the concept of discrete scale invariance and how it leads to complex critical exponents (or dimensions), i.e. to the log-periodic corrections to scaling. Proposed illustrations are diffusion-limited-aggregation clusters, rupture in heterogeneous systems, earthquakes, animals (a generalization of percolation) among several other systems. We refer to several papers and a recent review for further information, especially on the use for predictions.

1. Introduction

Notwithstanding the extraordinary large variety of disciplines addressed in this conference, at least one simple concept was recurrent in most if not all presentations, namely the existence of a hierarchical structure. Indeed, in most complex systems, many length scales are present but they do not play an equivalent role. Rather, there are discrete levels in a global hierarchy, each level being described by a set of laws and tools. Examples are found in the weather (dust devils, tor-

---

[1]Supported in part by NSF EAR96-15357 and NSF EAR97-06488

nadoes, cyclones, large scale motions), in the tectonic crust (joints, small faults, main faults, plate boundaries), in proteins (primary, secondary, tertiary, quaternary structures), in physiology (molecules, cells, tissues, systems), in the brain (neurons, functional regions, lobes, hemispheres), in economy (traders, companies, countries, currency blocks), in society (individual, family, local social group, town, region, country), etc.

However, the question of the origin of this hierarchy is poorly understood probably due to the fact that many detailled mechanisms come into play. Here, we would like to propose a robust "meta-mechanism" according to which hierarchies can naturally emerge from the *partial* breakdown of the most natural symmetry in systems with many length scales, namely the symmetry of continuous scale invariance.

The partial breakdown of continuous scale invariance corresponds to *discrete scale invariance*. We have urgued [1] that generic field theories are non-unitary and, as a consequence, have operators with complex dimensions. Complex dimensions or exponents is synonymous to log-periodic behavior, and therefore to a discrete scale invariance.

After their initial suggestion as formal solutions of renormalization group equations in the seventies, complex exponents have been studied in the eighties in relation to various problems of physics embedded in hierarchical systems. Only recently has it been realized that discrete scale invariance and its associated complex exponents may appear "spontaneously" in euclidean systems, i.e. without the need for a pre-existing hierarchy. We briefly list the known mechanisms for the spontaneous generation of discrete scale invariance and situations where complex exponents have been found. This is done in order to provide a basis for a better fundamental understanding of discrete scale invariance. The main motivation to study discrete scale invariance and its signatures is that it provides new insights in the underlying mechanisms of scale invariance. It may also be very interesting for prediction purposes.

We refer to [2] for a recent review.

2. What is 'discrete scale invariance'

Let us first recall what is the concept of (continuous) scale invariance : in a nutshell, it means reproducing itself on different time or space scales. More precisely, an observable $\mathcal{O}$ which depends on a "control" parameter x is scale invariant under the arbitrary change $x \rightarrow \lambda x$ [2], if there is a number $\mu(\lambda)$ such that

$$\mathcal{O}(x) = \mu \mathcal{O}(\lambda x) \quad . \tag{2.1}$$

---

[2]Here, we implicitly assume that a change of scale leads to a change of control parameter as in the renormalization group formalism. More directly, x can itself be a scale.

Eq.(2.1) defines a homogeneous function and is encountered in the theory of critical phenomena, in turbulence, etc. Its solution is simply a power law

$$\mathcal{O}(x) = Cx^\alpha \ , \tag{2.2}$$

with

$$\alpha = -\frac{\log \mu}{\log \lambda} \ , \tag{2.3}$$

which can be verified directly by insertion. Power laws are the hallmark of scale invariance as the ratio $\frac{\mathcal{O}(\lambda x)}{\mathcal{O}(x)} = \lambda^\alpha$ does not depend on x, i.e. the relative value of the observable at two different scales only depend on the *ratio* of the two scales [3]. This is the fundamental property that associates power laws to scale invariance, self-similarity, and criticality (Criticality refers to the state of a system which has scale invariant properties. The critical state is usually reached by tuning a control parameter as in liquid-gas and paramagnetic-ferromagnetic phase transitions. Many driven extended out-of-equilibrium systems seem also to exhibit a kind of dynamical criticality, that has been coined "self-organized criticality".

Discrete scale invariance (DSI) is a weaker kind of scale invariance according to which the system or the observable obeys scale invariance as defined above only for specific choices of λ (and therefore μ), which form in general an infinite but countable set of values $\lambda_1, \lambda_2, ...$ that can be written as $\lambda_n = \lambda^n$. λ is the fundamental scaling ratio. This property can be qualitatively seen to encode a *lacunarity* of the fractal structure.

Note that, since $x \to \lambda x$ and $\mathcal{O}(x) \to \mu\mathcal{O}(\lambda x)$ is equivalent to $y = \log x \to y + \log \lambda$ and $\log \mathcal{O}(y) \to \log \mathcal{O}(y + \log \lambda) + \log \mu$, a scale transformation is simply a translation of $\log x$ leading to a translation of $\mathcal{O}$. Continuous scale invariance is thus the same as continuous translational invariance expressed on the logarithms of the variables. DSI is then seen as the restriction of the continuous translational invariance to a *discrete* translational invariance : $\log \mathcal{O}$ is simply translated when translating y by a multiple of a fundamental "unit" size $\log \lambda$. Going from continuous scale invariance to DSI can thus be compared with (in logarithmic scales) going from the fluid state to the solid state in condensed matter physics! In other words, the symmetry group is no more the full set of translations but only those which are multiple of a fundamental discrete generator.

3. Properties

3.1. Signature of DSI

The hallmark of scale invariance is the existence of power laws. The signature of DSI is the presence of power laws with *complex* exponents α which manifests itself in data by log-periodic corrections to scaling.

---

[3]This is only true for a function of a single parameter. Homogeneous functions of several variables take a more complex form than (2.1).

3.2. Importance and usefulness of DSI

Log-periodic structures in data indicate that the system and/or the underlying physical mechanisms have characteristic length scales. This is extremely interesting as this provides important constraints on the underlying physics. Indeed, simple power law behaviors are found everywhere, as seen from the explosion of the concepts of fractals, criticality and self-organized-criticality. For instance, the power law distribution of earthquake energies which is known as the Gutenberg-Richter law can be obtained by many different mechanisms and a variety of models and is thus extremely limited in constraining the underlying physics. Its usefulness as a modelling constraint is even doubtful, in contradiction to the common belief held by physicists on the importance of this power law. In contrast, the presence of log-periodic features would teach us that important physical structures, that would be hidden in the fully scale invariant description, existed.

3.3. Prediction

It is important to stress the practical consequence of log-periodic structures. For prediction purposes, it is much more constrained and thus reliable to fit a part of an oscillating data than a simple power law which can be quite degenerate especially in the presence of noise. This remark has been used and is vigorously investigated in several applied domains, such as earthquakes, rupture prediction and financial crashes.

4. Mechanisms leading to DSI and examples

We list below the examples discussed in the litterature and refer to [2] for further informations.

- Built-in geometrical hierarchy;

- Programming and number theory;

- Diffusion in anisotropic quenched random lattices;

- Cascade of ultra-violet instabilities : growth processes and rupture;

- Cascades of sub-harmonic bifurcations in the transition to chaos;

- Animals;

- Quenched disordered systems;

- The bronchial tree;

- Turbulence;

- Titius-Bode law;

- Gravitational collapse and black hole formation;

- Rate of escape from stable attractors;

- Interface crack tip stress singularity;

- Eigenfunctions of the Laplace transform.

The references below refer to the work carried out in our team. Ref.[2] provides an extensive list of references covering most of the litterature on this subject.

Bibliography

[1] SALEUR H. and D. SORNETTE, Complex exponents and log-periodic corrections in frustrated systems, *J.Phys.I France* **6** (1996), 327–355.

[2] SORNETTE, D., Discrete scale invariance and complex dimensions, *Physics Reports*, in press (1997) (http://xxx.lanl.gov/abs/cond-mat/9707012).

[3] SORNETTE, D. and C.G. SAMMIS, Complex critical exponents from renormalization group theory of earthquakes : Implications for earthquake predictions, *J.Phys.I France* **5** (1995), 607–619 .

[4] SAMMIS, C.G., D. SORNETTE and H. SALEUR, Complexity and earthquake forecasting, Reduction & Predictability of Natural Disasters, eds. J. Rundle, Turcotte D.L. and W. Klein, Santa Fe Institute Studies in the Sciences of Complexity, vol. XXV (Addison-Wesley, 1995), 143–156.

[5] ANIFRANI, J.-C., C. LE FLOC'H, D. SORNETTE and B. SOUILLARD, Universal Log-periodic correction to renormalization group scaling for rupture stress prediction from acoustic emissions, *J.Phys.I France* **5** (1995), 631–638.

[6] SORNETTE D., A. JOHANSEN, A. ARNÉODO, J.-F. MUZY and H. SALEUR, Complex fractal dimensions describe the internal hierarchical structure of DLA, *Phys. Rev. Lett.* **76** (1996), 251–254.

[7] SORNETTE D., A. JOHANSEN, and J.-P. BOUCHAUD, Stock market crashes, Precursors and Replicas, *J.Phys.I France* **6** (1996), 167-175.

[8] SALEUR H., C.G. SAMMIS and D. SORNETTE, Renormalization group theory of earthquakes, *Nonlinear Processes in Geophysics* **3**, (1996), 102–109.

[9] OUILLON G., D.SORNETTE, A. GENTER and C. CASTAING, The imaginary part of rock jointing, *J. Phys. I France* **6** (1996), 1127–1139.

[10] A. JOHANSEN, D.SORNETTE, H. WAKITA, U. TSUNOGAI, W.I. NEWMAN and H. SALEUR, Discrete scaling in earthquake precursory phenomena : evidence in the Kobe earthquake, Japan, *J.Phys.I France* **6** (1996) 1391–1402.

518

[11] SALEUR H., C.G. SAMMIS and D. SORNETTE, Discrete scale invariance, complex fractal dimensions and log-periodic corrections in earthquakes, *Journal of Geophysical Research-Solid Earth* **101** (1996), 17661–17677.

[12] HUANG, Y., G. OUILLON, H. SALEUR and D. SORNETTE, Spontaneous generation of discrete scale invariance in growth models, *Phys. Rev. E* **55** (1997) 6433–6447.

[13] HUANG, Y., H. SALEUR, C.G. SAMMIS and D. SORNETTE, Precursors, aftershocks, criticality and self-organized criticality, Europhys. Lett. in press (http://xxx.lanl.gov/abs/cond-mat/9612065)

[14] JÖGI, P., D. SORNETTE and M. BLANK, Fine structure and complex exponents in power law distributions from random maps, *Phys. Rev. E*, in press (1997) (http://xxx.lanl.gov/abs/cond-mat/9708220).

Chapter 46

Emergence in earthquakes

Didier Sornette[1]

LPMC, CNRS and Université de Nice-Sophia Antipolis
06108 Nice Cedex 2, France
and IGPP and ESS, Box 951567
Los Angeles, CA 90095-1567, USA

We present a new theory of shallow earthquakes in which phase transformations allows the crust to store a fraction of the total released energy. These phase transformations also control the triggering nucleation phase. We briefly discuss consequences and comparisons with data.

1. Introduction

Research in earthquake prediction has a long history and is still very active in many parts of the world [1, 2, 3, 4]. Nourished by the fact that earthquake prediction has never really delivered [5], there is however a growing consensus [5, 6] that earthquake prediction might be utterly impossible in practice, due to the sheer complexity of the problem. From a scientific standpoint, this view is that of a Bayesian [7], for whom each failed attempt lowers the probability for the success of the next attempt.

Bayes'approach is warranted for stationary processes. Clearly, scientific progress is not in this class as testified by the long list of unexpected discoveries [8, 9]. According to standard wisdom, an earthquake is a sudden rupture in the earth's crust or mantle caused by tectonic stress. To understand the physics of earthquakes, it is thus necessary to know the state of stress in the crust. These premises are usually elaborated by models that attempt to account for seismological and geological data as well as constraints from laboratory experiments.

---
[1]Supported in part by NSF EAR96-15357 and NSF EAR97-06488

However, stress is probably the least known of all geophysical fields, the failure criterion is still largely unknown and we do not understand the conditions that make a fracture propagate into big earthquakes. Furthermore, the mechanical view of earthquakes lead to important paradoxes, the strain paradox, the stress paradox and the heat flow paradox, that are difficult to account for, either individually or when taken together. Their resolution usually calls for additional assumptions on the nature of the rupture process (such as novel modes of deformations and ruptures) prior to and/or during an earthquake, on the nature of the fault and on the effect of trapped fluids within the crust at seismogenic depths.

2. Role of water and phase transformations

What if some field other than stress was conducting the preparation of an earthquake? There is growing evidence [10, 11, 12, 13] of the importance of water as a key participant for decreasing normal lithostatic stress in the fault core and weakening rock materials by stress corrosion.

Here, we advance the novel concept that water has an additional role in the earthquake process, which is the alteration of minerals submitted to finite strains leading eventually to the phase transformation into other metastable minerals in out-of-equilibrium conditions. This transformation provides another route for storing energy during the earthquake cycle in the form of "chemical" energy. This new form of energy storage can be documented in details for the quartz to coesite transformation [14], shown to occur at seismogenic depth under finite deformations.

Under increasing strain, the transformed minerals eventually become unstable, as shown from Landau theory of structural phase transitions, and transform back explosively, creating a slightly supersonic shock wave propagating along the altered fault core leaving a wake of shaking fluidified fragments. As long as the resulting high-frequency waves remain of sufficient amplitude to fluidify the fault core, the mechanics of granular media teaches us that the elastic modulii and the shear strength of the granular fault gouge are substantially decreased in the presence of this shaking at acoustic pressure two order of magnitude smaller than the lithostatic value. This central zone is thus like a loose granular medium with low acoustic impedence, hence a trap of seismic waves. This weakening then triggers the rupture instability by unlocking the fault which starts to move and releases the elastic part of the stored energy. The healing of the fault is controlled by the duration of the shaking due to the trapped waves within the central zone. When the shaking dies off, the fault locks again and heals. This implies that the rupture propagation during the earthquake develops a pulse-like structure, with a pulse width increasing approximately proportionally with time since the inception of the earthquake.

3. Consequences and predictions

In this theory, the fault slip is not triggered by reaching a stress threshold (corresponding either to friction unlocking or rupture nucleation), but rather by a chemical instability: as a consequence, any level of stress will activate the fault slip when the explosive phase transformation occurs. This theory allows one to rationalize many observations within a coherent framework, including the strain, stress and heat flow paradoxes [15], the observation that the larger the recurrence time, the larger the stress drop [16], the shear heating observation [17], the seismic P-wave precursors [18], the fact that earthquakes occur preferentially in strong rocks [19], the existence of creeping fault sections and slow earthquakes [20], tilt anomalies [21], electric effects [3] and relation with deep earthquakes [22].

Bibliography

[1] V.I. Keilis-Borok, Phys. Earth and Planet. Interiors **61**, 1-139 (1990).

[2] MOGI, K., *J. Phys. Earth* **43** (1995), 533–561.

[3] DEBATE ON VAN, Special issue of *Geophys. Res. Lett.* **23** (1996), 1291–1452.

[4] WYSS, M., *Pure and Applied Geophysics* **149** (1997), 3–16.

[5] GELLER, R.J., D.D. JACKSON, Y.Y. KAGAN and F. MULARGIA, *Science* **275** (1997),1616–1617.

[6] MAIN, I., *Nature* **385** (1997), 19–20.

[7] EARMAN, J., Bayes or bust? : a critical examination of Bayesian confirmation theory, *Cambridge, Mass. : MIT Press* (1992).

[8] BAHCALL, J.N., *Physics Today* **44** (1991), 24–30.

[9] BROWNSTONE, D.M. and I. FRANCK, Timelines of the 20th century: a chronology of 7,500 key events, discoveries, and people that shaped our century, *Boston: Little, Brown*, (1996).

[10] KIRBY, S.H., *J. Geophys. Res.* **89** (1984), 3991–3995.

[11] HICKMAN, S., R. SIBSON and R. BRUHN, *J. Geophys. Res.* **100** (1995), 12831–12840.

[12] J.P. Evans and F.M. Chester, J. Geophys. Res. **100**, 13007-13020 (1995).

[13] THURBER, C., S. ROECKER, W. ELLSWORTH, Y. CHEN, W. LUTTER and R. SESSIONS, *Geophys. Res. Lett.* **24** (1997), 591–1594.

[14] SORNETTE, D., Earthquakes: a new way, *preprint* (1997).

[15] LACHENBRUCH, A.H. and J.H. SASS, *J. Geophys. Res.* **85**, (1980), 6185–6222.

[16] KANAMORI, H., *Ann. Rev. Earth Planet. Sci.* **22** (1994), 207–237.

[17] SCHOLZ, C.H., *J. Geophys. Res.* **85** (1980) 6174–6184.

[18] BEROZA, G.C. and W.L. ELLWORTH, *Tectonophysics* **261** (1996), 209–227.

[19] ZHAO, D.P. and H. KANAMORI, *Geophys. Res. Lett.* **22** (1995), 763–766.

[20] LINDE, A.T., M.T. GLADWIN, M.J.S. JOHNSTON, R.L. GWYTHER and R.G. BILHAM, *Nature* **383** (1996), 65–68.

[21] WYSS, M., *Pure and Applied Geophysics* **149** (1997), 17–20.

[22] GREEN, H.W. and H. HOUSTON, *Ann. Rev. Earth Sci.* **23** (1995), 169–213.

Chapter 47

Chemical oscillation in symbolic chemical system and its behavioral pattern

Yasuhiro Suzuki
Hiroshi Tanaka
Medical Research Institute, Tokyo Medical and Dental University
Yushima 1-5-45, Bunkyo Tokyo 113 JAPAN

One of the most essential temporal structures in life systems is a cycle, which can be observed in any hierarchy of living things, such as TCA cycle in cytoplasmic level, the cell cycle in cell division, and the life cycle of living things. The importance of this structure has been pointed out by several authors, such as Eigen's hypercycle, Maturana's autopoiesis, Kauffman's NK network, and Fontana's Algorithmic Chemistry. However, these researches do not systematically address the conditions under which such cycle structures will emerge and become stable. In this paper, abstract rewriting system on a multiset is introduced to a model chemical reaction as a symbolic rewriting system acting on a multiset, which can be viewed as a chemical reaction system in a test tube. By use of this model, we made experiments on Brusselator model (the Brusselator is a model of chemical oscillations which are found on the Belousov-Zabotinsky reaction). And we confirm that non-linear oscillation emerges.

1. Introduction

One of the most essential temporal structures in life systems is a cycle, which can be observed in any hierarchy of living things. For example, in metabolic pathways inside the cell, TCA cycle is used to extract energy from organic chemicals, which plays an important role in cell metabolism[17]. Also, the cell cycle in cell division is important for the growth of life systems, and the life cycle can be observed in all living things[17].

Thus, a cycle can be viewed as an universal structure in all the hierarchies of complex systems, especially in living things, whose importance has been pointed out by several researches on complex system, such as Eigen's hypercycle[2], Maturana's Autopoiesis[14], Kauffman's NK network[10], and Fontana's Algorithmic Chemistry[3]. However, these researches do not focus on the conditions under which such a cycle will emerge and the process in which it becomes stable.

In this paper, we introduce a new abstract rewriting system on multiple sets (ARMS) in order to examine the conditions of cycle emergence and the algebraic characteristics of the obtained cycle. In this model, a multiset [1] is taken as a role of a test tube, which contains "symbols", which correspond to chemical compounds. Then, rewriting rules, corresponding to chemical reaction formulae, act on this multiset, and the applied symbols are transformed into other symbols, corresponding to chemical reactions. By using this model, we made experiments on Brusselator model (the Brusselator is a model of chemical oscillations which are found on the Belousov-Zabotinsky reaction). And we confirm that non-linear oscillation emerges.

Furthermore, we suggest that confluence property and terminating property are two principal components to characterize the behavior of ARMS: confluence property, which indicates spatial characteristics of rewriting rules, is closely related with variabilities of emerging cycles. On the other hand, terminating property, which indicates temporal characteristics of rewriting rules, is related with the steps needed for the emergence of cycles. On the basis of these features, a table which classifies the ARMS behavior is obtained, where a system which is both weakly confluent and has non terminating property typically simulates not only the emergence of various cycles, but also their complex behavior, such as fusion of cycles.

2. Model

In this section, we describe an abstract chemical system in terms of an abstract rewriting system. Before describing the system in detail, we introduce abstract rewriting systems in general.

2.1. Abstract rewriting system (ARS)

An abstract rewriting system models the algebraic characteristics of calculation. By introducing this formal structure, several characteristics of calculation can be discussed in a common framework. This concept is applied to various formal methods in mathematics and computer science, for example in proof theory, the algebraic description of computer software, and automated deduction.

The principle of calculation within an ARS is simple. A calculation is performed by rewriting using rules as in formal grammar: $a \to Sa$.

---

[1] A multiset is defined as a set which is allowed to include the same elements. For example, $\{a, a, b, c\}$ is not a set, but a multiset, while $\{a, b, c\}$ is both a set and a multiset.

Definition [Abstract Rewriting System]

 An abstract rewriting system is defined as a pair $(A, \mathcal{R})$, where A and $\mathcal{R}$ denotes a finite alphabet and a finite set of pairs of words over the alphabet A, respectively. A rule which transforms $b \in A$ into $a \in A$ is written as $a \to b$ and we say that $a \to b$ is a "rewriting step".

An abstract rewriting system is a string-replacing system. If the left hand side of a rule matches a string, it is replaced with the right hand side of the rule. The final result of a calculation is called a normal form:

Definition, [Normal Form] *If there does not exist b such as $a \to b$ and $b \in A$, then $a \in A$ is called a normal form.*

Multiplication, for example, can be viewed as an abstract rewriting model. Let us define a set of rewriting rules, Ru as: $Ru = \{2 \times 2 \to 4, 4 \times 2 \to 8\}$. Using Ru, $2 \times 2 \times 2$ is calculated as shown in Fig. 2.1,

$$\mathbf{2 \times 4} \leftarrow 2 \times 2 \times 2 \to 4 \times 2 \to \mathbf{8}.$$

Figure 2.1: An example of rewriting calculus

In the first step, since only the rule $2 \times 2 \to 4$ can be used on $2 \times 2 \times 2$, the string $2 \times 2 \times 2$ is rewritten into 4×2, for example. This string on the other hand can be transformed into 8 using the rule $4 \times 2 \to 8$. Because there are no rules that apply to 8, the latter is a normal form. However, as the first step we can also transform $2 \times 2 \times 2$ into 2×4, which turns out to be another normal form. In this calculation, we see that two normal forms exist (see Fig. 2.1).

2.2. ARMS

Extending the concepts of the abstract rewriting system, we introduce an abstract rewriting system on multi-sets (ARMS). Intuitively, ARMS is like a chemical solution in which floating *molecules* can interact with each other according to reaction rules. Technically, a chemical solution is a finite multi-set of elements denoted $A^k = \{a, b, \dots, \}$; these elements correspond to *molecules*, and reaction rules are specified in terms of rewrite rules. As to the intuitive meaning of an ARMS, we refer to the study of chemical abstract machines [7]. In fact, rewrite rule systems can be thought of as reflecting an underlying *algorithmic chemistry*.

 We denote the empty set by ϕ, and the *base number* of a multi-set (size of a multi-set) by $|S|$ (where S is a multi-set), respectively. Then we define:

Definition [Multi-set]

 A "multi-set" is an element $t \in A^k$ $(1 \le k \le n) \in \Sigma$, where n is a finite number, and A^k is a Cartesian product $A_1 \dots A_k$. $A^k = A_1 \times A_2 \cdots \times A_k$, Σ denotes the set of multi-sets, and n is called the "maximal multi-set size."

The multi-sets correspond to possible states of *chemical solution*. The set of multi-sets corresponds to the space of transitions of an ARMS.

Definition [Rewriting rule]

A *"rewriting rule"* is a relation $l \, \mathcal{R} \, r$ $(l, \, r \in \Sigma)$, $|l|, |r| \leq$ *maximal multi-set size, n. A rewriting rule $l \, \mathcal{R} \, r$ is denoted as $l \rightarrow r$.*

A rewriting rule such as

$$a \rightarrow a \ldots b, \tag{2.1}$$

is called a *heating rule* and denoted as $r_{\Delta > 0}$; it is intended to contribute to the stirring solution. It breaks a complex *molecule* into smaller ones: *ions*. On the other hand, a rule such as

$$a \ldots c \rightarrow b, \tag{2.2}$$

is called a *cooling rule* and denoted as $r_{\Delta < 0}$; it rebuilds *molecules* from smaller ones. In this paper, reversible reactions, i.e., $S \rightleftharpoons T$, are not considered. We shall not formally introduce the refinement of *ions* and *molecules* though we use refinement informally to help intuition.

Definition [ARMS] An *"Abstract Rewriting System on Multi-sets"* (ARMS) is a pair (T, Ru) consisting of a multi-set T and a set Ru of rewriting rules.

Definition [Rewriting on ARMS]

Let (T, Ru) be an ARMS. We write $s \xrightarrow{Ru} t$ if there exists a rewriting rule $l \rightarrow r \in Ru$ such that $l \subseteq s$ and $t = (\mathbf{s} - l) \cup r$.

The ARMS can construct input by such a rule, for example, $\phi \rightarrow a$.

Definition [Normal Form in ARMS]

If no rule in Ru can be applied to a multi-set and no symbols can be inputted to the multi-set without the resulting base number exceeding the limit on the multi-set, then the multi-set is called Normal Form (final state).

Normal forms correspond to a steady state.

The reader will notice that the method of rewriting of ARMS is different from that of the abstract rewriting system. Since the abstract rewriting system is a string-replacing system, the string ab and the string ba are treated as different strings on rewriting. On the other hand, since ARMS is a multi-set replacing system, the system regards ab and ba as multi-sets of symbols, $\{ab\}$ and $\{ba\}$. Thus, they are treated the same. Hence, e.g., ARS cannot rewrite ab using the rule $ba \rightarrow c$, while ARMS, however, can rewrite ab into c using this rule.

2.3. How ARMS works

In ARMS, we assume that one randomly selected rule is applied in each rewriting step, unless no input is allowed. An algorithm for rewriting steps in ARMS is described in Fig. 2.2.

procedure ARMS (Rewriting Step)
begin
    count-step ← 0;
    while count-step ≠ n **do**
      begin
        if the multi-set reached Normal Form **then**
          count-step:= n;
        else
          Input string(s) to the multi-set;
          Select a rule;
          if the rule can rewrite the multi-set **then**
            Rewrite the multi-set;
            count-step := count-step + 1;
          end if
        end if
      end
    end while
end.

Figure 2.2: An algorithm for ARMS (for the first n steps)

Example In this example, we assume that a will be inputted on each rewriting step, the maximal multi-set size is 4 and the initial state is given by $\{a, a, f, a\}$. The set of the rewriting rules, Ru_1 is $\{r_1, r_2, r_3, r_4\}$, where each rule is described by the following:

$$aaa \rightarrow b : r_1, \ b \rightarrow a : r_2, \ b \rightarrow c : r_3, \ a \rightarrow bb : r_4.$$

In this example, we assume that rules are selected as following the order $\{r_4 \Rightarrow r_1 \Rightarrow r_3 \Rightarrow r_2\}$. Then, each rule is applied in the following way. First, r_4 is applied. Next, as steps 2 and 3, r_1 and r_3 are applied, respectively. Finally, as step 4, r_2 is applied.

$$
\begin{array}{ll}
\{aafa\} & \subseteq a \text{ (the left hand side of } r_4) \\
\downarrow & \text{.... can not input } a \text{ and can not apply } r_4, \\
\{aafa\} & \subseteq aaa \text{ (the left hand side of } r_1) \\
\downarrow & \text{.... can not input } a \text{ but can apply } r_1 \\
\{bf\} &
\end{array}
$$

Figure 2.3: Example of rewriting steps of ARMS

Fig. 2.3 illustrates two rewriting steps of the calculation from the initial state.

As the first step, since the base number of the multi-set is 4, the system can not input a. On the left hand side of r_4, a is included in $\{aafa\}$, however, r_4 can not be used. If a is replaced with bb, the base number of the multi-set becomes 5 and it exceeds the maximal multi-set size, 4.

In the next step, the system can not input a, however, r_1 can apply to the multi-set and $\{aafa\}$ is rewritten into $\{ba\}$ (because if aaa is replaced with b, the base number of the multi-set does not exceed the maximal multi-set size, see Fig. 2.3).

In step 3, ARMS inputs a to the multi-set and transforms it to $\{c, a, a\}$ with r_3.

$$\text{Step } 3 : \{c, a, a\}.$$

In step 4, the system inputs a, but r_2 can not apply to it. Thus $\{c, a, a\}$ becomes $\{c, a, a, a\}$.

$$\text{Step } 4 : \{c, a, a, a\}.$$

Typical Examples In this paragraph, we shall present two examples. Let us assume a set of rewriting rule Ru_1 and a maximal multi-set size of 4. The first example is a case where ARMS generates two cycles. This example has the following rule order:

$$\{r_4 \Rightarrow r_3 \Rightarrow r_2 \Rightarrow r_4 \Rightarrow r_1 \Rightarrow r_2 \Rightarrow r_1 \Rightarrow r_3 \Rightarrow r_4\},$$

whose state transition is shown in Fig. 2.4. After 8 steps, the system forms two cycles, whose periods are of 3 steps.

0.	$\{f\}$	
1.	$\{a, f\}$	
2.	$\{a, a, f\}$	
3.	$\{a, a, a, f\}$	
4.	$\{b, f\}$	↑
5.	$\{a, b, f\}$	a cycle
8.	$\{a, a, a, f\}$	↓
9.	$\{b, f\}$	↑
10.	$\{a, b, f\}$	a cycle
11.	$\{a, a, a, f\}$	↓
12.	$\{b, f\}$	

Figure 2.4: Example of a system that generates cycles

The next example is a case where ARMS terminates. Although ARMS applies the same rules, the obtained result is completely different (Fig. 2.5). This example has the following rule order:

$$\{r_4 \Rightarrow r_1 \Rightarrow r_2 \Rightarrow r_4 \Rightarrow r_3 \Rightarrow r_1 \Rightarrow r_2\}.$$

The state transition is shown in Fig. 2.5:

$$
\begin{aligned}
&0. \quad \{f\}\\
&1. \quad \{a, f\}\\
&2. \quad \{a, a, f\}\\
&3. \quad \{a, a, a, f\}\\
&4. \quad \{b, f\}\\
&5. \quad \{a, b, f\}\\
&6. \quad \{a, a, b, f\}\\
&7. \quad \{a, a, c, f\}.
\end{aligned}
$$

Figure 2.5: Example of a system that halts

2.4. Brusselator on ARMS

We performed an experiment about Brusselator[15] model by using an ARMS. The Brusselator is a model of chemical oscillation as the Belousov-Zabotinsky reaction (Fig. 2.6).

$$
\begin{aligned}
A &\xrightarrow{k_1} X\\
B + X &\xrightarrow{k_2} Y + D\\
2X + Y &\xrightarrow{k_3} 3X\\
X &\xrightarrow{k_4} E
\end{aligned}
$$

Figure 2.6: Abstract chemical model of Brusselator.

We regard these abstract chemical reaction equations as the rewriting rules as shown in Fig. 2.7.

$$
\begin{aligned}
A &\rightarrow X && : r_1\\
B \; X &\rightarrow Y \; D && : r_2\\
X \; X \; Y &\rightarrow X \; X \; X && : r_3\\
X &\rightarrow E && : r_4
\end{aligned}
$$

Figure 2.7: Rewriting rules for Brusselator.

In this simulation, the reaction rate corresponds to the frequency of rule application. If r_1 has the highest reaction rate, then r_1 is applied at the highest frequency.

Simulation of the Brusselator model Let us examine the relationship between the frequency of rule application (reaction rate) and the concentration of X and Y in the multi-set. The concentration of X and Y in the multi-set is indicated by the number of X and Y present in the multi-set.

530

As to the initial condition, we assume that the maximal multi-set size is equal to 5000 and the initial state of the multi-set is an empty multi-set. We assume that the system makes inputs A and B continually. Hence this model can be regarded as a continuously-fed stirred tank reactor (CSTR).

In this simulation, we confirmed that oscillations between the number of X and Y in the multi-set emerged. Furthermore, we discovered three types of oscillations as follows: (1) divergence and convergence (Fig. 2.8, Fig. 2.9), (2) quasi-stable oscillations (Fig. 2.10) and (3) unstable oscillations (Fig. 2.11).

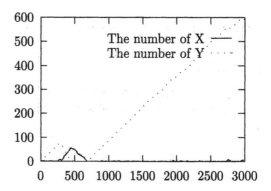

Figure 2.8: Example of Divergence

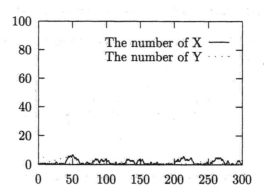

Figure 2.9: Example of Convergence

Oscillation on the number of between X and Y emerges as follows; at first, the number of X increases and as decreasing the number of X, the number of Y increases. We will call this type of oscillation as X-Y **oscillation**, and oscillations on the number of X or Y itself as X-**oscillation** and Y-**oscillation**, respectively, below.

Divergence

When the frequency of rule application of r_2 is much larger than r_1, the system becomes unstable and the system diverges in many cases.

In this case, a system generates X-Y oscillations, as shown in Fig. 2.8. The system diverges easily by perturbation and it is impossible to predict if the system generates oscillations or diverges. Fig. 2.8 shows this case (the frequency of r_1 and r_2 are 0.1, 0.45, respectively).

Convergence

When the frequency of rule application of r_1 is much larger than r_2, then the system converges. Fig. 2.9 illustrates this case (the frequency of rule application of r_1 and r_2 are 0.1, 0.04, respectively). In this case almost all the oscillations are X or Y-oscillation and only few X-Y oscillations can be found,

(quasi) stable oscillation

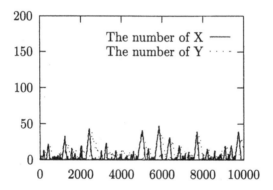

Figure 2.10: Example of (quasi) Stable oscillation

When the frequency of rule application of r_1 is slightly larger than r_2, the system is apt to generate X-Y oscillations, which resemble limit cycle. Even if we introduce perturbation to the system, it maintains this type of oscillation.

Unstable oscillation

When the frequency of r_2 is slightly larger than r_1, the system is apt to exhibit unstable oscillation. In this case, there are a few X or Y-oscillation and almost all the oscillations become X-Y oscillation. A great variety of X-Y oscillation can be found.

Fig. 2.11 shows unstable oscillations (the frequency of r_1 and r_2 are 0.1, 0.35, respectively).

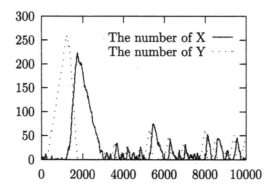

Figure 2.11: Example of Unstable oscillation

2.5. On behavioral pattern of the system

Through these experiments, we obtained the relationship between the frequency of rule application and system's behavior. It is summarized into Table 2.1.

r_1 is much larger than r_2	Convergence
r_1 is slightly larger than r_2	Stable Oscillation
r_1 is slightly less than r_2	Unstable Oscillation
r_1 is much less than r_2	Divergence

Table 2.1: Relationship between the frequency of rule application and system's behavior

The balance of r_1 and r_2 controls system's behavior. When a system applies r_1 in the high frequency, the system produces many Xs (by using $A \rightarrow X{:}r_1$). In this case, if the frequency of rule application of r_2 is quite low, the system hardly to generate Ys. Because of r_3 $(X\ X\ Y \rightarrow X\ X\ X)$ needs a Y, almost all the generated Xs will be used by r_4 $(X \rightarrow E)$. Hence the system generates X-oscillation.

In order to generate X-Y oscillation, a system must use r_2. r_2 plays a key role in generating X-Y oscillation. It generates Ys then r_3 $(X\ X\ Y \rightarrow X\ X\ X)$ is able to act. This r_3 is "auto-catalytic reaction " and it increases Xs 1.5-fold and consumes Ys. Hence, the system generates X-Y oscillation.

3. Classification of behavioral pattern of ARMS

We have been studying the formal properties of ARMS, based on our experimental results. Throughout the investigation we made a conjecture that the global behavior of ARMS is closely related to two essential properties: termination and confluence. Before entering into a detailed discussion of classifying the behavior patterns of ARMS, we shall describe the confluence termination properties.

Terminating Property

Terminating property is concerned with halting of computation. They are divided into following two types.

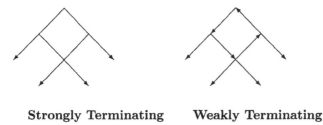

Strongly Terminating   **Weakly Terminating**

Figure 3.1: Terminating property

In these figures, each bold line illustrates a process of rewriting and each arrow illustrates the direction of process. Each fork denotes the case when a multiset is rewritten into two types of multiset. These figures illustrate examples that correspond to each property. Each fork, in fact, can corresponds to the case when a multiset is rewritten into more than two.

Weakly terminating   The meaning of weakly terminating property is that some rewrite sequences reach normal form and others do not reach normal form. The right side of Fig. 3.1 shows it.

Strongly Terminating   Strong terminating means that all rewrite sequences reach normal form. What has to be noticed this property does not ensure unique normal form. The left side of Fig. 3.1 indicates it.

Confluence property

The confluence property is related to pathway of rewriting calculus. Confluence is of two types as follows:

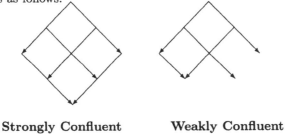

Strongly Confluent   **Weakly Confluent**

Figure 3.2: confluence property

534

Weakly Confluent Weakly confluent means that same rewriting sequences reach the same forms, but others do not meet other forms. The right side of Fig. 3.2 shows it.

Strongly Confluent The meaning of this property is that every rewriting sequences reach the same form. This property ensure that the system has one unique reduced form. The left side of Fig. 3.2 indicates it.

3.1. Classification table of ARMS behavior

We can classify behavior of ARMS using above characteristics of ARS. Table 3.1 illustrates the behavioral pattern with respect to terminating and confluence properties. The vertical array of the table indicates characteristics of confluence property, and the horizontal array indicates terminating property.

Terminating

Confluence	Strong (I)	Weak (II)	Non (III)
Strong	⊥	⊥ or ∞	∞
Weak	⊥	⊥ or ⊙	⊙
Non	⊥	⊥ or ⊙	⊙

Table 3.1: Classification table of ARMS behavior

In this table, ⊥ denotes the case where the system does not generates any cycles, ⊙, ⊙ and ∞ illustrate the case where the system generates cycles and variety of periods.

⊙ means the case where the system generates cycles of short period. ⊙ illustrated the case where the system generates cycles of short and middle period and ∞ denotes the case where the system generates cycles of short, middle and long period and *romans numerals* indicate its class, respectively. The most important feature of this table is that confluence property determines the complexity of cycles, and terminating property determines the number of cycles.

Effect of confluence property

The onfluence property concerns "spatial" behavior. If the system does not have this property, which corresponds to "non" in the table, once a rewriting sequence branches from a state, it never returns the state. Therefore, its trajectory resemble spiral. There are only a few interactions between cycles. ⊥ denotes this scheme.

If the system has weakly confluence property, once a rewriting sequence branches from a state some sequences can return the state, others can not return the state, as a result it brings about interactions between cycles such as fusion of cycles. As the property is getting stronger, some cycles come together and make colonies of cycles, ⊙ denotes this type of behavior.

When the system has strong confluence property, every branched sequences must return the former states. With the property becomes strong, these colonies getting together and create a complex cycle, so that the interactions get dense. As a result, cycles of short, middle and long period emerge and interact each other. ∞ denotes this scheme.

Effect of Terminating property

Terminating property is concerned with the number of cycles.

Class I (Strongly terminating) In this class, every rewriting sequence reaches the normal form. If the system has a cycle, it does not ensure that every rewrite sequence halts (c.f. Weakly Terminating, left side of Fig. 3.1), which contradicts the strong termination propriety. Therefore, in Class I, the system does not generates any cycles.

Class II (Weakly terminating) On class II, some rewriting sequences halt, others does not halt. Because of ARMS has finite states[2], if the system dose not halt, it must generate cycles. In the case of the system with perturbation, for example, when it decreases the frequency of input, introduce randomness in rule order and so on, we do not know the system halt or not. Because a perturbation sometimes makes the system halt, sometimes it revives normal form and prolongs computation. Class II (with perturbation) is analogous to undecidable class of computation[8] such as halting problem for universal Turing Machine, Hilbert's 10th problem[13] and so on.

Undecidable class of computation has provoked a great deal of controversy. Wolfram[18] and Langton[12] addresses this class on their study of Cellular Automaton (CA). They provide explanation qualitative classes of CA behavior. They classify the class as which CA capable of universal computation.

In [18], S. Wolfram discusses qualitative characterization of CA. He divides behavior pattern of CA into four types [18]. In his class 4, the behavior of CA is essentially unpredictable. And in this class, CA's of this class are capable of universal computation. Langton[12] investigates the phase transition of CA's, especially in the regime, between halting and non-halting computation (edge of chaos) where most complex computations are found. And he also refers to the regime which correspond to Wolflam's class 4. From these computational classification view, our Class II (with perturbation) corresponds to Wolfram's class 4, Langton's edge of chaos regime, respectively. Because they are all in undecidable computational class.

Class III (the system does not have terminating property) In this class, none of the rewriting sequences halt, leading to cycles.

---

[2]It is easy to show by definition of ARMS

4. Condition of cycles emergence

We investigated the formal properties of termination, and obtained a halting condition for ARMS computations. In order to discuss the condition precisely, we will introduce some mathematical background.

4.1. Mathematical background

Definition[Well-founded order]

  Let R be a relation on a set A. If there are no infinite descending sequences $a_1 R a_2 R a_3 \ldots$ of elements of A, we say that R is well-founded.

For example, if R is a relation $>$ on the set of natural number, then $>$ is well-founded, but on the set of integer, $>$ is not well-founded. A well-founded order is a powerful method to prove ARS's termination. We use lexicographic order as "order" between terms. It is defined as follows:

Definition[Lexicographic Order]

  Let $(A_i, R_i)(i = 1, 2, 3, \ldots)$ be sets equipped with total order R_i. We define a binary relation R on A,

$$A = \cup_{m=1}^{n}\{< a_1, a_2, \ldots a_n > | a_i \in A_i (i = 1, 2, \ldots, n)\}$$

(where n is a finite natural number, $< \cdots >$ is a element of Cartesian product of $A_1 \ldots A_n$), as follows:
$< a_1, a_2, \ldots a_n > R < b_1, b_2, \ldots b_m >$ if and only if $\exists i, 1 \le i \le m, 1 \le i \le n$, $(a_1 = b_1) \wedge \ldots (a_{i-1} = b_{i-1}) \wedge (a_i < b_i)$, or $m \le n$ and $(a_1 = b_1) \wedge \ldots (a_n = a_m)$. We denote lexicographic order $\succ_{lex}$.

For example, If we introduce a binary relation $" >"$ on *Alphabet*, such as, a $> b > c \ldots > z$, $\{aaac\} \succ_{lex} \{aaba\}$.

Definition

  A binary relation on terms is called a rewrite relation.

If $\{aaac\} \to_{Ru} \{aaba\}$, by $ac \to ab \in Ru$, then the rewrite relation between $\{aaac\}$ and $\{aaab\}$ is $\{aaac\} \succ_{lex} \{aaba\}$.

Definition

  A proper order on terms that is also a rewrite relation is called a rewrite order. A reduction order is a well-founded rewrite order.

There exists a proper order $\succ$ such that $s \succ t$ $(s, t \succ \Sigma)$, and if $\succ$ is well-founded order then $s \succ t$ is a reduction order. On lexicographic order, we cannot say that if
$< a_1, a_2 > \succ_{lex} < b_1, b_2 >$ then $< a_2, a_1 > \succ_{lex} < b_1, b_2 >$ (quasi order), then we introduce the concept of *sorted*.

Definition [Sorted]

  *In ARMS, if the all the alphabets in term t are sorted in descending order by an arbitrary total order $\succ$ on alphabets, then we say that the term is **sorted**.*

Lemma1

    ARMS is terminating if and only if there exists a well founded order R on terms.

Proof.

    (If) Immediate by definition of well founded order. (Only if) If there exists on infinite rewrite sequence, it contradicts the definition of well-founded order ∎.

Lemma2

    On ARMS, if $s \to_{Ru} t$, and $l \succ_{lex} r$ then the rewrite relation $s \succ t$.

Proof.

    Clear, by definition of rewriting ∎.

Theorem

    ARMS is terminating if and only if there exist a reduction order $\succ$ on terms such that $l \succ r$ for every rewrite rule $l \to r \in Ru$.

Proof.

    (If) Clear. (Only if) According to above lemma 1, it suffices to show $\{\to_R\} \subseteq \succ$. By assumption $l \succ r$ for every rewrite rule $l \to r \in Ru$. If we take $\succ_{lex}$ on sorted terms instead of $\succ$, then $l \succ r$ by lemma 2. Hence $\{\to_R\} \subseteq \succ$ ∎.

By above theorem, we obtain a following corollary as a condition of the system generates cycles.

Corollary

    ARMS generate cycles if and only if the set of terms Σ is finite set and there does not exist a reduction order $\succ$ on terms such that $l \succ r$ for any rewrite rule $l \to r \in Ru$.

Proof.

    By the contradiction of Theorem 1 (and Köning's tree lemma [6]) ∎.

5. Related work

L-systems   The most famous application of rewriting system to computational model of living things is L-systems[5], which model the morphogenesis of living things, such as the growth of trees. This system formalizes the process of morphogenesis as firing of rewriting rules and is defined as a formal language system $(\Sigma, P, \omega,$ "(", ")")$, where Σ, P, and ω, denotes a set of states, a set of rewriting rules, and a type of cell at initial state, respectively. For example, let Σ be equal to $\{1, 2, 3, 4\}$, and P be equal to: $\{P = \{1 \to 42, 2 \to 23, 3 \to 1(2)\}$ Then the process of morphogenesis can be viewed as the following rewriting process: $\{S1 = 1 \to S2 = 42 \to S3 = 423 \to S4 = 4231(2) \to S5 = 4231(2)42(23)\}$. The most important difference between L-systems and ARMS is that ARMS focuses on the temporal aspects of rewriting processes, rather than the spatial aspects.

Algorithmic Chemistry Fontana introduces an abstract model, called λ-gas[1], where a new chemical is generated by interactions between existing chemicals, using λ-calculus. This model is described by a set of functions which correspond to molecules. Two functions are randomly selected, and they interact with each other, which is represented as a compound of function $f(g)$. Although this model also focuses on characteristics of chemical reaction, the main difference between λ-gas and ARMS is that our interest is in temporal aspects of emergence of cycles.

Furthermore, Fontana also addresses the difference between λ-calculus and abstract rewriting system as follows[3];

With some ingenuity the observer will further derive all laws supplied by the uncovered group structure(Knuth and Bendix[11], 1970: Huet and Oppen[9], 1980). If read as rewrite rules, the equations thus obtained will enable the observer to exactly describe (and predict) each and every collision product in the system - without any knowledge about λ-calculus. The observer will, then, have discovered a perfectly valid theory of that organization, without reference to its underlying micro-mechanics. (from: [3])

Cellular Automata Using the concepts of abstract rewriting system, cellular automaton[18] can be formalized as rewriting system. For example, the following local rules,

000	001	010	100	101	110	111
0	1	0	1	0	1	0

can be viewed as rewriting rules such as $000 \rightarrow 0$.

The most important differences between cellular automata and ARMS are the following two points. First, ARMS does not use any spatial information. Rewriting rules are global, rather than local in cellular automata. Thus, although cellular automata focus on local interactions between neighbors, ARMS focuses on the interaction of rules with a given multiset. Second, our interest is in formal characteristics of rewriting calculus, rather than pattern formation in cellular automata.

In this paper, we introduce a new abstract rewriting system on multisets, and model chemical reactions as a symbolic rewriting system which acts on a multiset, which can be viewed as a test tube. Using this model, we made experiments on Brusselator model and we confirm non-linear oscillation emerge.

Furthermore, we suggest that confluence property and terminating property are two principal components to characterize the behavior of ARMS and we obtain the condition when ARMS generates cycle. These results, if properly translated in the terms of chemical reactions, might provide insightful analysis for prebiotic conditions where the proto-from of life emerges as a cycle of reactions.

Acknowledgments

This research is supported by Grants-in-Aid for Scientific Research No.07243203 from the Ministry of Education, Science and Culture in Japan.

Bibliography

[1] N. Dershowitz and J. P. Jouannaud, Rewrite Systems, in Handbook of theoretical computer science 245-309, Elsevier, 1990.

[2] M. Eigen and P. Schuster, The Hypercycle, Springer-Verlag, 1979.

[3] W. Fontana and L. W. Buss, The Arrival of the fittest: Toward a Theory of biological organization, Bulletin of Mathematical Biology, Vol.56, No.1, 1-64, 1994.

[4] W. Fontana, Algorithmic Chemistry, Artificial Life II, 160-209, Addison Wesley, 1994.

[5] D. Frijtyers and A. Lindenmayer, L systems, Lecture Notes In Computer Science, vol. 15, Springer Verlag, 1994.

[6] J. H. Gallier, Logic for Computer Science, p89, John Wiley & Sons, 1987.

[7] Bellin, G. and G. Boudol. 1992. The chemical abstract machine. *Theoretical Computer Science* 96: 217–248.

[8] J. E. Hopcroft and J. D. Ullman, Introduction to Automata theory, Languages and Computation, Addison-Wesley, 1979.

[9] G. Huet and D. S. Lankford, On the Uniform halting problem for Term Rewriting Systems, Rapport 359, INRIA, 1978.

[10] S. A. Kauffman, The Origins of Order, Oxford University Press, 1993.

[11] D. E. Knuth and P. B. Bendix, Simple word problems in universal algebras, North-Holland, 1985.

[12] C. G. Langton, Life at the Edge of Chaos, Artificial Life II, Addison & Wesley, 1991.

[13] Ju. Matijasevich, Diofantovo predstavlenie perechislimykh predikatov, Izv. Akad. Nauk SSSR, Ser. Matem., Vol.7, No5, pp.935-938, 1988.

[14] H. R. Maturana and F. J. Varela, Autopoiesis and Cognition, D. Reidel Publishing Company, 1980.

[15] G. Nicolis and I. Prigogine, Exploring Complexity, An Introduction, Freeman and Company, 1989.

[16] M. J. O'Donnell, Computing in system described by equations, Lecture Note in Computer Science, Vol.58, Springer Verlag, 1977.

[17] J. D. Watson, N. H. Hopkins at el, Molecular Biology of the Gene, The Benjamin/Cummings publishing Company, Inc, 1992.

[18] S. Wolfram, Cellular Automata and Complexity, Addison Wesley, 1994.

Chapter 48

Extinction dynamics in a large ecological system with random interspecies interactions

Kei Tokita[1]
Department of Chemistry and Chemical Biology
Harvard University, 12 Oxford Street, Cambridge MA 02138, USA
Ayumu Yasutomi
Suntory and Toyota International Centres for Economics and
Related Disciplines
London School of Economics and Political Science
London WC2A 2AE, UK

Introducing the effect of extinction into the so-called replicator equations (mathematically equivalent to the Lotka-Volterra equations) of population dynamics, we construct a general mathematical model of ecosystems. In the model, the diversity of species, i.e. the dimension of the equation, is a time-dependent variable. By the explicit introduction of the effect of extinction, the system shows very different behavior from the original replicator equation and leads to mass extinction when the system initially has high diversity. The present theory can serve as a mathematical foundation for the paleontologic theory for mass extinction.

[1]WWW: http://paradox.harvard.edu/~tokita/index.html
Permanent address: Condensed Matter Theory Group, Graduate School of Science, Osaka University, Toyonaka 560-0043, Japan.

1. Introduction

Mathematical biological models of evolution[1] have been a recent object of study in relation to complex systems. In particular, the problem of the extinction of species in ecosystem[2] has been discussed within the framework of physics[3, 4, 5, 6]. On the other hand, the mechanism of mass extinction has been a classical and controversial problem studied by a number of researchers in paleontology[7, 8, 9] and evolutional biology[10]. The conclusions based on the results of these studies can be divided into two categories, one emphasizing exogenous shocks (*bad luck*)[11, 7, 12] and the other, endogenous causes (*bad genes*)[13, 5]. Building on both views, we construct a general mathematical model of mass extinction starting from a traditional mathematical biological equation which describes the dynamics of populations of interacting species. This model reflects the former view, e.g., the situation where several biotas which have been separated from each other for a long time are suddenly integrated into a larger ecological network by some exogenous shock (*biotic fusion*)[14, 15]. (One example of this kind of large-scale extinction caused by such biotic fusion can be seen in a comparison of the number of families of land mammals in North America and South America before, during, and after the formation of the Panama land bridge between the two continents in the Pleistocene epoch about two million years ago[16].) We assume that the interaction coefficients for this newly produced ecosystem can be written in the form of a random matrix[17, 18, 19]. Meanwhile, following the latter view, we adopt the concept of an *extinction threshold*, which we introduce into the replicator equations[1] of the population dynamics. We refer to these large dimensional replicator equations with random interspecies interactions and the extinction threshold as *extinction dynamics*.

The extensive numerical simulations clarify the nature of the phase space of extinction dynamics, which in general has a vast number of multiple basins of attraction. One of the results of our study is that the distribution of such basin size is characterized by a *power law*. Moreover, the dependence of the results on parameters suggests that the original replicator equations without the extinction threshold should also follow the same law. Therefore, extinction dynamics can be a powerful tool for investigating the complex behavior of original replicator equations because extinction dynamics has rather simple attractors, while the original replicator equations often have complex attractors like chaos.

We also find several significant new features that characterize mass extinction. Defining the *diversity* as the number of existing species, we first find that final value of this quantity is largely independent of its initial value. Second, we find that mass extinction does not occur immediately after an environmental change, but begins after a number of *induction time*[20].

2. Model

2.1. Replicator equations with random interspecies interactions

We investigate the time development for populations of species, a development which obeys the following equations:

$$\mathrm{d}x_i(t)/\mathrm{d}t = x_i(t)(f_i(t) - \bar{f}) \qquad (i = 1, \cdots, N_I) \tag{2.1}$$

$$f_i(t) = \sum_{j=1}^{N_I} a_{ij} x_j(t) \tag{2.2}$$

$$\bar{f}(t) = \sum_{i=1}^{N_I} f_i(t) x_i(t) \tag{2.3}$$

$$\sum_{i=1}^{N_I} x_i(t) = 1 \qquad (0 \le x_i(t) \le 1). \tag{2.4}$$

These equations, the *replicator equations(RE)*[1], are generally used to describe the evolution of self-replicating entities, so-called replicators[21]. The replicator equations have become a well-established model in many fields, including sociobiology, studies of the prebiotic evolution of macromolecules, mathematical ecology, population genetics, game theory, and even economics. In particular, N_I dimensional RE is equivalent to the $N_I - 1$ dimensional general Lotka-Volterra equation[1], the analysis of which is one of the main subjects of mathematical biology.

In eqs. (2.1)-(2.4), the variable x_i denotes the *population density* of the replicator i. Hereafter, we will in general use the more general term ' species' in reference to these replicators. N_I denotes the initial number of species, that is, the initial value of the *diversity*. $f_i(t)$ and $\bar{f}(t)$ denote *fitness* of species i and their average at time t, respectively. The (i, j)-th element of the matrix $A = (a_{ij})$ determines the effect of species j on the growth rate of species i. Here we use $a_{ii} = -1$ to represent self-regulation, and we assign the interspecies interaction coefficients a_{ij} ($i \ne j$) as Gaussian random numbers with mean 0 and variance v. In general, this random asymmetric interaction matrix drives this system into a non-equilibrium state.

The assumptions we make for these asymmetric random interactions are based on the hypothesis that a *biotic fusion* reorganizes species relationships in a random fashion[15]. Here the biotic fusion is defined as a situation where a number of biotas which have been separated from each other for a long time are suddenly integrated into a larger ecological network by some exogenous shock[14]. This kind of ecosystem with random interaction also can be produced by a reduction in a habitat area, which paleontologists have asserted to be a trigger for mass extinction[9]. Because the reduction in a habitat area may confine many biologically isolated species to a narrow area, it drives them into competition and, eventually, brings biotic fusion. In this sense, a large-scale biotic fusion of many biotas can occur as well as a fusion of two biotas[15]. One of the purposes of the present study is to show that any large-scale biotic fusion and any subsequent random interspecies interactions may play a role in the bad luck effect for mass extinction.

Several pioneering studies of such a random interaction model have been carried out using the theory of random systems. However, these studies have dealt with limited cases such as the local stability condition for a linear version of RE[18], the replica variational theory for RE with symmetric random interactions which ensures equilibrium states[3], and the dynamic mean-field theory for noise-driven RE with asymmetric interactions[4] only in the parameter region where the asymmetry is weak and the system is ensured to approach a fixed point. On the other hand, the global behavior of RE with fully asymmetric random interaction is hardly treated analytically, because the equations are highly nonlinear and the dynamics often show not only convergence to a fixed point, but also complex behavior such as *heteroclinic orbits* [22, 23] or *chaos*[24, 25], even at low dimension ($N_I \geq 4$).

2.2. Introduction of the extinction threshold

We should note here that extinction is not well-defined in the RE model with large N_I and random interactions because such a model generally has heteroclinic orbits. When as heteroclinic orbit approaches a *saddle* where some species are extinct, the population densities exponentially approach zero. However, they never actually reach zero because the orbit is bound in the *interior* of the simplex (2.4). In the vicinity of the saddle, the values for these population densities are too small to cause underflow by naive numerical calculation. Nevertheless, some of these populations eventually begin to revive, causing the orbit to leave for another saddle. This transition among saddles continues cyclically or chaotically. The exponential approach of population to zero and its revival to the order $O(1)$ play a significant role in heteroclinic orbits. However, in the real world such a small population density cannot be sustained. In this sense, heteroclinic orbits have never been believed to be biologically significant.

Considering the above problem, we introduce the parameter δ to the dynamics to represent the extinction threshold (2.1)-(2.4) : at each discrete time step, the population density x_k is set to zero if this quantity becomes less than δ. The population densities of the surviving species $\{x_i\}$ ($i \neq k$) are then renormalized to satisfy $\sum_{i \neq k} x_i = 1$. This renormalization implies that the niche of an extinct species is divided among the survivors. The diversity decreases through the above process, and we denote its value by N. The introduction of δ is also nothing but a finite size effect on RE, because δ coincides with a minimum unit of reproduction for each species, and its reciprocal $1/\delta$ corresponds to the permissible population size of an ecosystem. We should note that the original replicator equations can be restored in the limit $\delta \to 0$ of no extinction.

It must be noted here that the present model belongs to a class of systems for which the dimension is a time-dependent variable. Although this time-dependence is inevitable not only in population dynamics[26, 27, 28, 29, 30] but in many other fields, such a highly nonlinear model has never been systematically analyzed.

Whenever there is a given set of parameters A, δ and an initial diversity $N_I \equiv N(t = 0)$, the initial state $\{x_i(0)\}$ evolves until a steady state is achieved. Extinction never occurs in this steady state, and there remains a stable subecosystem with a comparatively small number of surviving species (*core species*) N_F ($\leq N_I$). Although almost all orbits converge to an equilibrium point in this state, we also find periodic orbits. Chaotic orbits are very rare. Heteroclinic orbits are never achieved because the existence of the finite δ prohibits any orbit from approaching a saddle. Such stability is always achieved by any finite δ. This is a new type of destruction of a high dimensional attractor, which is, in general, called *crises* in the theory of chaos[31]. Therefore, let us refer to this kind of dynamics as *extinction dynamics* (ED). By a series of extensive numerical simulations, we investigate the novel features of ED, especially the dependence of ED on three parameters: N_I, v, and δ.

From the point of view of random system theory, it is important to observe typical behavior for ED by executing *random average* of quantities over samples of a random matrix A. Hereafter, we will in general write this average as $\langle ... \rangle_A$.

3. Results

3.1. Basin-size distribution

The first results presented here concern the basin-size distribution for ED with a large number of basins of attraction. Here, we identify each 'attractor' only by the composition of core species, not by its trajectory. In other words, even if several isolated attractors coexist in a system of core species, we do not discriminate between these attractors and we regard them to be in one basin of 'attraction'. The reason for this is that in ED such coexistence is rare, and this classification of basins of attraction also agrees with the classification of subecosystems created by ED.

In order to obtain the basin-size distribution, we (a) iterate ED starting from a sufficient number of random initial states in a system with the same parameters and the same random matrix A, (b) count basin size as the number of initial states which converge to each 'attractor', and (c) make a rank-size distribution $S(n)$, where the natural number n denotes the rank of each basin and can reach the total number of 'attractors' found in the simulation. Moreover, the above process is iterated for a sufficient number of random matrices A with the same v, and we finally obtain a basin-size distribution $\langle S(n) \rangle_A$ for a parameter set. $\langle S(n) \rangle_A$'s for various parameter sets are shown in Fig. 3.1.

It is clear that the basin-size distribution $\langle S(n) \rangle_A$ characteristically follows a power law. Moreover, each exponent of the power depends only on N_I, not on δ nor v. This exclusive dependence on N_I can be understood intuitively because the number of combinations of core species (the number of 'attractors') depends only on N_I. Therefore, a larger N_I provides a larger number of 'attractors' and hence, a smaller exponent. On the other hand, the independence of δ strongly suggests that the basin-size distribution of the original RE (ED in the limit $\delta \rightarrow 0$) also follows the power law. This conjecture is relevant to the hierarchical coexistence of infinitely many attractors in RE[32]. The power law of rank-size relationship with exponent near unity is often referred to as Zipf's law[33] in linguistics and other diverse fields[34].

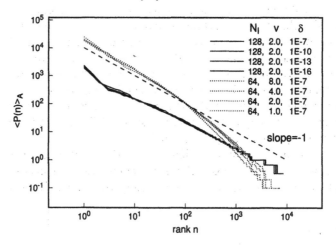

Figure 3.1: Basin-size distributions for (a) $N_I = 64$ sampled from 100000 initial states and averaged over 10 samples of A, (b) $N_I = 128$ from 20000 initial states and 3 samples of A.

3.2. Extinction curves

Fig. 3.2 shows $\langle N(t) \rangle_A$, the average diversity as a function of time(*the extinction curve*). Two significant characteristics can be observed from this figure. The first is that the average final diversity $\langle N_F \rangle_A$ is independent of N_I. This surprising result implies that no matter how large the diversity of initial species, the average diversity of species in the final state is small in comparison. That is $N_F \ll N_I$. In other words, when a large random ecosystem emerges as a result of the merging of many smaller ecosystems, a mass extinction will occur of 'size' $N_I - \langle N_F \rangle_A$. The other significant characteristic is that the avalanche of mass extinction begins after some *induction time* t_I and ends in each case at nearly the same time $t_R \sim 10^3 (\geq t_I)$. As N_I becomes larger, t_I becomes larger and approaches t_R. Therefore, for sufficiently large N_I, the extinction curve shows a sharp drop at t_I. Such an abrupt mass extinction occurring on a short time scale is highly relevant to the notion of 'punctuated equilibria[35]'.

The induction time and the abrupt drop in diversity at a large N_I is explained by the small rate of change for each x_i at $t = 0$, and a faster than exponential decay of x_i for extinct species. At time $t = 0$, the absolute value of the fitness $f_i(0) \equiv \sum_{j=1}^{N_I} a_{ij} x_j(0)$ (the first term in the parentheses of *r.h.s.* of the equation

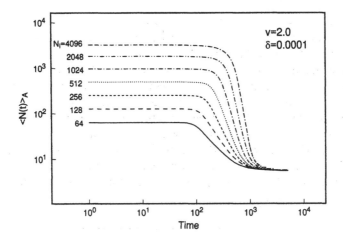

Figure 3.2: Extinction curves for several values of the initial diversity of species N_I with $\delta = 0.0001$ and $v = 2.0$. Each curve represents an average taken over 1000 samples of A.

(2.1)) for each species i is of estimated order $O(\sqrt{v/N_I})$ by a simple calculation. The absolute value of the average fitness $\bar{f}(0) \equiv \sum_{i=1}^{N_I} f_i(0)x_i(0)$ (the second term in the parentheses) has the same order. Therefore, as N_I becomes larger, the absolute value of the change rate $f_i - \bar{f}$ at $t = 0$ becomes smaller in proportion to $1/\sqrt{N_I}$, which makes the induction time larger because a smaller change rate makes populations change more slowly. However, the rapid decay eventually drives populations into extinction around the induction time. Therefore, almost all species, except for core species, are expected to become extinct synchronously in the limit of large N_I.

It should be noted here that and annihilations such as those caused by the impact of a meteor[11] or large-scale volcanic activity[12] are not necessarily required for mass extinction; biotic fusion triggered by some environmental change could be sufficient if it produces a large ecosystem with random interspecies interactions. It should also be noted here not only that the average diversity $\langle N_F \rangle_A$ of a core species is independent of N_I but also that the distribution $P(N_F)$ of N_F does not depend on N_I, as shown in Fig. 3.3.

Fig. 3.4 concerns the variation of extinction curves with v. As v becomes larger, the induction time t_I becomes shorter(Fig. 3.4a), and the average final diversity of surviving species $\langle N_F \rangle_A$ becomes smaller(Fig. 3.4b). Consequently, when the order of the interspecies interaction becomes large compared with the absolute value of the self-regulation ($\{a_{ii} = -1\}$), the avalanche of mass extinction begins earlier, and a smaller diversity of species survives. This explains why existing species seem to organize rather small ecosystems instead of large and complex relationships (at most, 5 species are inter-related, as suggested by work on the food webs[36]). This can be explained by existing ecosystems being formed as a consequence of extinction dynamics.

The extinction curves for several values of δ's are also shown in Fig. 3.5. We observe that $\langle N_F \rangle_A$ is also independent from δ. In other words, the result here means that no matter how small the extinction threshold, an extinction (of ' size' $N_I - \langle N_F \rangle_A$) is inevitable. We should note that the present result and the fact that there is no extinction in the original RE (ED in the limit $\delta \to 0$), are not inconsistent with each other. This phenomenon results from the simulation and the analytical estimation equally showing $t_I \sim -\log(\delta)$, and from t_I diverging to infinity in the limit $\delta \to 0$, that is, with no extinction occuring in finite time. From Fig. 3.2, Fig. 3.4 and Fig. 3.5, we can conclude that $\langle N_F \rangle_A$ depends only on v, but not on N_I nor δ, which is in contrast with the parameter dependence of $\langle S(n) \rangle_A$ only on N_I.

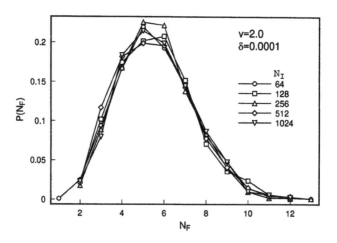

Figure 3.3: Distribution $P(N_F)$ of the diversity of core species for several values of N_I. The distribution was obtained using 1000 samples of random matrices A (1000 runs of extinction dynamics).

3.3. Average fitness and the nature of the shrink matrix

Here we discuss the time development of average fitness $\bar{f}$ through extinction dynamics, as depicted in Fig. 4.1. It should be noted that the average fitness takes on positive values, except during the short period at the beginning. The final value for average fitness $\bar{f} \sim 0.4$ is higher than what would be expected for a randomly generated ecosystem with the same diversity($N_F \sim 8$). Thus, more stable ecosystems are self-organized by ED. We also observe that $\langle \bar{f}(t) \rangle_A$ does not show a monotonic increase and reaches a maximum value at a time near t_I. This in general suggests that the average fitness shoots up in response to the avalanche of extinction of low-fitness species around the induction time and settles down to a final value via competition among core species.

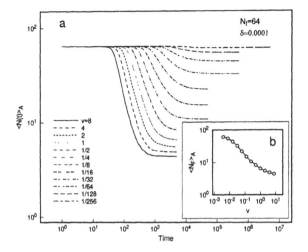

Figure 3.4: (a) Extinction curves for several values of the variance v of random interactions with $N_I = 64$ and $\delta = 0.0001$. (b) The average diversity of the final surviving species $\langle N_F \rangle_A$ is calculated as a function of v.

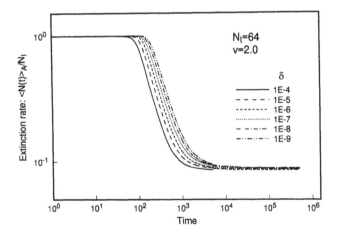

Figure 3.5: Extinction curves for several values of the extinction threshold δ with $N_I = 64$ and $v = 2.0$.

The time development for the distribution of elements of interaction matrices via extinction dynamics is depicted in Fig. 4.2. The average of a_{ij} shifts to a positive value, which means that the interaction matrix of the subecosystem becomes cooperative via extinction dynamics. This also contributes to an increase in the average fitness. It should be noted that the distribution continuously holds its gauss distribution shape. Therefore, the interspecies interaction coefficients of core species are still random, and various types of relationship among core species are realized by ED. The time development of $\langle a_{ij} \rangle_A$ is also shown in Fig. 4.3.

4. Estimation of induction time

Here let us clearly define the induction time t_I. It is convenient to define the induction time as the half-life of extinct species, as follows:

$$t_I \equiv \text{Min}\{t \,|\, N(t) < (N_I + N_F)/2\}. \tag{4.1}$$

In this section we estimate the t_I depending on parameters. At $t = 0$, the distribution of the elements of the interaction matrix $P(a_{ij})$ is Gaussian , with mean 0 and variance v. If we set each population density to the same value $x_i(0) = 1/N_I$ at $t = 0$, the variance of the average fitness can be estimated as

$$\langle f_i(0)^2 \rangle_A = \langle (\sum_j a_{ij} x_j(0))^2 \rangle_A \tag{4.2}$$

$$\simeq \frac{v}{N_I}. \tag{4.3}$$

Moreover, one can easily see that the distribution of the average fitness at $t = 0$ is again Gaussian, with mean 0 and variance v/N_I.

Meanwhile, it is useful to transform the variable $x_i(t)$ to $y_i(t) \equiv \log(x_i(t))$ and to deal with the replicator equations in the form of difference equations as

$$y_i(t+1) = y_i(t) + \Delta t(f_i(t) - \bar{f}) \tag{4.4}$$

$$f_i(t) = \sum_j a_{ij} e^{y_j(t)} \tag{4.5}$$

$$\bar{f}(t) = \sum_i e^{y_i(t)} f_i(t). \tag{4.6}$$

4.1. Approximation 1

We first estimate an induction time using a rather rough fitness approximation. Here we consider the standard deviation for the distribution of the fitness at $t = 0$ as the characteristic value for the negative fitness of a species:

$$f_i(t) \approx f_i(0) \sim -\sqrt{\frac{v}{N_I}} \tag{4.7}$$

$$\bar{f}(t) \approx \bar{f}(0) = 0. \tag{4.8}$$

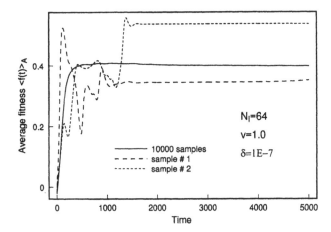

Figure 4.1: Time development of average fitness $\bar{f}(t)$ over 1000 samples of A with $N_I = 64$, $v = 2.0$ and $\delta = 10^{-7}$. Each dotted line represents one sample. The fitness $\bar{f}$ goes up and down and, in general, the final value is not the highest.

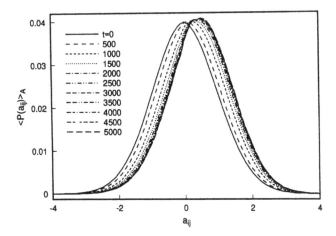

Figure 4.2: Time development of the distribution of elements of interaction matrices averaged over 2000 samples. $N_I = 64$, $v = 2.0$, and $\delta = 10^{-7}$.

Furthermore we assume that the induction time t_I is the time when this species i with characteristic value of negative fitness becomes extinct. By this approximation, we can write the following equation for t_I as

$$t_I \approx -\sqrt{\frac{N_I}{v}} \log(\delta N_I)\frac{1}{\Delta t}. \tag{4.9}$$

This approximate estimation (approx. 1) of the induction time is shown in

552

Fig. 4.4, Fig. 4.5 and Fig. 4.6 as a function of the parameters N_I, v and δ, respectively. This approximation is clearly inadequate in quantitatively estimating t_I as a function of N_I and δ. However, the log dependency of t_I on δ is qualitatively presented by this very rough approximation, and the v dependency is rather good.

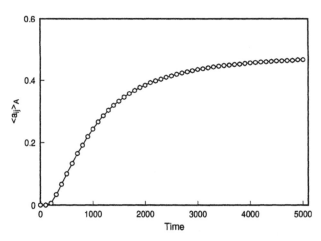

Figure 4.3: Time development of the average of elements of interaction matrices averaged over 2000 samples. $N_I = 64$, $v = 2.0$, and $\delta = 10^{-7}$.

4.2. Approximation 2

We next consider the time development of the fitness, although we again use the notion of the characteristic value of fitness. Here we regard the gaussian distribution of fitness at $t = 0$ as two delta functions at standard deviation. Therefore, we consider that $N_I/2$ species take fitness $f_+(0) = \sqrt{v/N_I}$, and the remaining $N_I/2$ species take fitness $f_-(0) = -\sqrt{v/N_I}$. By this approximation, it is possible to calculate the log-transformed population density $y_+(t)(y_-(t))$ of the species with positive (negative) fitness at arbitrary time t in the form of a difference equation as

$$y_\pm(t+1) = y_\pm(t) + \lambda_\pm(t) \tag{4.10}$$

$$\lambda_\pm(t) = \Delta_\pm \cosh\left(\sum_{t'=0}^{t-1} \lambda_\pm(t')\right) \tag{4.11}$$

$$\lambda_\pm(0) = \Delta_\pm \equiv \Delta t f_\pm(0). \tag{4.12}$$

This equation shows that a population density x_i with negative fitness decreases much faster than an exponential function in this approximation. Here t_I

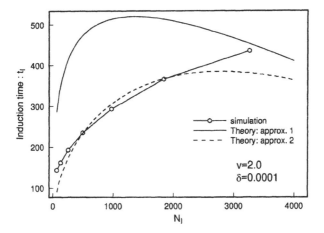

Figure 4.4: t_I vs. N_I.

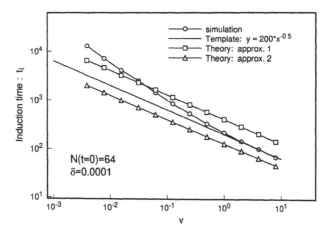

Figure 4.5: t_I vs. v.

is decided by the time when $y_-(t) < \log(\delta)$, half of the species, become extinct. Fig. 4.4 shows that N_I dependency is improved in this approximation.

5. Discussion

In this paper, we have ignored any effects of immigrants or invaders, which increase the diversity, and we have focused on global biotic fusion where no species ever come from the outside. Moreover, we did not consider any mutants, because the avalanche of mass extinction occurs so quickly that no significant evolution of mutants can occur. By neglecting these effects, the nature of extinction on

554

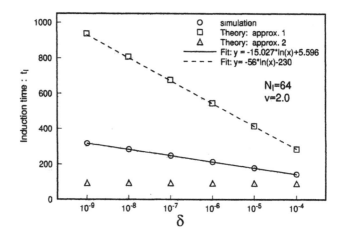

Figure 4.6: t_I vs. δ.

a rather short time scale was exclusively clarified. However, by introducing the effect of increasing diversity, we can study the nature of ED on a much larger time scale. Actually, an analysis of the interesting problem of whether ED shows *self-organized criticality*[5, 6] is in progress[37].

Our results suggests a new view of mass extinction in which an integration of several biotas that has been caused by some external shock will, after a certain induction time, cause a mass extinction through internal dynamics. Furthermore, these results clarify the importance of the finite size effect for replicator equations. One implication of our findings is that if some biotas are integrated by exogenous shocks, the diversity of species will be maintained for a period of time and then suddenly decrease to a certain value that is independent of the initial diversity. This alerts us to the fact that mass extinction may arise due to human activities if they cause some integration of biotas, and, moreover, when we apply some shock to an environment, we must continue observation over a long period of time, because its effects will not emerge at once.

Finally, it should be noted that the present results can be seen in a broad classes of general models of population dynamics because the replicator equations are accepted as a well-established model and the existence of an extinction threshold (a finite size effect of populations) seems inevitable in population dynamics.

Acknowledgements

The authors would like to thank A. Hallam, E. Shakhnovich, Y Aizawa, K. Kaneko, T Ikegami, and T. Chawanya for their fruitful discussions and for their encouragement, and G. Sheen for her careful reading of the manuscript. The present work is partially supported by the Japan Society for the Promotion of Science, a Grant-in-Aid from the Ministry of Education, Science, and Culture of Japan, and the Suntory foundation. Most of the numerical calculations were carried out on a Fujitsu VPP500/40 in the Supercomputer Center, Institute for Solid State Physics, University of Tokyo.

Bibliography

[1] Hofbauer, J. , and Sigmund K. , *The Theory of Evolution and Dynamical Systems*, Cambridge Univ. Press: Cambridge (1988)

[2] B. W. Roberts and M. E. J. Newman, J. Theor. Biol. **180**, 39 (1996)

[3] S. Diederich and M. Opper, Phys. Rev. A **39**, 4333 (1989)

[4] M. Opper and S. Diederich, Phys. Rev. Lett. **69**, 1616 (1992)

[5] Bak, P. , and Sneppen, K. , *Phys. Rev. Lett.* **71**, 4083-4086 (1993)

[6] P. Bak, K. Chen, and M. Creutz, Nature (London) **342**, 780 (1989)

[7] Hallam, A. , In *Causes of Evolution*(Ross, R. M. and Allmon, W. D. , eds.), Chicago: University of Chicago Press (1990)

[8] Raup, D. M. , *Extinction: Bad Genes or Bad Luck?*, New York:Norton (1991)

[9] A. Hallam, *An Outline of Phanerozoic Biogeography* (Oxford University Press, Oxford: New York, 1994)

[10] Maynard Smith, J. , *Phil. Trans. R. Soc. Lond. B*, **325**, 242 (1989)

[11] Alvarez, L. W. , Alvarez, W. , Asara, F. , and Michel, H. V. , *Science* **208**, 1095-1108 (1980)

[12] Coffin, M. F. , and Eldholm, O. , *Scientific American* **269**, 26-33 (1993)

[13] Van Valen, L. , *Evol. Theor.* **1**, 1-30 (1973)

[14] Vermeij, G. J. , *Science* **253**, 1099-1104 (1991)

[15] Gilpin, M. E. , *Proc. Natl. Acad. Sci. USA* **91**, 3252-3254 (1994)

[16] May, R. M. , *Scientific American* **239**, 119-132 (1978)

[17] Gardner, M. R. , and Ashby, W. R. , *Nature* **228**, 784-784 (1970)

[18] May, R. M. , *Nature* **238**, 413-414 (1972)

[19] Roberts, A. , *Nature* **251**, 607-608 (1974)

[20] Saitô, N., Ooyama, N., Aizawa Y., and Hirooka, H., *Prog. Theor. Phys. Suppl.* **45**, 209-230 (1970)

[21] Dawkins, R. , *The Extended Phenotype*, Oxford-San Francisco: Freeman (1982)

[22] R. M. May and W. J. Leonard, SIAM J. Appl. Math. **29**, 243 (1975)

[23] T. Chawanya, Prog. Theor. Phys. **94**, 163 (1995)

[24] M. E. Gilpin, Amer. Natl. **113**, 306 (1979)

[25] A. Arneodo, P. Coullet and C. Tresser, Phys. Lett. **79A**, 259 (1980)

[26] Tregonning, K. , and Roberts, A. , *Bull. Math. Biol.* **40**, 513-524 (1978)

[27] Tregonning, K. , and Roberts, A. , *Nature* **281**, 563-564 (1979)

[28] Roberts, A. , and Tregonning, K. , *Nature* **288**, 265-266 (1980)

[29] Ginzburg, L. R. , Akçakaya, H. R. , and Kim, J. , *J. theor. Biol.* **133**, 513-523 (1988)

[30] Pimm, S. L. , *The balance of Nature?* (The University of Chicago Press, Chicago and London, 1991)

[31] E. Ott, *Chaos in Dynamical Systems* (Cambridge University Press, Cambridge, 1993)

[32] T. Chawanya, Prog. Theor. Phys. **95**, 679 (1996)

[33] G. K. Zipf, *Human Behavior and the Pinciple of Least Effect* (Addison-Wesley, Cambridge, 1945)

[34] I. Kanter and D. A. Kessler, Phys. Rev. Lett **74**, 4559 (1995)

[35] Gould, S. J. , and Eldredge, N. , *Nature* **366**, 223-227 (1993)

[36] Sugihara, G. , Schoenly, K. , and Trombla, A. , *Science* **245**, 48-52 (1989)

[37] Yasutomi, A. and Tokita, K. , 1997, preprint

Chapter 49

Functional differentiation in developmental systems

Irina Trofimova

Keldysh Institute of Applied Mathematics RAS, Moscow, Russia

The study of functional differentiation involves understanding the processes whereby a population of individuals, whether humans or cells, become specialized in the behaviors which they express. Research into functional differentiation, across several levels, has suggested that this specialization occurs as a result of constraints on the flow of resources which pass through the individual from the environment (including other agents). These constraints are: 1) the maximum amount of resource that an individual can accept from outside, 2) the maximum amount of resource that an individual can give back to the population or other environment, and 3) the maximum amount of contacts that an individual can hold with such environment. This issue was explored through the Compatibility Model and Functional Differentiation model (FD-model).

1. Association and dissociation of system elements

An effect of cooperative behavior upon a system of multiple elements or agents is the emergence of novel systemic properties. We consider such effects to be the result of associations of elements and in so doing pay less attention to another system effect—that of the functional dissociation of elements within a system. Symmetry breaking is a sign of internal differentiation between different parts of a system or between a system and environment. There are three interesting aspects here that we would like to emphasize:

> The diversity of elements within naturally occurring systems is not random: a system cannot be an association of just any number or kind of elements nor can it appear from identical elements. The

emergence of such a system happens only given a particular diversity of elements, and this diversity is not arbitrary. As examples we can take ecosystems, social or economical systems. A central question then is: what types of elements, how many of each type and what conditions are necessary for the emergence of such systems?

The functional differentiation of elements occurs during and persists following the emergence of these systems: the subsequent dynamical activity of elements is also not simply random. Elements have differing abilities to carry out one or more functions, and the dissociation of elements into different functional classes, which is founded upon these basic abilities, results in a gradual specialization of elements. An element will happen to carry out one function or another based upon the ranking of its ability relative to the other elements, and it may change its function should it receive a better ranking for another ability, or should some other element appear with a better ranking than its own in this specialization. So another question is: what are the conditions necessary for such dynamical specialization?

The poly-functionality of many elements shows us that usually an element has a disposition to be effective in carrying out several functions as long as its configuration is compatible with such functions. Also the structure of connections between an element and other elements is based upon the compatibility of its configurations, and here degree of diversity of elements and their compatibilities, as well as population size play a role in the emergence of a system. How these formal parameters relate to each other?

In the analysis of functional differentiation we need to consider two processes—that of the association of elements in a system and that of their dissociation into different types.

2. Compatibility model

It was decided to first focus attention upon the third of our questions and so we analyzed the impact of diversity, size of population and sociability of its elements on system behavior.

Let us consider a model of a population in which agents possess an abstract set of characteristics, and seek to form connections with other agents according to the degree of conditional compatibility between these characteristics. Each connection carries with it a relative valuation on the part of the agent forming it, and the agents attempt to optimize their valuations over time. This results in a highly dynamic connection structure. The effect of cooperation depends upon not only compatibility of elements, but also upon a size of population, diversity and the number of contacts that a given element can form with other elements (sociability).

The Compatibility model[7] is a random graphical dynamical system with features reminiscent of spin glass models. We consider an ensemble of N cells, each of which possesses a "resource of life" R, and a k-dimensional "vector of traits" v, where each component can equal ± 1. Each cell forms connections with other cells, and both the maximal number of connections per cell S, termed the 'sociability' and the rate of connection attempts a are fixed and identical for all cells.

Time is discrete, and at every time step each cell i attempts a random connections, and its life resource R_i is adjusted according to the "quality" of its current connections. If $R_i > 0$, the cell dies, to be replaced by a new cell having the maximal life resource R and a random vector of traits. The quality of a contact between i-th and j-th cells is evaluated according to the traits of both cells:

$$q(i,j) = \sigma(T_{ij})(v_i, v_j) = \tanh(\gamma T_{ij}) \sum_{m=1}^{k} V_{im} \; V_{jm} \tag{2.1}$$

where $(\,,\,)$ denotes the inner product of two vectors, T_{ij} is the duration of the contact between the cells i and j, and $\sigma(T)$ is the "efficiency of the contact"—for small T it linearly increases, and after several time steps saturates at $\sigma = 1$. The quality varies between σk for aligned trait vectors, to $-\sigma k$ for anti-aligned trait vectors. For cell i having n_i connections and connection set $\{i_m\}$, the value of the life resource at the next step is

$$S_i(t+1) = S_i(t) - \delta(n_i, \{i_m\})$$

$$\delta(n_i, \{i_m\}) = \frac{\delta_0}{1 + \alpha n_i} - \sum_{m=1}^{n} q(i, i_m) \,. \tag{2.2}$$

Thus high quality connections result in a longer life.

Connections are adjusted as follows. At each time step, a cell i is randomly chosen and a possible contact cell j, which is not a member of the connection set of i is randomly selected. The possible profit of this connection for both i and j is determined by calculating the effect of this connection on the calculation of $\delta(n_i, \{i_m\})$ and $\delta(n_i, \{j_k\})$. If the effect is beneficial to both i and j then the connection is formed and added to the connection sets of both i and j. If the number of connections in either set is submaximal, it is simply added to that set. Otherwise the least profitable connection is deleted prior to adding the new connection. This process is repeated a times so that the number of new connections formed depends upon their profitability.

Populations of size 50, 100, 200, 300, 400, 500, 1000 and 2000 were simulated. The sociabilities ranged in increments of 20 contacts from 10 up to 130. The trait vectors were of dimension 2, 4, and 8, that gave us correspondingly 4, 16 and 256 types of elements. The model was simulated up to 10000 time steps. In general, as the population size increased, the number of simulation steps required to achieve stationary statistics also increased.

The object of study was the formation of connected components. In the compatibility model, these components proved to be highly volatile, with large fluctuations in connection structure occurring for all parameter values. An order parameter A, termed the affiliation of the group, was defined as the ratio of the mode of the distribution to the maximum generated cluster size. This order parameter is close to 0 when the distribution is highly skewed towards small clusters, and close to 1 when it is skewed towards big clusters. Using A, it was found that the compatibility model exhibits a stochastic phase transition in the distribution of cluster sizes as function of the sociability. For low populations, this transition bears many of the features of a continuous phase transition. There is a gradual increase in the order parameter across the transition. The entropy of the distributions gradually increases, reflecting a uniformization of the distribution, and thus a less ordered state, and then gradually declines again, reflecting a transition to a more ordered state. For large populations, the transition becomes more abrupt. The affiliation shows a very sharp transition from a level near 0 below a critical value, to a level close to 1 above this critical value. We were able to estimate the critical sociability as $S_c = P^{0,6}$, where P is the population size. The discontinuous nature of the change in affiliation suggests a first order transition, yet the entropy continues to show the rise and fall typical of the continuous transition.

Continuous phase transitions have been extensively studied and two well known models, percolation models[2] and random graph models[4] bear a striking similarity to the compatibility model. In all of these models, connections are established between a collection of vertices. The difference between the models lies mostly in the fact that in percolation and random graph models, the connections, once established, are fixed, as are the vertices. Thus the percolation and random graph models simulate equilibrium conditions. In addition, those models yield a value for the critical sociability of $S_c = 2$, in contrast to our model. The connections in the compatibility model are highly dynamic with large fluctuations occurring for all values of the sociability. In addition, the apparent critical state of our model is characterized by a probability distribution of high entropy, a close approximation to a uniform distribution. Such a uniform distribution reflects a complete lack of structure in the system, with clusters of all scales appearing with more or less equal frequency. Such a state is maintained so long as the sociability remains constant at the critical value. This situation is reminiscent of that observed in so-called self organized criticality models[1].

Usually we see in the appearance of a system a significant degree of redundancy as regards the details of its construction. The normal natural organization can be presented as some organized structure together with a grab bag of garbage, representing leftovers from the construction process. If a system permits a large sociability, and there is enough diversity, then the system "can find a place for everybody": the system allows its elements to be poly-functional and adaptive to different tasks and neighbors.

Thus the sociability and population size appeared to be a more critical factor of this transition then the diversity of elements. Given the small range of

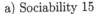

a) Sociability 15

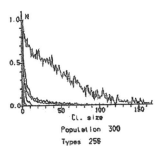

b) Sociability 40

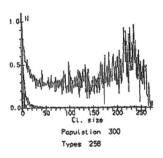

q) Sociability 90

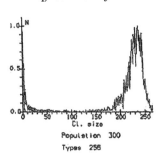

Figure 2.1: Cluster Distribution Functions. The y-axis represents number of clusters normalized against the mode.

population studied it is still too early to draw any definitions or final conclusions about the effect of diversity on system dynamics.

However what determines the process of clusterization? What constitutes compatibility? What configuration of elements make them compatible or incompatible with one another and to the tasks of the environment? What parameters

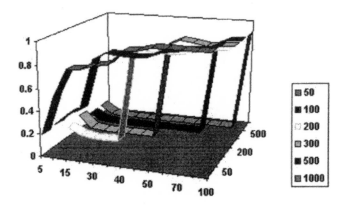

Figure 2.2: Affiliation Landscape. The lines demonstrate the affiliation as a function of populations and sociability.

describe this?

3. What parameters describe functions?—Life on the flow

Functionality is the property of system's units to execute special functions within a system's community. Thus, the study of functional differentiation involves understanding the processes whereby a population of individuals, whether humans or cells, economical institutions or species of animals, become specialized in the behaviors which they express. In the case of cells, we see differentiation into blood forming elements, repair elements, control elements, structural elements, and so forth. In the case of humans, we find ourselves drifting into different roles—leader, teacher, healer, mediator, worker, couch potato, etc. Such a specialization reflects certain needs of a system. So if we want to understand and classify specialization of elements inside a system, we have to find parameters that describe these needs.

One can begin the search for such parameters from the following general principle. The complexity of natural systems is connected with their property to pass some energetic flow through themselves: such systems are both open and dissipative, they accept and lose different kinds of resources. This fact allows us to analyse the activity of a system using the concept of flow parameters.

A natural system is always an organized structure through which the incoming flow is processed. It possesses a set of attractors reflective of its dealings with this flow, resulting in a set of possible states which this structure might adopt throughout its dynamical life. A system is an association of multiple elements which, individually and in groups, possess a structure of connections which also evolves during their life. They support each other dynamically and eventually establish a kind of homeostasis, preventing each other from going too far from the states that are appropriate for keeping this ensemble together.

This means that we can apply the concept of homeostasis here in a very dynamic sense—as the tendency of a system to keep itself in a certain set of nonequilibrium states. We can consider at least two possible tendencies in the dynamics of any system which form the directions (and eventually functions) of its elements: (1) to keep the characteristics of the incoming flow within optimal limits, and (2) to keep open the possibility to organize the flow from a system. Under such conditions the incoming flow can "play" inside the system under the control of its internal structure.

We can extract two basic parameters: "ability to accept incoming resource flow" (I) and "ability to spend or expel output resource" (O). A natural system is thus likened to a conduit, and our parameters are similar to regulators on the different ends of this pipe. In general, systems must regulate the flows of many differing types of resources. In addition, resource is not merely passed from one element to another. In general it will undergo some form of transformation. The transformation of resource within a system can also be described by means of a set of formal parameters.

The existence of a set of such parameters presupposes some form of partial ordering on the set of possible manifestations of the resource. This might be a material quantification or a functional valuation. Inhomogeneity in the temporal and spatial properties and in the density of the flow induce the differentiation of elements within the system. Thus differentiation of function is a result of the character of a flow passing through a system. Inhomogeneity in the flow results in a spreading of properties along selected parameters, resulting in a polarization of abilities within the population of elements.

Table 3.1 and Table 3.2 demonstrate this polarization for the first three levels of functional differentiation in our input-output example. The first division corresponds to the establishment of roles, that is who can play the role of "input end" and who the role of "output end". The second division is based upon a differentiation of input and output elements. For example, if an element lacks appropriate complementary neighbours, a further differentiation of role may take place. Consider the case of a spinster living alone. Given the need to expend resource in the form of nurturing, and lacking human companions, they may instead expend this resource on the nurturing of a pet.

The third division is connected with the process of association and cooperation of elements. This plays an important role, especially in maintaining some semblance of homeostasis in an uncertain environment (for example to prevent the disappearance of a vital resource). As we mentioned above, cooperation depends upon the number of contacts that a given element can form with other elements (sociability). Thus the maximum number of contacts that an element can establish and keep becomes the critical characteristic, and contacts itself become a type of resource. Moreover the exchange of resources at its most primitive level requires, at the very least, a formal, albeit transient, connection between the elements which participate in this exchange.

Table 3.1: Extreme types of elements by 3 parameters.

Parameter	I(nput)		O(utput)		Sociability	
Rank of element by parameter	High	Low	High	Low	High	Low
Flow	Takes much	Takes few	Gives much	Gives few	Many contacts	Few contacts
Index	1**	0**	*1*	*0*	**1	**0

It is important to emphasize that in a population of interconnected elements personal traits do not completely define the functions of these elements. For example, J. S. Hale and L. J. Eaves[5] created a dynamic model of the relationships between behavioral strategies in evolution. They showed that a real evolutionarily stable strategy includes some dominant behavioral strategies and certain balances between them. If the density of agents possessing one of these dominant strategies declines, breaking the correlation between all of the strategies, then the density of agents with alternative strategies also changes. There will be shift towards the use of other strategies. As a consequence, the predisposition of animals to different types of behaviour depends upon the characteristics of the other members of the population. The environment resonantly excites one or another feature possessed by each individual and these features could be opposite to each other yet cohere together in that one individual.

How does this differentiation develops further?

4. Development of a system is a specialization of its elements

Formally speaking, the structure of the connections between the elements of a system is a major determinant of the changes that characterize the development of the system. This structure represents the internal organization of a system as regards the process of resource flow. Elements such as 011 create a possibility of taking the pattern of real structure and using it in their activity. Such a separation of a virtual pattern of structure from the real structure produced in the course of evolution another type of resource—information.

Thus, systems could make transformations between 3 types of resource: energy, connections and information.

It seems likely that in the case in which an element is unable to accept all input, it needs some other element with which to share it (and so sometimes even save itself). In this case if it has a resource *connection*, it can cope (with some limitations) with all of the tribulations of unpredictable flow and save some energy through effective use of this structure of connections. A most significant accomplishment for such elements is the shift from merely distributing and re-

Table **3.2**: Three divisions of functional differentiation.

Index:

I O S   Function

1 Acceptor

   1 0   Condenser

   1 0 1   Condenser flexible—selector, relaxator, leader

   1 0 0   Condenser rigid—condenser itself, limiter, inhibitor, warehouse

   1 1   Conductor

   1 1 1   Conductor flexible—socially active element, *controller the rightness*, regulator

   1 1 0   Conductor rigid—conductor itself, deliver

0 Extractor

   0 1   Intensifier of input, producer, cleaner

   0 1 1   Intensifier and cleaner flexible—reflector and self-reflector of structure

   0 1 0   Intensifier and cleaner rigid—extractor itself, producer and worker by instruction

   0 0   Shell and stable structures, connections structures

   0 0 1   Shell, stable structures, connections flexible—external and internal soft shells

   0 0 0   Shell, stable structures, connections rigid—bonds, hard shells, defend

distributing resource, to actually carryign out some modification of the resource before it is passed on to the next elements. Self-regulated movement provides an example of this in the context of energy flow. Perception, evaluation, reasoning provide examples within the domain of information flow. Such transformations of resource create new possibilities for a system in its interaction with its environment.

All natural systems have a period of *childhood* that is a period of intensive primary internal structuring and sensitivity to initial conditions. We can probably say that specialization of elements inside a system characterize its development, and even more—such differentiation and change of functional activity of elements is the obligatory sign of any development. Probably the degree of change and the sensitivity of internal structure to the character of the flow reduces at the end of *childhood*. The *Adult* state of a system implies the existence of an optimal distribution of elements and interconnections between their groups, that is why it is already more or less stable, or metastable. Metastability of a system derives from the fact that subsystems stimulate each other: even with just 2 functions ("to maintain input" and "to maintain output"), each of these corresponding subsystems stimulates the other: "output" needs something to put out and so "input" is stimulated to maximize that which they bring in.

These ideas are well recognized in developmental biology. I posit that a similar basic division of functions will arise within any naturally occurring morphological or organizational structure. I have begun to explore this issue through the use of a computer Functional Differentiation model (FD-model). A population of cells is allowed to adapt through local interactions with neighboring cells. Each cell possesses a set of possible action (such as reproduction, division, creation of contacts, dissolution of contacts, secretion or absorption of resource, death) and each action is given a probability of selection. Each time an action is performed, its probability is augmented while those of the other actions is decremented. Each cell possesses a quantity of resource, which fluctuates according to the nature of the local contacts with neighbors, plus internal factors. Selection of an action also depends upon the level of resource. For example, low levels bias in favor of only movement or death, while high levels permit more complex actions such as division. Resource levels reflect environmental contributions to action selection while dispositions reflect internal factors.

The promising results obtained through the use of the Compatibility model suggest that the use of such computer simulations provides us with a powerful tool through which we can forge an understanding of the deep and complex questions concerning the origin of functional differentiation.

Bibliography

[1] Bak P., Tang C. and Wiesenfeld K. Self Organized Criticality, Phys Rev A 38 (1). (1988), 364–374.

[2] Grimmett G. Percolation. Berlin, Springer-Verlag. (1989).

[3] Huberman B.A. and Glance N.S. Diversity and collective action. In: H.Haken and A. Mikhailov (Eds.), Interdisciplinary approaches to nonlinear systems. New York, Springer-Verlag. (1993), 44–64.

[4] Palmer E. Graphical Evolution. New York, Wiley Interscience. (1985).

[5] Dawkins R. The Selfish Gene. Oxford University Press, England. (1976).

[6] Trofimova I., Mitin N., Potapov A., Malinetzky G. Description of Ensembles with Variable Structure. New Models of Mathematical Psychology. Preprint N 34 of KIAM RAS. (1997).

[7] Trofimova I., Potapov A., Sulis W. Collective Effects on Individual Behavior: In Search of Universality. Chaos Theory and Applications. (1997). In press.

Chapter 50

Tuning complexity on randomly occupied lattices

Tsang Ing Jyh
Tsang Ing Ren
VisionLab - Department of Physics
RUCA - University of Antwerp
Groenenborgerlaan 171, Antwerp B-2020 Belgium

Using diversity of cluster size (mass) as a measurement of structural complexity on randomly occupied lattices, we describe a tuning effect in complexity by parameters L, size of the lattice, and p, probability of occupation. We also show the behavior of the number of fragments (clusters) in relation to the probability of occupation p. Another kind of pattern used to define diversity was the form (shape) of the cluster. A behavior similar to the cluster size diversity was observed. Showing that diversity as a measurement of complexity has an inherent subjective constraint. Scaling relations between the variables measured show some aspects of the complexity of the system and can give more insights into the tuning effect observed.

1. Introduction

The notion of complexity can be associated with a variety of properties of a system. In various situations, complexity is associated with the diversity of size scales. Even though, there is so far no generally accepted definition of a complexity measure, several studies[1, 2, 3, 4, 5, 6] suggest that there should exist a complexity measure which attains its maximum value for configurations in between a completely ordered and disordered state.

The concept of diversity appears in various problems in biology[7], evolution[8, 9], self-organization and cellular automata[10, 11], fractals[12, 13] and non-equilibrium phenomena[14]. In the last few years, diversity of size or mass has been studied in several dissipative processes and cellular automata which generate a distribution of clusters and are of interest in physics, chemistry, biology and ecology [15, 16, 17]. Recently, Gomes et al.[5, 6] proposed the diversity of size of fragments as a measurement for the complexity in aggregation and fragmentation processes. These studies are of interest since it captures the idea of a complex configuration in between a completely connected structure (initial state, ordered) and completely disconnected state with no structure (disordered).

Here we concentrate our effort in understanding complexity on randomly occupied lattices. Note the similarity with the problem of fragmentation processes, even though in the system considered there are no dynamics involved. Also randomly occupied lattice is used as initial configuration in most of the studies on cellular automata. This paper is organized in the following way. In §2 we describe quantitative measurements that characterize complexity. In §3, we describe the experiment, the tuning effect in complexity, and the critical probability associated with each measurement. In §4, we show various scaling relationships between the variables which are taken into consideration. And in §5 some concluding remarks.

2. Diversity and complexity

In nature, diversity is one of the most important characteristics of biological phenomena and it manifests in several ways. Diversity is characterized by the amount of different kinds of patterns that can be identified or delimited. The types of patterns can be obtained by measurement of different physical properties of the system. So diversity also attains a subjective view as a measurement.

On randomly occupied lattices diversity can be obtained by differentiation of cluster sizes or, in another way, by distinction between the forms of the clusters. Here a cluster is defined as a collection of occupied sites connected by nearest neighbors relationship. We define cluster form as being an animal lattice that can assume different orientations, i.e. rotation will not generate a different animal. Since the system in consideration has no dynamics and no initial correlations among the sites, the structural complexity can be well defined as the amount of different cluster sizes or forms.

The mathematical definition for the total number of fragments is given by:

$$N(p) = \left\langle \sum_s N(s,p) \right\rangle, \qquad (2.1)$$

for the diversity of mass of fragment,

$$D_s(p) = \left\langle \sum_s \Theta[N(s,p)] \right\rangle. \qquad (2.2)$$

and for the diversity of form of fragment,

$$D_f(p) = \left\langle \sum_f \Theta[N(f,p)] \right\rangle.$$ (2.3)

In these expressions, $N(s,p)$ is the number of fragments of size s and $N(f,p)$ is the number of fragments with the same form, in a single experiment and for occupation probability p; $\Theta(x) = 1$ if $x > 0$ and $\Theta(x) = 0$ otherwise and the averaging $\langle \rangle$ is over different experiments.

3. Tuning effect and critical probabilities

We performed Monte Carlo simulations on square lattices with size ranging from L=60 to L=4000, with averages taken on 6000 to 50 experiments. The lattices were randomly occupied with probability p ranging from 0.01 to 0.99 with steps of 0.01. To obtain a more accurate value of the critical probabilities, we used steps of 0.002 in the maximum region of N and Ds. These variables were measured as functions of both L and p.

From the point of view of clusters generated on the lattice, the probability of occupation p determines the structure of the system. Here, with a low value of p various small and equal clusters without connection are generated, while with p close to 1 a totally connected single cluster appears. An intermediate value of p gives a configuration with maximum diversity (complexity).

The size of the lattices L determines the possible amount of cluster diversity, since increasing the lattice size means an increase in the range of the clusters size scales. As we tune the size of the lattice and the probability of occupation, it is possible to achieve not only a configuration with great diversity, but also a probability in which the spanning cluster percolates the lattice. Thus relating the measurement of complexity on randomly occupied square lattice with the percolation problem[19].

Fig. 3.1 shows the tuning effect of the cluster size diversity by parameters L and p. For the two-dimensional square lattice at the thermodynamic limit, the critical probability associated with maximum diversity, $P_c(Ds_{max})$, has to be tuned to the value of 0.57 ± 0.02. Here we improve the results of our previous work [18], performing more computer simulations with larger lattice sizes. The critical probability $P_c(Ds_{max})$ has a value closer to the percolation threshold ($P_c = 0.5961$). This suggests that further studies are necessary to clarify the relationship between $P_c(Ds_{max})$ and P_c.

The behavior of the cluster form diversity curve is similar to Fig. 3.1. However the range of probabilities associated with the maximum of diversity is larger than the previous case. We can observe that the maximum of complexity, in this case, is obtained at a lower value of p, as shown in Fig. 3.2. This is due to the fact that as the probability increases, less space is left for different forms to appear. The critical probability, $P_c(Df_{max})$, is obtained at the value of 0.45 ± 0.01 at the limit $L\rightarrow \infty$. For the cluster form diversity the simulations were performed on

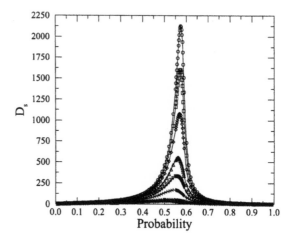

Figure 3.1: This figure shows the behavior of diversity of size as function of p for various values of L. The tuning effect on diversity (complexity) is observed.

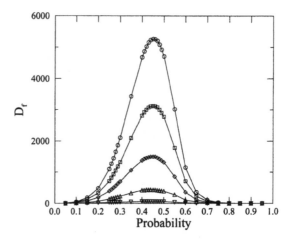

Figure 3.2: The behavior of diversity of form as function of p for various values of L.

lattices with size ranging from L=50 to L=650 and with averages taken on 200 to 50 experiments.

Another interesting variable measured is the number of fragments N. In Fig. 3.3 the number of fragments is normalized by L^2 so that all curves collapse independent of lattice size. Here the variable measured also increases with p, reaches a maximum and decreases afterwards. The maximum number of fragments is obtained at $P_c(N_{max})=0.27\pm0.01$ for an infinite size square lattice.

Finite size scaling analysis was used to determine the critical probabilities

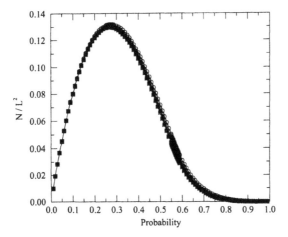

Figure 3.3: The collapsing curves of N normalized by L^2 against the probability of occupation.

associated with the number of clusters, diversity of size and diversity of form at the thermodynamic limit as shown in Fig. 3.4. The values obtained are summarized in the table below.

	$P_c(N_{max})$	$P_c(Df_{max})$	$P_c(Ds_{max})$
d = 2	0.27	0.45	0.57

4. Scaling relations

The robust scaling relation $N_{max} \sim Ds_{max}^{2.04}$, reported in different fragmentation and aggregation dynamics[5, 6, 14, 17] was also observed in this system, as shown in Fig. 4.1. This figure also shows the plots of the scaling relations $Ds_{max} \sim L^{0.98}$ and $N_{max} \sim L^2$ observed in the simulations.

By fixing a probability p and taking the scaling relation of N versus L and Ds versus L we obtain the exponents α and β in function of p. Following the scaling:

$$N(p) \sim L^{\alpha(p)}. \tag{4.1}$$

and

$$D(p) \sim L^{\beta(p)}. \tag{4.2}$$

One derives

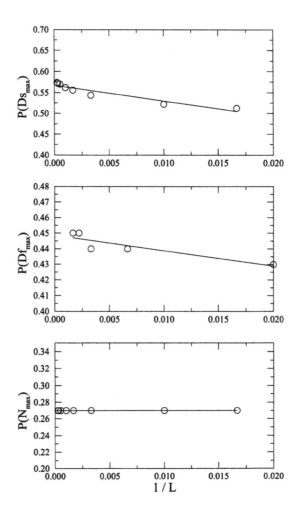

Figure 3.4: Plot of the critical probabilities as a function of 1/L.

$$D(p) \sim N(p)^{\gamma(p)}. \qquad (4.3)$$

where

$$\gamma(p) = \beta(p)/\alpha(p). \qquad (4.4)$$

In Fig. 4.2, we have the behavior of the exponent α versus p. Note that the exponent α maintains constant around the value 2 for almost all the values of p until the system attains a configuration where no scaling relation is observed since just one cluster appears with a large value of p.

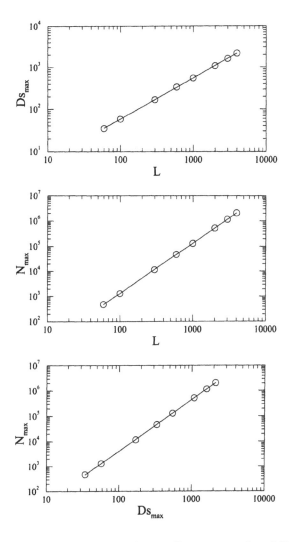

Figure 4.1: Log-log plot of N_{max} versus Ds_{max}, Ds_{max} versus L and N_{max} versus L. Giving the scaling exponents between these variables.

The same scaling relation is obtained for diversity. In Fig. 4.3, we show the plot of these values. The behavior of β unlike α does not follow a constant. β increases with p until it attains a maximum and decreases afterwards, similarly to the diversity plot. This shows that the rate of increase in diversity for different Ls is higher in the maximum region.

In Fig. 4.4, we have a plot of γ versus p. This graphic is similar to β, since for almost all values of p α assumes a constant. The exponent γ is of interest because it relates the number of fragments (population of cluster) with

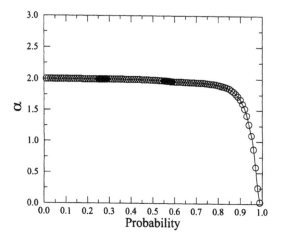

Figure 4.2: Graphic of the exponent α versus p.

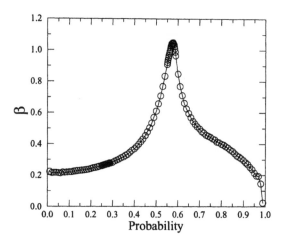

Figure 4.3: Plot of β versus p

the diversity (complexity) of the system.

5. Conclusion

We described a measurement of complexity suitable for systems that generate a statistical distribution of clusters. A tuning effect in complexity on randomly occupied lattices was observed. However some analytical work is necessary to fully understand this effect. In this system the robust scaling relation between $N_{max} \sim Ds_{max}^2$ present in fragmentation processes is also reported. Only the two-

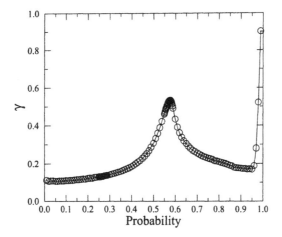

Figure 4.4: Plot of γ versus p

dimensional lattice was considered here although extension to higher dimensions is subject to future studies.

Acknowledgments

The authors gratefully acknowledge financial support by CAPES (Brazilian Government Agency)

Bibliography

[1] HUBERMAN, B.A., and T. HOGG, "Complexity and Adaptation", *Physica* **22D** (1986), 376–384.

[2] CRUTCHFIELD, James P., "The Calculi of Emergence: Computation, Dynamics, and Induction", *Physica D* **75** (1994), 11–54.

[3] CRUTCHFIELD, James P., and K. YOUNG, "Computation at the Onset of Chaos", *Entropy, Complexity, and Physics of Information* (W. ZUREK ed.), SFI Studies in the Science of Complexity, VIII, Addison-Wesley (1990), 223–269.

[4] GELL-MANN, Murray, *The Quark and the Jaguar*, Abacus (1994).

[5] GOMES, M.A.F., F.A.O. DE SOUZA, and K. ADHIKARI, "Formation and Maintenance of Complex Systems", *J. Phys. A: Math. Gen.* **28** (1995), L613–L618.

[6] GARCIA, J.B.C., M.A.F. GOMES, T.I. JYH, T.I. REN, and T.R.M. SALES, "Diversity and Complexity: Two Sides of the Same Coin?", *The Evolution of Complexity* (F. HEYLIGHEN ed.), Kluwer Academic, (1997) (in press).

[7] THOMPSON, D.W., *On Growth and Form*, Cambridge University Press (1971).

[8] RAUP, David M., Stephen Jay GOULD, Thomas J.M. SCHOPF, and Daniel S. SIMBERLOFF, "Stochastic Models of Phylogeny and the Evolution of Diversity", *The Journal of Geology* **81** (1973), 525–542.

[9] KAUFFMAN, Stuart A., *The Origins of Order*, Oxford University Press (1993).

[10] WOLFRAM, Stephen, *Theory and Applications of Cellular Automata*, World Scientific (1986).

[11] WOLFRAM, Stephen, *Cellular Automata and Complexity: Collect Papers*, Addison-Wesley (1994).

[12] MANDELBROT, Benoit B., *The Fractal Geometry of Nature*, Freeman (1983).

[13] PEITGEN, Heinz-Otto, and Peter H. RICHTER, *The Beauty of Fractals: Images of Complex Dynamical Systems*, Springer-Verlag (1986).

[14] COUTINHO, K., M.A.F. GOMES, and S.K. ADHIKARI, "Robust Scaling in Fragmentation from d=1 to 5", *Europhys. Lett.* **18** (1992), 119–124.

[15] GARCIA, J.B.C., M.A.F. GOMES, T.I. JYH, T.I. REN, and T.R.M. SALES, "Nonlinear Dynamics of the Cellular-Automaton "Game of Life"", *Physical Review E* **48** (1993), 3345–3351.

[16] SALES, Tasso R.M., "Unidimensional Games, Propitious Environments, and Maximum Diversity", *Physical Review E* **48** (1993), 2418–2421.

[17] COUTINHO, K.R., M.D. COUTINHO-FILHO, M.A.F. *Gomes*, and A.M. *Nemirovsky*, "Partial and Random Lattice Covering Times in Two Dimensions", *Physical Review Letters* **72** (1994), 3745–3749.

[18] TSANG I. R., and TSANG I. J., "Critical Probabilities for Diversity and Number of Clusters in Randomly Occupied Square Lattices", *J. Phys. A: Math. Gen.* **30** (1997), L239–L243.

[19] STAUFFER, D., *Introduction to Percolation Theory*, Taylor and Francis (1985).

Chapter 51

Socioeconomic organization on directional resource landscapes

James G. Uber
Department of Civil and Environmental Engineering
University of Cincinnati, Cincinnati, OH 45221-0071
Ali A. Minai
Dept. Elec. & Comp. Eng., and Computer Science
University of Cincinnati, Cincinnati, OH 45221-0030

We describe a spatially distributed model with adaptive agents that live on a landscape and interact via a simple economy. This economy consists of a single good that is produced and consumed by the agents, plus a market mechanism that distributes this good according to the agents' bids. Most importantly, production of the good depends on consumption of a spatially distributed resource flow described by an acyclic directed graph that connects the potential production sites. Thus, resource use at one production site implies less resource flow to other connected sites, establishing an implicit hierarchy of production sites. We investigate the influence of resource flow and distribution on the emergent behavior of the agent economy in terms of price dynamics, wealth and health distribution, and population size.

The motivation for the above abstraction derives from water resources flowing in river networks. The topology of such networks has been studied extensively, and has been described by fractals as well as various empirical "laws." Our agents can be thought of as households participating in an agrarian economy where water is input to the production process and food is output. Each agent behaves according to relatively simple production and consumption strategies, but the system of agents produces emergent economies that exhibit intermittent and large price fluctuations, periods of rapid wealth transfer, and social hierarchy.

1. Introduction and motivation

Although the importance of water as a production resource is taken for granted, there has been no systematic study of the coupling between this important resource flow and socioeconomic adaptation. This remains true despite significant evidence pointing to the importance of water resources and its directional nature on the evolution of civilizations from pre-agrarian times to the modern, post-industrial age[16, 21, 24]. Of all resources critical to human survival (water, food, land, building material, etc.), water is virtually unique in its strongly directional character. Lakes and swamps notwithstanding, most water used by human populations comes directly or indirectly from streams and rivers — flowing inexorably under the influence of gravity over a landscape that resists major alteration, and replenished by the wholly uncontrollable mechanisms of precipitation and snowmelt. As such, until very recent times, water was the least controllable and the most unpredictable of all basic resources. Even today, establishing a reliable and controlled supply of water on a significant scale requires major investment and has consequences on transcontinental scales (cf. the effect of dams in the Colorado River basin).

An important and distinct attribute of flowing water is that, due to its directionality, it produces explicit connections and dependencies between agents in a society. Socioeconomic gradients in a community seem to be correlated with hydraulic gradients in a river basin; downstream riparians have historically been at a disadvantage compared to their upstream neighbors because they drank water that was comprised, in part, of their neighbors wastes. However, the consequences of living at a particular point in a watershed are not simple to determine. Upstream dwellers might have first use of water — and the consequent freedom of choosing their usage — but those who live downstream can often get more water from a larger number of sources than their upstream counterparts. While this might make them more dependent on these upstream dwellers in theory, the dependence may not matter if those upstream have no way to control water flow, or if they fail to act in concert. It is precisely to study and understand such complex interactions that we have developed the preliminary model described here. It is similar in many ways to other distributed models of societal self-organization[10], but differs fundamentally in its focus on the *directional* and *connective* aspect introduced by the stream network.

The remainder of this paper is organized as follows. The basic methodology is outlined in §2. Section §3 describes a dimensional model of the system, followed by a non-dimensionalized model. Finally, simulation results are presented in §4 for some configurations of the dimensionless model.

2. Background and methodology

The relatively recent synthesis of nonlinear dynamics, game theory, adaptive systems theory, and statistical physics into a science of complex systems[8] has, for the first time, made it possible to model societal dynamics in a distributed way[10] rather than with standard lumped models. Our approach is to develop a spatially distributed (cellular) model of human adaptation in the watershed environment. The model comprises two parts: 1) A cellular model of the landscape where the resource is distributed, and on which the population resides; and 2) A cellular societal model with a population of locally interacting adaptive agents that occupy land, engage in the production and consumption of a single good, perish if unsuccessful and, in the process, compete for water and land. The landscape/resource model is based on well developed hypotheses regarding the spatial nature of drainage basins [14, 22, 23]. These hypotheses, which are supported by extensive field data, lend themselves to a multi-scale, cellular representation of the watershed landscape with purely local interactions[19].

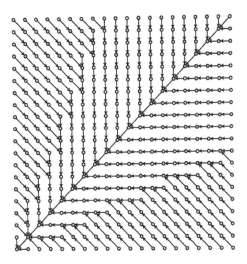

Figure 2.1: A simple symmetric river network.

The societal model is patterned after multi-agent schemes developed in economics [2, 11, 13, 18] to model decision-making based on inductive reasoning and bounded rationality. Multi-agent models evolved out of game-theory as it became computationally tractable to simulate large numbers of interacting agents in a simple economy [5, 9]. However, many of these models have been based on the classical economic formalism of perfect rationality, i.e., the agents act with perfect knowledge of all other agents' strategies, and the system as a whole is in equilibrium. Real economic agents, however, are not omniscient, which is precisely why markets never reach equilibrium, and some investors make money while others do not. In any real situation, human agents act based on their "best predictions" of others' behavior, subjective judgement, and non-stationary information, which distinguishes the savvy from the rest. Such agents are said to possess *bounded rationality*[20], and may be able to capture the essential complexity of real economic (or other) decision-making. Recently, multi-agent bounded-rationality models have been integrated with the evolutionary adaptation formalism developed by Holland [12] to study economies and ecologies with adaptive agents[4, 5, 17]. Such adaptive mechanisms are also incorporated into our model.

The choice of existing, simple models for landscape and socioeconomic behavior reflects a desire to minimize ad-hoc assumptions. At the same time, the combination of the two models represents a major step forward, in that it focuses on the interaction of landscape and society rather than on each in isolation.

3. Spatially distributed agent model

In this section, we describe the mathematical model of agents and environment used in our simulations. The agent model is an extension of those studied in [15, 18] to the case where resource and strategies are explicit and commodity production is endogenous to the system. Production decisions evolve in a complex dynamics coupled with the market dynamics, with an emergent — and generally nonstationary — commodity distribution.

The system evolves in discrete time steps, k, each corresponding to one production cycle. The simulated environment consists of an $m = M \times M$ lattice, with each cell representing a potential production/habitation site. There is a total of $N^k \leq m$ immmobile agents in the system at time step k, and each agent, i, occupies a distinct site, j, on the lattice. This is denoted by the relations $Loc(i) = j$ and $Occ(j) = i$ where $Occ(j) = 0$ if site j is unoccupied. The landscape is drained by a *stream channel network*, which is an acyclic, directed graph, $G(V, E)$ whose vertices, V, correspond to the cells of the lattice, and the edges, E, connect neighboring cells. Thus, G forms a spanning tree for the lattice.

Water flows from the leaves to the root of G, which is the outlet for the system, and is always in the lower left corner (see Fig. 2.1). Each cell (except the root) drains to exactly one of its eight neighboring cells. Each cell, j, also receives an exogenous resource input, s_j^k, at step k, so the total resource available in the system is $R^k = \sum_{j=1}^m s_j^k$; this input can be seen as a simple model for precipitation. The *watershed*, W_j, of location j is defined as the set of all cells in the subtree rooted at j (i.e., all locations upstream of j), while its *discharge path*, D_j, is defined as the set of all cells in the path from j to the root cell, including the root (i.e., all locations downstream of j).

Every location is assumed to represent a land area suitable for the production of a commodity through use of the resource. The economics of commodity production depends on the water resource available to and controlled by each agent, as well as that used by each agent in production. These quantities are endogenous to the system and are governed by the stream channel network and the economic activities of the agents. The resource *used* by agent i at time k is denoted by u_i^k, which must not exceed the resource *available* to the agent,

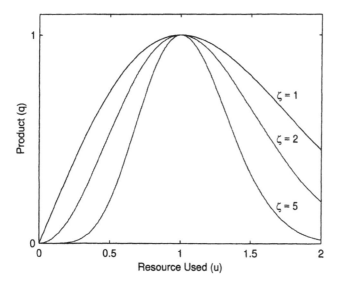

Figure 3.1: The dimensionless production function, $\eta = 1$.

$\hat{r}^k_{Loc(i)}$. The resource available at location j is given recursively by:

$$\hat{r}^k_j = s^k_j + \sum_{l \in I_j}(\hat{r}^k_l - u^k_{Occ(l)})$$

where I_j is defined as the set of all neighboring cells which have a directed edge into j, and $u^k_0 = 0$ (*i.e.*, unoccupied sites do not use resource). The *controllable* resource, $\hat{\rho}^k_j$, at cell j is defined as that resource which arrives at j without passing any other occupied cells; thus, an agent at j would have full control of this resource. Controllable resource is calculated recursively as

$$\hat{\rho}^k_j = s^k_j + \sum_{l \in I_j}\hat{\rho}^k_l(1 - \delta^k_l)$$

where $\delta^k_l = 1$ whenever $Occ(l) \neq 0$ and $\delta^k_l = 0$ otherwise. The resource quantity available to agent i at time period k is denoted by $r^k_i \equiv \hat{r}^k_{Loc(i)}$, and the quantity controllable by $\rho^k_i = \hat{\rho}^k_{Loc(i)}$. Essentially, the controllable resource for an agent is that part of its available resource which is not affected by changes in resource usage by other agents.

Each agent is characterised intrinsically by *wealth*, w^k_i, which is fiat money used to guard against environmental fluctuations, and *health*, h^k_i, which reflects the agent's level of satisfaction from commodity consumption, as well as its physical well-being. Agent health is a function of its commodity consumption history, and a certain level of health must be maintained for survival.

At time k, Agent i uses an amount $u^k_i \leq r^k_i$ of the resource to produce a quantity q^k_i of the commodity. Given usage u^k_i, production is determined by a *production function*, $P(u)$, embodying the Malthusian assumption that increasing resource usage eventually leads to a decrease in productivity (e.g., corresponding to inundation of the land). The particular function we used is:

$$P(u) = a_1 u^{a_2} exp(-a_3 u^2/2)$$

where a_1, a_2, and a_3 are constant parameters (Fig. 3.1 shows this function in dimensionless form — see below).

Since commodity consumption is both desirable and essential for survival, each agent also bids a portion, $b_i^k \leq w_i^k$ of its wealth to purchase the commodity. The price of the commodity is determined by a simple global market as:

$$p_q^k = B^k / Q^k$$

where $B^k = \sum_{i=1}^{N^k} b_i^k$ is the total bid and $Q^k = \sum_{i=1}^{N^k} q_i^k$ is the total commodity produced[18]. The commodity is then distributed to agents in proportion to their bid, with agent i receiving amount

$$x_i^k = b_i^k / p_q^k$$

The commodity is assumed to be perishable, and is consumed immediately by the agent. The agent's health is then updated by

$$h_i^k = \sum_{\tau=0}^{k} \gamma^{k-\tau} B(x_i^\tau) = \gamma h_i^{k-1} + B(x_i^k)$$

where $0 \leq \gamma \leq 1$ represents a "discount factor" for past consumption — reflecting the agent's metabolic rate or otherwise a decay in satisfaction derived from consumption —, and $B(x)$ quantifies the immediate benefit accrued from consumption x. The benefit function obeys $B(0) = 0$, and is strictly increasing and concave. The particular function used here is:

$$B(x) = b_1 \left[1 - e^{-b_2 x} \right]$$

where b_1 and b_2 are constant parameters.

Agent survival depends on maintaining health above a critical health level, h_d; death occurs at time k if $h_i^k < h_d$. Since there is no mechanism of birth in the current model, the wealth of a dead agent is simply distributed equally among living agents. The introduction of a birth model will be accompanied by an inheritance mechanism in future versions.

Aside from distributions after agent death, there are two primary types of transactions that determine the wealth dynamics in the system: 1) Sale of commodity in the market; and 2) Payments by users of resource to those who control it. The latter represents one mechanism through which the resource network influences the economy, another being the indirect influence of resource availability on production. The wealth for agent i after time step k is updated as

$$w_i^{k+1} = w_i^k - p_q^k [x_i^k - q_i^k] - p_r [y_i^k - z_i^k]$$

where p_r is a fixed, exogenously set price of the resource, y_i^k is the *excess resource* agent i purchased from other agents to satisfy its resource usage, and z_i^k is the *surplus resource* it sold to other *customer* agents (see below). In general, wealth is conserved in the system, though the amount of positive wealth in the system may exceed the total initial wealth, reflecting a credit subeconomy. Runaway borrowing is controlled by preventing debtor agents (those with negative wealth) from bidding for the commodity. Thus debtor agents will perish if their debt remains unpaid.

We define $y_i^k = MAX(0, u_i^k - \rho_i^k)$. Thus, each agent can use resource up to the amount it controls without payment. Any excess is purchased from other agents who are its *suppliers*. Conversely, any part of the controllable resource that an agent does *not* use is a source of income, if used by other agents who are its *customers*. This *surplus supply* is $\sigma_i^k = MAX(0, \rho_i^k - u_i^k)$. The potential suppliers, S_i^k, for i are defined as all agents living on $W_{Loc(i)}$ — the watershed of i's location, while its potential customers, C_i^k, are defined as all agents living on $D_{Loc(i)}$ — the discharge path for i's site. The excess resource actually purchased by agent i from potential supplier l, $z_{l_i}^k$, is calculated in proportion to the surplus supplies:

$$z_{l_i}^k = y_i^k \frac{\sigma_l^k}{\sum_{j \in S_i^k} \sigma_j^k}$$

which means that the total resource sold by agent l is

$$z_l^k = \sum_{i \in C_l^k} z_{l_i}^k.$$

3.1. Production and consumption strategies

Each agent must make two economic decisions at every step: 1) How much resource to commit to production; and 2) How much commodity to purchase and consume. In terms of modeled quantities, these production and consumption decisions are mapped into *desired* usage and consumption, $\tilde{u}_i^k$ and $\tilde{x}_i^k$. A desired bid is then calculated using an estimate of commodity price and the desired consumption, $\tilde{b}_i^k = \tilde{p}_q^k \tilde{x}_i^k$, where we simply determine $\tilde{p}_q^k = p_q^{k-1}$. The agent may not be able to satisfy its desires if they exceed the available resource or wealth. Thus,

$$u_i^k = \min(\tilde{u}_i^k, r_i^k), \quad b_i^k = \min(\tilde{b}_i^k, w_i^k)$$

Strategic adaptation of the desired usage and consumption — and hence of the agent's economic personality — is modeled using a framework based on evolution[12, 13]. Here agents learn the usage and consumption that is best for them, based on a reward received. Arthur[2, 3] described a general framework for a parameterized learning automaton which improves its strategies for making simple decisions. As Arthur notes, such adaptive strategies are interesting

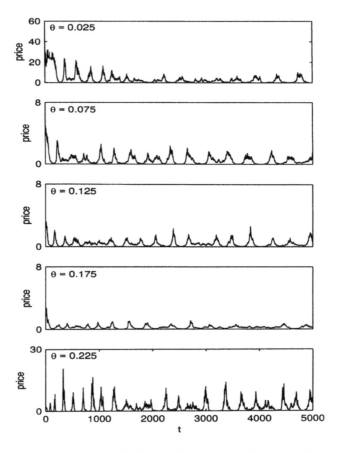

Figure 3.2: Price dynamics for the symmetric network. Resource price parameter, λ is set to 10.

because they allow agents to learn inductively from sequences of experiences rather than from a complete and rational analysis of choices.

In our context an agent *strategy* is defined as a combination of desired usage and consumption values. The possible strategies are defined by a discrete global strategy space $\mathcal{S} = \mathcal{S}_u \times \mathcal{S}_x$, where $\mathcal{S}_u$ is a set of n_u desired usage values, $\mathcal{S}_u = \{u_1, u_2, ..., u_{n_u}\}$, and $\mathcal{S}_x$ is a set of n_x desired consumption values, $\mathcal{S}_x = \{x_1, x_2, ..., x_{n_x}\}$. Agent i in time period k selects randomly a strategy $j \in \mathcal{S}$, with probabilities that are updated to reflect rewards received over time. Let the *strength* of strategy j for agent i in time period k be $S_i^k(j)$, where the initial strength $S_i^0(j) > 0$ for all j. The probabilities associated with each strategy are calculated in proportion to their strengths

$$P_i^k(j) = \frac{S_i^k(j)}{T_i^k}$$

where $T_i^k = \sum_{j=1}^m S_i^k(j)$. The agent then selects a strategy j at time period k consistent with the above probabilities.

Implementation of strategy j yields a *reward* in time period $k+1$, $\mathcal{R}_i^{k+1}(j) = \mathcal{U}_i^{k+1} - \mathcal{U}_i^k$ where $\mathcal{U}_i^k(h_i^k, w_i^k)$ is the agent *utility* associated with the health and wealth at time k

$$\mathcal{U}_i^k = \sqrt{v(h_i^k - h_d)w_i^k}$$

and v is a constant utility parameter. Thus utility exhibits a decreasing marginal rate of substitution of health for wealth as health is progressively substituted for wealth, and approaches zero as health approaches the critical level h_d. This concept of utility expresses the preferences of the agent vis-a-vis immediate commodity consumption and long-term wealth accumulation. These preferences essentially represent lifestyle choices: An agent may equally select a frugal lifestyle focused on staying alive and conserving wealth, an extravagant lifestyle of living at the limit of one's means with no savings, or something in between.

Given the reward $\mathcal{R}_i^{k+1}(j)$ the strength of strategy j is updated

$$S_i^{k+1}(j) = S_i^k(j) + \mathcal{R}_i^{k+1}(j)$$

after which the strengths are *renormalized* to sum to $T_i^{k+1} = Tk^\nu$. This renormalization allows the *rate of learning* of the agents to be controlled by the learning parameters T and ν. Arthur[1] showed that the probability of selecting any one action grows at a proportional to $1/Tk^\nu$, for the simplified case where the rewards are exogenous and obtained from a stationary probability distribution. Thus if ν is "small," the sum of the strengths T_i^k grows relatively slowly with time, and the agent learns its best strategies at a relatively slow rate (allowing for more experiences to influence the strategy selection).

3.2. Dimensional analysis

While the model described above could be simulated as described, it would be difficult to generalize because the parameters lack a simple and intuitive meaning. The results are more meaningful, however, once the model has been expressed in dimensionless form. The dimensionless model is mathematically equivalent to the original, but is expressed in terms of dimensionless health, wealth, resource, and commodity, and dependent on a smaller — and more meaningful — set of dimensionless parameter groups.

Derivation of an equivalent dimensionless model requires that characteristic quantities for the fundamental units of health, wealth, resource, and commodity be specified. The dimensionless model variables are then scaled to their respective characteristic values. We have defined the following characteristic quantities for our model: $h^* = b_1/(1 - \gamma) =$ the asymptotic maximum health obtained by setting $B(x_i^k) = b_1$ during each period k; $w^* = \sum_{i=1}^{N^0} w_i^0/N^0 =$ the average initial wealth; $u^* = (a_2/a_3)^{1/2} =$ the resource usage leading to maximum production; $q^* = -[ln(1 - h_d(1 - \gamma)/b_1)]/b_2 =$ the commodity quantity which, if

588

consumed indefinitely, leads to the critical health, h_d. Given these character-istic quantities, the corresponding scaled (dimensionless) values, $h_i'^k = h_i^k/h^*$, $w_i'^k = w_i^k/w^*$, $u_i'^k = u_i^k/u^*$, and $q_i'^k = q_i^k/q^*$, are relatively easy to interpret: the maximum dimensionless health, $h_i'^k$, is unity, a unit value of dimensionless wealth, $w_i'^k$, implies exactly average initial wealth, a unit value of dimensionless resource, $u_i'^k$, leads to maximum commodity production, and a unit value of dimensionless commodity, $q_i'^k$, will just allow survival, if consumed indefinitely.

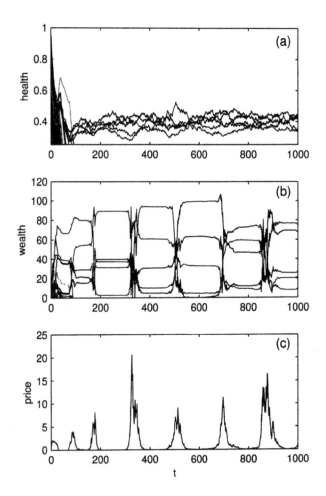

Figure 3.3: Wealth and health dynamics for the symmetric network ($\lambda = 10$, $\theta = 0.225$).

When the model is expressed using the above dimensionless variables, five dimensionless parameter groups emerge, each allowing a simple interpretation. Specifically, the production function, and the health and wealth dynamics become:

$$P(u_i'^k) = \eta(u_i'^k)^\zeta exp(\zeta(1 - (u_i'^k)^2)/2$$

$$h_i'^k = \gamma h_i'^{k-1} + (1 - \gamma)[1 - (1 - \xi)^{x_i'^k}]$$

$$w_i'^{k+1} = w_i'^k - p_q'^k[x_i'^k - q_i'^k] - \lambda[y_i'^k - z_i'^k]$$

where dimensionless commodity purchased $x_i'^k = x_i^k/q^*$, dimensionless bid $b_i'^k = b_i^k/w^*$, dimensionless commodity price $p_q'^k = B'^k/Q'^k = (B^k/w^*)/(Q^k/q^*)$, and dimensionless controllable resource $\rho_i'^k = \rho_i^k/u^*$. The dimensionless parameter groups γ, ζ, η, λ, and ξ are the principle determinants of model behavior, and are defined in terms of the dimensional parameters:

$$\gamma = \gamma, \quad \zeta = a_2, \quad \lambda = \frac{N^0 p_r (a_2/a_3)^{1/2}}{W^0}$$

$$\eta = \frac{a_1(a_2/a_3)^{a_2/2} exp(-a_2/2)}{-[ln(1 - \xi)]/b_2}, \quad \xi = \frac{h_d}{b_1/(1 - \gamma)}$$

Thus the health discount factor γ appears also in the dimensionless model, ζ is a factor that impacts production efficiency by determining its "width" (see Fig. 3.1), η is the ratio of maximum commodity production to minimum commodity consumption (for survival), λ is the fraction of total system wealth required to purchase sufficient resource for maximum commodity production, and ξ is the critical health level expressed as a fraction of the maximum asymptotic health level, h^*, and hence also equal to the dimensionless critical health ($h_i'^k \geq \xi$ is required for survival). In addition to these dimensionless quantities, we also define the dimensionless resource input at each node of the stream channel network, $\theta_j^k = s_j^k/u^*$. Given the dimensionless quantities γ, ζ, η, λ, ξ, and θ_j^k, we now have a logical basis with which to specify model input and thus to interpret model output. The results in the next section are discussed using the terminology of the dimensionless model.

4. Results and discussion

Here we present a brief set of simulation results to illustrate some interesting features of commodity price and agent wealth and health dynamics, for various environmental conditions. We do not attempt a detailed explanation of these results here, but rather present them as useful pedagogical examples that evoke images of real economic dynamics.

In general, the dynamic model behavior under adaptive strategies exhibits varying degrees of intermittency, with occasional large fluctuations in commodity price and wealth, and economies that exhibit emergent class structures in terms

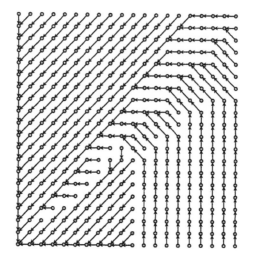

Figure 4.1: A simple asymmetric river network.

of agent wealth and health. Further, such features appear fairly robust for a range of dimensionless resource supply and price. A change in the resource network is shown to affect the qualitative character of these dynamics, and although a quantitative description of this affect is not currently available, such knowledge would clearly be an important component of future work.

Interestingly, in the absence of adaptive agent migration strategies, it appears to be quite challenging to make a living and survive as an economic agent on our current resource landscapes. As a result, solutions obtained for immortal agents are quite different than those where agents live or die according to their health (and thus commodity consumption); in the latter case, large numbers of agents perish during a fairly brief early time period, leaving a relatively fewer number of surviving agents that presumably benefit from a better position on the resource landscape. This massive agent extinction occurs largely because agents are implicitly allowed to issue "I.O.U.s" for resource purchased for production (i.e. to borrow money from the resource suppliers). A small number of agents borrow heavily to purchase resource but are doomed; when these agents inevitably die and their debt is (unfairly) distributed over all agents, a large number of agents living at the margins of success never recover. The weakest segment of society thus pays the price for the irresponsible excesses of a few. Obviously, very different behavior could result if borrowing was not allowed or was attributed to specific "creditor" agents. Also, although not shown below, the dynamics obtained for static (non-adaptive) strategies are very different than those for the current model where agents adapt their usage and consumption strategies based on a utility reward. In the non-adaptive case, solutions tend quickly toward equilibrium, with commodity price either collapsing to a stable zero value, or approaching a stable non-zero fixed point. We can explain very precisely this price behavior based solely on whether the desired agent consumption levels are above or below a critical threshold, which depends on dimensionless critical health and maximum resource production. In contrast, the adaptive strategy dynamics never exhibit stable fixed point behavior.

All results below are obtained for a 20×20 resource landscape. The stream channel networks, such as that in Fig. 2.1, were developed using a search algorithm that constructs a local cost-minimizing spanning tree of the resource landscape, where edge "costs" are functionally related to surface slope for a defined topographic mapping — higher slopes leading to lower costs. The symmetric stream channel network in Fig. 2.1 was generated using a quadratic topography $z = c_1 x^2 + c_2 y^2 + c_3 xy$ and $c_1 = 1$, $c_2 = 1$, $c_3 = 0$, where the (x, y) origin (and thus the stream channel network outlet) is at the southwest corner node.

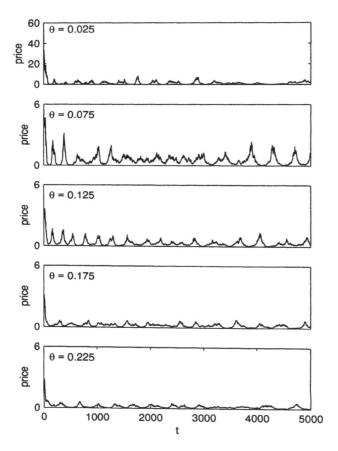

Figure 4.2: Price dynamics for the asymmetric network. Resource price parameter, λ is set to 10.

An initial population of 200 agents is randomly distributed on the stream channel network graphs, and seeded with a random initial distribution of wealth and health. The landscape is supplied with a steady and homogeneous resource input $\theta_j^k = \theta$ that is distributed to the agents via the stream channel network. The dimensionless desired usage strategies $\mathcal{S}_u = \{0, 1/3, 2/3, 1\}$ and the dimensionless desired consumption strategies $\mathcal{S}_x = \{1.25, 2, 2.75, 3.5\}$; thus the 4×4 strategy set ranges from zero to maximum production (recall that a unit usage results in maximum production), and from relatively small to relatively large commodity consumption (recall that a unit consumption is the minimum continuous consumption required to maintain health above the critical level). The strategy strengths at the initial step are set uniformly to $\mathcal{S}_i^1(j) = T/16$ for all agents i and strategies j, so that all sixteen strategies begin with equal probability of selection. The strategy adaptation parameters are $T = 0.025$ and $\nu = 0.5$.

Price dynamics for the symmetric stream channel network of Fig. 2.1 are shown in Fig. 3.2 for a resource input θ ranging from relatively scarce (0.025) to relatively plentiful (0.225), and resource price $\lambda = 10$. The values of other dimensionless parameter are $\gamma = 0.95$, $\zeta = 2.0$, $\eta = 2.0$, and $\xi = 0.25$. The commodity price exhibits intermittency with larger swings in price generally occuring as resource supply dwindles, although paradoxically large swings are also seen at the very high resource supply level (though, other results show, not when resource price is decreased). The exact events initiating the intermittent price jumps remain unidentified at this time, but informal analysis of individual agent dynamics suggests that a change of strategy by a successful agent might produce destabilization of price leading to catastrophic (revolutionary?) shifts in wealth. It would be interesting to explore this phenomenon within the framework of critical phenomena [7, 6].

Fig. 3.3 shows agent health and wealth for the same case along with price, for the first 1000 iterations and resource supply $\theta = 0.225$. Here one notices the dramatic numbers of agents that perish before the first 100 iterations (shown by their plummeting health levels). The agents that remain engage in a very interesting competition for resource and commodity, as seen by the wealth dynamics. During periods when price is relatively stable (and low) there exists a defined economic hierarchy, with both very rich and very poor agents. Associated with the intermittent periods of rapid inflation, rich agents often lose their status and transfer this wealth to poorer agents in an interesting switching of economic roles. The health dynamics are also related to the punctuated fluctuations in price, with health variance generally increasing significantly during periods of price instability.

To briefly investigate the effect of resource network topology, results were obtained for the highly asymmetric network shown in Fig. 4.1, which was generated using the same quadratic topography function but with parameters $c_1 = 1$, $c_2 = 1$, $c_3 = -1.5$. Other parameters were identical to the above symmetric case. Fig. 4.2 shows the price dynamics for the asymmetric network, again for varying resource supply θ and resource price $\lambda = 10.0$. There are distinct qualitative differences in these dynamics as compared to the symmetric case, and in general the magnitude of price swings is damped in the asymmetric network (this obser-

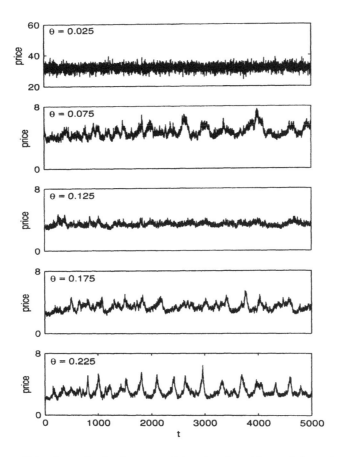

Figure 4.3: Price dynamics for the symmetric network and immortal agents. Resource price parameter, λ is set to 10.

vation is further supported by results at different resource prices). Although not discernible from the price dynamics, the asymmetric network supports a much larger number of agents in the long run, and although wealth volatility is still associated with price volatility, the magnitude of the wealth swings is greatly reduced.

Price dynamics for immortal agents living on the symmetric resource network are shown in Fig. 4.3, all other parameters remaining identical with the above symmetric case. Here we see again qualitatively different behavior; although price intermittently exhibits wide fluctuations, the floor of the price dynamics is non-zero. Quantitative analysis of price dynamics in a small number of cases reveals three distinct types of behavior: 1) White noise-like (for very low resource only); 2) 1-over-f (for low to moderate resource); and 3) Fractional Brownian motion with spectral exponents between 1.6 and 2 (for moderate to high resource). Finding a theoretical basis for this behavior would be of great interest

594

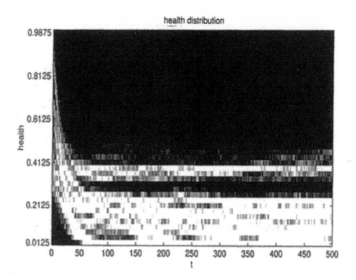

Figure 4.4. Dynamics of health distribution for the symmetric network and immortal agents ($\lambda = 10$, $\theta = 0.225$). Each column of the plot represents the histogram of health values at a time step. White indicates large values and black small values.

and this issue will be addressed in future papers.

The dynamics of the distribution of health values for immortal agents are shown in Fig. 4.4 for $\theta = 0.225$, where white values indicate large numbers of agents and black values small numbers. Although the initial health distribution is uniform, a distinct bimodal health distribution emerges, with a clear minority of agents enjoying relatively large commodity consumption. This bimodal distribution appears to be stable despite fairly large price swings, suggesting that large numbers of agents in this economy will never be able to satisfy their desired consumption. (Note that even for an agent consistently selecting the minimum desired consumption $S_x = 1.25$, that agent's health would be greater than $\xi = 0.25$.)

The preliminary model described here includes many of the basic processes we consider necessary for a model of socioeconomic organization on directional resource landscapes: economic agents that produce and consume, a limiting resource distributed along a network, adaptive decision strategies, and boundedly rational agents attempting to maximize their utility. The only spatial metric in the current model is induced by the stream network. Many economic interactions, however, occur in Euclidean space (e.g., migration), and it will be necessary to define a correspondence between network space and Euclidean space in extensions of the model.

Bibliography

[1] ARTHUR, W. Brian, "A Learning Algorithm that Mimics Human Learning", *Santafe Institute Working Paper* **90-026** (1990).

[2] ARTHUR, W. Brian, "Designing Economic Agents That Act Like Human Agents: A Behavioral Approach to Bounded Rationality", *Amer. Econ. Rev.* **81** (1991), 353–359.

[3] ARTHUR, W. Brian, "Inductive Behavior and Bounded Rationality", *Amer. Econ. Rev.* **84** (1994), 406–411.

[4] ARTHUR, W. Brian, "Complexity in Economic and Financial Markets", *Complexity* **1** (1995), 20–25.

[5] ARTHUR, W. Brian, John H. HOLLAND, Blake LeBARON, Richard PALMER, and Paul TAYLER, "Asset Pricing Under Endogenous Expectations in an Artificial Stock Market", *The Economy as an Evolving Complex System II* (W. Brian ARTHUR et al.ed.), Addison-Wesley (1997), 15-44.

[6] BAK, Per, *How Nature Works*, Copernicus (1996).

[7] BAK, Per, Chao TANG, and Kurt WIESENFELD, "Self-Organized Criticality", *Phys. Rev. A* **38** (1988), 364–374.

[8] BAR-YAM, Yaneer, *Dynamics of Complex Systems*, Addison-Wesley (1997).

[9] DARLEY, V.M., and Stuart .A. KAUFFMAN, "Natural Rationality", *The Economy as an Evolving Complex System II* (W. Brian ARTHUR et al.ed.), Addison-Wesley (1997), 45-79.

[10] EPSTEIN, Joshua M., and Robert AXTELL, *Growing Artificial Societies: Social Science from the Bottom Up*, MIT Press (1996).

[11] HOLLAND, John H., and J. MILLER, "Artificial Adaptive Agents in Economic Theory", *Amer. Econ. Rev.* **81** (1991), 365–370.

[12] HOLLAND, John H., *Adaptation in Natural and Artificial Systems*, MIT Press (1992).

[13] HOLLAND, John H., *Hidden Order: How Adaptation Builds Complexity*, Addison-Wesley (1995).

[14] HORTON, R.E., "Erosional Development of Streams and Their Drainage Basins: Hydrophysical Approach to Quantitative Morphology", *Geol. Soc. Am. Bull.* **56** (1945), 275–370.

[15] KARATZAS, I., M. SHUBIK, and W.D. SUDDERTH, "Construction of Stationary Markov Equilibria in a Strategic Market Game", *Math. Op. Res.* **19** (1994), 975–1006.

[16] LANSING, J.S., and J.N. KRAMER, "Emergent Properties of Balinese Water Temple Networks: Coadaptation on a Rugged Fitness Landscape", *American Anthropologist* **95(1)** (1993), 97-114.

[17] LINDGREN, Kristian, "Evolutionary Dynamics in Game-Theoretic Models", *The Economy as an Evolving Complex System II* (W. Brian ARTHUR et al.ed.), Addison-Wesley (1997), 337-367.

[18] MILLER, J.H., and M. SHUBIK, "Some Dynamics of a Strategic Market Game with a Large Number of Agents", *J. Econ.* **60** (1994), 1–28.

[19] RIGON, R., A. RINALDO, and I. RODRIGUEZ-ITURBE, "On Landscape Self-Organization", *J. Geophys. Res.* **99** (1994), 11971–11993.

[20] SARGENT, T., *Bounded Rationality in Macroeconomics*, Clarendon Press (1993).

[21] SCARBOROUGH, V.L., "Ancient Maya Water Management in the Southern Maya Lowlands", *National Geographic Research and Exploration* (1994).

[22] SHREVE, R.L., "Statistical Law of Stream Numbers", *J. Geol.* **74** (1966), 17–37.

[23] TARBOTON, D.G., R.L. BRAS, and I. RODRIGUEZ-ITURBE, "The Fractal Nature of River Networks", *Water Resources Res.* 24 (1988), 1317–1322.

[24] WORSTER, D., *Rivers of Empire: Water, Aridity, and the Growth of the American West*, Pantheon Books (1985).

Chapter 52

"Continuous time" in Feigenbaum's model

Alexander Yu. Vlasov

FRC/IRH, St.–Petersburg, Russia

The sequence $X_{n+1} = \lambda X_n(1 - X_n)$ is well known and it is often used as a model of complex behavior in economics, biology etc.. The universality of this model was shown by M.Feigenbaum. This is model with *discrete time*. It is interesting to consider some analogues with *continuous time* i.e., some functions of *real* argument with recurrent formula $F(x + 1) = \lambda F(x)(1 - F(x))$.

The simple example for $\lambda = 4$ is $F_4(x) = \sin^2(A\,2^x)$. It is interesting what such a simple formula describes behavior of the model what usually called chaotic. To find the particular sequence with some X_1 we should set $A = \arcsin(\sqrt{X_1})/2$, when $X_n = F_4(n)$. It is possible also write simple solution for $\lambda = 2$ when model show regular behavior - sequence coverage to $1/2$ very fast. Really for $\lambda = 2$ it is possible to choose $F_2(x) = \left(1 - \exp(-B\,2^x)\right)/2$.

These examples are show that such functions are useful for understanding of behavior of the model. For other k also can be found an analytical functions, but they do not have such simple expression. Author is described one of class of such functions $\mathbb{F}_\lambda(x)$ (1993). They can be effective calculated due to *series* whose converge very fast in an interval $1 \leqslant x \leqslant 2$ and recursive formula above. For particular values $\lambda = 2$ and $\lambda = 4$ the functions are coincide with analytical expression above. Here is presented the expressions and further description of properties of the functions $\mathbb{F}_\lambda(x)$ for other values of λ.

1. Introduction

Let us consider a sequence:

$$X_{n+1} = \lambda X_n(1 - X_n)$$

$$2 \leqslant \lambda \leqslant 4 \qquad (1.1)$$

Here we are not discussing applications of the sequence, there are huge amount of different work devoted to it and it could not be reviewed in a short article. We consider question about possibility of using some *real* (or complex) value x instead of natural number n.

It could help to know more about the initial model. For example, sequence(1.1) for $\lambda = 4$ produces behavior that usually is called *chaotic*, but it can be written as "exact expression":

$$X_n = \sin^2(A\,2^n) \qquad (1.2)$$

with parameter $A = \arcsin(\sqrt{X_1})/2$. It can be simply checked:

$$
\begin{aligned}
X_{n+1} &= \sin^2(A\,2^{n+1}) = \sin^2(2\,A\,2^n) = \\
&= \left(2\sin(A\,2^n)\cos(A\,2^n)\right)^2 = 4\sin^2(A\,2^n)\cos^2(A\,2^n) = \\
&= 4\sin^2(A\,2^n)\left(1 - \sin^2(A\,2^n)\right) = 4X_n(1 - X_n)
\end{aligned}
$$

For $\lambda = 2$ it is also possible to build similar formula

$$X_n = \left(1 - \exp(-B\,2^n)\right)/2 \qquad (1.3)$$

with $B = -\ln(1 - 2X_1)/2$. The solution converge to $1/2$ very fast due to e^{2^n} in (1.3).

2. Expressions with continuous parameter

It is possible to use expressions (1.2), (1.3) with real parameters x instead of n. Let us try to find solution for other λ. It is a function of real argument x that mets the following equation:

$$\mathbb{F}_\lambda(x + 1) = \lambda\,\mathbb{F}_\lambda(x)\left(1 - \mathbb{F}_\lambda(x)\right) \qquad (2.1)$$

It should be mentioned what the function (2.1) is not unique. For a function $F(x)$ that meets (2.1) and a periodic function[1] $g(x + 1) = g(x)$ the function $F_g(x) = F(x + g(x))$ also meets (2.1). To minimize such kind of ambiguity let us try to look for functions similar to (1.2), (1.3). We can introduce auxiliary function $\varphi_\lambda(x)$:

$$\mathbb{F}_\lambda(x) = \varphi_\lambda(\alpha^x) \qquad (2.2)$$

and write instead of (2.1) an equation for $\varphi_\lambda(x)$:

$$\varphi_\lambda(\alpha x) = \lambda\varphi_\lambda(x)\left(1 - \varphi_\lambda(x)\right) \qquad (2.3)$$

---

[1] For example $g(x) = \sin(2\pi x)$

For resolving the equation let us write the Teilor's series for function $\varphi_\lambda(x)$ near $x = 0$:

$$\varphi_\lambda(x) = a_0 + a_1 x + a_2 x^2 + \cdots \qquad (2.4)$$

We can rewrite (2.3) as:

$$\begin{aligned}(a_0 + \alpha a_1 x + \alpha^2 a_2 x^2 + \cdots) &= \\ = \lambda(a_0 + a_1 x + a_2 x^2 + \cdots) &\times \\ \times \left((1 - a_0) - a_1 x - a_2 x^2 - \cdots\right)\end{aligned} \qquad (2.5)$$

Coefficients of x^k in both parts of the equation must be equal and for a_0 :

$$a_0 + \lambda a_0(1 - a_0) \Rightarrow \begin{cases} a_0 = 0 \\ a_0 = 1 - \frac{1}{\lambda} \end{cases} \qquad (2.6)$$

For a_1 :

$$\alpha a_1 = \lambda\left(-a_0 a_1 + a_1(1 - a_0)\right) = \lambda a_1(1 - 2a_0) \qquad (2.7)$$

If $\varphi_\lambda(x)$ is solution of (2.2) and $a_1 \neq 0$, the $\varphi_\lambda(x/a_1)$ is also solution, so it is possible to let $a_1 = 1$.

We have two sets of solutions:

$$\begin{cases} a_0 = 0, & \alpha = \lambda \\ a_0 = 1 - \frac{1}{\lambda}, & \alpha = 2 - \lambda \end{cases} \qquad (2.8)$$

For a_k, $k \geqslant 2$ we have:

$$\alpha^k a_k = \lambda(a_k \underbrace{(1 - 2a_0)}_{\alpha/\lambda} - a_1 a_{k-1} - $$
$$-a_2 a_{k-2} - \cdots - a_{k-1} a_1) \Rightarrow$$
$$(\alpha^k - \alpha)a_k = -\lambda \sum_{j=1}^{k-1} a_j a_{k-j} \Rightarrow$$
$$a_k = \frac{\lambda}{\alpha - \alpha^k} \sum_{j=1}^{k-1} a_j a_{k-j} \qquad (2.9)$$

To use the same Teilor's series for both solutions, let us substitute $x \to \lambda x$, then we can write:

$\alpha = \lambda$	$\alpha = 2 - \lambda$
$a_0 = 0$	$a_0 = \lambda - 1$
$a_1 = 1$	
$a_{\{k\|k>1\}} = \dfrac{-\sum_{j=1}^{k-1} a_j a_{k-j}}{\alpha^k - \alpha}$	
$\varphi_\lambda(x) = \frac{1}{\lambda} \sum_{k=0}^{\infty} a_k x^k$	
$\mathbb{F}_\lambda(x) = \varphi_\lambda(\alpha^x)$	

(2.10)

The a_k in (2.10) depends only on one parameter α. We can use the first solution with $\alpha = \lambda$ and write second one as:

$$\tilde{\varphi}_{2-\lambda}(x) = \frac{1 - \lambda + \lambda \varphi_\lambda(x)}{2 - \lambda} \tag{2.11}$$

3. The functions $\mathbb{F}_\lambda$ for $\lambda = 2, 4$

Let us consider properties of the functions $\mathbb{F}_\lambda$, φ_λ for particular values of λ.

$\lambda = 2$:
For the $\lambda = 2$ the coefficients a_k in formulae (2.10) are:

$$
\begin{aligned}
a_0 &= 0 \\
a_1 &= 1 \\
a_2 &= -\frac{1}{2} \\
a_3 &= \frac{1}{6} \\
&\cdots
\end{aligned}
$$

It can be proven what:

$$a_k = -\frac{(-1)^k}{k!}; \quad k > 1 \tag{3.1}$$

i.e.

$$
\begin{aligned}
\varphi_2(x) &= \frac{1 - \exp(-x)}{2} \\
\mathbb{F}_2(x) &= \frac{1 - \exp(-2^x)}{2}
\end{aligned} \tag{3.2}
$$

It corresponds to formula (1.3).

$\lambda = 4$:
For the $\lambda = 4$ the coefficients a_k in formulae (2.10) are:

$$
\begin{aligned}
a_0 &= 0 \\
a_1 &= 1 \\
a_2 &= -\frac{1}{3 \cdot 4} \\
a_3 &= \frac{1}{3 \cdot 4 \cdot 5 \cdot 6} \\
&\cdots
\end{aligned}
$$

It can be proven what:

$$a_k = -\frac{2 \cdot (-1)^k}{2k!}; \quad k > 1 \tag{3.3}$$

i.e.

$$
\begin{aligned}
\varphi_4(x^2) &= \frac{1 - \cos(x)}{2} = \sin^2\left(\frac{x}{2}\right) \\
\mathbb{F}_4(x) &= \sin^2(2^x / \sqrt{2})
\end{aligned} \tag{3.4}
$$

It corresponds to formula (1.2).

4. An application to fractals: Mandelbrot set

The functions $\mathbb{F}_\lambda$ could be simply rewritten for other sequences with quadratic terms. For example, let us consider the sequence:

$$Z_{n+1} = Z_n^2 + c \tag{4.1}$$

Where Z_n and c are complex numbers. For some complex c (and Z_0) the sequence of Z_n is bounded. Set of possible c (let $Z_0 = 0$)[2] is called *Mandelbrot set* and it is a simple example of fractal. The sequences (4.1) and (1.1) are simply related for real c, $-2 \leqslant c \leqslant 1/4$:

$$Z_n = \lambda\left(\tfrac{1}{2} - X_n\right) = \lambda\left(\tfrac{1}{2} - \mathbb{F}_\lambda(n)\right) \equiv \mathcal{M}_c(n) \tag{4.2}$$

$$c = \tfrac{\lambda(2-\lambda)}{4}, \quad \lambda = 1 \pm \sqrt{1 - 4c} \tag{4.3}$$

It make possible to use analytic continuation of $\mathbb{F}_\lambda(x)$ for extension the formulae (4.1), (4.2) for complex z:

$$\mathcal{M}_c(z + 1) = \mathcal{M}_c^2(z) + c \tag{4.4}$$

Additional property of the sequence (4.1) is that equations (4.2) for both functions (2.10) with $\alpha = \lambda$ and $\alpha = 2 - \lambda$ coincide due to (4.3), (2.11).

5. Conclusion and possible applications

The continuous functions presented in the work are demonstrative because they disprove possible suggestion about necessity of strong relation of chaotic behavior with discrete, recursive description of the models. It is already clear from formulae (1.2), (1.3) and extension to arbitrary values of parameter λ make the demonstration more general.

There is known useful example of similar continuation of another recursive function, the factorial $n! = n \cdot (n - 1)!$, i.e. $\Gamma(z) = \int_0^\infty t^{z-1}e^{-t}\, dt$, $n! = \Gamma(n+1)$.

A straightforward application of continuous functions is possibility to calculate derivatives. For example for (4.1) with given $Z_0 = \mathcal{M}_c(0) \neq 0$, there is simple expression for product of n elements of the sequence with using derivative of the function (4.4), $\mathcal{M}_c'(z)$:

$$Z_k \cdot Z_{k+1} \cdots Z_{k+n-1} = \frac{\mathcal{M}_c'(n + k)}{2^n \mathcal{M}_c'(k)}, \quad k \geq 0 \tag{5.1}$$

It follows from differentiation of the expression (4.4):

$$\mathcal{M}_c'(z + 1) = 2\mathcal{M}_c(z)\mathcal{M}_c'(z) \tag{5.2}$$

---

[2]Set of possible Z_0 for given c is called *Julia set*

With other initial $Z_0 = a$ it is possible to apply *translation* $\mathcal{M}_{c,a}(z) \equiv \mathcal{M}(z + \mu_a)$, where $\mathcal{M}(\mu_a) = a$. For standard choice $Z_0 = 0$ and so $\mathcal{M}'_{c,0}(k) = 0$ it is possible to write instead of (5.1) equation with second derivative:

$$Z_k \cdot Z_{k+1} \cdots Z_{k+n-1} = \frac{\mathcal{M}''_{c,0}(n+k)}{2^n \mathcal{M}''_{c,0}(k)}, \quad k \geq 1 \tag{5.3}$$

Chapter 53

Ordering chaos in a neural network with linear feedback

Lipo Wang
School of Computing and Mathematics, Deakin University
662 Blackburn Road, Clayton, Victoria 3168 Australia
lwang@deakin.edu.au
http://www.cm.deakin.edu.au/~lwang

There have been extensive, multi-discipline research interests in ordering chaos recently; however, most work has been done on systems with few degrees of freedom. In the case of neural networks, numerical simulations are usually relied upon to study networks with a small number of neurons. Based on our earlier work [Wang et al, *Proc. Natl. Acad. Sci. USA,* vol.87, 9467, 1990], we present ordering of chaos in an exactly solvable neural network with arbitrarily large number of neurons using a simple linear feedback.

In this network, N binary McCulloch-Pitts neurons are connected with both first-order and second-order Hebbian synapses, which are disconnected at random in the spirit of the sparse and asymmetric connectivities found in the real brain. In the absence of the linear feedback, depending on the network parameters, such as the number of stored patterns and the relative strength of first- and second-order interactions, the network can exhibit a variety of dynamics such as static fixed points, periodic oscillations, and chaos.

When the output of each neuron, multiplied by a positive constant, is fedback to this neuron itself, we show that chaos and oscillations in this modified Hopfield neural network are suppressed or even eliminated.

When a negative feedback is used, "crisis" can occur: all fixed points, oscillations, and chaos with positive overlaps between the network state and the initial attracting pattern disappear.

1. Introduction

Recently many researchers have studied ways to order chaos in various areas of science and technology (see e.g., [15], [14], [16], [13]). For example, Ott, Grebogi, and Yorke [15] studied controlling chaos in the context of a physical system. Ogorzalek [14] discussed ordering chaos in electronic circuits. Sepulchre and Babloyantz [16] studied controlling chaos in a network of neuron-like oscillators, whereas Lourenco and Babloyantz [13] explored ordering chaos in a neural network with time delays. However, the systems studied so far have relatively small numbers of degrees of freedom. For example, there are 16 excitatory neurons and 16 inhibitory neurons in the neural network studied in [13].

Artificial neural networks attempt to capture various information processing mechanisms of the brain [5]. In particular, oscillatory and chaotic neural networks have attracted much research attention lately (e.g., [17], [26], [27], [18]-[25]). In an earlier work, we [24] proposed a higher-order Hopfield-type neural networks [10] [11]. The main advantage of this network is that it has an arbitrarily large number of neurons and its dynamics is exactly solvable. We showed that this network exhibits a wide spectrum of dynamical behaviors, including stable fixed points, periodic oscillations, and chaos. Oscillatory and chaotic behaviors can be useful; however, many practical applications require the system to be non-chaotic. In the present paper we demonstrate how to order chaos in this higher-order neural network by simply adding a linear feedback to each neuron in the network.

2. System and analysis

In this section we first describe the system and then analyze its dynamics. It consists of $N \gg 1$ binary neurons, i.e.,

$$S_i(t+1) = \text{sgn}[h_i(t)] = \pm 1$$

where S_i is the state of neuron i and

$$i = 1, 2, ..., N .$$

The neurons interact with both first-order and second-order Hebbian synaptic weights [7], [4], [2], [1], [10], some of which are randomly disconnected [6] to model the asymmetric and sparse connectivity found in the real brain. The total input for neuron i is

$$h_i(t) = \sum_{j=1}^{N} T_{ij} S_j(t) - \sum_{j,k=1}^{N} T_{ijk} S_j(t) S_k(t)$$

$$+ \kappa S_i(t) + \eta_i ,$$

where we have introduced a linear feedback for each neuron with a strength κ (see Fig. 2.1). In addition,

$$T_{ij} = C_{ij} \sum_{\mu=1}^{p} S_i^\mu S_j^\mu$$

and

$$T_{ijk} = C_{ijk} \sum_{\mu=1}^{p} S_i^\mu S_j^\mu S_k^\mu$$

are the *modified* Hebbian [7], [4], [2], [1], [10] synaptic weights, $\vec{S}^\mu$ is the μ-th stored random pattern, and p is the number of patterns stored. We have introduced synaptic disruption in the weights T_{ij} and T_{ijk} by choosing random variables C_{ij} and C_{ijk} as follows: C_{ij} is 1 with a probability (C/N), C_{ijk} is 1 with a probability $(2C/N^2)$, C_{ij} and C_{ijk} are zero otherwise. Hence after synaptic disruption, each neuron interacts with about C neurons with first-order interactions and C pairs of neurons with second-order interactions [24] (Fig. 2.1). We have also included a background Gaussian noise η_i with a standard deviation σ_o to take into account the presence of signal transmission noise. Since the random numbers C_{ij} and C_{ji} for $i \neq j$ are independently chosen, we have, in general

$$T_{ij} \neq T_{ji} \quad , \quad \text{for} \quad i \neq j \ .$$

Similarly

$$T_{ijk} \neq T_{ikj} \neq T_{jik} \neq T_{jki} \neq T_{kij} \neq T_{kji} \ ,$$

$$\text{for} \quad i \neq j \neq k \ .$$

Hence in the presence of synaptic disruption, our system does not have an energy (Lyapunov) function as the Hopfield network does [10].

We also assume that the network is initially in the neighborhood of the memory state ν, i.e.,

$$m(t = 0) \equiv m^\nu(t = 0) \gg m^\mu(t = 0) \, , \mu \neq \nu \quad , \tag{1}$$

where

$$m^\lambda(t) = \frac{1}{N} < \vec{S}(t) \cdot \vec{S}^\lambda > \tag{2}$$

is a statistical average of the overlap between the state of the network at time t and the stored pattern λ ($\lambda = 1, 2, ..., N$). For parallel updating, i.e., when the states of all neurons are updated synchronously, the dynamics of the network can be shown in a way similar to [24]:

$$m(t+1) = [\frac{1+m(t)}{2}] \, \text{erf}\{\frac{m(t) - [m(t)]^2 + \kappa_c}{\sqrt{2}\sigma}\}$$

$$+ [\frac{1-m(t)}{2}] \, \text{erf}\{\frac{m(t) - [m(t)]^2 - \kappa_c}{\sqrt{2}\sigma}\} \ , \tag{3}$$

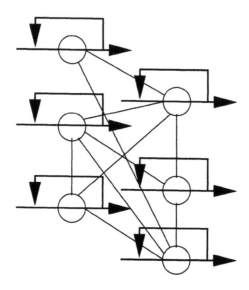

Figure 2.1: A schematic diagram of the system studied in the present work.

where

$$\kappa_c = \kappa/C \tag{4}$$

is a rescaled strength of the linear feedback, and

$$\mathrm{erf}(x) \equiv \frac{2}{\sqrt{\pi}} \int_0^x dz \, e^{-z^2}$$

is the standard error function. Furthermore,

$$\sigma \equiv \sqrt{2(p-1)/C + (\sigma_o/C)^2} \tag{5}$$

is a rescaled noise level that represents the combined effects of the random synaptic disruption, interference between stored patterns, and additional background noise.

In the absence of linear feedback, i.e., $\kappa_c = 0$, the dynamics of the network was studied by Wang *et al* [24]. A wide variety of dynamic behaviors, such as stable fixed points, periodic oscillations, and chaos, were found. Wang *et al* [24] identified a bifurcation parameter σ, which represents the combined effects of first- and second-order interactions, interference between stored patterns, random synaptic disconnection, and a background noise. For example, with $\sigma = 0.1$ and initial conditions $m(t = 0) = 0.15$ and $m(t = 0) = 0.3$, where $m(t)$ is the overlap between the state of the network and the initial attracting memory pattern, the dynamics of the network is chaotic, as shown in Fig. 2.2.

When the linear feedback for the neurons is applied, for example, $\kappa_c = 0.02$, chaos is ordered into periodic oscillations (Fig. 2.3). As the strength of the

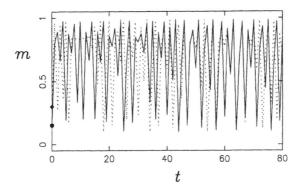

Figure 2.2: The average overlap m between the network state and the initial attracting memory pattern as a function of time t, according to eq.(3), in the absence of linear feedback ($\kappa_c = 0$) and with $\sigma = 0.1$. The solid line has an initial condition $m(0) = 0.15$ (denoted by the solid hexagon), whereas the dashed line has an initial condition of $m(0) = 0.3$ (denoted by the solid diamond). Despite the small difference in the two initial conditions, the resulting dynamics are very different: a hallmark of chaos.

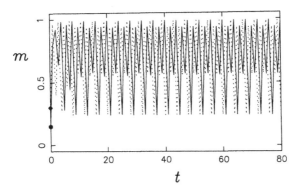

Figure 2.3: The average overlap m between the network state and the initial attracting memory pattern as a function of time t, according to eq.(3), in the presence of weak linear feedback $\kappa_c = 0.02$ and with $\sigma = 0.1$. Chaos are ordered into periodic oscillations (with a period-4). The solid line has an initial condition $m(0) = 0.15$ (denoted by the solid hexagon), whereas the dashed line has an initial condition of $m(0) = 0.3$ (denoted by the solid diamond). The difference in the two initial conditions causes a phase difference in the dynamics.

linear feedback increases, e.g., $\kappa_c = 0.5$, the amplitude of the periodic oscillations decreases (Fig. 2.4). If the interactions between the networks become even stronger, for example, $\kappa_c = 0.9$, the network becomes stable (Fig. 2.5). If *negative* linear feedback are used ($\kappa_c < 0$), "crisis" can occur: the network evolves towards a negative fixed point -1 even if it starts with $m(0) > 0$ (Fig. 2.6). We have thus demonstrated that one can order chaos using a linear feedback for each neuron in a higher-order neural networks.

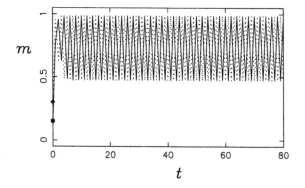

Figure 2.4: The average overlap m between the network state and the initial attracting memory pattern as a function of time t, according to eq.(3), in the presence of medium linear feedback $\kappa_c = 0.5$ and with $\sigma = 0.1$. The oscillations have a smaller amplitude and a period-2. The solid line has an initial condition $m(0) = 0.15$ (denoted by the solid hexagon), whereas the dashed line has an initial condition of $m(0) = 0.3$ (denoted by the solid diamond). The difference in the two initial conditions causes a phase difference in the dynamics.

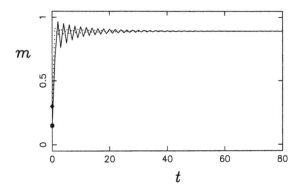

Figure 2.5: The average overlap m between the network state and the initial attracting memory pattern as a function of time t, according to eq.(3), in the presence of strong linear feedback $\kappa_c = 0.9$ and with $\sigma = 0.1$. The system becomes stable. The solid line has an initial condition $m(0) = 0.15$ (denoted by the solid hexagon), whereas the dashed line has an initial condition of $m(0) = 0.3$ (denoted by the solid diamond). The difference in the two initial conditions causes little difference in the dynamics.

3. Summary and conclusions

We have showed ordering chaos in modified Hopfield neural network. Specifically, we investigated the dynamical behaviors of a chaotic higher-order neural network when a linear feedback is applied to each neuron in the network. We showed that this type of simple linear feedback can serve as an effective means to order chaos.

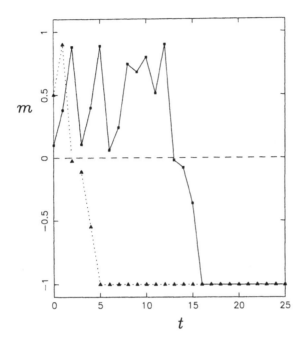

Figure 2.6: The average overlap m between the network state and the initial attracting memory pattern as a function of time t, according to eq.(3), in the presence of *negative* linear feedback $\kappa_c = -0.2$ and with $\sigma = 0.1$. All trajectories are attracted to the negative fixed point -1, even if the initial overlaps are positive ("crisis"). The solid line has an initial condition $m(0) = 0.15$, whereas the dashed line has an initial condition of $m(0) = 0.3$.

The dynamics of the system are described in terms of the statistical average overlap $m(t)$ given by eq.(3). $m(t)$ is the overlap with the stored random pattern $\vec{S}^\nu$ which the system is initially close to (eqs.(1) and (2)). Fig. 2.2–Fig. 2.6 indicate that although the control mechanism does not depend on the stored patterns, the state of the system does depend on its initial condition.

The second order synapses are necessary for the system to exhibit chaos at the level of the overlap $m(t)$, since in the absence of the second order synapses, $m(t)$ has a stable dynamics [6] [24]. In addition, $m(t)$ shows chaos only for some values of σ (eq.(5)) [24], for example, $\sigma = 0.1$ used in Fig. 2.2–Fig. 2.6. The control mechanism discussed in the present paper is robust in terms of the choices of σ, since in 10 additional tests with different choices of σ, Fig. 2.2– Fig. 2.6 remain qualitatively the same. Similarly, as shown in Fig. 2.2–Fig. 2.6, the system dynamics depends on the initial overlap $m(0)$ only *quantitatively*, but not *qualitatively*, that is, for a given set of σ and feedback strength κ_c, whether the system is chaotic, periodic, or stable does not depend on the precise value of $m(0)$.

Acknowledgements

This work was supported by the Australian Research Council and Deakin University. We thank Michael Doebeli for helpful suggestions.

Bibliography

[1] S.-I. Amari, "Learning patterns and pattern sequences by self-organizing nets of threshold elements," *IEEE Trans. Computers*, vol. 21, no. 11, pp. 1197, 1972.

[2] H.D. Block, B.W. Knight, Jr., and F. Rosenblatt, "Analysis of a four-layer series-coupled perceptron. II," *Rev. Mod. Phys.*, vol. 34, 135-142, Jan., 1962.

[3] D.G. Bounds, "New optimization methods from physics and biology," *Nature*, vol. 329, pp.215-219, September, 1987.

[4] E.R. Caianiello, "Outline of a theory of thought processes and thinking machines," *J. Theor. Biol.*, vol. 2, pp. 204-235, April, 1961.

[5] P. S. Churchland and T. J. Sejnowski, *The Computational Brain*, MIT Press, Cambridge, MA, 1989.

[6] B. Derrida, E. Gardner, and A. Zippelius, "An exactly solvable asymmetric neural network model," *Europhys. Lett.*, vol. 4, no. 2, pp. 167-173, July, 1987.

[7] D.O. Hebb, *The Organization of Behavior*, New York: John Wiley, 1949, p.44.

[8] R. Hecht-Nielsen, *Neurocomputing*, Reading, MA: Addison-Wesley, 1990, p.98.

[9] J. Hertz, A. Krogh, and R. G. Palmer, *Introduction to the Theory of Neural Computation*, Addison-Wesley, 1990.

[10] J. J. Hopfield, "Neural networks and physical systems with emergent collective computational abilities," *Proc. Natl. Acad. Sci. USA* vol. 79, pp. 2554-2558, April, 1982.

[11] J. J. Hopfield, "Neurons with graded response have collective computational properties like those of two-state neurons," *Proc. Natl. Acad. Sci. USA* vol. 81, pp. 3088-3092, May, 1984.

[12] J. J. Hopfield and D. Tank, "Neural computation of decisions in optimization problems," *Biol. Cybern.* vol. 52, pp. 141-152, 1985.

[13] C. Lourenco and A. Babloyantz, "Control of chaos in networks with delay: a model for synchronization of cortical tissue", *Neural Computation*, vol. 6, pp.1141-1154, 1994.

[14] EM.J. Ogorzalek, "Taming chaos", *IEEE Trans. on Circuits and Systems I: Fundamental Theory and Applications,* vol.40, no.10, pp. 693-706, 1993.

[15] E. Ott, C. Grebogi, and J.A. Yorke, "Controlling chaos," *Phys. Rev. Lett.,* vol. 64, no. 11, pp. 1196-1199, March, 1990.

[16] J. A. Sepulchre and A. Babloyantz, "Controlling chaos in a network of oscillators", *Phys. Rev. E,* vol. 48, no. 2, pp. 945-950, August, 1993.

[17] C. A. Skarda and W. J. Freeman, "How brains make chaos in order to make sense of the world," *Brain and behavioral Science,* vol. 10, pp. 161-195, 1987.

[18] L. Wang, "Oscillatory and chaotic dynamics in neural networks under varying operating conditions," *IEEE Trans. Neural Networks,* vol. 7, no. 6, pp. 1382-1388, November, 1996.

[19] L. Wang, "Suppressing chaos with hysteresis in a higher-order neural network," *IEEE Trans. on Circuit and Systems-II: Analog and Digital Signal Processing,* vol. 43, no. 12, pp. 845-846, December, 1996.

[20] L. Wang, "Discrete-time convergence theory and updating rules for neural networks with energy functions," *IEEE Trans. on Neural Networks,* vol. 8, no. 2, pp. 445-447, March, 1997.

[21] L. Wang, "Noise injection into inputs in sparsely-connected Hopfield and winner-take-all neural networks," *IEEE Trans. on Systems, Man, and Cybernetics,* vol. 27, no.5, October, 1997, accepted.

[22] L. Wang, "Processing spatio-temporal sequences with any static associative neural network" *IEEE Trans. on Circuit and Systems-II: Analog and Digital Signal Processing,* accepted.

[23] L. Wang, "On the dynamics of discrete-time, continuous-state Hopfield neural networks," *IEEE Trans. on Circuit and Systems-II: Analog and Digital Signal Processing,* accepted.

[24] L. Wang, E. E. Pichler, and J. Ross, "Oscillations and chaos in neural networks: an exactly solvable model," *Proc. Natl. Acad. Sci. USA,* vol. 87, pp. 9467 - 9471, December, 1990.

[25] L. Wang and J. Ross, "Chaos, multiplicity, crisis, and synchronicity in higher order neural networks," *Phys. Rev. A* vol. 44, no. 4, pp. R2259 - 2262, August, 1991.

[26] Y. Yao and W. J. Freeman, "Model of biological pattern recognition with spatially chaotic dynamics," *Neural Networks,* vol. 3, pp. 153-170, 1990.

[27] Y. Yao, W.J. Freeman, B. Burke, and Q. Yang, "Pattern recognition by a distributed neural network: an industrial application," *Neural Networks,* vol. 4, pp. 103-121, 1991.

Chapter 54

Self-organisation and information-carrying capacity of collectively autocatalytic sets of polymers: Ligation systems

Peter Wills[1]
Santa Fe Institute
1399 Hyde Park Road, Santa Fe, New Mexico 87501
Leah Henderson
Department of Physics, University of Auckland
Private Bag 92019, Auckland, New Zealand

We derive a stability criterion, analogous to Eigen's replication error threshold, for the information carrying capacity of a collectively autocatalytic set of polymers. We consider a specific system comprised of polymer molecules in which certain general ligation (and cleavage) reactions occur. We describe the physical constraints that must be fulfilled for the system to be capable of increasing in functional complexity as a result of progressive dynamic selection. These constraints relate to the embedding of catalytic functions in the space of possible polymer structures as dictated by the laws of physics and chemistry. Functional complexity is defined in terms of the degree of catalytic specificity displayed by the population of molecules in the system. Our results are likely to find application in the study of catalytic nucleic acids and peptides and should help to clarify how the differentiated roles of information carriers (nucleic

[1]Supported in part by NASA-AMES grant no. NAG 2-1091

acids) and functional catalysts (proteins) emerged in prebiotic molecular biology.

1. Introduction

The concept of collectively autocatalytic systems of molecules can perhaps be attributed to Calvin [1], but detailed consideration of the possibility was left to others, notably Kauffman [2], Rössler [3] and Eigen [4], who independently and virtually simultaneously discussed how such systems might function. Related ideas have been expounded by Dyson [5] and Cohen [6], but the most detailed analyses of collectively autocatalytic sets has been carried out by Kauffman and co-workers [7] [8] [9] who have established that in sufficiently complex chemical reaction systems whose components can serve simultaneously as substrates products and catalysts, the "crystallization" of collectively autocatalytic sets is a highly probable phenomenon. The system must of course be maintained away from equilibrium for the autocatalytic set to survive and potentially evolve. Bagley *et al.* [9] have studied the specific case of collective autocatalytic networks comprised of RNA sequences which catalyse ligation and cleavage reactions. Their model has much in common with the earlier system of Anderson *et al.* [10] [11].

Although complex autocatalytic systems have not been realised experimentally, both nucleic acid and protein molecules have been found able to replicate by means of autocatalytic processes. It is more than a decade since von Kiedrowski [12] first demonstrated that a hexanucleotide could catalyse its own synthesis by ligating two trinucleotide fragments. More recently Lee *et al.* [13] have shown that a very similar reaction occurs in a peptide system; a peptide comprised of 32 amino acids catalyses its own synthesis from two fragments comprised of 15 and 17 amino acids. The much greater complexity of the peptide system offers scope for the experimental investigation of competition and selection among closely related sets of templates and reactants [14]. In the light of these new developments it seems timely to take a general look at the formal and dynamic conditions which constrain the selection and maintenance of collectively autocatalytic sets of polymers.

2. Dynamics of autocatalysis

The dynamics of collective autocatalysis can be modeled in the same way as coding systems [15] [16].

Consider a set of reactions $\mathbf{R} = \{R_1, R_2, \ldots R_N\}$ which are catalysed at rates w_{ij} by polymers taken from a set $\{P_1, P_2, \ldots P_M\}$. The total rate of catalysis of reaction R_i is then

$$w_i = \sum_j w_{ij} x_j$$

where x_j is the population number of polymer sequence P_j. Let $x_\mathbf{s}$ represent the population number of catalysts in the system which catalyse reactions chosen from a subset $S \subset R$ and let $x_{\bar{S}}$ represent the population of polymers which catalyse reactions not in S. (There is a third population of molecules which are not catalytically active and which may be made by reactions in S, $\bar{S}$ or a combination of these.) If the set S is autocatalytic, then catalysts for all the reactions in S must be able to be formed through series of reactions belonging exclusively to S.

We denote by p the probability that any reaction taking place in the system is a member of S. We consider polymers comprised of ν monomers and we make the simplifying assumptions (a) that any polymer either catalyses a reaction at a specified rate w^o or not at all, (b) that all reactions are of the same order, (c) that there is equal availability of substrates for all reactions, and (d) that any polymer is a catalyst for at most one reaction. Under these approximations

$$p = \frac{x_S}{x_S + x_{\bar{S}}}$$

With $w_0 = \sum_i w_i$ representing the total rate at which reactions are carried out in the system, we can calculate the time evolution of x_S on the basis of two assumptions: (i) that the synthesis of any molecule which is comprised of ν monomers and catalyses a reaction in S requires a series of $\nu - 1$ ligation reactions chosen from S, and (ii) no series including a reaction from $\bar{S}$ creates a catalyst for a reaction in S. Thus,

$$\frac{dx_S}{dt} = w_0 f p^{\nu - 1}$$

and

$$\frac{dx_{\bar{S}}}{dt} = w_0 f'(1 - p^{\nu - 1})$$

where f is the catalytically active fraction of the set of polymers from sequence space which can be made by reactions from S alone, and f' is the catalytically active fraction of the set of polymers which require at least one reaction from $\bar{S}$ for their synthesis.

The time evolution of p is given by

$$
\begin{aligned}
\frac{dp}{dt} &= \frac{1}{x_S + x_{\bar{S}}}[(1 - p)\frac{dx_S}{dt} - p\frac{dx_{\bar{S}}}{dt}] \\
&= w^o[f(1 - p)p^{\nu - 1} - f'p(1 - p^{\nu - 1})] \\
&= w^o p(1 - p)[fp^{\nu - 2} - f'(1 + p + p^2 + \ldots + p^{\nu - 2})]
\end{aligned}
$$

To find the steady states of the system we look for the roots when $dp/dt = 0$. There are obviously roots at $p = 1$, corresponding to the situation when the only polymers present are those made by catalysts of reactions in S, and at $p = 0$, correponding to the situation when no such polymers are synthesized in the system. For $f/f' > 1/(\nu - 1)$ there is a third root in the range $0 < p < 1$, corresponding to a situation in which all reactions in $R = S + \bar{S}$ are catalysed to some degree. This state of "disorganised" synthesis cannot necessarily be described as truly "random" because it depends on the manner in which catalysis of the reactions in **R** is embedded in the polymer sequence space.

The stability of these steady states can be investigated by examining when the inequality

$$\frac{d}{dp}\left(\frac{dp}{dt}\right) = w^{\circ}[\nu(f' - f)p^{\nu-1} + f(\nu - 1)p^{\nu-2} - f'] < 0$$

is satisfied. Of interest is the criterion

$$\nu < \frac{f + f'}{f'}$$

for the stability of the steady state at p=1. This represents a general *selection threshold criterion* for the existence of a collectively autocatalytic set of polymers able to catalyse the reactions needed for their self-construction. It dictates the maximum number of monomer residues which can be maintained in polymers belonging to a collectively autocatalytic set and it therefore also represents an *information threshold criterion* for such systems, similar to that derived by Eigen [4] for the Darwinian selection of a quasispecies in simple replicative systems. While the Eigen criterion depends primarily on the specificity of the means of replication, ranging from the enzyme-free replication of RNA to the replication of DNA in eukaryotic cells [17], the criterion for collectively autocatalytic sets refers specifically to systems for which the means of polymeric construction are completely internal to the system itself. It is relevant therefore to the hypothetical prebiotic "RNA world" and may be used to estimate, based on knowledge of the sequence requirements for catalysis in RNA systems, how much information could have accumulated prior to the advent of protein synthesis. (The determinants of the "usefulness" of such information for the nucleation of genetic coding have been discussed previously [18].)

3. Ligation/cleavage systems

We consider a set of λ monomers $A = \{a_1, a_2, ...a_\lambda\}$ which can be joined together to form (directed) linear sequences of length ν as a result of reactions,

$$... a_k + a_l ... \rightarrow ... a_k a_l ...$$

denoted by $a_k - a_l$. There are $M = \lambda^2$ possible reactions. Without loss of generality, the reactions can be considered reversible; the monomer a_l to be added to the terminal monomer a_k of a polymeric sequence may itself be the initial monomer of another polymer. In discussing collectively autocatalytic sets, we are interested in the sequences that can be formed from a chosen proper subset **S** comprised of $\mu : 0 < \mu < M$ of the reactions. The composition of the collectively autocatalytic set depends not only on how many reactions there are in the subset, but also which particular reactions are chosen.

The formal condition for autocatalytic closure of a set of reactions $S \subset R$ can be stated quite simply: the sequence of any catalyst which carries out any reaction in $\bar{S}$ must contain an element $\ldots a_k a_l \ldots$ formed as a result of the reaction $a_k - a_l \in \bar{S}$ and none of the reactions in S may require a catalyst containing any such structural element. This condition can be understood as follows. The subset S is formed from R by eliminating one or more reactions $a_k - a_l$, and thereby the capability of forming corresponding structural elements $\ldots a_k a_l \ldots$. No polymers formed as a result of the reactions S are allowed to catalyse reactions $a_k - a_l \in \bar{S}$, and therefore the structural elements $\ldots a_k a_l \ldots$ must be obligatory for catalysis of reactions in $\bar{S}$. Thus, the possibility of collective autocatalysis is critically dependent on the "structure-function relationship", i.e., the correspondence between polymer sequences and their catalytic properties. We call this relationship the *functional embedding* [15] [16] [18]. Even for a small number of functions embedded in a polymer sequence space of low dimension, the number of mathematically possible embeddings is very large and rapidly becomes hyper-astronomical as the number of separate functions and the dimension of the sequence space increase [15] [18]. On the other hand, at most one of all the possible embeddings can represent the actual catalytic properties of a set of polymers comprised of real monomers like amino acids. Therefore, it is only worthwhile discussing embeddings which are based on what can be argued are "natural principles".

3.1. Binary systems

It is instructive to consider the polymer sequences which can be constructed for different subsets of reactions chosen from sets which involve the making (or breaking) of bonds between monomers chosen from a set of size $\lambda = 2$. For notational simplicity, we represent the binary monomer alphabet as $A = \{a, b\}$ and the full reaction set as $R = \{a\text{-}a, a\text{-}b, b\text{-}a, b\text{-}b\}$. The proper subsets S of reactions which could potentially lead to collectively autocatalytic sets and the characteristic polymers belonging to those sets are classified in Table 3.1.

In understanding how autocatalysis might be possible, the problem is to conceive, in chemical terms, how polymers associated with any reaction subset could plausibly be expected to have features which allowed them to catalyse the specific reactions in the relevant subset, but not other reactions. Our guiding principle comes from what is known about the structure-function relationship in protein sequence space: very specific structural features are usually required to build good catalysts which specifically differentiate different substrates and thus selectively catalyse just one or a few members of a class of reactions.

In the case of the uniform polymers, a potential mechanism is clear. If poly-a catalysed the reaction $..a + a.. \rightarrow ..aa..$ and did not recognise b as a substrate, then it could replicate; similarly for poly-b. Alternatively, poly-a and poly-b could play complementary roles and the two moleclues could form a collectively autocatalytic set. It is more difficult to conceive of how a tagged molecule might replicate by carrying out the relevant pair of reactions needed unless it somehow catalysed general sequential copying. Such a mechanism would have a threshold

Reaction Subset	Characteristic Polymer	Category of Polymer
{a-a}	aaaaa....aa	uniform
{b-b}	bbbbb....bb	
{a-a, a-b}	aaaaa....ab	tagged
{b-a, b-b}	bbbbb....ba	
{a-a, b-a}	baaaa....aa	
{a-b, b-b}	abbbb....bb	
{a-b, b-a}	..abababa..	alternating
{a-a, a-b, b-a}	..aaababa..	punctuated
{a-b, b-a, b-b}	..babbabb..	
{a-a, a-b, b-b}	..aaaabbb..	stepped
{a-a, b-a, b-b}	..bbbbaaa..	

Table 3.1: Possible autocatalytic sets of polymers comprised of monomers from the binary alphabet $A=\{a, b\}$.

accuracy defining the specificity of substrate recognition which would have to be exceeded by a tagged polymer which replicated in a dynamically stable fashion [4]. More generally, the presence of a single different monomer at the end of a polymer is unlikely to confer on the polymer the special catalytic properties needed. An alternating polymer is a plausible replicator of slightly greater complexity than a uniform polymer and could possibly replicate by direct or, more likely, complementary copying. The punctuated polymers form the most plausible collectively autocatalytic sets, because of the structural variation in the polymers which the associated subset of reactions produces. Increasing with the length of the polymer, structures of greater and greater sequence complexity can be produced. If the structural features needed to build a catalyst were quite specific to the reaction, then the complexity of sequences that can be synthesized by any one of these reaction subsets may make autocatalytic closure possible. The complexity of stepped polymers is similar to that of tagged polymers and it is more difficult to conceive of how they might form autocatalytic sets.

A set of punctuated polymers made from a binary alphabet of monomers {a, b} could form a collectively autocatalytic set if a structural dimer ..aa.. (or alternatively ..bb..) was an obligatory sequence element for the construction of any catalyst of the reaction a-a (or b-b) and if none of the reactions in the set **S** excluding a-a (or b-b) required catalysts containing the element ..aa.. (or ..bb..). To put the discussion in a realistic context, let us imagine that the polymers constituting the collectively autocatalytic sets are peptides and that the alphabet {a, b} represents not two particular amino acids that are distinguished by their specific chemical structures but a binary partition over the class of the available amino acids. Such a partition would divide the amino acids into two classes, e.g., hydrophobic and hydrophilic, or polar and nonpolar. A collectively autocatalytic set of punctuated polymers could appear in such a system if all catalysts capable of joining hydrophobic (or alternatively hydrophilic) amino acids required two adjacent hydrophobic (or hydrophilic) amino acids somewhere in their sequence, and if catalysts for all the other reactions could be constructed without such a dimeric element. Given satisfaction of these conditions, one could imagine that, starting from a random mixture of peptides in which all reactions occurred with equal probability, the first stage of self-organisation might be the selection of a collectively autocatalytic set of polymers on the basis of their ability to selectively recognise monomers as either hydrophobic or hydrophilic depending on the presence or absence of homologous dimeric elements in the sequences of the relevant catalysts.

3.2. Functional decomposition

The selection of a collectively autocatalytic set of polymers comprised of differentiated hodrophobic and hydrophylic amino acids would correspond to an increase in the functional value of λ from 1 to 2, because the truly random synthesis of peptides, in which the joining of any two amino acids is of identical probability, is a situation in which all amino acids are functionally equivalent to one another. It should be emphasised that by defining the monomer alphabet as a partition over chemical species, we have allowed a single polymer sequence, even a uniform sequence like poly-a, to represent a large number of different polymers conceived in ordinary chemical terms. However, this functional definition of structural variation is no different in principle from the approach taken in chemistry whereby the different quantum states of groups of atoms are classed together according to the arrangement of covalent bonds between the atoms — those are the distinctions which are relevant to understanding chemical structures and processes.

We have discussed the formal conditions which must be satisfied before it is possible for dynamic processes to lead to the selection of autocatalytic sets of polymers comprised of monomers from a binary alphabet. We now turn our attention to the formal prerequisites for an increase from 2 to 3 in the functional value of λ. The general idea is that collective selection among the polymer species, with sequences defined at the level of binary recognition of monomers, should lead to the refined differentiation of one of the classes of monomers into two subclasses, raising the number of recognisable classes of monomers from 2 to 3. Let us represent the ternary alphabet as $\{\alpha, \beta, \gamma\}$, to avoid the confusion that a new chemically different monomer, c, has been introduced into the system to augment the binary alphabet {a, b}. We shall consider the decomposition

$$a \to \{\alpha\}$$

$$b \rightarrow \{\beta, \gamma\}.$$

(The symmetries in Table 3.1 make it unnecessary, for our purposes, to consider the alternative decomposition explicitly.) The decomposition of the reaction set follows the pattern

$$\begin{aligned}
\text{a-a} &\rightarrow \{\alpha\text{-}\alpha\} \\
\text{a-b} &\rightarrow \{\alpha\text{-}\beta,\ \alpha\text{-}\gamma\} \\
\text{b-a} &\rightarrow \{\beta\text{-}\alpha,\ \gamma\text{-}\alpha\} \\
\text{b-b} &\rightarrow \{\beta\text{-}\beta,\ \beta\text{-}\gamma,\ \gamma\text{-}\beta,\ \gamma\text{-}\gamma\}
\end{aligned}$$

and Table 3.2 shows the decomposition of the autocatalytic sets listed in Table 3.1.

Binary Reactions		Ternary Reactions
{a-a}	$\rightarrow$	$\{\alpha\text{-}\alpha\}$
{b-b}	$\rightarrow$	$\{\beta\text{-}\beta,\ \beta\text{-}\gamma,\ \gamma\text{-}\beta,\ \gamma\text{-}\gamma\}$
{a-a, a-b}	$\rightarrow$	$\{\alpha\text{-}\alpha,\ \alpha\text{-}\beta,\ \alpha\text{-}\gamma\}$
{b-a, b-b}	$\rightarrow$	$\{\beta\text{-}\alpha,\ \gamma\text{-}\alpha,\ \beta\text{-}\beta,\ \beta\text{-}\gamma,\ \gamma\text{-}\beta,\ \gamma\text{-}\gamma\}$
{a-a, b-a}	$\rightarrow$	$\{\alpha\text{-}\alpha,\ \beta\text{-}\alpha,\ \gamma\text{-}\alpha\}$
{a-b, b-b}	$\rightarrow$	$\{\alpha\text{-}\beta,\ \alpha\text{-}\gamma,\ \beta\text{-}\beta,\ \beta\text{-}\gamma,\ \gamma\text{-}\beta,\ \gamma\text{-}\gamma\}$
{a-b, b-a}	$\rightarrow$	$\{\alpha\text{-}\beta,\ \alpha\text{-}\gamma,\ \beta\text{-}\alpha,\ \gamma\text{-}\alpha\}$
{a-a, a-b, b-a}	$\rightarrow$	$\{\alpha\text{-}\alpha,\ \alpha\text{-}\beta,\ \alpha\text{-}\gamma,\ \beta\text{-}\alpha,\ \gamma\text{-}\alpha\}$
{a-b, b-a, b-b}	$\rightarrow$	$\{\alpha\text{-}\beta,\ \alpha\text{-}\gamma,\ \beta\text{-}\alpha,\ \gamma\text{-}\alpha,\ \beta\text{-}\beta,\ \beta\text{-}\gamma,\ \gamma\text{-}\beta,\ \gamma\text{-}\gamma\}$
{a-a, a-b, b-b}	$\rightarrow$	$\{\alpha\text{-}\alpha,\ \alpha\text{-}\beta,\ \alpha\text{-}\gamma,\ \beta\text{-}\beta,\ \beta\text{-}\gamma,\ \gamma\text{-}\beta,\ \gamma\text{-}\gamma\}$
{a-a, b-a, b-b}	$\rightarrow$	$\{\alpha\text{-}\alpha,\ \beta\text{-}\alpha,\ \gamma\text{-}\alpha,\ \beta\text{-}\beta,\ \beta\text{-}\gamma,\ \gamma\text{-}\beta,\ \gamma\text{-}\gamma\}$

Table 3.2: Decomposition of collectively autocatalyic sets based on the alphabet decomposition a $\rightarrow$ $\{\alpha\}$, b $\rightarrow$ $\{\beta, \gamma\}$.

Sets of reactions based on the ternary alphabet and differing from those based on the binary alphabet can be found by eliminating reactions from the sets in the right hand column of Table 3.2. It is possible to categorise the polymers made by each such set of reactions in much the same way as the binary-alphabet polymers were categorised in Table 3.1 (results not shown). What is striking about the categorisation of these potentially autocatalytic sets of ternary-alphabet polymers is that they are, with only one exception, of similar complexity to the binary-alphabet polymers listed in Table 3.1. Virtually all of the ternary-alphabet polymers comprising collectively autocatalytic sets which can be derived by decomposition from binary-alphabet autocatalytic sets either have the same character as the binary-alphabet sets but in a different alphabet (such as $\{\beta, \gamma\}$ instead of $\{a, b\}$) or a simple combination of two such characteristics (e.g., poly-α tagged with β or γ, or poly-α in stepped combination with a punctuated β-γ polymer). Only those derived from the second "punctuated" category (9th column in Table 3.1) have sequences of markedly different complexity from the binary-alphabet polymers. This is not entirely unexpected since, under the decomposition, the set $\{a\text{-}b, b\text{-}a, b\text{-}b\}$ gives rise to 8 out of the 9 possible ternary-alphabet reactions, from which 8 of the 72 possible 7-member subsets of reactions can be generated.

Two general conclusions emerge from this analysis. The first is that the collectively autocatalytic sets that can evolve through progressive decomposition of the monomer alphabet A and the corresponding reaction set $\mathbf{R}$ comprise a specialized fraction of all the possible subsets of the enlarged reaction set. The second conclusion represents the reemergence of the general rule of autocatalysis at the level of functional detail: selection of refined functions is only possible if the structures which carry out the refined functions are differentiated in specialized ways through the presence or absence of the refined structural features which the refined functions produce. The evolutionary operation of decomposition maintains, at the level of functional and structural detail, the obligatory relationships between structure and function which make autocatalysis possible.

4. Conclusion

We have shown how the information-carrying capacity of a collectively autocatalytic set of polymers can be calculated from knowledge of the relationship between polymer structure, specified in terms of the operations needed to construct different molecules, and polymer function, specified in terms of the polymers' catalytic capabilities that allow the constructive operations to be performed. The information-carrying capacity of an autocatalytic system depends on a selection threshold criterion derived from the dynamics of polymer synthesis. The criterion is an analogue of that derived by Eigen [4] for systems of replicating molecules. The importance of this new result is the demonstration that the stepwise replication of polymers is not an absolute prerequisite for information storage in non-equilibrium chemical systems. Thus, there could have been the stable preservation of information in the putative RNA world prior to the emergence of general replicases, provided the ribozymal structure-function relationship allowed self-constructive collectively autocatalytic sets to be synthesized. Eigen [19] draws a distinction between "inherent" reproductive capacity, like that possessed by nucleic acids on account of base-pairing complementarity, and "incidental" reproductivity, like that displayed more generally by autocatalytic systems. Eigen suggests that "inherent" reproductivity is required for evolutionary selection. Our analysis of autocatalysis in ligation/cleavage systems indicates that this is not the case. Collectively autocatalytic sets of polymers can both carry information and undergo joint selection. Although we have not investigated the possibility here, there is no reason why this phenomenon could not be stochastically driven, as it is for "inherently" reproductive systems in which mutation occurs.

Our second conclusion is that the selection of progressively more complex collectively autocatalytic sets of polymers is possible in systems whose structure-function relationship satisfies certain constraints. By examining simple ligation/cleavage systems we illustrate what is likely to be a general precondition for the evolutionary emergence of refined biological functions: structures which carry out refined functions should be differentiated in specialized ways through the presence or absence of the refined structural features which the refined functions selectively produce. In contrast with the linear logic of the Central Dogma of molecular biology [20] and the Darwinian selection of genetic information [4], the logic of functional evolution is strangely circular and even seems Lamarckian [21].

Bibliography

[1] CALVIN, Melvin, *Chemical Evolution*, Oxford University Press (1969).

[2] KAUFFMAN, Stuart A., "Cellular homeostasis, epigenesis and replication in randomly aggregated macromolecular systems" *J. Cybernetics* 1 (1971), 71–96.

[3] RÖSSLER, Otto E., "A system-theoretic model of biogenesis" *Zeit. f. Natur-forsch. B* 266 (1971), 741–746.

[4] EIGEN, Manfred, "Selforganisation of Matter and the Evolution of Biological Macromolecules", *Naturwissenschaften* 58 (1971), 465–522.

[5] DYSON, Freeman, "A model for the origin of life" *J. Mol. Evol.* 18 (1982), 344–350.

[6] COHEN, Joel E., "Threshold Phenomena in Random Structures", *Discrete Applied Math.* 19 (1988), 113–128.

[7] KAUFFMAN, Stuart A., "Autocatalytic sets of proteins" *J. Theor. Biol.* 119 (1986), 1–24.

[8] FARMER, J. Doyne, Stuart A. KAUFFMAN and Norman H. PACKARD "Autocatalytic replication of polymers" *Physica D* 22 (1986), 50–67.

[9] BAGLEY, Richard J., J. Doyne FARMER, Stuart A. KAUFFMAN, Norman H. PACKARD, Alan S. PERELSON and Irene. M. STADNYK, "Modeling adaptive biological systems" *Biosystems* 23 (1989), 113–137.

[10] ANDERSON, Philip W., "Suggested model for prebiotic evolution: The use of chaos" *Proc. Natl. Acad. Sci. USA* 80 (1983), 3386–3390.

[11] ROKSHAR Daniel. S., Philip W. ANDERSON, Daniel L. STEIN, "Self-Organization in Prebiological Systems: Simulations of a Model for the Origin of Genetic Information", *J. Mol. Evol* **23** (1986), 119–126.

[12] VON KIEDROWSKI, Günther, "A Self-Replicating Hexadeoxynucleotide", *Angewandte Chemie Intl. Eng. Ed.* **25** (1986), 932–935.

[13] LEE, David H., Juan R. GRANJA, Jose A. MARTINEZ, Kay SEVERIN and M. Reza GHADIRI, "A self-replicating peptide", *Nature* **382** (1996), 525–528.

[14] WILLS, Peter R., Stuart A. KAUFFMAN, Bärbel M. R. STADLER and Peter F. STADLER, "Selection Dynamics in Autocatalytic Systems: Templates Replicating Through Binary Ligation", *Working Paper no. 97-07-065*, Santa Fe Institute, (Jul. 1997).

[15] WILLS, Peter R., "Self-organization of Genetic Coding", *J. Theor. Biol.* **162** (1993), 267–287.

[16] WILLS, Peter R., "Does Information Acquire Meaning Naturally?", *Ber. Bunsenges. Phys. Chem.* **98** (1994), 1129–1134.

[17] EIGEN, Manfred and Peter SCHUSTER, *The Hypercycle*, Springer-Verlag (1979).

[18] NIESELT-STRUWE, Kay and Peter R. WILLS, "The Emergence of Genetic Coding in Physical Systems", *J. Theor. Biol.* **187** (1997), 1–14.

[19] EIGEN, Manfred, "Prionics, or the kinetic basis of prion diseases", *Biophys. Chem.* **63** (1996), A1-A18.

[20] CRICK, Francis H. C., "Central dogma of molecular biology", *Nature* **227** (1970), 561-563.

[21] WILLS, Peter R., "Genetic information and the determination of functional organization in biological systems", *Systems Res.* **6** (1989), 219–216.

Chapter 55

Self-dissimilarity: An empirically observable complexity measure

David H. Wolpert

NASA Ames Research Center

MS269-2, Moffett Field, CA, 94035

William G. Macready

Bios Group LP

317 Paseo de Peralta

Santa Fe, NM, 87501

For many systems characterized as "complex/living/intelligent" the spatio-temporal patterns exhibited on different scales differ markedly from one another. For example the biomass distribution of a human body "looks very different" depending on the spatial scale at which one examines that biomass. Conversely, the density patterns at different scales in "dead/simple" systems (e.g., gases, mountains, crystals) do not vary significantly from one another. Accordingly, we argue that the degrees of self-*dis*similarity between the various scales with which a system is examined constitute a complexity "signature" of that system. Such signatures can be empirically measured for many real-world data sets concerning spatio-temporal densities, be they mass densities, species densities, or symbol densities. This allows one to compare the complexity signatures of wholly different kinds of systems (e.g., systems involving information density in a digital computer, vs. species densities in a rain-forest, vs. capital density in an economy, *etc.*). Such signatures can also be clustered, to provide an empirically determined taxonomy of "kinds of systems" that share organizational traits. The precise measure of dissimilarity between scales that we propose is the amount of extra information on one scale beyond that which exists on a different scale. This "added information" is perhaps most naturally determined using a maximum entropy inference of the distribution of patterns at the second scale, based on the provided distribution at the first scale. We briefly discuss using our measure with other inference mechanisms (e.g., Kolmogorov complexity-based inference).

1. Introduction

Historically, the concepts of life, intelligence, culture, and complexity have resisted all attempts at formal scientific analysis. Indeed, there are not even widely agreed-upon formal definitions of those terms [6, 3]. Why is this?

We argue that the underlying problem is that many of the attempted analyses have constructed an extensive formal model before considering any experimental data. For example, some proposed definitions of complexity are founded on statistical mechanics [7], while others use computer science abstractions like finite automata [5] or universal Turing machines [4, 8, 2]. None of these models arose from consideration of any particular experimental data.

This contrasts with the more empirical approach that characterized the (astonishingly successful) growth of the natural sciences. This approach begins with the specification of readily measurable "attributes of interest" of real-world phenomena followed by observation of the inter-relationships of those attributes in real-world systems. *Then* there is an attempt to explain those inter-relationships via a theoretical model. For the most part, the natural sciences were born of raw experimental data and a need to explain it, rather than from theoretical musing.

It is not difficult to see why data-driven approaches may be more successful in general. In many respects, before a model-driven approach can be used to assign a complexity to a system, one must already fully understand that system (to the point that the system is formally encapsulated in terms of one's model class). So only once most of the work in analyzing the system has already been done can one investigate that system using these proposed measures of complexity. Another major problem with model-driven approaches is that they are prone to degeneration into theorizing and simulating, in isolation from the real world. This lack of coupling to experimental data vitiates the most important means by which theoretical models can be compared, refuted, and modified.

In this paper we follow a more data-driven approach, in which we start with an attribute of interest. Our choice for attribute of interest is based on the observation that most systems that people characterize as complex/living/intelligent have the following property: *over different space and time scales, the patterns exhibited by a complex system vary greatly, and in ways that are unexpected given the patterns on the other scales.* Accordingly, a system's self-*dissimilarity* is the attribute of interest we propose be measured — completely devoid of the context of any formal model at this point. (Bar Yam also proposes a complexity profile which is based on the characteristics of a system at different scales — see [1].)

The human body is a familiar example of such self-dissimilarity; as one changes the scale of the spatio-temporal microscope with which one observes the body, the pattern one sees varies tremendously. Other examples from biology are how, as one changes the scale of observation, the internal structures of a biological cell, or of an ecosystem, differ greatly from one another. By measuring patterns in quantities other than the mass distribution (*e.g.*, in information distributions), one can also argue that the patterns in economies and other cultural institutions vary enormously with scale. It may also be that as one changes the scale of observation there are also large variations in the charge density patterns inside the human brain.

In contrast, simple systems like crystals and ideal gases may exhibit some variation in pattern over a small range of scales, but invariably when viewed over broad ranges of scales the amount of variation falls away. Similarly, viewed over a broad range of spatio-temporal scales (approximately the scales from complexes of several hundred molecules on up to microns), a mountain, or a chair, would appear to exhibit relatively little variation in mass density patterns. As an extreme example, relative to its state when alive, a creature that has died and decomposed exhibits no variation over temporal scales. Such a creature also exhibits far less variation over spatial scales than it did when alive.

Our thesis is that variation in a system's spatio-temporal patterns as one changes scales is not simply a side-effect of what is "really going on" in a complex system. Rather it is a crucial aspect of the system's complexity. We propose that it is only after we have measured such self-dissimilar aspects of real-world systems, when we have gone on to construct formal models explaining those data, that we will have models that "get at the heart" of complex systems.

There are a number of apparent contrasts between our proposed approach and much previous work on complexity. In particular, fractals have often been characterized as being incredibly complex due to their possessing nontrivial structure at all different scales; in our approach they are instead viewed as relatively simple objects since the structure found at different scales is in many respects the *same*.

Similarly, a cottage industry exists in finding self-similar degrees of freedom in all kinds of real-world systems, some of which can properly be described as complex systems. Our thesis is that independent of such self-similar degrees of freedom, it is the alternative self-dissimilar degrees of freedom which are more directly important for analyzing a system's complexity. We hypothesize that, in large measure, to concentrate on self-similar degrees of freedom of a complex system is to concentrate on the degrees of freedom that can be very compactly encoded, and therefore are not fundamental aspects of that system's complexity.

As an example, consider a successful, flexible, modern corporation, a system that is "self-similar" in certain variables ([9]). Consider such a corporation that specializes in an information processing service of some sort, so that its interaction with its environment can be characterized primarily in terms of such processing rather than in terms of gross physical manipulation of that environment. Now hypothesize that in *all* important regards that corporation is self-similar. Then the behavior of that corporation — and in particular its effective dynamic adaptation to and interaction with its environment — is specified using the extremely small amount of information determining the scaling behavior. In such a situation, one could replace that adaptive corporation with a very small computer program based on that scaling information, and the interaction with the environment would be unchanged. The patent absurdity of this claim demonstrates that *what is most important* about a corporation is not captured by those variables that are self-similar.

More generally, even if one could find a system commonly viewed as complex that was clearly self-similar in all important regards, it is hard to see how the same system wouldn't be considered even more "complex" if it were self-dissimilar. Indeed, it is hard to imagine a system that is highly self-dissimilar in both space and time that wouldn't be considered complex. Self-dissimilarity would appear to be a sufficient condition for a system to be complex, even if it is not a necessary condition.

In §2 we further motivate why self-dissimilarity is a good measure of complexity. Section §3 then takes up the challenge of formalizing some of these vague notions. The essence of our approach is the comparison of spatio-temporal structure at different scales. Since we adopt a strongly empirical perspective, how to infer structure on one scale from structure on another is a central issue. This naturally leads to the probabilistic measure we propose in this section. Finally, in §4 we discuss some of the general attributes of our measure and how to estimate it from data. In future work we plan to apply those estimation schemes to real-world data sets.

It is worth emphasizing that we make no claim whatsoever that self-dissimilarity captures all that is important in complex systems. Nor do we even wish to identify self-dissimilarity with complexity. We only suggest that self-dissimilarity is an important component of complexity, one with the novel advantage that it can actually be evaluating for real-world systems.

2. Self-dissimilarity

In the real world, one analyzes a system by first being provided information (e.g., some experimental data) in one space, and then from that information making inferences about the full system living in a broader space. The essence of our approach is to characterize a system's complexity in terms of how the inferences about that broader space differ from one another as one varies the information-gathering spaces. In other words, our approach is concerned with characterizing how readily the full system can be inferred from incomplete measurements of it. Violent swings in such inferences as one changes what is measured — large self-dissimilarity — constitute complexity for us.

2.1. Why might complex systems be self-dissimilar?

Before turning to formal definitions of self-dissimilarity we speculate on why self-dissimilarity might be an important indicator of complexity. Certainly self-dissimilar systems will be *interesting*, but why should they also coincide with what are commonly considered to be complex systems?

Most systems commonly viewed as complex/interesting have been constructed by an evolutionary process (*e.g.* life, culture, intelligence). If we assume that there is some selective advantage in such systems for maximizing the amount of information processing within the system's volume, then we are led to consider systems which are able to process information in many different ways on many spatio-temporal scales, with those different processes all communicating with one another. By exploiting different scales to run different information processing, such systems are in a certain sense maximally dense with respect to how much information processing they achieve in a given volume. Systems processing information similarly on different scales, or even worse not exploiting different scales at all, are simply inefficient in their information-processing capabilities.

To make maximal use of the different information processes at different scales, presumably there must be efficient communication between those processes. Such inter-scale communication is common in systems usually viewed as complex. For example, typically the effects of large scale occurrences (like broken bones in organisms) propagate to the smallest levels (stimulating bone cell growth) in complex systems. Similarly, slight changes at small scales (the bankruptcy of a firm, or the mutation of a gene) can have marked large-scale (industry-wide, or body-wide) effects.

Despite the clear potential benefits of multi-scale information processing, explicitly constructing a system which engages in such behavior seems to be a formidable challenge. Even specifying the necessary dynamical conditions (e.g., a Hamiltonian) for a system to be able to support multi-scale information processing appears difficult. (Tellingly, it is also difficult to explicitly construct a physical system that engages in what most researchers would consider "life-like" behavior, or one that engages in "intelligent" behavior; our hypothesis is that this is not a coincidence, but reflects the fact that such systems engage in multi-scale information processing.) In this paper, rather than try to construct systems that engage in multi-scale information processing, we merely assume that nature has stumbled upon ways to do so. Our present goal is only to determine how to recognize and quantify such multi-scale information processing in the first place, and then to measure such processing in real-world systems.

This perspective of communication between scales suggests that there are upper bounds on how self-dissimilar a viable complex system can be. Since the structure at one scale must have meaning at another scale to allow communication between the two, presumably those structures cannot be *too* different. Also for a complex system to be stable it must be robust with respect to changes in its environment. This suggests that the effects of random perturbations on a particular scale should be isolated to one or a few scales lest the full system be prone to collapse. To this extent scales must be insulated from each other. Accordingly, as a function of the noise inherent in an environment, there may be very precise and constrained ways in which scales can interact in robust systems. If so it would be hoped that when applied to real-world complex systems a self-dissimilarity measure would uncover such a modularity of multi-scale information processing.

This perspective also gives rise to some interesting conjectures concerning the concept of intelligence. It is generally agreed that any "intelligent" organism has a huge amount of extra-genetic information-processing concerning the outside world, in its brain. (If all the processing could take place directly via genome-directed mechanisms, there would be no need for an adaptive structure like a brain.) In other words, the information processing in the brain of an intelligent organism is tightly and extensively coupled to the information processing of the outside world. So to an intelligent organism, the outside world — which is physically a scale up from the organism — has the same kind of information coupling with the organism that living, complex organisms have between the various scales within their own bodies.

So what is intelligence? This perspective suggests a definition. An intelligence is a system that is coupled to the broader external world exactly as though it were a subsystem of a living body consisting of that broader world. In other words, it is a system whose relationship with the outside world is similar to its relationship with its own internal subsystems. An intelligence is a system configured so that the border of what-is-living/complex extends beyond the system, to the surrounding environment.

2.2. Advantages of the approach

The reliance on self-dissimilarity as a starting point for a science of complexity has many advantages beyond its being part of a data-driven approach. For example, puzzles like how to decide whether a system "is alive" are rendered mute under such an approach. We argue that such difficulties arise from trying to squeeze physical phenomena into pre-existing theoretical models (e.g., for models concerning "life" one must identify the atomic units of the physical system, define what is meant for them to reproduce, *etc.*). Taking our purely empiricist approach, life is instead a characteristic signature of a system's self-dissimilarity over a range of spatio-temporal scales. Presumably highly complex living systems exhibit highly detailed, large self-dissimilarity signatures, while less complex, more dead systems exhibit shallower signatures with less fine detail. We argue that life is more than a yes/no bit, and even more than a real number signifying a degree—it is an entire signature. In addition to superseding sterile semantic arguments, adopting this point of view opens entirely new fields of research. For example, one can meaningfully consider questions like how the life-signature of the biosphere changes as one species (*e.g.*, humans) takes over that biosphere.

More generally, self-dissimilarity signatures can be used to compare entirely different kinds of systems (*e.g.*, information densities in human organizations versus mass distributions in galaxies). With this complexity measure we can, in theory at least, meaningfully address questions like the following: How does a modern economy's complexity signature compare to that of the organelles inside a prokaryotic cell? What naturally occurring ecology is most like that of a modern city? Most like that of the charge densities moving across the internet?

Can cultures be distinguished according to their self-dissimilarity measure? Can one reliably distinguish between different kinds of text streams, like poetry and prose, in terms of their complexity?

By concentrating on self-dissimilarity signatures we can compare systems over different regions of scales, thereby investigating how the complexity character itself changes as one varies the scale. This allows us to address questions like: For what range of scales is the associated self-dissimilarity signature of a transportation system most like the signature of the current densities inside a computer? How much is the self-dissimilarity signature of the mass density of the astronomy-scale universe like that of an ideal gas when examined on mesoscopic scales, *etc.*?

In fact, by applying the statistical technique of clustering to self-dissimilarity signatures, we should be able to create empirically-defined taxonomies ranging over broad classes of real-world systems. For example, self-dissimilarity signatures certainly will separate marine environments (where the mass density within organisms is similar to the mass density of the environment) from terrestrial environments (where the mass densities within organisms is quite different from that of their environment). One might also hope that such signatures would divide marine creatures from terrestrial ones, since the bodily processes of marine creatures observe broad commonalities not present in terrestrial creatures (and vice-versa). Certainly one would expect that such signatures could separate prokaryotes from eukaryotes, plants from animals, *etc.* In short, statistical clustering of self-dissimilarity signatures may provide a purely data-driven (rather than model-driven or — worse still — subjective) means of generating a biological taxonomy. Moreover, we can extend the set of signatures being clustered far beyond biological systems, thereby creating, in theory at least, a taxonomy of all natural phenomena. For example, not only could we cluster cultural institutions (do Eastern and Western socio-economic institutions break up into distinct clusters?); we could also cluster the signatures of such institutions together with those of insect colonies (do hives fall in the same cluster as human feudal societies, or are they more like democracies?).

The self-dissimilarity concept also leads to many interesting conjectures. For example, in the spirit of the Church-Turing thesis, one might posit that any naturally-occurring system with sufficiently complex yet non-random behavior at some scale s must have a relatively large and detailed self-dissimilarity signature at scales finer than s. If this hypothesis holds, then (for example) due to the fact that its large-scale physical behavior (i.e., the dynamics of its intelligent actions) is complex, the human mind *necessarily* has a large and detailed self-dissimilarity signature at scales smaller than that of the brain. Such a scenario suggests that the different dynamical patterns on different scales within the human brain is not some side-effect of how nature happened to solve the question of how to build an intelligence, given its constraints of noisy carbon-based life. Rather it is fundamental, being required for any (naturally occurring) intelligence. This would in turn suggest that (for example) work on artificial neural nets will have difficulty creating convincing mimics of human beings until those nets are built on several different scales at once.

3. Probabilistic measures of self-dissimilarity

We begin by noting that any physical system is a realization of a stochastic process, and it is the properties of that underlying process that are fundamentally important. This leads us to consider an explicitly probabilistic setting for measuring self-dissimilarity, in which we are comparing the probability distributions over the various scale s patterns that the process can generate.

By incorporating probabilistic concerns into its foundations in this way, the proposed measure explicitly reflects the fundamental role that statistical inference (for example of patterns at one scale from patterns at another scale) plays in complexity. It also means that the framework will involve the quantities that are of direct interest physically. In addition, via information theory, it provides us with some very natural candidate measures for the amount of dissimilarity between structures at two different scales (*e.g.*, the Kullback-Leibler [10] distance between those structures). The implicit viewpoint of such measures is that "how dissimilar" two structures at different scales are is how much information is provided in the larger-scale structure that is absent in the smaller-scale structure. (The exploration of other, non-information-theoretic measures of self-dissimilarity is the subject of future research.)

To formalize the proposed measure of self-dissimilarity, we begin with a definition of a scale's "stochastic structure". Then we specify how to convert structures on different scales to the same scale by using statistical inference. As the final step, we specify how to quantify the difference between two structures on the same scale. Applied on a scale s_c to a pair of structures converted from scales s_1 and s_2, this quantity will be our measure of the self-dissimilarity exhibited by scales s_1 and s_2.

3.1. Defining the structure at a scale

Assume an integer-indexed set of spaces, Ω_s. The indices on the spaces are called *scales*. For any two scales s_1 and $s_2 > s_1$, assume also that we have a set of mappings $\{\rho_{s_1 \leftarrow s_2}^{(i)}\}$ labeled by i, each taking elements of Ω_{s_2} to elements of the smaller scale space Ω_{s_1}.

In this paper, "scales" will be akin to the widths of the translatable masking windows with which a system is examined, rather than to different levels of precision with which it is examined. The index i labeling the mapping set specifies the location of the masking window through which the system is examined (colloquially, i tells us where we are pointing our microscope). The fact that we have a full mapping *set* simply reflects the multitude of such locations.

Two elaborations of window-based scales are provided by the following two examples. Both examples involve one-dimensional sequences of characters as the objects under study

Example 1: The members of Ω_{s_2} are the sequences of s_2 successive characters. Indicate such a sequence as $\omega_{s_2}(k)$, with $1 \leq k \leq s_2$ indexing the characters. $\rho_{s_1 \leftarrow s_2}^{(i)}$ is the projective mapping taking any ω_{s_2} to the sequence of s_1 characters ω_{s_1} where $\omega_{s_1}(j) = \omega_{s_2}(j + i)$ for $1 \leq j \leq s_1$, and $0 \leq i \leq s_2 - s_1$. So the $\rho_{s_1 \leftarrow s_2}^{(i)}$ are translations of a simple masking operation creating a subsequence of s_1 characters, with i indicating the translation.

Example 2: This is a modification of example 1 so that the mapping sets and spaces Ω_s are as scale-invariant as possible, and therefore introduce minimal *a priori* bias into the self-dissimilarity measure. We require that s_1 must equal $a2^{k_1}$ and s_2 must equal $b2^{k_2}$, for some integer constants $b > a > 0$ and $k_2 \geq k_1 \geq 0$. We then have $\omega_{s_1}(j) = \omega_{s_2}(j + i2^{k_1})$, where $1 \leq j \leq a2^{k_1}$, and $0 \leq i \leq b2^{k_2 - k_1} - a$. So for example if b and a are fixed and $k_2 = k_1$, then for all (k_1-indexed) pairs of a small scale and a large scale, the kinds of of overlaps among the small scale windows appear the same, "from the perspective" of the large scale.

If we are given a probability distribution π_{s_2} over Ω_{s_2} and any single member of the mapping set $\{\rho_{s_1 \leftarrow s_2}^{(i)}\}$, we obtain an induced probability distribution over Ω_{s_1} in the usual way. Call that distribution $\rho_{s_1 \leftarrow s_2}^{(i)}(\pi_{s_2})$, or just $\pi_{s_1 \leftarrow s_2}^{(i)}$ for short. It will often be convenient to construct a quantitative synopsis of the set of all of these scale s_1 distributions. If that synopsis is a single probability distribution, then forming this synopsis puts Ω_{s_1} and Ω_{s_2} on equal footing, in that they are both associated with a single distribution. In this paper, we use the average $\rho_{s_1 \leftarrow s_2}(\pi_{s_2}) \equiv \pi_{s_1 \leftarrow s_2} \equiv \sum_i \pi_{s_1 \leftarrow s_2}^{(i)} / \sum_i 1$ as the synopsis of $\{\pi_{s_1 \leftarrow s_2}^{(i)}\}$.

We would like to be able to talk about the probabilistic structure at scale s (i.e. a distribution describing the kinds of patterns seen at scale s). This structure may characterize the statistical regularities of a single object or the regularities of an ensemble of the objects. Either way though, we would like this distribution to be independent of quantities at scales other than s.

Accordingly, we restrict attention to mapping sets such that for some fixed generating scale s_g, for any $s_1 < s_2 < s_g$, the set $\{\rho_{s_1 \leftarrow s_g}^{(k)}\}$ is the set of all compositions $\rho_{s_1 \leftarrow s_2}^{(i)} \rho_{s_2 \leftarrow s_g}^{(j)}$. We call this restriction *composability of mapping sets*. By itself, composability of mapping sets does not quite force $\rho_{s_1 \leftarrow s_g}(\pi_{s_g})$ to equal $\rho_{s_1 \leftarrow s_2}(\rho_{s_2 \leftarrow s_g}(\pi_{s_g}))$.[1] In this paper though we focus on mapping sets such that for the scales of interest $\pi_{s_1 \leftarrow s_g} \approx \rho_{s_1 \leftarrow s_2}(\rho_{s_2 \leftarrow s_g}(\pi_{s_g}))$. Under this restriction we can, with small error, just write π_s for any scale of interest s, without specifying how it is generated from π_{s_g}. For situations where this restriction holds we will say that we have (approximate) *composability of distributions*. Given such composability, we adopt the sitribution π_s as our definition of the *stochastic structure at scale s*.

---

[1]The problem is that the ratio of the number of times a particular mapping $\rho_{s_1 \leftarrow s_g}^{(k^*)}$ occurs in the set $\{\rho_{s_1 \leftarrow s_g}^{(k)}\}$, divided by the number of times it can be created by compositions $\rho_{s_1 \leftarrow s_2}^{(i)} \rho_{s_2 \leftarrow s_g}^{(j)}$, may not be the same for all k^*.

Example 1 continued: Here $\pi^{(i)}_{s_1 \leftarrow s_2}(\omega_{s_1})$ is the probability that a sequence randomly sampled from Ω_{s_2} (according to π_{s_2}) will have the subsequence ω_{s_1} starting at its i'th character. So $\pi_{s_1 \leftarrow s_2}(\omega_{s_1})$ is the probability that a sequence randomly sampled from Ω_{s_2} will, when sampled starting at a random character i, have the sequence ω_{s_1}.

In this example, although we have composability of mapping sets, in general we do not have composability of distributions unless s_g/s_2 is quite large. The problem arises from edge effects due to the finite extent of Ω_{s_g}. Say $\pi_{s_g}(\omega_{s_g}) = 1$ for some particular ω_{s_g}; all other elements of Ω_{s_g} are disallowed. Then a subsequence of s_1 characters occurring only once in ω_{s_g} will occur just once in $\{\rho^{(k)}_{s_1 \leftarrow s_g}(\omega_{s_g})\}$, and accordingly is assigned the value $1/(s_g - s_1)$ by $\pi_{s_1 \leftarrow s_g}$, regardless of where it occurs in ω_{s_g}. If that subsequence arises at the end of ω_{s_g} and nowhere else it will also occur just once in the set $\{\rho^{(i)}_{s_1 \leftarrow s_2}\rho^{(j)}_{s_2 \leftarrow s_g}(\omega_{s_g})\}$. However if it occurs just once in ω_{s_g}, but away from the ends of ω_{s_g}, it will occur more than once in the set $\{\rho^{(i)}_{s_1 \leftarrow s_2}\rho^{(j)}_{s_2 \leftarrow s_g}(\omega_{s_g})\}$. Accordingly, its value under $\rho_{s_1 \leftarrow s_2}(\rho_{s_2 \leftarrow s_g}(\pi_{s_g}))$ is dependent on its position in ω_{s_g}, in contrast to its value under $\rho_{s_1 \leftarrow s_g}(\pi_{s_g})$.

Fortunately, so long as s_g/s_2 is large, we would expect that any sequence of s_1 characters in ω_{s_g} that has a significantly non-zero probability will occur many times in ω_{s_g}, and in particular will occur many times in regions far enough away from the edges of ω_{s_g} so that the edges are effectively invisible. Accordingly, we would expect that the edge effects are negligible under those conditions, and therefore that we have approximate composability of distributions.

The fact that they are generated via mappings $\rho_{s_1 \leftarrow s_g}$ and $\rho_{s_2 \leftarrow s_g}$ imposes some restrictions relating the stochastic structures π_{s_1} and π_{s_2}. Firstly, note that the mapping from the space of possible π_{s_2} to the space of possible π_{s_1} given by a particular $\rho_{s_1 \leftarrow s_2}(\cdot)$ usually will not be one-to-one. In addition, it need not be onto, i.e. there may be π_{s_1}'s that do not live in the space of possible $\pi_{s_1 \leftarrow s_2}$. In particular, consider example 1 above, where the character set is binary. Say that $s_1 = 2$. Then $\pi_{s_1}(\omega_{s_1}) = \delta_{\omega_{s_1},(0,1)}$ is not an allowed $\pi_{s_1 \leftarrow s_2}$. For such a distribution to exist in the set of possible $\pi_{s_1 \leftarrow s_2}$ would require that there be sequences ω_{s_2} for which any successive pair of bits is the sequence (0, 1). Clearly this is impossible for there must necessarily be successive pairs of bits in ω_{s_2} consisting of (1,0).

Accordingly, for any $s < s_g$, in general not all π_s are possible, due solely to the mapping set $\rho_{s \leftarrow s_g}$. Therefore for any $s_1 < s_2$, the posterior probability[2] $P(\pi_{s_2}|\pi_{s_1})$ must reflect a mapping set concerned with a scale other than s_1 or s_2, namely $\rho_{s_2 \leftarrow s_g}$. This is in addition to reflecting $\rho_{s_1 \leftarrow s_2}$, and holds even for composable distributions.

Also due to this fact that (depending on the mapping set) not all π_s are possible in general, the functional form of any $P(\pi_{s_g})$ will often not be "consis-

---

[2]$P(\pi_{s_2}|\pi_{s_1})$ is the probability of stochastic structure π_{s_2} at scale s_2 given a stochastic structure π_{s_1} at scale s_1.

tent" with the associated induced functional form of $P(\pi_s) = \int d\pi_s P(\pi_{s_g})\delta(\pi_s - \rho_{s \leftarrow s_g}(\pi_{s_g}))$ (the integral is implicitly restricted to the unit s-dimensional simplex). When this happens, we cannot employ first-principles arguments to set a functional form for a prior probability distribution over structures π_s and then apply that prior to all scales s simultaneously. In particular, a $P(\pi_{s_g})$ that assigns non-zero weight to all possible π_{s_g} will not assign non-zero weight to all possible π_s in general, and in this sense the functional forms on the two scales are not consistent.

3.2. Comparison to traditional methods of scaling

It is worth taking a brief aside to discuss the numerous alternative ways one might define the structure at a particular scale. In particular, one could imagine modifying any of the several different methods that have been used for studying self-similarity. Although we plan to investigate those methods in future work, it is important to note that they often have aspects that make them appear problematic for the study of self-dissimilarity. For example, one potential approach would start by decomposing the full pattern at the largest scale into a linear combination of patterns over smaller scales, as in wavelet analysis for example. One could then measure the "weight" of the combining coefficients for each scale, to ascertain how much the various scales contribute to the full pattern. However such an approach has the difficulty that comparing the weight associated with the patterns at a pair of scales in no sense directly compares the patterns at those scales. At best, it reflects — in a non-information-theoretic sense — how much is "left over" and still needs to be explained in the small scale pattern, once the full scale pattern is taken into account.

Many of the other traditional methods for studying self-similarity rely on scale-indexed blurring functions (*e.g.* convolution functions, or even scaled and translated mother wavelets) B_s that wash out detail at scales finer than s (for example by forming convolutions of the distribution with such blurring functions). With all such approaches one compares some aspect of the pattern one gets after applying B_s to one's underlying distribution, to the pattern one gets after applying $B_{s' \neq s}$. If after appropriate rescaling those patterns are the same for all s and s' then the underlying system is self-similar.

There are certain respects shared by our approach and these alternatives. For example, usually a set of spaces $\{\rho^{(i)}_{s_1 \leftarrow s_2} \Omega_{s_2}\}$ are used by those alternative approaches in defining the structure at a particular scale. (Often those spaces are translations of one another, corresponding to translations of the blurring function.)

However unlike these traditional approaches our approach makes no use of a blurring function. This is important since there are a number of difficulties with using a blurring function to characterize self-dissimilarity. One obvious problem is how to choose the blurring function, a problem that is especially vexing if one wishes to apply the same (or at least closely-related) self-dissimilarity measure to a broad range of systems, including both systems made up of symbols and systems that are numeric. Indeed, for symbolic spaces how even to define blurring functions in general is problematic. This is because the essence of a blurring function B_s is that for any point x, applying B_s reduces the pattern over a neighborhood of width s about x to a single value. There is some form of average or integration involving that blurring function that produces the pattern at the new scale — this is how information on smaller scales than s is washed out. But what general rule should one use to reduce a symbol sequence of width s to a single symbol?

More generally, even for numeric spaces, how should one deal with the statistical artifacts that arise from the fact that the probability distribution of possible values at a point x will differ before and after application of blurring at x? In traditional approaches, for numeric spaces, this issue is addressed by dividing by the variance of the distribution. But that leaves higher order moments unaccounted for, an oversight that can be crucial if one is quantifying how patterns at two different scales differ from one another.

Such artifacts reflect two dangers that should be avoided by any candidate self-dissimilarity measure:

1. The possibility of changes in the underlying statistical process that don't affect how we view the process's self-dissimilarity, but that do modify the value the candidate self-dissimilarity measure assigns to that process.

2. The possibility of changes in the underlying process that modify how we view the self-dissimilarity of the process but not the value assigned to that process by our candidate measure.

In general, unless the measure is derived in a first principles fashion directly from the concept of self-dissimilarity, we can never be sure that the measure is free of such artifacts.

Our current focus is on approaches that are based on mapping sets, and in which rather than directly compare two scale-indexed structures that live in different spaces (as in the traditional approaches), one first performs statistical inference to map the structures to the same space. There will always be the possibility of artifacts when making comparisons between systems that are different in kind (e.g., that live in non-isomorphic spaces). However properly done, an inference-based approach should at least avoid hidden statistical artifacts in comparisons between scales within a single system since the statistical aspects are explicit.[3] In particular, with such an inference-based approach there is no need for a blurring function, and the problems inherent in careless use of such functions can be avoided. Intuitively, the inference-based approach achieves this by having the information at scale s_2 be a superset of the information at any scale $s_1 < s_2$. This is clarified in the following discussion.

---

[3]Indeed, it may even prove possible to combine such inference-based mappings — and the associated lack of unforeseen statistical artifacts — with the structures used in the traditional approaches (e.g., blurring-based structures). This is the subject of future research.

3.3. Converting structures on different scales to the same scale

It will be convenient to introduce yet a fourth "comparison scale", s_c, at which to compare our (inferences based on) our structures. Often s_c is set in some manner by the problem at hand, and in particular, we can have $s_c = s_2$, and/or $s_c = s_g$. But this is not required by the general formulation. For the rest of this paper, we will always take $s_c \geq \max[s_1, s_2]$, where s_1 and s_2 are the two scales whose structures are being compared.

Suppose we are interested in the scale s_c structure, π_{s_c}, and are given the structure on scale s. Then via Bayes' theorem, that scale s structure fixes a posterior distribution over the elements of $\omega_{s_c} \in \Omega_{s_c}$, i.e., it fixes an estimate of the scale s_c structure:

$$
\begin{aligned}
P(\omega_{s_c} \mid \pi_s) &= \int d\pi_{s_c} \, P(\omega_{s_c} \mid \pi_{s_c}) P(\pi_{s_c} \mid \pi_s) \\
&= \int d\pi_{s_c} \, \pi_{s_c}(\omega_{s_c}) P(\pi_{s_c} \mid \pi_s) \\
&= \frac{\int d\pi_{s_c} \, \pi_{s_c}(\omega_{s_c}) P(\pi_s \mid \pi_{s_c}) P(\pi_{s_c})}{\int d\pi'_{s_c} \, P(\pi_s \mid \pi'_{s_c}) P(\pi'_{s_c})}
\end{aligned}
\tag{3.1}
$$

where π'_{s_c} is a dummy argument π_{s_c}, and in the usual Bayesian way

$$
\begin{aligned}
P(\pi_{s_c}) &= \int d\pi_{s_g} P(\pi_{s_c} \mid \pi_{s_g}) P(\pi_{s_g}) \\
&= \int d\pi_{s_g} \delta(\pi_{s_c} - \rho_{s_c \leftarrow s_g} \pi_{s_g}) P(\pi_{s_g}),
\end{aligned}
$$

where $P(\pi_{s_g})$ is a prior over the real-valued multi-dimensional vector π_{s_g}.

The implicit model here is that π_{s_c} is formed by first sampling $P(\pi_{s_g})$ to get a π_{s_g}, and then having the mapping set $\rho_{s_c \leftarrow s_g}$ generate π_{s_c} from that π_{s_g}. Then ω_{s_c} is formed by sampling that π_{s_c}. To generate π_s, one applies the mapping $\rho_{s \leftarrow s_g}$ to π_{s_g} directly.

As an example, by composability $\pi_s = \rho_{s \leftarrow s_c} \pi_{s_c}$, and therefore

$$
P(\omega_{s_c} \mid \pi_s) = \frac{\int d\pi_{s_c} \, \pi_{s_c}(\omega_{s_c}) \delta(\pi_s - \rho_{s \leftarrow s_c}(\pi_{s_c})) P(\pi_{s_c})}{\int d\pi'_{s_c} \, \delta(\pi_s - \rho_{s \leftarrow s_c}(\pi'_{s_c})) P(\pi'_{s_c})}
$$

(As always, sums replace integrals if appropriate.) In this situation, $P(\omega_{s_c} \mid \pi_s)$ may not even be an allowed distribution, in the sense that $P(\pi_{s_c})$ assigns zero probability to the distribution whose ω_{s_c}-dependence is given by $P(\omega_{s_c} \mid \pi_s)$. As an alternative decomposition, we can write

$$
\begin{aligned}
P(\omega_{s_c} \mid \pi_s) &= \frac{P(\pi_s \mid \omega_{s_c}) P(\omega_{s_c})}{P(\pi_s)} \\
&= \frac{P(\pi_s \mid \omega_{s_c}) \int d\pi_{s_c} \pi_{s_c}(\omega_{s_c}) P(\pi_{s_c})}{\sum_{\omega_{s_c}} \text{numerator}}.
\end{aligned}
\tag{3.2}
$$

In practice, rather than set the prior $P(\pi_{s_c})$ and try to evaluate the integrals in equations (3.1) and (3.2), one might approximate the fully Bayesian approach of equations (3.1) and (3.2), for example via MAXENT [11], MDL [12], or by minimizing algorithmic complexity [14]. Indeed, even if we were to restrict ourselves to analyses relying on Bayes' theorem and even if $s_g \neq s_c$, we might (for example) wish to "pretend" that s_c is our generating scale, and therefore measure the dissimilarity between $P(\omega_{s_g} \mid \pi_{s_1})|_{s_g=s_c}$ and $P(\omega_{s_g} \mid \pi_{s_2})|_{s_g=s_c}$, rather than the dissimilarity between $P(\omega_{s_c} \mid \pi_{s_1})$ and $P(\omega_{s_c} \mid \pi_{s_2})$.

To allow full generality then, for each pair of scales s_2 and $s_1 < s_2$, introduce the random variable $\pi_{s_2}^{s_1}$ to indicate a distribution over Ω_{s_2} that is inferred from the structure at scale s_1. Indicate an element sampled from $\pi_{s_2}^{s_1}$ by $\omega_{s_2}^{s_1}$. Given a structure at scale s_1, π_{s_1}, we call the rule taking π_{s_1} to a distribution $\pi_{s_2}^{s_1}$ the *inference mechanism* for going from that scale-s_1 structure to a guess for the distribution at scale s_2, and indicate the action of the inference mechanism by writing $\pi_{s_2}^{s_1} = \pi_{s_2}^{s_1}(\pi_{s_1})$. As examples, equations (3.1) and (3.2) provide two formulations of a Bayesian inference mechanism.

Once we have calculated both $\pi_{s_c}^{s_1}(\pi_{s_1})$, the scale-$s_1$-inferred distribution over Ω_{s_c}, and $\pi_{s_c}^{s_2}(\pi_{s_2})$, the scale-$s_2$-inferred distribution over Ω_{s_c}, we have translated both our structures at scale s_1 and s_2 into two new structures, both of which are in the same space, Ω_{s_c}. We can now directly compare the two new structures that were generated by the structures at scales s_1 and s_2. In this way we can quantify how dissimilar the structures over s_1 and s_2 are. In this paper, we will concentrate on quantifications that can be viewed as the amount of information (concerning scale s_c) inferable from the structure at scale s_2 that goes beyond what is inferrable from the structure at scale s_1.

3.4. Comparing structures on the same scale

To define a complexity measure we must next choose a scalar-valued function Δ_{s_c} that measures a distance between probability distributions over Ω_{s_c}.[4] Intuitively, $\Delta_s(Q_s, Q'_s)$ should reflect the information-theoretic similarity between the two distributions over Ω_s given by Q_s and Q'_s. Accordingly Δ_{s_c} should satisfy some simple requirements. It is reasonable to require that for a fixed π_s, $\Delta_s(\pi_s, Q_s)$ is minimized by setting Q_s to equal π_s. Also, in some circumstances it might be appropriate to require that for any s_2, $s_1 < s_2$, π_{s_2}, and Q_{s_2}, $\Delta_{s_2}(\pi_{s_2}, Q_{s_2}) \geq \Delta_{s_1}(\rho_{s_1 \leftarrow s_2}(\pi_{s_2}), \rho_{s_1 \leftarrow s_2}(Q_{s_2}))$. In this paper we will not impose a rigid set of requirements on Δ_s, but rather as we discuss various candidate Δ_s we will note how they are related to such desiderata.

As an example $\Delta_s(Q_s, Q'_s)$ might be $|KL(\pi_s, Q_s) - KL(\pi_s, Q'_s)|$, where $KL(\cdot, \cdot)$ is the Kullback-Leibler (KL) distance [10] and π_{s_c} is the implicit true distribution over Ω_{s_c}.[5] One nice aspect of $\Delta_{s_c}^{KL}$ is that it can be viewed as a quantification of the amount of extra information concerning Ω_{s_c} that exists in

---

[4] We use the word "distance" advisedly, since we do not require that Δ_{s_c} obey the properties of a metric in general.

[5] When, as in this case, specification of π_{s_c} is needed, we should properly write $\Delta_s^{KL}(Q_s, Q_{s'}; \pi_s)$.

Q_s but not in $Q_{s'}$. I.e., it is the amount of extra information in Q_s beyond that in $Q_{s'}$.

Consider $s_c = s_2$. In this case $\pi^{s_2}_{s_c}(\cdot)$ is the identity function and $\Delta^{KL}_{s_c}(\pi^{s_2}_{s_c}(\pi_{s_2}), \pi^{s_1}_{s_c}(\pi_{s_1}); \pi_{s_c}) = KL(\pi_{s_c}, \pi^{s_1}_{s_c}(\pi_{s_1}))$. I.e., in this scenario, $\Delta^{KL}_{s_c}$ is the KL distance between π_{s_c} and the inference for π_{s_c} based on π_{s_1}. This suggests another natural choice for $\Delta_s(Q_s, Q'_s)$, which is to set it to $KL(Q_s, Q'_s)$ always, regardless of the scale s_c distribution or of whether $s_c = s_2$. However this choice for Δ_s could be misleading if neither Q_s nor Q'_s is "well-aligned" with the true π_s; in such a case the two distributions may appear very similar according to Δ_s, but that similarity is specious. In contrast, Δ^{KL}_s forces the inference mechanisms to be "honest", as far as the resultant value of dissimilarity is concerned. In addition, $\Delta^{KL}_s(Q_s, Q'_s)$ obeys the triangle inequality, and unlike $KL(Q_s, Q'_s)$, $\Delta^{KL}_s(Q_s, Q'_s)$ is symmetric in its arguments. Unfortunately though, $\Delta^{KL}_s(Q_s, Q'_s) = 0$ does not imply that $Q_s = Q'_s$. So Δ^{KL}_s is not ideal, and there may be situations where $KL(\cdot, \cdot)$ is preferable.

4. Discussion

In this section we discuss how to estimate our self-dissimilarity measure from finite data and discuss some of the broad features of our measure.

4.1. Comparing structures when information is limited

In the previous section we saw that to measure how dissimilar two structures π_{s_1} and π_{s_2} are we translate both to a distribution over the common space Ω_{s_c} and then measure how dissimilar those two distributions are. Unless we know the structures π_{s_1}, π_{s_2}, and π_{s_c} though, rather than evaluate Δ_{s_c}, we have to be content with the expected value of Δ_{s_c} conditioned on our provided information, $\mathcal{I}$. We indicate such an expectation in its full generality as follows:

$$I_{s_1, s_2; s_c}(\mathcal{I}) \equiv \int d\pi_{s_1} d\pi_{s_2} d\pi_{s_c} \Delta_{s_c}(\pi^{s_1}_{s_c}(\pi_{s_1}), \pi^{s_2}_{s_c}(\pi_{s_2}); \pi_{s_c})$$
$$\times P(\pi_{s_1}, \pi_{s_2}, \pi_{s_c} \mid \mathcal{I}), \qquad (4.1)$$

where in turn

$$P(\pi_{s_1}, \pi_{s_2}, \pi_{s_c} \mid \mathcal{I}) = P(\pi_{s_c} \mid \pi_{s_1}, \pi_{s_2}, \mathcal{I})$$
$$\times P(\pi_{s_1}, \pi_{s_2} \mid \pi_{s_c}, \mathcal{I}).$$

In this last equation, the last term on the right-hand side is the likelihood function for generating the structures at scales s_1 and s_2.

As an example, if the provided information is π_{s_1} and π_{s_2}, then we can write the expected distance as

$$I_{s_1, s_2; s_c}(\pi_{s_1}, \pi_{s_2}) = \int d\pi_{s_c} \Delta_{s_c}(\pi^{s_1}_{s_c}(\pi_{s_1}), \pi^{s_2}_{s_c}(\pi_{s_2}); \pi_{s_c})$$
$$\times P(\pi_{s_c} \mid \pi_{s_1}, \pi_{s_2}), \qquad (4.2)$$

where by Bayes' theorem

$$P(\pi_{s_c} \mid \pi_{s_1}, \pi_{s_2}) \propto$$
$$\delta[\pi_{s_1} - \rho_{s_1 \leftarrow s_c}(\pi_{s_c})]\delta[\pi_{s_2} - \rho_{s_2 \leftarrow s_c}(\pi_{s_c})]P(\pi_{s_c}),$$

$$(4.3)$$

with the proportionality constant set by normalization.

$I_{s_1, s_2; s_c}$ is a quantification of how dissimilar the structures at scales s_1 and s_2 are. The dissimilarity signature of a system is the upper-triangular matrix $\Delta_{s_1, s_2} = I_{s_1, s_2; s_c}$. Large matrix elements correspond to unanticipated new structure between scales.

In light of the foregoing, there are a number of restrictions we might impose on our inference mechanism, in addition to the possible restrictions on the distance measure. For example, it is reasonable to expect that for scales $i < j < k$ that $I_{i,k;s_c} \geq I_{i,j;s_c}$. Plugging in equation (4.2) with $\rho_{i \leftarrow k}$ set equal to $\rho_{i \leftarrow j}\rho_{j \leftarrow k}$ translates this inequality into a restriction on allowed inference mechanisms π_k^i and π_k^j. As with a full investigation of restrictions on distance measures, an investigation of restrictions on inference mechanisms is the subject of future research.

4.2. Features of the measure

Although we are primarily interested in cases where the indices s correspond to physical scales and the Ω_s to versions of physical spaces observed on those scales, our proposed self-dissimilarity measure does not require this, especially if one allows for non-composable mapping sets. Rather our measure simply acknowledges that in the real world information is gathered in one space, and from that information inferences are made about the full system. The essence of our measure is to characterize a system's complexity in terms of how those inferences change as one varies the information-gathering spaces.

Accordingly, there are three elements involved in specifying $I_{s_1, s_2; s_c}(\pi_{s_1}, \pi_{s_2})$:

1. A set of mapping sets $\{\rho_{s \leftarrow s';i}^{(i)}\}$ relating various scales s and s', to define the "structure" at a particular scale;

2. An inference mechanism to estimate structure on one scale from a structure on another scale;

3. A measure of how alike two same-scale structures are (potentially based on a third structure on that scale).

The choice of these elements can often be made in an axiomatic manner. First, the measure in (3) can often be uniquely determined based on information theory and the issues under investigation. Next, assuming one has a prior probability distribution over the possible states of the system, then for any provided mapping set, one can combine that prior with the measure of (3) to fix the unique "Bayes-optimal" inference mechanism: The optimal inference mechanism is the one that produces the minimal expected value of the measure in (3) given the information provided by application of the mapping set. For $s_2 = s_c$, $I_{s_1, s_2; s_c}(\pi_{s_1}, \pi_{s_2}) = \Delta_{s_2}(\pi_{s_2}, \pi_{s_2}^{s_1}(\pi_{s_1}))$, and for example for the Kullback-Leibler Δ, the Bayes-optimal $\pi_{s_2}^{s_1}(\pi_{s_1})$ is $P(\omega_{s_2} \mid \pi_{s_1})$, as given in equation 3.1. (This solution for the Bayes-optimal inference mechanism holds for many natural choices of Δ; see the discussion on scoring and density estimation in ([13]).)

Finally, given the mapping-set-indexed Bayes-optimal inference mechanisms, and given the measure of (3), one can axiomatically choose the mapping set itself: The optimal mapping set of size K from Ω_s to $\Omega_{s' \neq s}$ is the set of K mappings that *minimizes* the expected value of the self-dissimilarity of the system. In other words, one can choose the mapping set so that the expected result of applying it to a particular Ω_s results in a distribution over $\Omega_{s'}$ that is maximally informative concerning the distribution over Ω_s, in the sense of inducing a small expected value of the measure in (3). At this point all three components of I are specified. The only input from the researcher was what issues they wish to investigate concerning the system, and their prior knowledge concerning the system.

In practice, one might not wish to pursue such a full axiomatization of (1,2,3). We view the ease with which our measure allows one to slot in portions of such an alternative non-axiomatic approach to be one of the measure's strengths. For example, one could fix (1) and (3), perhaps without much concern for *a priori* justifiability, and then choose the inference mechanism in a more axiomatic manner. In particular, if we know that the system has certain symmetries (e.g., translational invariance), then those symmetries can be made part of the inference mechanism. This would allow us to incorporate our prior knowledge concerning the system directly into our analysis of its complexity without following the fully axiomatic approach.

Another advantage of allowing various inference mechanisms is that it allows us to create more refined versions of some of the traditional measures of complexity. For example, consider a real-world scheme for estimating the algorithmic information complexity of a particular infinite real-world system. Such a scheme would involve gathering a finite amount of data about the system (e.g., data from a finite window), and then finding small Turing machines that can account for that data [14]. The size of the smallest such machine is an upper bound on the algorithmic complexity of the data. In addition, the appropriately weighted distribution of the full patterns these Turing machines would produce if allowed to run forever can be taken as a probabilistic inference for the full underlying system. Self-dissimilarity then measures how this inference for the full system varies as one gathers data in more and more refined spaces. Systems with small algorithmic complexity should be quite self-similar according to such a measure, since once a certain quality of data has been gathered, refining the data further (*i.e.*, increasing the window size) will not affect the set of minimal Turing machines that could have produced that data. Accordingly, such refining will not significantly affect the inference for the full underlying system, and therefore will result in low dissimilarity values. Conversely, algorithmically complex systems should possess large amounts of self-dissimilarity. Note also that rather than characterize a system with just a single number, as the traditional use of algorithmic complexity does, this proposed variant yields a more nuanced signature (the set $\{I_{s_i,s_j}\}$).

The self-dissimilarity measure can even be made to closely approximate traditional, blurring-function-based measures of similarity by an appropriate choice of the inference mechanism. This would be the case if for example the inference mechanism worked by estimating the fractal character of the pattern at scale s_1, and extrapolated that character upward to scales $s_2 > s_1$.

Acknowledgements We would like to thank Tony Begg, Liane Gabora, Isaac Saias, and Kevin Wheeler for insightful discussions.

Bibliography

[1] BAR-YAM, Y. *Dynamics of Complex Systems*, Addison-Wesley (1997).

[2] BENNETT, C. H. *Found. Phys.*, **16**, (1986), 585.

[3] CASTI, J. L., "What if", *New Scientist* **151** (1996), 36–40.

[4] CHAITIN, G. *Algorithmic Information Theory*, Cambridge University Press, 1987

[5] CRUTCHFIELD, J. P., "The calculi of emergence", *Physica D*, **75** (1994), 11–54.

[6] LLOYD, S, "Physical Measures of Complexity", *1989 Lectures in Complex Systems*, (E. Jen ed), Addison-Wesley, 1990.

[7] LLOYD, S. and H. PAGELS, "Complexity as thermodynamic depth", *Annals of Physics*, (1988), 186–213.

[8] SOLOMONOFF, R. J., *Inform. Control*, **7**, (1964), 1.

[9] STANLEY, M. "Scaling Behaviour in the Growth of Companies", *Nature*, **379**, (1996), 804–806.

[10] COVER, T. M., and J. A. THOMAS, *Elements of information theory*, John Wiley & Sons (1991).

[11] JAYNES E. T., *Probability theory: the logic of science*, fragmentary edition available at ftp://bayes.wustl.edu/pub/Jaynes/book.probability.theory

[12] BUNTINE, W. "Bayesian back-propagation", *Complex Systems*, **5**, (1991), 603–643.

[13] BERNARDO, and SMITH, *Bayesian Theory*, John Wiley & Sons (1995).

[14] SCHMIDHUBER, J. "Discovering solutions with low Kolmogorov complexity and high generalization ability", *The Twelfth International Conference on Machine Learning*, (Prieditis and Russel Eds.), Morgan Kauffman, 1995.

Chapter 56

Complexity and order in chemical and biological systems

Gad Yagil

Dept. of Molecular Cell Biology, The Weizmann Institute of
Science
Rehovot, Israel 76100
lcyagil@weizmann.ac.il

The complexity of structurally defined objects in the chemical and biological context is calculated using a previously defined formalism. Structural complexity is defined as the minimal number of numerical and compositional specifications needed to describe the ordered features of an object . The formalism is applied to a set of simple one to four atomic molecules, incl. H_2, HCl, H_2O and H_2O_2. H_2O_2 is shown to be the prototype of partially ordered systems. Order is defined as the ratio of the ordered to the total coordinates of an object or sequence. Ordered coordinates are discernible from non-ordered ones by a reproducibility criterion proposed. Partial order is the dominant feature in most bio- and man-made systems so that the proposed formalism is capable of treating such systems. This is demonstrated in the last part, where the formalism is applied to a biostructure, the wing eyespot patterns of an African butterfly $B.$ *anynana*, and is shown to correctly predict the number of genes specifying that structure.

1. Introduction

Biosystems differ from simpler chemical and physical systems in that they contain a huge arsenal of coded instructions embodied in their genetic material.

The complex behavior of biosystems is a direct result of this special feature, and makes it difficult to predict responses of a bioentity to changes in external conditions based merely on the local state of its different non-instructive components. To understand the way the genetic material is decoded and expressed in response to a change in external conditions, is essential to the understanding of cellular and organismic behavior.

How can genetic instruction processing be incorporated into a quantitative description of biosystems? On the experimental level, the detailed interactions between the genetic material and surrounding components have to be studied and quantified (see e.g. Yagil, 1975, Schneider, 1991, Westerhoff, 1994). On the conceptual level, a formalism which is capable of incorporating the special properties of template encoded instructions and of their mode of implementation must be built. To serve as templates, molecules have to be highly ordered, in the sense that each coding element (nucleotide base) has to be exactly at its specified position, with any change being potentially lethal. DNA and the bioproperties directed by it are, therefore, unlikely to be amenable to the techniques of nonlinear dynamics or of simple cellular automata. The concept of Algorithmic Complexity (AC, Chaitin, 1990), which has the capability for dealing with highly ordered (or 'structured', Crutchfield, 1994) systems offers a more suitable approach to the understanding of these instructive systems. In this paper we extend a formalism previously proposed for assessing the structural complexity of molecular systems (Yagil, 1985,1993a,b,1995). Simple molecules are treated here for the first time and serve to illustrate the method; an application to a particular biopattern is described in a later part, demonstrating the utility of the concept. A similar concept of effective complexity has recently been offered (Gell-Mann, 1994, Gell-Mann and Lloyd, 1996).

2. Order

The definition of biocomplexity is tightly connected with the concept of order. An *ordered system* can be defined as a *system (string, object) in which each element occupies the same coordinates in every specimen of the system.* A structure most often associated with order is that of a crystal. A crystal is, however, not a good paradigm for order, because it is also highly *regular*, which is to say that its elements (atoms) are not only located at specified locations but also that these locations manifest *repetitious spatial relations between them.* These regular, repetitious features can be specified by a short list of numerical statements. Regularity is however not a necessary condition for a structure to be ordered. Here are three examples of ordered sequences devoid of any regularity, but which are still highly ordered, in the sense that displacing any of their elements could completely annihilate their meaning or function:

(a) The sequence of a DNA molecule.

(b) The First Verse of the Bible: "Bereshit Bara Elohim et Hashamaim ve'et Haarets" (In the beginning God created earth and heaven).

(c) The often mentioned second hundred digits of π.

What is common to the three strings, or sequences, is that all three have a governing principle or design behind them. The necessity to keep the prescribed order is clear to any person who knows the code employed and understands the principle or design which governs each sequence. The knowledge of the principle or design makes these ordered strings *reproducible*, which is to say that the strings can be reproduced either by invoking the principle behind them, or by reconstructing them according to the known design. In the case of π, its generating function will serve as the guiding principle; the story which the original writer of the Bible had in mind can be considered as the design of the First Verse and the complementary strand of the DNA (perfected by that blind watchmaker, Dawkins 1986) can serve as the design or template for the DNA sequence.

DNA is a particularly good paradigm for *ordered systems*, because there are very few regularities in DNA. I hope that even a non-biologist realizes that whenever a DNA region is sampled from any tissue of a certain human, or even from different humans (ignoring about 1% "polymorphisms"), one and the same sequence will be obtained. This is in sharp contrast to an ideal gas, where the positions and moments of each molecule have no fixed values; so are the letters in the proverbial text printed by a monkey. Both systems have no design behind them, their coordinates are *irreproducible* and both are consequently *non ordered* systems. I am avoiding the word "random", because it is currently used to characterize both ordered and non-ordered strings. It seems nevertheless somehow inappropriate to call the sequence of any DNA "a random string" by the daily meaning of that word.

The property of reproducibility offers a criterion to determine whether a system (string, object) is ordered or not. This can be done by sampling an *ensemble* of the strings or objects, and to examine the positions/coordinates of each sampled member (specimen). If the position of every coordinate is the same in every (or most) sampled members, then the system members are ordered. For our three strings, the reproducibility test can be done by picking up another Bible, or by consulting a Bible scholar, in the case of the First Verse; by performing a DNA sequencing analysis from a kin, or by regenerating π by the well known series. Only a sequence or object which passes this *reproducibility criterion* can be considered as ordered (partly ordered systems will be considered later below). This was articulated in detail, because it is a basic tenet of our formalism that *structural complexity can be attributed only to the ordered coordinates of a system*. For non ordered coordinates, only macro variables like entropy can be evaluated. In many disciplines, ranging from molecular biology to man made systems, it is the ordered micro-structure which is responsible for the inner working of these systems and which is responsible for their complex features.

3. Structural complexity of point systems

Chemical molecules provide a class of systems the structural complexity of which can be readily calculated. Molecules can be considered as typed point systems, i.e. systems or objects the elements of which consist of points with a different type or "color" for each point. The spatial position of each atom within a molecule is customarily specified by numerical values in a suitable coordinate system. Three spatial coordinates + one color coordinate define a 4 dimensioned space. In dynamical situations, time can be added as a fifth coordinate.

The coordinates of each element can now be divided into two classes: *Ordered coordinates*, i.e. coordinates which do obey the reproducibility criterion defined above, i.e. do assume the same numerical values for every member of an ensemble of these molecules, and *non-ordered coordinates*, which do not have a fixed value, and therefore can not contribute to the evaluated complexity. The ordered coordinates can be redivided into *uniquely specified* ones and *regular* ones. The uniquely specified coordinates (e.g. the bases in the DNA example, or the atoms in a simple chiral molecule) contribute to the complexity a full unit each. The regular coordinates can be compressed by a mathematical or logical expression and will consequently contribute less then a unit, in analogy to the shorter program of algorithmic complexity. *Structural complexity* may thus be written:

$$\mathbf{C} = \min\{\sum_k [c_k/k] - c'\}$$

where c_k is the number of ordered coordinates sharing a k fold regularity and c' is the number of specifications necessary to place the system in the external framework (normally 6).

The next step in the analysis of a molecule is to identify regularities present, each regularity contributing in inverse proportion to the number of elements sharing that regularity (The more regularities, the less complex the system is). This procedure can be repeated in other coordinate systems. The complexity value to be assigned is of course that in the coordinate system yielding the minimal value of $\mathbf{C}$. There is at present no systematic way to determine the minimal set, and there seems to be no general solution (Chaitin 1990, Crutchfield 1994). We have however formulated a set of rules, shown in the Appendix, which enables a consistent identification of regularities present in typed point systems (Yagil, 1985,1993a). We scan now proceed to illustrate the procedure on a set of simple molecules.

4. The simple molecules:

In Table 4.1 the complexity of several of the simplest molecules is presented. The simplest molecular system is obviously a single atom, represented by helium.

Once placed in the external world ($c' = 3$) only its color ϵ ("He") needs to be specified. Consequently both the *maximal complexity* $\mathbf{C}_{max}$ And the actual complexity $\mathbf{C}$ have a value of unity.

Molecule	c	c'	no.	ε	r	φ	z	New	C	Cmax	Cr
1 1 He	4	3	1	He	0*	0*	0*	1	1	1	1
H–H	8	5	1	H	0*	0*	0*	1			
			2	H	R	0*	0*	1	2	3	2/3
H–Cl	8	5	1	H	0*	0*	0*	1			
			2	Cl	R	0*	0*	2	3	3	1
O=C=O	12	5	1	C	0*	0*	0*	1			
			2	O	R_{CO}	0*	0*	2			
			3	O	R_{CO}	π	0	2	5	7	5/7
N	12	5	1	H	0*	0*	0*	1			
H–C≡N	12	5	1	H	0*	0*	0*	1			
			2	C	R_{CH}	0*	0*	2			
			3	N	R_{CN}	π	0	4	7	7	1
H–O	12	6	1	O	0*	0*	0*	1			
\			2	H	.98**	0*	0*	2			
H			3	H	.98	108.4	0*	1	4	6	2/3
H–O	12	6	1	H	0*	0*	0	1			
\			2	O	R_{OH}	0*	0*	2			
Cl			3	Cl	R_{OCl}	Θ	0*	3	6	6	1

c - Total number of coordinates of the molecule (no. of atoms × 4)

c' - Placement coordinates, i.e. coordinates placing the molecule in an external framework (= number of translations + rotations).

ε - type (color) coordinate

r, ϕ, z - cylindrical position coordinates

New - Number of new, independent numerical and symbolic specifications in this row, i.e. specifications not counted in a previous row (*excluded)

C - Complexity - sum of independent specifications

Cmax - Maximal complexity possible. Cmax is equal to the total number of coordinates, minus the placement coordinates, minus disordered coordinates, if any

Cr - Relative Complexity, C/Cmax

* - Placement coordinates

** - Numerical values are shown to illustrate that numerical values are meant

Table 4.1: Simple Molecules – Specification tables

For hydrogen, a single color specification and a single spatial specification, the interatomic distance (R_{HH}), are all that is needed to completely describe the molecule. The complexity of H_2 is consequently $C = 2$. The maximal complexity $C_{max} = c - c'$ is nevertheless 3, because 2 colored points have have 8 coordinates (2nd column) in a 4 dimensional space, of which $c' = 5$ (3rd column) are placement coordinates. The complexity is not even 3 because of a twofold regularity (both atoms are H), yielding a *relative complexity* of $C_r = C/C_{max} = 2/3$ (last column).

In the next molecule, hydrogen chloride, the single regularity of hydrogen is removed. To specify HCl, two colors and one distance, R_{HCl} , are needed. The Complexity, $C = 3$, is therefore equal to maximal complexity attainable in a two atom system, with $C_r = 1$

For three atomic molecules both linear ($c' = 5$) and nonlinear structures ($c' = 6$) are possible. Several molecular regularities are manifested, some of which are represented in the Table. I hope the way complexity is derived is obvious by now. The maximally complex triatomic molecule (no regularities; $C_r = 1$) is represented by HCN for linear molecules and by HOCl for the nonlinear ones.

No	ϵ	r	ϕ	z	spcf.
1	O	0^*	0^*	0^*	1
2	O	0^*	0^*	Roo	1
3	H	R	$\phi 1^*$	-r	3
4	H	R	Any	r+Roo	0

$C_{max} = 9$; $C = 5$
$C_r = 5/9$; $\Omega = 9/10 = 0.9$
$R = R_{OH} \sin \theta$; $r = R_{OH} \cos \theta$
$*$ - placement coordinate

Table 4.2: Complexity and Order of Hydrogen Peroxide (H_2O_2); 4 atoms (elements), 16 coordinates: $c' = 6; c_{ord} = 9; c_{nord} = 1$

All the molecules treated so far adhere to the reproducibility criterion formulated above – whenever we observe an HOCl molecule its inner coordinates will have the same numerical values as stated in Table 4.1 (ignoring excited states). HOCl is therefore a completely ordered structure. We shall proceed now to the next molecule, which has already a non-ordered feature.

In Table 4.2 the specifications of the four atomic hydrogen peroxide molecule are shown. In H_2O_2 one of the O–H bonds can rotate freely against the other O–H bond. In the cylindrical coordinate system chosen (the z axis passing through the two oxygen atoms) this means that the ϕ coordinate of atom H2 can assume any value between 0 and 360 degrees. This is expressed in Table 4.2 by assigning the value of "Any" to the $\phi 4$ angle. In other words, only nine of the $4 \times 4 - 6 = 10$ spatial and color coordinates are ordered and do contribute to the complexity, while the tenth one is non-ordered. H_2O_2 would therefore not obey the reproducibility criterion, or, to be exact, nine out of its ten coordinates would pass the reproducibility test ; only the tenth coordinate, which assumes a different value in each molecule in an ensemble of H_2O_2 molecules, would not pass the test.

Hydrogen peroxide can serve thus as a prototype of the wide spread phenomenon of partly ordered systems. Most biosystems are partly ordered systems - free rotating side chains of proteins or flopping wings of a bird are typical examples. Returning to our prototype, the recognition that only nine out of ten coordinates are ordered permits a quantitative definition of order, namely *The Order of a system is defined as the ratio of its ordered to its total non placement coordinates*:

$$\Omega = (c_{ord})/(c_{tot} - c');$$

$(c_{tot} - c' = c_{ord} + c_{nord};$ nord =non-ordered) in H_2O_2: $\Omega = 9/10 = 0.9$

i.e. the hydrogen peroxide molecule is a 90% ordered system. The proposed formalism provides thus a simple way to discern between ordered on non-ordered parts of a system, at least in those cases where a separation is possible.

5. Wing patterns of the butterfly *Bicyclus anynana*

Biosystems are, as said, instructed systems. The complexity of a particular subsystem, like a metabolic pathway or morphogenetic pattern, should be related to the number of instructions required to specify that pathway or pattern, i.e. give an indication on the number of genes involved. Thus, it was previously (Yagil, 1985) pointed out that the high complexity of adenine ($C = 59$) reflects the large number of genes involved in the purine biosynthesis pathway (13 genes, one of the longest pathways in metabolic map). An example of a morphogentic process in a higher organism will be given here:

The genetics of wing pattern formation by this butterfly have been intensively studied by Bakerfield and colleagues (1996). The wings of *B. anynana* exhibit "eyespot" patterns, consisting of three concentric rings, each of a different color. To specify the size of each ring, 3 radial specifications are required. Three further specifications spell the color of the different rings; 2 more specifications determine the location of the center of the spot relative to a defined origin on the wing, altogether 8 specifications. Genetic crossing experiments by Bakerfield et al. revealed the eyespot patterns are multigene traits; the number of genes involved could be estimated from the crossing experiments and were found to be 4.8–9.3 for males and 7.5–10.8 for females (Bakerfield et al., 1996). There seems thus to be a good correspondence between the number predicted from complexity analysis and the much more tedious genetic analysis. While direct experimental analysis is certainly required in any actual situation, complexity analysis can give directions as to anticipated results, and help thus in setting up and interpreting the experimental results. The formalism proposed here has therefore the potential to be helpful in understanding biopattern generation.

In summary, the procedure and examples described here, as well as previously treated molecules and structures (Yagil, 1985, 1993a,b, 1995) offer a practical way to assess complexities in a wide range of systems of interest, in which ordered features predominate. At present an algorithm to provide systematically minimal complexity is not available, and is considered impossible in the general case (Chaitin 1990). So far, consistency with other techniques, such as normal mode analysis ensured reliable results for the simple molecules treated here. Other treatments of physical complexity (Crutchfield, 1994; Gell-Mann, 1994) proposed that complexity reaches a maximum somewhere between complete order and complete disorder. The present procedure sees no continuum between order and disorder, but rather divides coordinate space into ordered and non-ordered realms. Structural complexity is defined only in the ordered realm and then increases continuously with declining regularity up to complete order, i.e. up to $\Omega = 1$, in a uniquely specified system. It is thus anticipated that the formalism described here may help to assess complexity in a wide range of systems.

The inputs and criticisms of my colleagues Shneior Lifson, David Mukamel, David Harel and Uri Feige are gratefully acknowledged.

Appendix: Rules for determining structural complexity

To compute structural complexity, the following set of rules was adopted, the criterion being the extent they lead to consistent descriptions using different coordinate and numbering systems:

1. The unit of structural complexity $\mathbf{C}$ is the specification.

2. The assignment of a numerical value to a single spatial coordinate in a system, or the declaration of the color of a point, count as one specification. A color may be a chemical element, a nucleotide base, a cell type, or any other compositional element, depending on the hierarchical levels chosen.

3. A mathematical or logical statement relating the specifications of several points (a regularity) is counted as a single specification, only if a single numerical value is specified; else it contributes as many specifications as the independent numerical values that are present.

4. A numerical value appearing in more than one statement is counted only once, except if clearly repeated by coincidence.

5. An ordinal number is not counted as a specification.

6. A range statement is not counted as a specification

7. A simple numerical coefficient like $(-1)^i$ (alternation) is not counted as a specification.

8. A transformation of the coordinates adds to the complexity as many statements as new constants are included.

9. A function $(\sin\phi, \log r)$ is not counted separately of its argument.

Bibliography

[1] BAKERFIELD, P.M., GATES, J., KEYS, D., KESBEKE, F., WIJNGAARDEN, P.J., MONTEIRO, A., and CARROL, S.B. "Development, plasticity and evolution of butterfly eyespot patterns" *Nature* **384** (1996) 236–242.

[2] CHAITIN, G. J., " Algorithmic information theory", Cambridge University Press (1990).

[3] CRUTCHFIELD, J. P. " The calculi of emergence: Computation, dynamics and induction" *Physica D* (1994) **75** 11–54.

[4] DAWKINS, " The blind Watchmaker", Norton and Cpy. Inc. (1986).

[5] GELL-MANN, M., "The Quark and the Jaguar", Abacus (1994), 3–120.

[6] GELL-MANN, M. and LLOYD, S., "Information measures, Effective complexity and total information", *Complexity* **2** (1996) 44–53.

[7] SCHNEIDER, T.D., "The theory of molecular machines parts I and II", *J. Theor. Biol.* **148** (1991) 83-137.

[8] WESTERHOFF, H.V., " Biothermokinetics", Andover publ. (1997).

[9] YAGIL, G., " Quantitiative aspects of protein induction", *Curr. Topics. Cell. Regul.* **9** (1975) 183–236.

[10] YAGIL, G., " On the structural complexity of simple biosystems." *J. Theor. Biol.*, **112** (1985) 1–23.

[11] YAGIL, G., "On the structural complexity of templated systems", *1992 lectures in complex systems*, L. Nadel and D. Stein, Eds., The Santa Fe Institute and Addison-Wesley, (1993a), 519–530.

[12] YAGIL, G., " Complexity analysis of a protein molecule." In: *Mathematics in Biology and Medicine*, (J. DEMONGEOT, ed.), Wuertz Publ., Winnipeg, Canada (1993b), 303–313.

[13] YAGIL, G., " Complexity analysis of a self-organizing versus a template directed system." *Lectures in artificial intelligence*, **929** (1995), 179–187.

Authors Index

Milton Keynes UK
Ingram Content Group UK Ltd.
UKHW040712141024
449569UK00012B/603